Neonatal and Paediatric Clinical Neurophysiology

For Elsevier Churchill Livingstone:

Commissioning Editor: Alison Taylor
Development Editor: Kim Benson
Production Manager: Susan Stuart
Project Management and Production: Helius
Cover Design: Kneath Associates

Neonatal and Paediatric Clinical Neurophysiology

EDITED BY

Ronit M Pressler MD PhD MRCPCH

Specialist Registrar in Clinical Neurophysiology, National Hospital for Neurology and Neurosurgery and Great Ormond Street Hospital for Children, London, UK

Colin D Binnie MD FRCP

Emeritus Professor of Clinical Neurophysiology, King's College Hospital, London, UK

Raymond Cooper BSc PhD

Former Director, Burden Neurological Institute, Bristol, UK

Richard Robinson FRCP FRCPCH

Emeritus Professor of Paediatric Neurology, Guy's Hospital, London, UK

FOREWORD BY: Professor Jean Aicardi MD FRCP FRCPCH

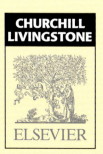

EDINBURGH   LONDON   NEW YORK   OXFORD   PHILADELPHIA   ST LOUIS   SYDNEY   TORONTO   2007

CHURCHILL
LIVINGSTONE
ELSEVIER

ISBN-13: 978-0-443-10173-1
ISBN-10: 0-443-10173-6

British Library Cataloguing in Publication Data
A catalogue record for this book is available from the British Library.

Library of Congress Cataloging in Publication Data
A catalog record for this book is available from the Library of Congress.

Note
Knowledge and best practice in this field are constantly changing. As new research and experience
broaden our knowledge, changes in practice, treatment and drug therapy may become necessary or
appropriate. Readers are advised to check the most current information provided (i) on procedures
featured or (ii) by the manufacturer of each product to be administered, to verify the recommended dose
or formula, the method and duration of administration, and contraindications. It is the responsibility
of the practitioner, relying on their own experience and knowledge of the patient, to make diagnoses,
to determine dosages and the best treatment for each individual patient, and to take all appropriate safety
precautions. To the fullest extent of the law, neither the Publisher nor the Editors assumes any liability
for any injury and/or damage to persons or property arising out or related to any use of the material
contained in this book.

The Publisher

Printed in China

Editors and Contributors

Editors

Ronit M Pressler MD PhD MRCPCH

Colin D Binnie MD FRCP

Raymond Cooper BSc PhD

Richard Robinson FRCP FRCPCH

Contributors

This is a list of contributors to the original work in the parent volumes. In the present book there has been much updating and the responsible authors are named in the contents pages.

Bernadette Bady MD (deceased)
Chef de Laboratoire d'Electromyographie
Hôpital Neurologique
Lyon, France

Colin Binnie MD FRCP
Emeritus Professor of Clinical Neurophysiology
King's College Hospital
London, UK

Stewart G Boyd MD FRCPCH
Consultant Clinical Neurophysiologist
Department of Clinical Neurophysiology
Great Ormond Street Hospital for Children
London, UK

Geraldine B Boylan PhD
Lecturer, Department of Paediatrics and Child Health
Cork University Hospital
Cork, Ireland

John A Connell MA (Oxon) BM (Oxon) BCh FRCP
Consultant Paediatrician
Wexham Park Hospital
Slough, UK

Raymond Cooper BSc PhD
Former Director, Burden Neurological Institute
Bristol, UK

Walter van Emde Boas MD PhD
Neurologist and Clinical Neurophysiologist
Epilepsy Clinics "Meer & Bosch"
"Heemstede" & "Heemstaete"
Zwolle Stichting Epilepsie Instellingen Nederland
Heemstede, The Netherlands

Graham E Holder PhD
Director of Electrophysiology
Moorfields Eye Hospital
London, UK

Jürg Lütschg MD
Facharzt FMH für Pädiatrie und Neuropädiatrie
Universitäts Kinderspital beider Basel
Standort Bruderholz
Basel, Switzerland

François Mauguière MD PhD
Professor of Neurology
Department of Functional Neurology and Epileptology
Neurological Hospital Pierre Wertheimer
Lyon, France

Roweena C Oozeer
Former Chief Physiological Measurement Technician
Department of Clinical Neurology
The Royal Postgraduate Medical School
Hammersmith Hospital
London, UK

John W Osselton BSc
Former Lecturer in Clinical Neurophysiology
University of Newcastle upon Tyne
Newcastle upon Tyne, UK

Matthew Pitt MD FRCP
Consultant Clinical Neurophysiologist
Department of Clinical Neurophysiology
Great Ormond Street Hospital for Children
London, UK

Ronit M Pressler MD PhD MRCPCH
Specialist Registrar in Clinical Neurophysiology
National Hospital for Neurology and Neurosurgery and
Great Ormond Street Hospital for Children
London, UK

Pamela F Prior MD FRCP
Consultant Clinical Neurophysiologist
Department of Clinical Neurophysiology
St. Bartholomew's and the Royal London Hospitals
London, UK

Richard Robinson FRCP FRCPCH
Emeritus Professor of Paediatric Neurology
Guy's Hospital
London, UK

Dieter Scheffner MD
Professor and Head of the Neuropaediatric Department
Children's Hospital, Free University and
Humboldt University
Berlin, Germany

John C Shaw BSc PhD CEng MIEE (deceased)
Formerly of the Medical Research Council
Clinical Psychiatry Research Unit
Chichester, UK

Hermann Stefan MD
Professor and Chairman Epilepsy Centre,
Department of Neurology
University of Erlangen-Nürnberg
Erlangen, Germany

Agnese Suppiej MD PhD
Specialist in Paediatrics and Paediatric Neuropsychiatry
Department of Clinical Neurophysiology
Children's Hospital of Padua
Padua, Italy

Brian M Tedman BSc BMedSci BM BS PhD FRCP
Consultant Clinical Neurophysiologist
The Walton Centre for Neurology and Neurosurgery
Liverpool, UK

Tom Wisman
Product Manager
Tefa-Portanje
Woerden, The Netherlands

Foreword

Over the past few decades clinical neurophysiology has made enormous contributions to child neurology, and has indeed become an essential support of this discipline, leading to major improvements in the study and diagnosis of a host of paediatric disorders. It is hardly necessary to remind paediatricians of the revolution represented by the EEG, which has totally transformed our understanding of epilepsy and its treatment in clinical practice. Electromyography and the study of nerve conduction velocities have also contributed considerably to the diagnosis of other disorders, especially of the neuromuscular diseases. The more recently introduced technique of evoked potential measurement appears promising both in clinical practice – for example for the assessment of vision and hearing, and in monitoring spinal cord function during surgery on the vertebral column – and as a research tool, where it has permitted new insight into the mechanisms of brain function, thus opening new avenues to the scientific study of neuropsychology and the processing of information of perception in normal and pathological situations. Recent technical improvements have made neurophysiological techniques easier to apply in children and infants, improved the quality and reliability of these investigations, and enlarged their range of indications.

It is therefore imperative not only that clinical neurophysiologists but also all those interested in the study of brain function and its abnormalities in childhood, primarily child neurologists and paediatricians, become familiar with the contributions that neurophysiology can offer and their overall significance. This book meets this need. It presents an up to date and in-depth review of the basic and practical aspects of the methods and results of neurophysiological investigations in infancy and childhood, which all too often are poorly understood by the clinicians who request them. In addition, clinicians are often unaware of the problems of performance and interpretation these investigations pose, and of the significance of the information that they can be expected to yield. This book offers an extensive coverage of all the electrophysiological techniques currently applied in infants and children, and brings together an enormous amount of information that usually has to be looked for in several different places. For each technique, the physiological basis, normal results and the maturation of results with age are fully discussed. The text is profusely illustra-

ted and completed by useful tables summarising the most important data.

The text also includes a considerable amount of clinical information concerning the multiple aspects of neurological disease in childhood; and this is a novel and welcome feature. The clinical manifestations of epilepsy are presented in great detail, using the now classical classifications of seizures and syndromes. In addition, many other neurological conditions, from neuromuscular diseases to malformations, chromosomal abnormalities, tumours and psychiatric diseases, are described in detail. The causes of neurological diseases are also discussed at length. This book is thus not only an up to date textbook of child neurophysiology that reviews all the fields of the discipline, but it extends beyond these narrow limits, thus responding to the needs of both neurophysiologists and child neurologists.

Neonatal and Paediatric Clinical Neurophysiology is the result of a joint effort of distinguished neurophysiologists and child neurologists from the UK and all over Europe. This collaboration explains the prominence given to the clinical aspects of the conditions studied. Because the book deals with infants and children, it dedicates much attention to the practical specificities of the neurophysiological testing of children, and emphasises such practical details as the safety and comfort of patients, which are absolutely necessary to perform a satisfactory study. Dealing with young patients, who are growing and maturing, also implies the need for a completely different perspective from that applied in adults, not only in the performance of examinations but also in the interpretation of results, which must take into account the essential factor of age. This requires that doctors, technicians and nurses be fully trained to deal with infants and children in a paediatric environment, and necessitates the integration of neurophysiological units in children's hospitals or specialised centres.

Careful attention during testing is also indispensable, and demands a precise knowledge of what to look for during tests and the technical and clinical knowledge necessary to interpret the events correctly as a function of the context. The detection and appreciation of the significance of minimal symptoms and signs, such as myoclonic jerks, responsiveness, and minimal changes in attitude and memory, during and immediately after testing by physiologists, neurophysiologists or child neurologists are essential, and cannot be replaced even by recent techniques such as video recording or telemetry. Again this demands a good knowledge of clinical neurology.

Over one hundred pages are dedicated to the description of the clinical manifestations of specific disorders, underlining the importance attached by the authors to this aspect. Even more importantly, this clinical information is presented within an integrated perspective, the clinical, EEG, EMG and evoked potential results being described for each of the clinical disorders covered. As a clinician, I find extremely important and encouraging the willingness on the part of neurophysiologists to take into account the clinical perspective, and on the part of clinicians to realise fully not only the importance of 'ancillary' findings but also that these are not 'routine' or systematic investigations but have specific indications, and that both the indications for such tests and the interpretation of the results are best discussed with their neurophysiological colleagues. It seems that we have now reached the point where both clinicians and neurophysiologists are willing not only to take advantage of each other's contribution but fully realise the need for some sort of institutionally organised cooperation between neurophysiologists and child neurologists. With full implementation of such cooperation, the future of both disciplines will certainly be bright.

Jean Aicardi MD FRCP FRCPCH

Preface

This book is the last of a trilogy derived from the parent volumes on *Clinical Neurophysiology* (Binnie et al 2003, 2004). The first of the three (Cooper et al 2005) is concerned with the technical aspects of neurophysiology; the second (Smith et al 2006) the application of these techniques to neurophysiological monitoring during intensive care and surgery; and this last book describes application of the techniques in neonatal and paediatric practice.

Children present particular challenges to the clinician, the neurophysiologist and the technologist who has to collect data in difficult circumstances. We hope that this book, describing as it does the practice and role of EEG, EMG and evoked potentials in this difficult area, will help in the treatment of these vulnerable patients.

The diagnosis of epilepsy in children is not always straightforward – and the value of the EEG is enhanced by simultaneous video. Whilst DNA analysis for the diagnosis of spinal muscular atrophy types I and II and Duchenne muscular dystrophy, for example, plays a more predominant role in the assessment of the floppy infant and the weak child, EMG nevertheless remains indispensable. As with difficulties in the diagnosis of epilepsy, it is not always easy for the clinician to determine whether weakness is neurogenic or myopathic, for example. The new and increasing use of evoked potentials in the neonate show changes in the brain as the newborn acquires information about the sudden, strange new environment.

The rapid developments of the definition of epilepsy syndromes (some of which are confined to childhood), the myriad conditions that may mimic epilepsy and the increasing understanding of the role of EEG in the neonatal period (to name but three) now embodies a sufficiently large body of knowledge to justify amply the creation of posts for neurophysiologists with particular expertise in children. We are firmly of the view that the results of neurophysiological investigations should be discussed at regular meetings at which the technologist(s), the neurophysiologist(s) and those requesting the investigations are present. It is at this interface that a more complete understanding of the problem is achieved and plans laid for further investigations.

This book is not intended to cover all aspects of EEG, EMG and EPs – the parent volumes do this – but rather to give paediatricians an insight into the value of clinical neurophysiology, and neurophysiologists an insight into paediatric practice, as well as giving practical help with the collection of data by the technologist. We believe that neurophysiology should be firmly established within the interdisciplinary team and hope that this book will help to bring this about.

Ronit M Pressler
Colin D Binnie
Raymond Cooper
Richard Robinson

REFERENCES

Binnie CD, Cooper R, Mauguière F, Osselton JW, Prior P, Tedman B 2003 *Clinical Neurophysiology, EEG, Paediatric Neurophysiology, Special Techniques and Applications,* Vol. 2. Elsevier, Amsterdam.

Binnie CD, Cooper R, Mauguière F, Osselton JW, Prior P, Tedman B 2004 *Clinical Neurophysiology, EMG, Nerve Conduction and Evoked Potentials,* Vol. 1. Elsevier, Amsterdam.

Cooper R, Binnie C, Billings R 2005 *Techniques in Clinical Neurophysiology; A Practical Manual.* Elsevier, Oxford.

Smith NJ, van Gills M, Prior PF 2006 *Neurophysiological Monitoring during Intensive Care and Surgery.* Elsevier, Oxford.

Acknowledgements

We are grateful to many colleagues and friends for their generous help that has contributed in various ways to the making of this book. In particular we wish to thank Alison Taylor and Kim Benson of Elsevier, Oxford, for their patience, and Drs Pam Prior, Shelagh Smith and Stewart Boyd for much practical help in all kinds of ways.

We are also deeply indebted to our expert technologists, our secretaries and our families, who have all contributed to making this work possible. Jane Sweetland, librarian of the Burden Neurological Institute, has been particularly helpful. We are also indebted to Noah (age 6 years), Joshua (4 years), Luke (6 years) and Nada (9 years) for the unique artwork at the head of each chapter.

Abbreviations

A-ABR	automated auditory brainstem response (test)
Ac	contralateral earlobe (electrode)
ACh	acetylcholine
AChE	acetylcholine esterase
AChR	acetylcholine receptor
ACTH	adrenocorticotrophin hormone
ADEM	acute disseminated encephalomyelitis
AED	anti-epileptic drugs
AEP	auditory evoked potential
Ag/AgCl	silver/silver chloride (electrode)
Ai	ipsilateral earlobe (electrode)
AIDS	acquired immune deficiency syndrome
APGAR	activity, pulse, grimace, appearance, respiration
AR	autoregressive
AS	active sleep
ASSD	arginosuccinate synthetase deficiency
ATP	adenosine triphosphate
AVM	arteriovenous malformation
BAEP	brainstem auditory evoked potential
BECTS	benign epilepsy with centrotemporal spikes
BETS	benign epileptiform transients of sleep
BIRD	brief intermittent rhythmic discharge
C	central (electrode site)
CA	conceptual (conceptional) age
CAE	childhood absence epilepsy
CAP	compound action potential
CCT	central conduction time
CFAM	cerebral function analysing monitor
CFM	Cerebral Function Monitor
CH	congenital hypomyelination
CHB	complete heart block
CIDP	chronic inflammatory demyelinating polyneuropathy
CJD	Creutzfeldt–Jakob disease
CLEMS	congenital Lambert–Eaton syndrome
CM	cochlear microphonic
CMAP	compound muscle action potential
CMS	congenital myasthenic syndrome
CMT	Charcot–Marie–Tooth (disease)
CMV	cytomegalovirus
CNE	concentric needle electrode
CNS	central nervous system

CP	cerebral palsy
CSA	compressed spectral array
CSNB	congenital stationary night blindness
CSWS	continuous spike and wave during slow sleep
CT	computerised tomography
Cv7	seventh cervical vertebra (electrode)
dB	decibel
DC	direct current
DC	dorsal column
DNA	deoxyribonucleic acid
DSD	Dejerine–Sottas disease
DTL	Dawson–Trick–Litzkow (ERG electrodes)
ECG	electrocardiogram
ECI	electrocerebral inactivity
ECMO	extracorporeal membrane oxygenation
ECN	electroencephalography and clinical neurophysiology
ECochG	electrocochleogram
ECoG	electrocorticogram
ECS	electrocerebral silence
EDC	extensor digitorum communis
EEG	electroencephalogram, electroencephalograph, electroencephalography
ELAE	episodic low amplitude event
EMG	electromyogram, electromyograph, electromyographic
EOG	electro-oculogram
EP	evoked potential
ERG	electroretinogram, electroretinographic, electroretinography
ERP	event related potential
ESES	electrical status epilepticus during slow wave sleep
F	frontal (electrode site)
FC	febrile convulsion
FERG	flash ERG
FIM	familial infantile myasthenia
FIRDA	frontal intermittent rhythmic delta activity
fMRI	functional MRI
f/s	flashes per second
FVEP	flash (evoked) VEP

g	gram
GA	gestational age
GABA	γ-aminobutyric acid
GBS	Guillain–Barré syndrome
GEFS+	generalised epilepsy with febrile seizures plus
HIE	hypoxic–ischaemic encephalopathy
HIV	human immunodeficiency virus
HL	hearing level
HMN	hereditary motor neuropathy
HMSN	hereditary motor and sensory neuropathy
HNPP	hereditary neuropathy with liability to pressure palsy
HSAN	hereditary sensory and autonomic neuropathy
HV	hyperventilation
Hz	hertz, cycles per second (frequency)
IBI	interburst interval
IFCN	International Federation for Clinical Neurophysiology
IFSECN	International Federation of Societies for Electroencephalography and Clinical Neurophysiology
IGE	idiopathic generalised epilepsies
ILAE	International League Against Epilepsy
IPI	interpeak interval (evoked potentials)
IPS	intermittent photic stimulation
ISCEV	International Society for Clinical Electrophysiology of Vision
ISI	interstimulus interval
ITU	intensive therapy unit
IVH	intraventricular haemorrhage
JMA	juvenile myoclonic absence
JME	juvenile myoclonic epilepsy
kΩ	kilohm
LED	light-emitting diode
LF	low frequency
LIF	latency–intensity function
LKS	Landau–Kleffner syndrome
LLAEP	long latency auditory evoked potential
LOC	left outer canthus
LVI	low voltage intermittent (irregular)
Mc	contralateral mastoid (electrode)
MCD	mean consecutive difference
MCV	motor conduction velocity
MELAS	mitochondrial encephalopathy with lactic acidosis and stroke-like episodes
MERRF	myoclonus epilepsy with ragged red fibres
Mi	ipsilateral mastoid (electrode)
MLAEP	middle latency auditory evoked potential
MLD	metachromatic leucodystrophy

MN-SEP	median nerve stimulation SEP
MRI	magnetic resonance imaging
MRSA	methicillin-resistant *Staphylococcus aureus*
ms	millisecond
MS	multiple sclerosis
mt-DNA	maternally transmitted DNA
MUAP	motor unit action potential
MuSK	muscle-specific tyrosine kinase
MΩ	megohm
NARP	neuropathy, ataxia and retinitis pigmentosa
NCL	neuronal ceroid lipofuscinosis
NCS	nerve conduction study
NCV	nerve conduction velocity
NES	non-epileptic seizures
nHL	normal hearing level
NICE	National Institute for Health and Clinical Excellence
NICU	neonatal intensive care unit
NMR	nuclear magnetic resonance
NREM	non-REM (of sleep)
O	occipital (electrode site)
OAE	otoacoustic emission
OCTD	ornithine carbamyl transferase deficiency
OIRDA	occipital intermittent rhythmic delta activity
P	parietal (electrode site)
Pa	pascal
P_aCO_2	arterial blood carbon dioxide tension
P_aO_2	arterial blood oxygen tension
PCA	postconceptional (postconceptual) age
PDS	paroxysmal depolarisation shift
PERG	pattern ERG
PET	positron emission tomography
PHB	partial heart block
PHR	photic high-frequency response
PKU	phenylketonuria
PLED	periodic lateralised epileptiform discharge
PLP	proteolipid protein
PMA	postmenstrual age
PMP	peripheral myelin protein
POSTs	positive occipital sharp transients (of sleep)
ppm	parts per million
PPR	photoparoxysmal response
PPS	prolonged photoconvulsive response
PRSW	positive Rolandic sharp wave
psi	pounds per square inch
PSP	postsynaptic potential
PTN-SEP	posterior tibial nerve stimulation SEP
PTθ	premature temporal theta
PVEP	pattern (evoked) VEP
PVL	periventricular leucomalacia

QS	quiet sleep
RBBB	right bundle branch block
R-CMAP	repetitive compound motor action potential
REM	rapid eye movement
rms	root mean square
RNS	repetitive nerve stimulation
ROC	right outer canthus
s	second
SAP	sensory action potential
SCBU	special care baby unit
SD	standard deviation
SEEG	stereo EEG
SEP	somatosensory evoked potential
SFEMG	single fibre EMG
SIDS	sudden infant death syndrome
SL	sensation level (of sound)
SMA	spinal muscular atrophy
SMARD	spinal muscular atrophy with respiratory distress
SMN	survival motor neuron (gene)
SMNc	centromeric SMN
SMNt	telomeric SMN
SNAP	sensory nerve action potential
SPECT	single proton emission computerised tomography
SPL	sound pressure level
SREDA	subclinical rhythmic epileptiform discharge of adults
SSEP	somatosensory evoked potential (= SEP)
SSPE	subacute sclerosing panencephalitis
SSS	small sharp spikes
STOPs	sharp theta on the occipitals of prematures
SW	spike and wave
SWS	slow wave sleep
T	temporal (electrode site)
TC	time constant
TCI	transitory cognitive impairment
V	vertex (electrode site)
VDU	video/visual display unit
VEP	visual evoked potential
μV	microvolt (= 10^{-6} volt)

Contents

Contents

Chapter 1

General Introduction

THE NEED FOR SPECIAL EXPERTISE

Neonatal and paediatric applications of clinical neurophysiology form an important and steadily increasing part of the practice of many departments of clinical physiology, typically constituting 40% of workload. Some 30% of epilepsy commences in childhood, often presenting with specific syndromes in the investigation of which electroencephalography (EEG) plays a major role.

In the UK the National Institute for Health and Clinical Excellence (NICE) was implemented to improve patient care and provide quality assurance in the form of clinical guidelines. The Epilepsy NICE guidelines published in 2004 (http://www.nice.org.uk) recommend that an EEG should be performed within 4 weeks after it has been requested in a child-centred environment. They also specify that an EEG should not be used in isolation to diagnose epilepsy but rather to determine seizure type and epilepsy syndrome. Equally an EEG should not be used to exclude epilepsy in a child with probable syncope or non-epileptic events. No NICE guidelines exist so far for the recording of electromyograms (EMGs) or evoked potentials (EPs) in children.

The importance of neonatal neurophysiology is becoming increasingly recognised. Nevertheless, neonatal and paediatric neurophysiology is very different from that in adults and represents a highly specialised area of practice, the more so because neurophysiologists tend to be trained mainly on adult material and may wrongly attempt to extrapolate this experience to children. Apart from the practicalities of investigating very young patients and the wide spectrum of age dependent disorders of infancy and childhood, there is also the problem of continuous maturation of the developing nervous system. Thus, when evaluating the neurophysiological recordings of a child, one must be constantly mindful of the fact that the effects of pathology, on the EEG in particular, are age dependent. Hence the characteristics may often be much more a reflection of the stage of structural maturation (sprouting of dendrites and myelination, etc.) of that particular child's nervous system than of any specific disease.

The normal development of EEG, EP, nerve conduction and EMG features are dealt with in general terms in the appropriate chapters. Here we are concerned with the specific technological and interpretative aspects of clinical neurophysiological recordings that are peculiar to the child from birth to adolescence. In this general introduction, details of some general departmental procedures are given that are helpful when running a service for children. The neurophysiology of the neonatal period is covered in Chapter 5 and that for the child in Chapter 6. After some special details on methodology for EEG, EP and EMG investigations in infants and children, each main section then attempts to provide a disease-orientated approach to the applications of the techniques. This is because a much more integrated use of neurophysiological investigation is required for diseases in infants and children than is the case in adults, problem solving often only being possible by combinations of different neurophysiological techniques. Some applications of EPs and EMG, particular to these age groups, complete each main section.

It is beyond the scope of this book to provide a comprehensive account of the whole field; indeed, the newcomer to this work should always seek formal training with properly qualified experts. The reader is referred to specialist texts (e.g. Eyre 1992, Brett 1997, Volpe 2000, Levene et al 2001, Binnie et al 2003, Holmes et al 2005) for more comprehensive and detailed accounts of neonatal and paediatric neurology and clinical neurophysiology.

GENERAL STRATEGIES FOR NEUROPHYSIOLOGICAL INVESTIGATION IN INFANTS AND CHILDREN

To obtain properly any type of neurophysiological recording in an infant or child is time consuming and requires much more than just skill in placing electrodes. Both patience and inventiveness are needed to maintain the child's interest – by talking, singing and playing – as well as the ability to give comfort and a feeling of relaxed security, which inspires the confidence of the child and parents. The investigation of the small neonate requires yet other skills to enable appropriate neurophysiological information to be expeditiously obtained without undue stress to the infant, family or staff.

General considerations about the planning and recording arrangements for infants and children tend to vary between the general department undertaking some neonatal and paediatric work and the specialised children's hospital. It cannot be emphasised too strongly that miniaturisation of adult facilities is not what is required, but rather a different approach, equipment and skills. Planning considerations should extend to the need for a higher than usual ratio of staff to patients and for longer times allocated for individual investigations than with adults. This should be reflected in statistical returns and budgetary forecasts.

Planning investigations

Any investigation in a young child is urgent for the family concerned. Nearly all neurophysiological investigations in newborn babies have to be considered as an emergency. Appointments for outpatient visits always have to be prioritised, but children should be fitted in as soon as possible and a clear list of indications for emergency, urgent and routine appointments should be made available to all the staff. Telephone liaison with the paediatrician can often be helpful. For example, a personal referral by the clinician direct from the clinic and an immediate investigation and discussion of results may help a prompt decision regarding the need for inpatient treatment. Moreover, the paediatric team will be familiar with the problems and stresses peculiar to any individual family and can often prevent difficulties during recording by their advice as to how the situation may best be handled.

The neonatal EEG is, as with all investigations regardless of the patient's age, most valuable when there is a specific clinical question and when the findings are correlated with the patient's clinical state and the results of other investigations. Moreover, it is impossible to interpret the EEG of a preterm infant without knowledge of gestational age, birth history or adequate information about the medication, as all these factors can influence the appearance of the EEG. Close interaction with the neonatologist is always important and will result in a mutually valuable exchange of relevant information.

Parents and children are helped by adequate information in advance of the examination: simple illustrated leaflets are a useful accompaniment to the appointment and the same pictures can be reproduced in the entrance and waiting areas. Thoughtful planning should allow for an outpatient appointment to be timed so that the family with a young child can arrive when a postprandial nap is due (if necessary feeding a baby or toddler in the department whilst electrodes are applied). The written instructions should mention the advantages of bringing a favourite drink and a familiar toy or comforter for the child. They should also indicate the need to stay for as long as required for the child to settle down and for an adequate recording to be obtained – for EEG this will, ideally, mean with the child awake, during natural drowsiness and in sleep. A clear indication of the possible maximum duration of the test will enable parents to make any necessary arrangements, for instance for the collection of siblings from school.

Recording environment

A neurophysiology team with experience in working with children will have built up many areas of expertise which result in a calm and friendly atmosphere. Staff may replace traditional uniforms with something more friendly and identify themselves with photos on a board in the waiting area. A display board for drawings in the reception area and an invitation to send a picture afterwards may allow children to express their impressions of a visit to the neurophysiology department. This may highlight both good and bad points about the service and encourage new patients. It is helpful to have appropriately decorated, homely rooms with toys and activities, facilities such as child-sized lavatories and, most importantly, staff with time and skills to answer the child's, parents' or siblings' questions. Experienced staff will talk to, and play with, a child to encourage cooperation, but must have an authoritative, orderly and unemotional approach to recording. Greetings and explanations of what is to be done should precede any electrode placement; preliminary demonstration on a toy may be helpful. Specific safety rules for paediatric work must be devised to take into account the inquisitive and exploratory nature of most toddlers and children.

Recording procedures

Strategies for obtaining recordings need some thought. It is rarely possible to prolong or repeat investigations to the extent that one might with an adult. Many experienced technologists find that it may be best to start with a basic minimum, even a technically inadequate recording with a limited electrode array, and then try to improve on this once some initial data have been captured, rather than lose the child's cooperation. Even a fractious child often falls asleep during the recording and further electrodes can then be applied. Combined electrode contact and adhesive pastes are useful in all but the most restless children, and have the advantage of ease of removal. Leads and sticky tape should be kept away from the face as far as possible. The technologist experienced in working with children learns many useful ploys: devoting time to play before attempting to apply electrodes, recording a doll's 'EEG' at the same time as the patient's, playing peek-a-boo to obtain eye-closure, encouraging hyperventilation by inviting the child to blow a tissue or a windmill. Early use of photic stimulation or even allowing the patient to watch the EEG as it is recorded for a short period may attract the child's interest and cooperation. Some children find it very difficult to relax in an unfamiliar environment and thus will not produce alpha rhythm on eye closure. Television may prove the ultimate panacea but is not conducive to recording with eyes closed. Often hyperventilation can help here as children become involved with the procedure and then are tired afterwards. A room with good sound insulation is sensible if sleep recording is envisaged or an unduly noisy or upset child is to be examined.

Working together with the parents is essential and their advice on how to get a toddler or difficult child to cooperate may make all the difference. Usually it is helpful if the parents stay with the child during the recording, but the experienced technologist will also identify situations where the interaction between child and parents is unhelpful and distressing to both. In this case, the parents should be invited to leave, as if this were routine practice, and to return towards the end of the investigation.

Adolescents may prefer to be without their parents during the investigation, and may confide to the technologist important clinical information previously withheld. Thus, one has to remember to talk to the adolescent rather than to the parents when explaining the procedures and ask whether he or she wants the parent to stay.

For visual evoked potentials, stimulators may utilise a combination of video cartoons with specific stimuli. Children often tolerate somatosensory stimulation or nerve conduction studies better than adults, but great care must be taken when such techniques are used.

The EMG investigation of the child presents a particular challenge and considerable experience is essential both in EMG and in working with children. The objective is to obtain the necessary information with minimum stress to the child. Of all the investigations that are performed on children, EMG is the one that should be explained only at the time it is done. Information that is sent out before, however it is phrased, almost always alarms and frightens children and their parents. Once in the EMG laboratory various techniques can be used for the different elements of the investigation. For example, one can use the changes on the screen, the movements that the stimuli produce and the sound of EMG to involve patients and parents, and it is rare to need sedation. If sedation is needed it should only be used after discussion with the paediatrician and with informed parental assent; caution in prescribing should be exercised – too little may increase difficult behaviour and too much will make it impossible to obtain active cooperation with the EMG requirements.

Health and safety of the patient

Of prime regard in any neurophysiological investigation is the well-being of the patient, particularly when investigating neonates and young children. No procedure is without risk and it is incumbent on the neurophysiological team to minimise possible injury. There are two main sources of risk – electrical due to malfunction of equipment and infection from the use and re-use of electrodes.

Electrical safety

Fortunately modern equipment is extremely reliable and the risk of injury in its use is extremely small. The danger is greater during intensive care monitoring and recording

in the operating theatre, or in a special baby unit when several electrical instruments may be connected to the patient for long periods. However, although the risk is very small, in view of the expensive legal proceedings that can be initiated for even trivial accidents, it is essential to take 'reasonable precautions'. Safety specifications for electromedical equipment are detailed in the International Electrotechnical Commission Standard IEC 601-1 (1988) (in the UK the British Standard BS 5724 (1989) applies). Particular requirements for EEG machines were published as IEC 60601-2-26: 2002 and adopted as European Standard 60601-2-26: 2003. American standards are described in American EEG Society Guideline Nine (1994).

The main hazard, as with any piece of household electrical equipment, is that arising from the inadvertent passage of electrical current through the body. Normally the high resistance of the skin offers sufficient protection, provided that the body does not touch those parts of the apparatus that are at high voltage. Low-voltage sources, such as torch batteries, can be handled without harm. However, during electrophysiological recording the situation is very different, as equipment is deliberately connected to the body with low contact resistance.

Electrical currents passing through the body may cause pain, burns, respiratory failure or ventricular fibrillation. The current depends upon the magnitude of the applied voltage and the impedance of the tissue through which it flows. Considerable current has to flow (tens of milliamps) through skin to produce a burn, but less than 100 μA of 50 Hz current applied directly to the right ventricle is sufficient to disturb the cardiac rhythm.

IEC 601-1 states: 'Equipment shall be so designed that the electric shock in normal use and in single fault condition is obviated as far as practicable'. There are two types of faults that may give rise to injury. In the first an electrode may become connected to a relatively high voltage source (with respect to earth) so that there is current flow through the electrode on the body to the earth electrode. This could be caused by a fault in the equipment. The second kind of fault is due to the earth connection becoming disconnected within the apparatus, at the mains outlet or in the electrical wiring system. This defect can easily escape detection, unless routine safety checks are performed. If a second fault then develops it is possible that the equipment casing or chassis or patient 'earth' connection could be at mains voltage without the fuse blowing. If the patient (or operator) then touches items of equipment or water pipes that are earthed, lethal current can flow. An additional hazard exists if a possible path to earth of especially low resistance is created when the skin is breached by the use of intravenous catheters, needles, etc.

In most modern electrophysiological equipment such accidents are prevented by isolating the electrodes and preamplifiers from the main amplifiers, displays and power supplies (where high voltages exist) by optical or high-frequency transformer coupling. IEC 601-1 recommends that the maximum current that leaks to earth should not exceed 100 µA at 50 Hz in normal circumstances, and not more than 10 µA where the current could pass directly through the heart. A single fault, such as an interruption of one of the supply conductors or a protective earth conductor, may increase the leakage current. The leakage current should be checked *and recorded* at regular intervals by an authorised person.

When two or more items of mains-operated equipment are connected to a patient they should be plugged into the same mains supply. This is because different earth points within a building may not be at the same potential and because of the risk of an accidental shorting of one of the mains supplies to earth. Apart from the safety aspect, this single point earth helps to minimise the interference that is developed from currents in the 'earth loops'. Only one earth electrode should be on the patient, and the cot or bed should be positioned so that the patient cannot reach out and touch water pipes or the metal cases of equipment.

Pulsed electrical stimulation as used in nerve conduction studies (Chapter 4) and in the evocation of somatosensory evoked potentials (Chapter 3) could affect the operation of cardiac pacemakers, especially those that detect atrial and ventricular excitation separately and those that detect and correct tachyarrhythmias (Smith 2000).

For further reading see Bruner and Leonard (1989).

Infection control

The acquired immune deficiency syndrome (AIDS) epidemic has served to highlight the longstanding problem of the risk of transmitting infection to patients or personnel through the use, and particularly the re-use, of electrodes. Apart from the human immunodeficiency virus (HIV), the infections most likely to be transmitted are Creutzfeldt–Jakob disease (CJD) and viral hepatitis. More recently, methicillin-resistant *Staphylococcus aureus* (MRSA) has become a major concern among hospital infections, as it is difficult to treat and can be transmitted directly from patient to patient, or via staff or infected electrodes. Detailed recommendations for the sterilisation of electrodes (and other measures) have been published by the American Electroencephalographic Society (1984) and by the Association of British Clinical Neurophysiologists and the British Society for Clinical Neurophysiology (Evans et al 1993). Approved practices are likely to change and to differ between countries and institutions and (not least for medico-legal reasons) the reader is advised to determine and follow local regulations. Only some *very* general principles and typical practices are summarised here.

Although some special precautions may be taken in dealing with known carriers of infection, the only safe practice is to assume that a risk is always present. Electrodes that are intended to penetrate the skin expose the technologist to the risk of infection by needle-stick injury or by blood coming in contact with broken skin. There is also obviously a risk of patient-to-patient transmission by the electrodes themselves. Skin preparation before use of surface electrodes may result in bleeding or oozing of serum, which may contaminate both the electrodes and the hands of the operator. Needles used for applying jelly under scalp electrodes are particularly likely to become contaminated.

Any lesions (cuts, abrasions, burns, eczema) on the operator's hands should be covered with a waterproof dressing. Disposable gloves should be worn during electrode application, removal and cleaning. Particular care should be taken to avoid touching other equipment when wearing contaminated gloves. If possible, sinks used to clean electrodes should not be used for hand-washing. Electrode application to patients known (or suspected) to be infected with MRSA should be carried out with full physical protection: mask, gown, plastic gloves and plastic apron. EMG needle electrodes should be cleaned in an ultrasonic cleaning bath and autoclaved (e.g. at 121°C for 15 min) after every use. Needles for introducing jelly require similar precautions and should be flushed with clean water before sterilisation, or should be disposable. The use of EEG needle electrodes is not recommended. Scalp disc electrodes should be cleaned in hot detergent after every use, including brushing to remove fragments of jelly or adhesive, and the contacts, but not the leads, immersed in 10,000 ppm sodium hypochlorite for 10 min after each use. After use in patients with known or suspected infection with CJD or MRSA, disc electrodes should be autoclaved at 134°C at 30 psi for 18 min or for six cycles of 3 min each. This is practicable only if the insulation of the leads has a high melting point (e.g. Teflon); it may be simpler to discard the leads into a suitable container for incineration. All electrodes should be washed thoroughly after immersion in disinfectant.

Equipment should be swabbed with 10,000 ppm sodium hypochlorite if contaminated by 'high-risk' biological fluids (blood, semen, female genital tract secretions, cerebrospinal fluid, amniotic fluid, pericardial fluid, synovial fluid, etc.); saliva, urine and faeces are not normally considered a risk for blood-borne viruses unless visibly blood-stained, although other infections such as *Salmonella* and tuberculosis could, of course, be present. As sodium hypochlorite corrodes metal it should be washed off after 10 min. Two per cent glutaraldehyde is a potent and useful disinfectant for all except CJD, but is not suitable for routine swabbing as it is likely to sensitise staff. Contaminated bedding, etc. should be handled according to locally recommended procedures.

Finally, the importance of following infection precautions in a manner that preserves the dignity of the patient cannot be overemphasised. This is more likely to be achieved where the measures are routine, rather than when they are adopted only exceptionally, when the perceived risk is high.

REFERENCES

American Electroencephalographic Society 1984 Report of the committee on infectious diseases. J Clin Neurophysiol 3(Suppl 1): 38–42.

American EEG Society Guideline Nine 1994 Guidelines on evoked potentials. J Clin Neurophysiol 11: 40–73.

Binnie CD, Cooper R, Mauguière F, et al (eds) 2003 *Clinical Neurophysiology*, Vol. 2. Elsevier, Amsterdam.

Brett EM (ed) 1997 *Paediatric Neurology*, 3rd edition. Churchill Livingstone, Edinburgh.

British Standards Institution 1989 *British Standard Specification for Safety of Medical Electrical Equipment*. Document BS 5724, Part I, General Requirements. British Standards Institution, London.

Bruner JMR, Leonard PF 1989 *Electricity, Safety and the Patient*. Year Book Medical Publishers, Chicago, IL.

Evans B, Kriss A, Jeffries D, et al 1993 British Society for Clinical Neurophysiology guidelines for preventing transmission of infective agents and toxic substances by clinical neurophysiology procedures: an update. J Electrophysiol Tech 19: 129–135.

Eyre A (ed) 1992 *The Neurophysiological Examination of the Newborn Infant*. Clinics in Developmental Medicine, No. 120. MacKeith Press (distributed by Cambridge University Press), London.

Holmes GL, Moshe SL, Jones HR (eds) 2005 *Clinical Neurophysiology of Infancy, Childhood, and Adolescence*. Butterworth-Heinemann, Boston, MA.

International Electrotechnical Commission (IEC) 1988 *Safety of Medical Electrical Equipment*. Publication 601, Part 1, General Requirements. IEC Secretariat, Geneva.

International Electrotechnical Commission (IEC) 2002 *Particular Requirements for Encephalographs*. Publication 60601, Part 2–26. IEC Secretariat, Geneva.

Levene MI, Bennett MJ, Punt J (eds) 2001 *Fetal and Neonatal Neurology and Neurosurgery*, 3rd edition. Churchill Livingstone, Edinburgh.

Smith NJ 2000 Patient safety in NCS clinics and theatres. Electrical nerve stimulation in patients with cardiac pacemakers. J Electrophysiol Technol 26: 177–180.

Volpe JJ (ed) 2000 *Neurology of the Newborn*, 4th edition. WB Saunders, Philadelphia.

Chapter 2

General Characteristics of the EEG

ORIGIN OF THE EEG

The electroencephalogram (EEG) is a signal recorded from the scalp and derived from the electrical activity of cortical neurons. A similar recording made directly from the surface of the cerebral cortex is termed the electrocorticogram (ECoG). Specifically, the signal is a spatiotemporal average of excitatory and inhibitory postsynaptic potentials. This description requires some explanation.

If neurons fire asynchronously, the averaged activity recorded at relatively distant and large electrodes will show no signal. Only when neurons undergo similar and simultaneous electrical changes does the temporal averaging process produce any recordable activity. Similarly, unless the electrical fields generated by the neurons are in alignment they will be cancelled out by spatial averaging. Practically, the EEG represents the synchronous activity of neurons arranged at right angles to the surface of the cerebral cortex and of the scalp. In a group of pyramidal neurons aligned in parallel and at right angles to the surface of the cortex, depolarisation of apical dendrites by excitatory postsynaptic potentials will render the cortical surface negative with respect to deeper layers (*Fig. 2.1*). Conversely, depolarisation of the somas produces a positive change on the surface with respect to the depth. Hyperpolarisation by inhibitory postsynaptic potentials will have the opposite effects. Some 6 cm² (1 inch by 1 inch) of cortex must be synchronously active to generate signals detectable in the EEG (Cooper et al 1965). For further details of the underlying physiological principles, see Cooper et al (2005) or any standard textbook of neurophysiology.

It may be noted that, as two-thirds of the cortex is folded into the sulci, only the one-third on the crests of the gyri contributes directly to the EEG. It is also an interesting paradox that neuronal populations mostly fire asynchronously when transmitting and processing information and tend to go into synchronous oscillation when not otherwise active; hence, the EEG largely reflects apparently non-productive brain activity. During arousal and concentration, EEG activity is fast and of low amplitude; it increases and slows during relaxation, and becomes yet slower and of maximum amplitude in sleep or coma. The inverse relationship between evident adaptive activity and the EEG may explain the failure of the latter to give any profound insights into psychophysiology and why clinical EEG interpretation remains an empirical study with little basis in known physiology.

EEG PHENOMENOLOGY

Describing EEG phenomena

The first stage in the clinical interpretation of an EEG is the identification of its various components and the recognition of any possibly abnormal features. The normal findings vary with age, consciousness, state of health and between individuals. The EEG changes throughout childhood, approaching the adult pattern by the end of the second decade. For didactic purposes, this

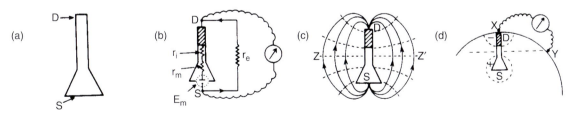

Fig. 2.1 The current flow due to local changes of membrane characteristics can produce potential differences on the cortical surface. The soma S and apical dendrite D are shown diagrammatically in the resting state in (a). The membrane of the entire cell is uniformly polarised, the inside being 80 mV negative with respect to the outside. Suppose an excitatory input causes the tip of the apical dendrite to depolarise completely; that is, the membrane potential in this region becomes zero. Ionic current will now flow through the cell and external fluid. This is shown in (b) as the membrane potential E_m causing current flow in the internal resistance r_i, membrane resistance r_m, and external resistance r_e. The external current will flow in all regions surrounding the cell and in the case of an isolated neuron gives rise to equipotential lines, shown dotted in (c). Potential differences can be measured from X to a distant point Y (d). (After Kiloh et al (1980), by permission.)

adult picture is used here as an archetype and the evolution from infantile to adult patterns is then described.

A brief excerpt from a typical EEG recording from a healthy awake adult is shown in *Fig. 2.2* and illustrates the main features used to describe EEG phenomena. These are:

1. *Wave shape (morphology).* When the eyes are closed, there is a rhythmical, though not sinusoidal, waveform of regularly fluctuating amplitude. This is an example of the alpha rhythm (see p. 14). Activity that persists for much of the time during a particular state (as here, awake with eyes closed) is described as 'ongoing'. Other examples of wave shape are shown in *Fig. 2.7* (see p. 11).

2. *Rhythmicity.* The regular fluctuation about the baseline gives a rhythmical nature to the activity, which is typical of the EEG. More complex, pathological waveforms may also be rhythmical. *Figure 2.3* shows the regular spike and wave (SW) pattern typically seen during an absence seizure. In some conditions, the wave pattern has no obvious rhythmicity, as in *Fig. 2.4*, and is described as irregular.

3. *Spatial distribution.* The amplitude of the alpha rhythm in *Fig. 2.2* is larger towards the occipital area, decreasing anteriorly, but is approximately symmetrical. The SW discharge in *Fig. 2.3* is also symmetrical but the amplitude is greatest anteriorly. In *Fig. 2.4*, the activity is localised to a particular part of the head. Such a distribution is often due to local brain pathology. More specific localising features are discussed on page 47.

4. *Transients.* Against a background of ongoing activities, discrete transient waves, or brief stereotyped sequences of two or more waves (complexes) may occur. Some are apparently spontaneous; others are evoked by extrinsic or intrinsic stimuli. Those elicited by known stimuli or cognitive events are designated evoked, or event related, potentials (see Chapter 3).

Some transients are normal, others pathological; it is therefore important to identify the different types. This is helped by the annotation of the record by the technologist, who should note intrinsic (e.g. patient moving) or extrinsic (e.g. telephone ringing) events.

There are two examples of transient signals in *Fig. 2.2*. First, a few minor fluctuations occur followed by several major ones, mainly in channels 1 and 5, coincident with eye movements. These are not

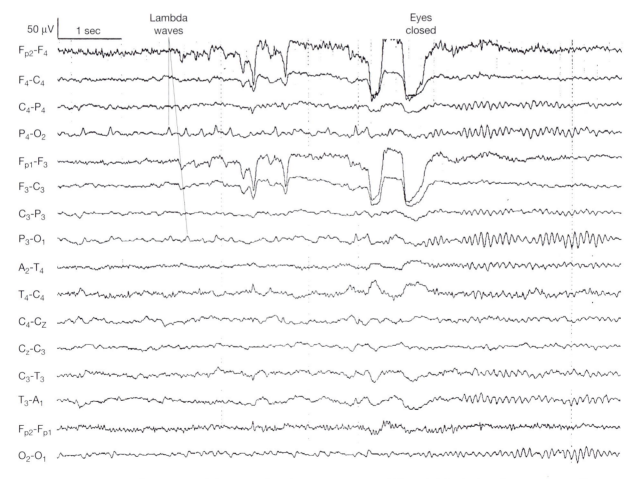

Fig. 2.2   Normal adult EEG showing lambda waves when the eyes are open and bilateral alpha rhythm after eye closure. (Courtesy of Dr NM Kane.)

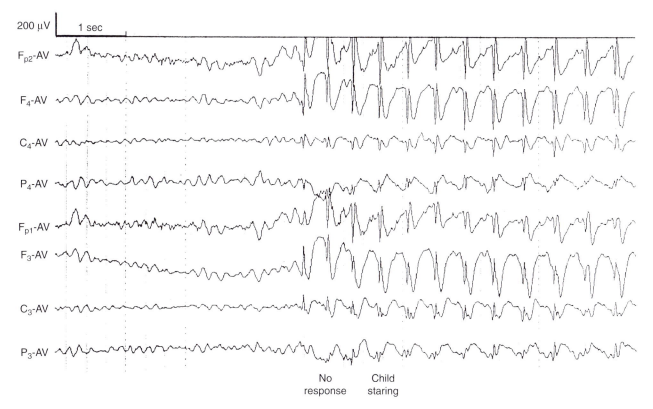

200 μV | 1 sec

Fp2-AV

F4-AV

C4-AV

P4-AV

Fp1-AV

F3-AV

C3-AV

P3-AV

No response   Child staring

Fig. 2.3 Regular SW activity at the beginning of an attack in a child with absence seizures. Average reference recording. (Courtesy of Dr NM Kane.)

of cerebral origin but artefacts due to movement of the eyeballs and eyelids on blinking and eye closure (see Barry and Jones 1965, Westmoreland, 1975). They have a very specific spatial distribution, with higher amplitude towards the front of the head.

Artefacts are described in more detail on page 22. Secondly, small transient waves of cerebral origin are seen, each lasting about 0.2 s in the occipital derivations (channels 4 and 8) when the subject's eyes are open. These particular transients are lambda

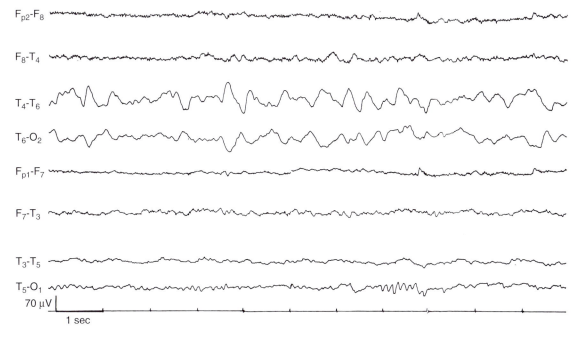

Fp2-F8

F8-T4

T4-T6

T6-O2

Fp1-F7

F7-T3

T3-T5

T5-O1

70 μV

1 sec

Fig. 2.4 Localised slow activity in the right posterior temporal region due to a cerebral tumour. (From Binnie et al (2003), by permission.)

waves. They also have a characteristic distribution and responsiveness as described on page 17.

Transient potentials may particularly be evoked by internal or external events if the subject is drowsy or asleep (e.g. the K-complex, see p. 36). An important, usually abnormal, transient that occurs spontaneously is the spike, a wave with a pointed peak clearly distinguished from background activity with a duration of 20–70 ms (see *Fig. 2.7(d)*).

5. *Reactivity*. Various EEG activities, both normal and abnormal, alter or react in response to many factors, both intrinsic and extrinsic. A common reaction is illustrated in *Fig. 2.2*, where the alpha rhythm occurs when the eyes are closed but is not seen in the initial eyes-open period. This attenuation, or 'blocking', of the alpha rhythm when the eyes are opened is also illustrated in *Fig. 2.8(a)* where the blocking is not sustained, the alpha activity returning after several seconds. Blocking is generally symmetrical, and marked asymmetry would be regarded as abnormal. Sometimes the reactivity is paradoxical. For example, alpha activity usually disappears during extreme drowsiness, but returns when the subject is alerted and opens his or her eyes (see p. 36).

6. *Frequency of repetition*. An ongoing EEG signal can usually be described in terms of the number of times the waveform repeats itself in 1 s. This is termed the frequency of the signal. Frequency can apply to a rhythmical or to an irregular signal.

There is an arcane discussion amongst neurophysiologists as to whether repetition rate should be measured in hertz (Hz), or 'cycles per second' (c/s). Both are used here, but for transients, which are discontinuous and not necessarily regular, it may be appropriate just to state the rate, e.g. as 'spikes per second'.

The frequencies seen in the ongoing EEG are classified into four arbitrary bands. Waves with frequencies below 4 Hz are delta waves; those from 4 Hz to less than 8 Hz are called theta waves; the 8–13 Hz range is the alpha band; and frequencies from more than 13 Hz to 40 Hz are called beta waves. Examples of activity in these frequency bands are shown in *Fig. 2.5*. A fifth range, above 40 Hz, called gamma, is of interest to neurophysiologists, but it has no established clinical significance (Bressler 1990), except at the onset of some epileptic seizures.

When an ongoing signal is irregular or 'polymorphic' (i.e. contains a mixture of frequencies), it may be difficult to assign it to a particular frequency band. An example of irregular delta activity is shown in *Fig. 2.6*. Isolated non-rhythmic events may also be described in terms of equivalent frequency. Thus, a wave of 300 ms duration is considered to have a frequency of 3 Hz and is described as a delta wave.

7. *Amplitude*. EEG signal amplitude may range from very small, a few microvolts, up to several hundred microvolts. The amplitude with which a phenom-

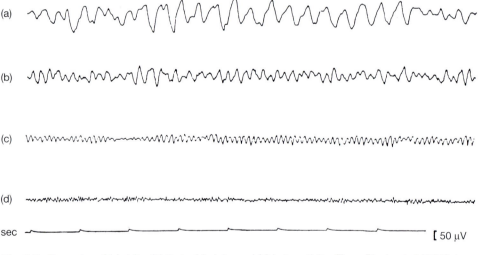

(a)

(b)

(c)

(d)

sec [50 μV

Fig. 2.5 Examples of (a) delta, (b) theta, (c) alpha and (d) beta activity. (From Binnie et al (2003), by permission.)

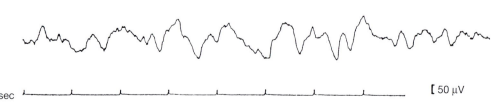

sec [50 μV

Fig. 2.6 Irregular (polymorphic) delta activity. (From Binnie et al (2003), by permission.)

enon is displayed depends on its distribution relative to the electrodes and the method of derivation (see p. 18). When visually interpreting an EEG, it is usual to estimate the average peak-to-peak amplitude of the waves, which results in a somewhat larger value than is mathematically correct.

The waxing and waning in amplitude of ongoing activity is often described as 'spindling' (see *Fig. 2.7(g)*). A group of waves that appears and disappears abruptly and is distinguished from background activity by differences in frequency, form and/or amplitude, is called a 'burst' (see *Fig. 2.7(i)*).

8. *Spatiotemporal patterns*. A particular activity may be characterised by a combination of frequency and spatial distribution. For example, apart from the alpha rhythm, there is another alpha frequency activity with a more central site of origin. This is the mu rhythm discussed below. It is therefore important to be able to recognise particular spatiotemporal patterns of EEG activity.

When there is brain pathology, the spatiotemporal pattern may help to localise the brain area involved. The recognition of specific spatial patterns, and their localisation, requires the use of both bipolar and reference electrode derivations (see p. 18).

9. *Symmetry and synchrony*. If an EEG feature such as a spike apparently occurs simultaneously in two locations, it is described as synchronous. There may, however, still be a small time difference between the two deflections, detected only by expanding the time scale or the use of instrumental analysis, which indicates that they are not strictly synchronous. Such subtle asynchronies may be of clinical importance, for instance when an abnormal activity is rapidly propagated from one hemisphere to the other.

Similarly, a rhythmical component that occurs bilaterally may have a small time displacement between comparable waves on either side. Special measurement techniques usually show that the degree of synchrony between the alpha peaks in the two hemispheres varies continually, with one side leading for a few seconds, then the other (Hoovey et al 1972). More extremely, if the rhythm on one side is of different frequency from that on the other (measured over an interval of about 1 s), then the two are unequivocally asynchronous. On the other hand, over a greater time interval, say several seconds, runs or spindles of activity such as mu rhythm (see below) may loosely be called 'synchronous' if they occur more or less simultaneously on each side.

Similar considerations apply to amplitude asymmetry. The peak-to-peak amplitudes of a transient recorded in two or more channels can be measured and compared, and for an ongoing activity, such as the alpha rhythm, average amplitudes on either side can be estimated to decide whether the record is significantly asymmetrical. However, sleep spindles,

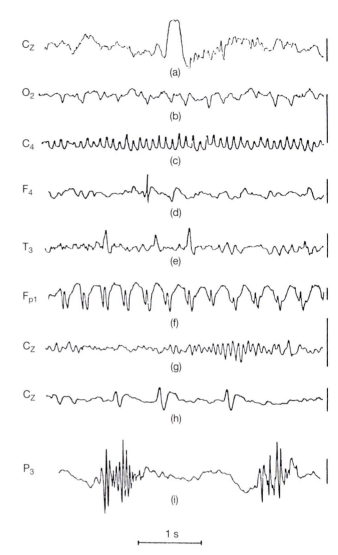

Fig. 2.7 Examples of waveforms referred to in the text, each recorded from the electrode indicated with respect to an electrode on the chin: (a) K-complex; (b) lambda wave; (c) mu rhythm; (d) spike; (e) sharp waves; (f) spike and wave; (g) sleep spindle; (h) vertex sharp waves; (i) multiple spikes. In all examples an upward deflection corresponds to negativity of the specified electrode with respect to the chin. Calibration marks on the right = 100 µV. (From Binnie et al (2003), by permission.)

which have a tendency to occur in discrete, non-simultaneous runs on either side, may still be regarded as 'symmetrical' if the overall amounts of activity are substantially the same.

The alpha rhythm commonly shows a slight asymmetry, with the higher amplitude over the right hemisphere. This may be related to handedness but is a subject of lasting controversy (Butler 1988, Davidson 1988, Hoptman and Davidson 1998). All normal rhythms may show some asymmetry, but asymmetries exceeding 50% may be abnormal.

An apparent asymmetry of the EEG may, of course, be due to asymmetrical electrode placement or spacing, or other technical errors, and this possibility should be checked by the technologist during the recording.

To summarise, these characteristics of waveform (wave shape or morphology), spatial distribution and responsiveness, frequency, amplitude, symmetry and synchrony, shown in the figures, are typical of the features used to describe the EEG. Some other waveform examples are illustrated in *Fig. 2.7* and are discussed below.

Because of the very wide variation in the appearance of records that are classed as normal, it is not possible to give an illustrated catalogue of all patterns likely to occur. However, the extremes can be typified by a few examples, and these are illustrated in *Fig. 2.8*. The dominant

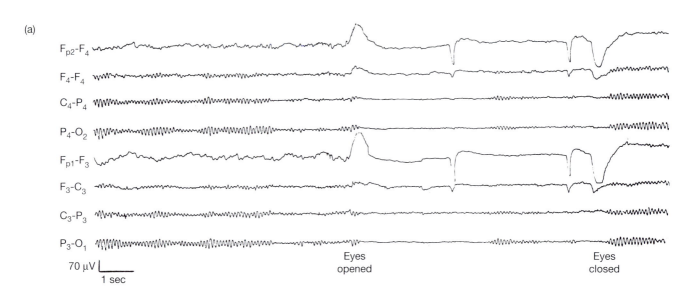

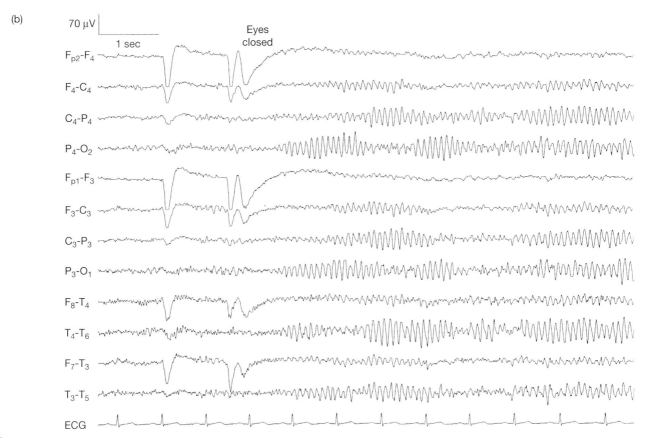

Fig. 2.8 Types of alpha rhythm. (a) Monorhythmic alpha rhythm, initially blocking on eye opening and reappearing on eye closure. (b) Asymmetrical alpha spindling (see text). (Courtesy of Dr RV Johnston.)

activity in (a) of this figure, while the eyes are closed, is an alpha rhythm with a constant frequency and wave-form, with maximum amplitude in the posterior areas and appearing simultaneously and more or less synchron-ously over the two hemispheres. In (b) the alpha rhythm, when the eyes are closed, has a more complex pattern with spindles, i.e. with groups of waves characterised by an amplitude that progressively waxes and wanes. In this example the spindles do not occur simultaneously over the channels. For instance, the figure shows that 2–3 s after eye closure the amplitude is increasing in the C_4–P_4 and C_3–P_3 channels, whereas it is decreasing in the P_4–O_2 and P_3–O_1 channels. The third example (c) shows a low amplitude record with alpha activity only just visible, but enhanced for a second or two when the eyes are first closed. Such low amplitude records may be seen in adolescents and are discussed on page 16.

IFCN recommendations and definitions
The International Federation for Clinical Neurophysiol-ogy (IFCN) exists to promote a high standard of technical and clinical expertise and understanding in electro-encephalography and clinical neurophysiology (ECN). The IFCN committees report on particular aspects of ECN to further these ends. Their recommendations are

published in the journal *Clinical Neurophysiology* (formerly *Electroencephalography and Clinical Neurophysiology*). Of particular value is a glossary of terms defining specific EEG phenomena; these are used in the descriptions of those phenomena seen in the EEG of the normal awake adult discussed below (Noachtar et al 1999). The use of recommended terminology is advocated to avoid ambi-guities of description. Such ambiguities, for example whether a particular transient should be classified as a spike, are a source of unreliability in interpretation (see p. 18).

The decisions of the International Federation of Soci-eties for Electroencephalography and Clinical Neuro-physiology (IFSECN) committees current in 1981, together with a glossary, were published in book form (IFSECN 1983), based in turn on earlier reports (IFSECN 1958, 1974). A volume of more recent updates was published in 1999 (Deuschl and Eisen 1999).

Ongoing activities
The term 'ongoing activity' introduced above has not been defined by the IFCN. It is often used as a synonym for 'background activity', to which the IFCN does attach a specific, but slightly different, meaning ('any EEG activity representing the setting in which a given normal

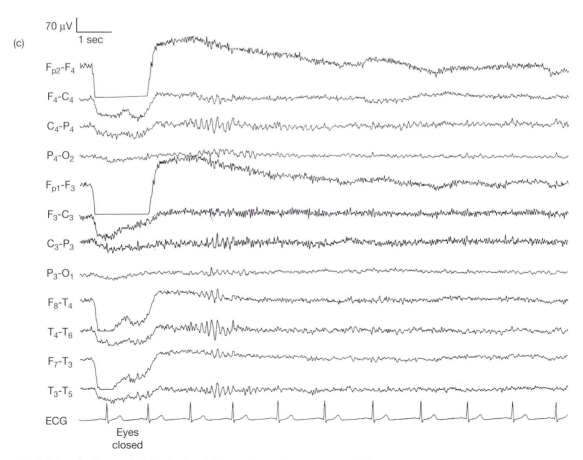

Fig. 2.8 (*cont.*)   Types of alpha rhythm. (c) Low voltage, beta dominant EEG showing only a brief run of alpha waves 2 s after eye closure. (Courtesy of Dr RV Johnston.)

or abnormal pattern appears'). A transient EEG event is 'any isolated wave or complex, distinguished from background activity'. Here we define several ongoing activities that are recognised by their rhythmical nature and the fact that they are easily classified into one of the established frequency bands. Transient activities are considered on page 16.

Alpha activity The range of frequencies from 8 to 13 Hz is called the alpha band and any activity within this band is called alpha. The alpha rhythm is a more specific phenomenon defined by the IFCN as a:

> rhythm at 8–13 Hz occurring during wakefulness over the posterior regions of the head, generally with maximum amplitudes over the occipital areas. Amplitude varies but is mostly below 50 μV in the adult. Best seen with the eyes closed and during physical relaxation and relative mental inactivity. Blocked or attenuated by attention, especially visual, and mental effort.

The alpha rhythm is the dominant EEG activity in the healthy awake adult. In children it is usually seen from around 3 years of age, and absence of alpha in an alert child aged 8 years or over has to be considered abnormal. The amplitude of alpha in children is usually higher than in adults, being between 50 and 100 μV.

Activities in the alpha band that differ from the alpha rhythm with regard to their topography and/or reactivity

either have specific names (e.g. mu rhythm) or should be described generically as 'alpha activity'.

Beta activity Activities with frequencies in the range > 13–40 Hz are referred to as beta, defined by the IFCN as:

> In general: any EEG rhythm between 14 and 40 Hz. Most characteristically: a rhythm from 14 to 40 Hz recorded over the fronto-central regions of the head during wakefulness. Amplitude of fronto-central beta rhythm is variable but is mostly below 30 μV. Other beta rhythms are most prominent in other locations or are diffuse.

Beta activity is usually a normal EEG phenomenon (see Fig. 2.8(c)), often increasing with drowsiness or light sleep.

Some drugs (e.g. barbiturates and benzodiazepines) increase the amplitude of beta activity, but it can be quite prominent in some normal individuals free from drugs (Fig. 2.9). Amplitudes above 30 μV are rare in normals but can occur. Beta activity may also be asymmetrical and, because it is usually asynchronous between the two hemispheres, its amplitude is often greater in transverse interhemispheric than in intrahemispheric derivations.

Beta activity is often regarded as replacing alpha activity when the latter is blocked by eye opening or by mental arousal. However, there is evidence that beta activity is present all the time, but is normally masked by the much larger amplitude of alpha activity (Daniel 1966,

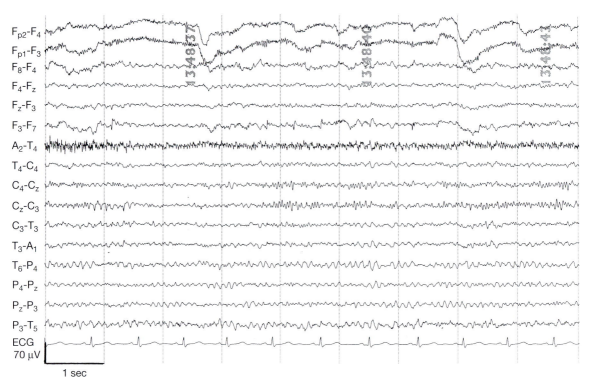

Fig. 2.9 Beta activity in anterior derivations (channels 1 and 2) and in transverse derivations about the vertex (channels 9 and 10). Note difference from muscle activity in channel 7. (Courtesy of Dr WL Merton.)

Gengerelli and Parker 1966, Goncharova and Barlow 1990). (See also 'fast alpha variant' below.)

Theta activity   EEG activity in the frequency range of 4 Hz to less than 8 Hz is called theta. The term originates from Grey Walter, who associated activity at theta frequencies with tumours in the region of the thalamus (Walter and Dovey 1944).

Like the alpha rhythm, its incidence and amplitude vary with age and with states of consciousness and health. In pre-school children, theta activity, mostly rhythmic, dominates the EEG. In school children and adults irregular low amplitude theta activity is a usual feature, and in the awake state is usually of greatest amplitude in the posterior temporal regions. It is often difficult to decide whether prominent theta activity is excessive in a particular individual, and there is a danger of giving it more clinical significance than is warranted, if the activity is not obviously focal or of high amplitude. Shinomiya et al (1994) have attempted to differentiate normal midfrontal theta rhythm from similar activity associated with pathology.

High amplitude rhythmic theta (hypnagogic hypersynchrony) is seen during drowsiness in younger children (see p. 38). Lower amplitude theta activity occurs naturally with the onset of drowsiness and light sleep in older children and adults. If its development during a recording is accompanied by a decrease in alpha activity, the onset of sleep should be suspected (*Fig. 2.10*). The technologist should observe this and annotate the recording appropriately.

Delta activity   Delta designates activity with a frequency below 4 Hz. This term also was originally coined by Grey Walter (1936) for all EEG rhythms with a frequency below that of alpha rhythm. With the recognition of theta rhythm as a functionally independent activity, the term delta was assigned to frequencies below 4 Hz.

There is no defined lower limit to the delta frequency band but the low frequency response of conventional EEG amplifiers limits the lowest frequencies that are recorded. Examples of delta activity are shown in *Figs 2.4* and *2.6*.

Activity at delta frequency is prominent in normal children during the first year of life and discrete delta waves are conspicuous posteriorly throughout maturation. Delta activity occurs in deep sleep. In the awake adult, rhythmical delta activity is usually an abnormal sign.

Instrumental frequency analysis of the EEG usually shows more delta activity than is seen by visual inspection. This is probably because small, slow fluctuations of the baseline tend to be masked by other activity.

Alpha variants   A small proportion of children and adults with no known brain disease have activity at theta or, more rarely, beta frequency arising from posterior regions in their EEG. The term 'alpha variant rhythms' or 'alpha variants' has been given to this activity, defined in the IFCN glossary as:

> *certain characteristic EEG rhythms which are recorded most prominently over the posterior regions of the head and differ in frequency but resemble in reactivity the alpha rhythm.*

Both slow and fast alpha variants are defined as being blocked by attention, especially visual and mental effort.

Slow alpha variant is a normal finding in the age range from 4 to about 23 years. The activity is usually at a subharmonic of the alpha frequency (*Fig. 2.11*). Fast alpha variant is also regarded as a normal phenomenon (Goodwin 1947), but is very rare in children. It can be induced by benzodiazepine drugs. Like the slow variant, it has a topographic distribution resembling that of the alpha rhythm. If a coexisting alpha rhythm is present, fast alpha variant is at a frequency twice that of the alpha activity.

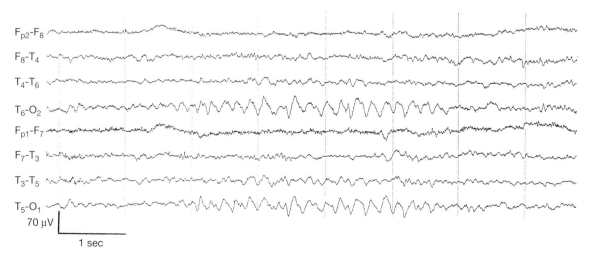

Fig. 2.10   Bilateral occipital theta activity during drowsiness. (Courtesy of Dr WL Merton.)

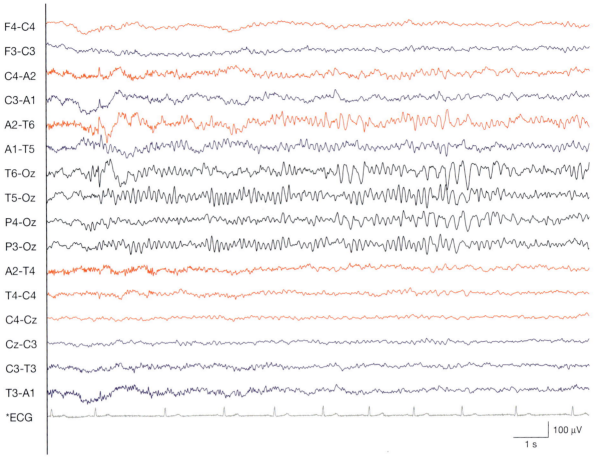

Fig. 2.11   Slow alpha variant: 10-year-old boy, recorded with the eyes closed. The basic alpha frequency of 10 Hz predominates but is combined with a subharmonic at 5 Hz in the occipital regions.

The low voltage EEG   An EEG having a mean peak-to-peak amplitude of 20 µV or less is classed as of low voltage (see *Fig. 2.8(c)*). This is a normal, often genetically determined EEG trait in adults. It is very rare in children under 15 years old. Eeg-Olofsson et al (1971) recorded routine EEGs in 743 healthy children aged 1–15 years and reported not one EEG with single low voltage alpha. It starts to appear after the age of 13 years (Gibbs and Gibbs 1950) and is seen in 7% of 20–39 year olds (Adams 1959). In children under the age of 10 years a low voltage EEG should be considered abnormal, particularly if hyperventilation and sleep do not change the amplitude. The description of such EEGs as 'flat' is best avoided, not only because of the imprecision of this description, but also because of possible confusion with an isoelectric state seen in irreversible coma when no EEG activity may be recordable. However, subjects with a habitual pattern of normal amplitude may produce a low voltage record if anxious or if the conduct of the investigation is not conducive to relaxation.

Very slow and very fast brain potentials   Developments in amplifier and electrode technology have enabled an extension of the measurement of slow brain potentials down to very low frequencies. Some of these potentials are so slow as to constitute changes in direct current (DC) levels. They have been related to both behaviour and psychopathology (Rockstroh et al 1989, Birbaumer et al 1990, McCallum and Curry 1993, Vitouch et al 1997), and have been found in premature infants (Vanhatelo et al 2002).

At the other end of the frequency scale, it is now recognised that oscillations above 40 Hz, called gamma activity, play an important part in cognitive brain physiology. Gamma frequencies are not seen in traditional EEG records and are not known to be of clinical significance, except during seizures (Medvedev 2001, Wheless and Kim 2002).

Non-pathological localised or transient phenomena

Transient waveforms in the EEG, even if apparently spontaneous, are usually the response to intrinsic or extrinsic (environmental) changes (see evoked potentials, Chapter 3). It is important to recognise transients that are normal phenomena. Some examples are shown in *Fig. 2.7*.

Lambda waves The transient waves in *Fig. 2.2* have already been designated as lambda waves. Characteristically, they consist of isolated electropositive sharp waves of up to 30 μV in amplitude and 200 ms in duration, arising at the occiput but sometimes extending into the parietal regions. The waveform is saw-toothed rather than spike-like. Lambda waves are elicited by scanning a patterned visual field. Any abrupt change of contrast in the field, such as a line, acts as an evoking stimulus. These waves are most commonly seen in children aged 2–15 years, but may be seen in younger children.

Mu rhythm Many EEG records show activity at alpha frequency arising from central regions, often with a specific waveform and with a reactivity differing from the occipital alpha. This is called mu rhythm (Gastaut 1952, Gastaut et al 1952). An analogous phenomenon, the sensorimotor rhythm, is seen in animals (Kulman 1980). Mu rhythm is defined by the IFCN as:

rhythm at 7–11 Hz, composed of arch-shaped waves occurring over the central or centroparietal regions of the scalp during wakefulness. Amplitude varies but is mostly below 50 μV. Blocked or attenuated most clearly by contralateral movement, thought of movement, readiness to move or tactile stimulation.

Mu rhythm usually has a sharp negative peak to each wave (showing the presence of a second harmonic) and giving it a characteristic wave shape, resembling the Greek letter μ. It was originally called 'comb' or 'wicket' rhythm (American usage, as in 'wicket fence'), or *rythme en arceau*. *Figure 2.12* shows an example of mu rhythm with the effect of fist clenching.

Mu activity may be asymmetrical in amplitude and asynchronous in the two hemispheres without the presence of brain pathology. However, a frequency asymmetry of as little as 0.5 Hz is usually abnormal. Pathological variants, often related to previous craniotomy and designated 'breach' rhythms, are discussed on page 57.

Posterior temporal slow waves The EEG records of children and young adults contain posterior slow activity, both slow alpha rhythms and more localised, isolated delta waves. These may present interpretative difficulties, as the occurrence of such activity in older subjects, or its presence in excessive amounts in the young, would be regarded as abnormal. It may take a variety of waveforms and, when mixed with alpha activity, the result is a complex wave pattern, which may be misinterpreted as a spike-and-slow wave complex. The slow waves themselves appear larger and sharper with common reference

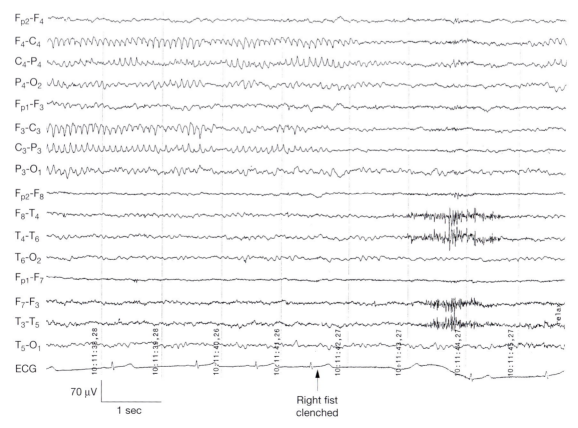

Fig. 2.12 Mu rhythm (phase-reversed about electrodes C3 and C4) responding to right-fist clenching, particularly on the left side. (Courtesy of Dr NM Kane.)

than with bipolar derivation. An example of posterior slow activity is shown in *Fig. 2.13*.

Early reports of this activity regarded it as an abnormal sign (Jasper et al 1938, Lindsley and Cutts 1940), and it does appear to be more prevalent in children with behaviour disorders and in adults with psychopathology. However, it may simply be a reflection of an immature brain and is now regarded as normal up to the early twenties, unless excessive for the age of the patient.

Non-pathological spiky waveforms   Mention has already been made of spikes and SW activities, which are of particular clinical significance in relation to epilepsy. Certain spiky waveforms, particularly those accompanying drowsiness and sleep, occur in healthy subjects. They have various characteristics by which they can, and must, be clearly distinguished from those spiky transients associated with epilepsy. In order to highlight the distinction, they are described, together with abnormal spikes, in the section on pathological EEG phenomena related to epilepsy (see pp. 51–57).

DERIVATION AND MONTAGES

Introduction

The EEG consists of the changes with time of the electrical field on the scalp. It is thus a phenomenon existing in five dimensions, three of space, and one each of electrical potential and time. Fortunately perhaps, for those who have to interpret EEGs, five-dimensional display systems are not available and it is necessary to represent this information by means of two-dimensional charts or graphics.

By the use of a stylised head outline, the three dimensions of the scalp can be represented in two, and the electrical field shown by means of isoelectric contour lines (*Fig. 2.14*). This is a convenient way of displaying the field at any instant in time and will be used repeatedly for didactic purposes in this chapter.

The EEG is usually displayed (formerly on a chart recorder but nowadays on a high resolution visual display unit) as an array of plots of potential difference against time. Topographical information is provided by the use of multiple traces and the human observer must learn to interpret these to determine the salient characteristics of the underlying electrical field. Each voltage/time tracing, or 'channel', presents the potential difference between two electrodes (or sometimes between one electrode and a reference point in the EEG machine; see p. 22). There are various systems or 'methods of derivation' for making these connections, which may facilitate (or hinder) the task of the interpreter. Unfortunately, the method of derivation used so markedly affects the appearance of the trace that to understand EEG traces at any but a most superficial level it is necessary to have some understanding of methods of derivation and their consequences. This chapter attempts to provide this information on a need-to-know basis, for users rather than providers of EEG services. For a comprehensive, detailed account see Cooper et al (2005).

When Berger first recorded the human EEG with a single channel he demonstrated only that a changing potential difference existed between an electrode at the front of the head and another at the back (*Fig. 2.15(a)*); this simple display gave no indication of the topography of the underlying electrical field. Using two channels with a moveable, exploring electrode common to both,

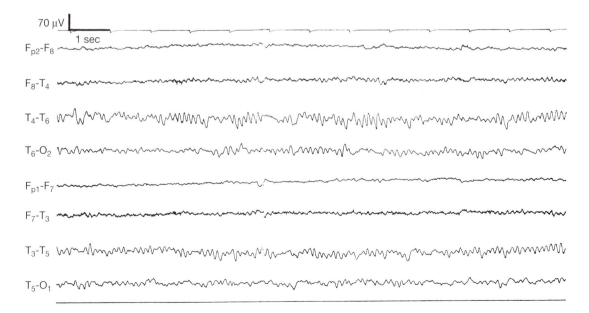

Fig. 2.13   Irregular posterior temporal slow activity mixed with alpha rhythm. (From Binnie et al (2003), by permission.)

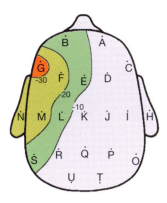

Reference: H

Fig. 2.14 Representation as a 'brain map' of the electrical field over the scalp at an instant in time. A focal feature is present at electrode G in the left frontal region. To display this focal event each electrode has been referred to the right ear. The contour map shows here, as elsewhere, the electrical field at the peak of the transient waveform. (From Binnie et al (2003), by permission.)

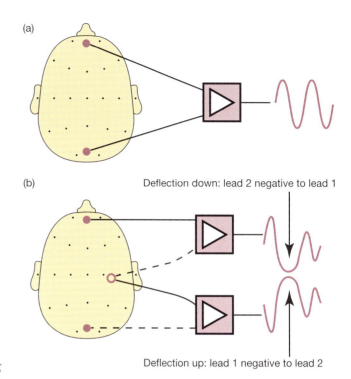

Fig. 2.15 Simple EEG displays. (a) Single channel recording demonstrates only that a changing potential difference exists between the front and back of the head. (b) Localisation of a wave by 'phase reversal' at the electrode common to two recording channels. (From Binnie et al (2003), by permission.)

Grey Walter in 1936 showed that when the exploring electrode was overlying a cerebral tumour, slow waves were recorded on both channels (*Fig. 2.15(b)*).

Grey Walter adopted a convention whereby the two connections to each channel were designated as 'lead 1' and 'lead 2' (shown in *Fig. 2.15(b)* and subsequent figures as full and broken lines, respectively), such that when lead 1 was negative with respect to lead 2 the channel concerned deflected upwards. In terms of digital recording systems, each trace displays the difference between the numerical potential values recorded (with respect to an internal reference) from two electrodes, such that when the absolute value at the 'lead 1 electrode' is less than that at the 'lead 2 electrode' the trace is deflected upwards. Grey Walter connected the exploring electrode to lead 1 of one channel and lead 2 of the other, so that the slow waves picked up from the tumour were rendered conspicuous as they caused the two channels to be deflected in opposite directions (so-called 'phase reversal', a phrase that often appears in EEG reports describing a localised feature).

Common reference derivation

The most obvious way of displaying an electrical field on a chart recorder is to connect lead 1 of every channel to a different recording electrode and to connect lead 2 to a common reference point; this method is termed 'common reference derivation'. The use of a reference for a topographic display finds an analogy in cartography, where heights are mapped with respect to mean sea level. In *Fig. 2.16(a)* an EEG event is shown schematically in a 21-channel common reference recording. An electrical field develops such that the left frontal region becomes negative with respect to the rest of the scalp, electrode G corresponding to the most negative region; this field then collapses. The potential distribution at the peak of

this brief event, or 'transient', is also illustrated as a contour map. Such a phenomenon, which causes a small part of the scalp to assume a potential difference with respect to the greater part (or, more strictly, which causes a minority of electrodes to assume a potential difference with respect to the majority), is described as 'focal'. Electrode G is most strongly negative with respect to the larger, unaffected area of the scalp and is thus at or near the site of the focus. The physiological generator producing such an electrical field is usually, but not always, located under the focus on the scalp. If the activity in question is abnormal then the focus often provides an indication of the site of physiological dysfunction.

Note that the isoelectric contour lines in *Fig. 2.16(a)* are labelled in microvolts (μV) with respect to electrode H, which is also used as the reference for the chart display. H, the right ear, was a fortunate choice of reference point, as it lies within a region of the scalp where there are no potential gradients (or contour lines) as it is unaffected by the physiological event producing the focal negative transient at electrode G.

The choice of different reference points (*Fig. 2.16*) would not change the appearance of the contour map, although the numbering of the isopotential lines would be different. It could, however, make a dramatic difference to the appearance of the chart recording. Part (b) of *Fig. 2.16* shows the effect of choosing a midfrontal reference, E, in a region of potential gradients (closely spaced

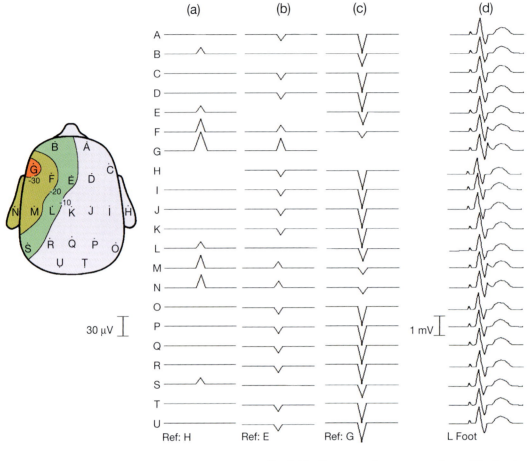

(a)   (b)   (c)   (d)

A
B
C
D
E
F
G
H
I
J
K
L
M
N
O
P
Q
R
S
T
U

30 µV

Ref: H   Ref: E   Ref: G   1 mV   L Foot

(e)

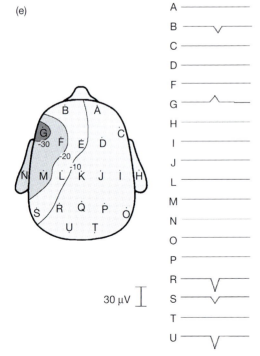

A
B
C
D
F
G
H
I
J
L
M
N
O
P
R
S
T
U

30 µV

Ref: Right-sided electrodes to H, left-sided to N

Fig. 2.16   Common reference recordings of the focal event in *Fig. 2.14* referred to: (a) right-ear electrode H, remote from focus; (b) midfrontal region, electrode E, adjacent to focus; (c) electrode G at the site of the focus; (d) the left foot (see text); (e) common reference recordings of focal event, referred to ipsilateral ears; (f) (see opposite page) comparison of common reference recordings of ongoing activity, referred to ipsilateral ear (middle trace), or nose (left). The greater involvement of the right ear than the left in this activity results in a reversal of the apparent asymmetry at the posterior temporal electrodes. The use of averaged joined ears (below and right) does not solve the problem. (From Binnie et al (2003), by permission.)

contour lines) produced by the left frontal transient. The focus at G remains negative with respect to the reference. There are also upward deflections in the channels corresponding to electrodes F, M and N, which are again negative with respect to the reference on lead 2, and the biggest upward deflection still corresponds to the focus at electrode G. However, as the reference is itself electronegative with respect to a large area on the right of the head, the corresponding channels show small downward deflections (leads 1 positive with respect to reference on leads 2). Why are the channels corresponding to B, L and S flat? Obviously because these electrodes are equipotential with the reference. Once one has grasped how a single EEG event causes some channels to deflect upwards, some downwards and some not at all, it is not difficult to work out its topography. The upward deflecting channels are still a minority, and the most atypical in its behaviour is that corresponding to electrode G which can, therefore, again be identified as the site of the focus.

In part (c) of Fig. 2.16, electrode G itself has been chosen as the reference. As this electrode is negative with respect to all others in the display, all channels deflect downwards (all leads 1 are positive with respect to the reference). Again, once it is understood how this can come about, it is not difficult to identify electrode G, the reference, as being the site of the focal electronegative event.

A deflection common to many channels is, however, not necessarily either an indicator of a focus or an aid to localisation. In part (d) of Fig. 2.16, the left foot has been chosen as a reference point (in the misguided belief, let us suppose, that a reference not sited on the head would give a more easily interpreted display). The heart, which generates its own electrical field, the electrocardiogram (ECG), is located between the reference point and the other recording electrodes. The ECG therefore produces a common potential difference between the reference and all scalp electrodes and appears on all channels of the recording as a signal of relatively high amplitude (note the calibration bar), totally obscuring the EEG.

Common average reference derivation

The difficulties of finding an ideal 'inactive' reference may in part be overcome by referring each electrode to the average of all. This is the 'common average reference' (Goldman 1950, Offner 1950), similar to the 'Wilson' electrode used in electrocardiography (Wilson et al 1934), which assumes the mean potential of all the electrodes in use. A property of the arithmetic mean of a set of numbers is that the sum of differences of those values from their mean is zero. Consequently, there will always be some upward and some downward deflections (i.e. some electrodes negative and some positive with respect to the reference) and the instantaneous deflections will be found to add up to zero. This produces a display that is predictable and will have an appearance (Fig. 2.17) similar to that in Fig. 2.16(b). In the extreme hypothetical case of a field causing one electrode to assume a potential difference of $P\,\mu V$ with respect to all the others in use (Fig. 2.17), the channel recording from the electrode in question will produce a deflection equivalent to $P(1 - 1/n)\,\mu V$ (where n is the number of electrodes), whereas all the other channels will show deflections of opposite polarity equivalent to $P/n\,\mu V$.

Bipolar derivation

Particularly where the number of available channels is limited, topographic features of the EEG may be

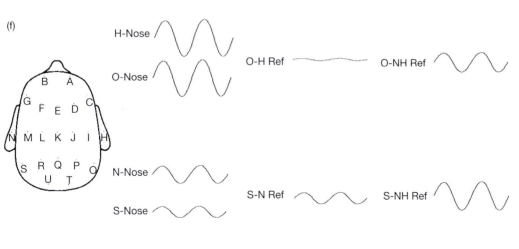

Ave-of H & N

Fig. 2.16 (*cont.*) See caption on opposite page.

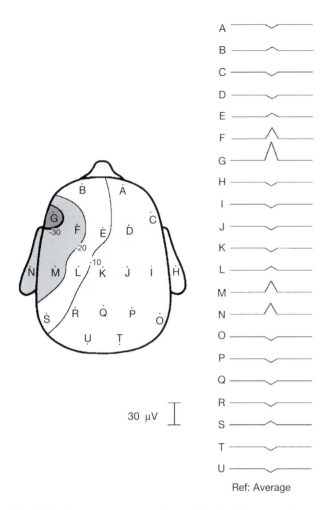

30 µV

Ref: Average

Fig. 2.17   Common average reference display of the event shown in *Fig. 2.16*. (From Binnie et al (2003), by permission.)

at electrode G'. The advantages of this type of display may not be apparent from the highly schematic figures used here, but when a focal event of relatively low amplitude appears against a background of higher voltage components, phase reversal may help to render it conspicuous.

Source reference derivation

There also exists a method termed 'source reference derivation'. Here the electrode on lead 1 of each channel is referred to its own unique reference, which is a weighted average derived from the adjacent electrodes. In fact there are complex underlying mathematical principles and the method is intended to represent the underlying sources of the EEG as the radial currents producing the surface potential field, and appears to give a sharper localisation. It has characteristics intermediate between those of the bipolar and common reference methods. It both highlights local potential gradients after the manner of bipolar recording and causes focal features to stand out as waveforms limited uniquely to a few channels. The overall appearance of source derivation records is very similar to that of those using the common average reference, although there is less tendency for focal components to appear with reduced amplitude and inverted polarity on channels recording from distant electrodes.

Montages

A recording montage is a specific pattern of connections of electrodes to the recording channels of an EEG machine and each usually employs only one method of derivation. Every method of derivation has advantages for displaying different features of the EEG. Their various features are summarised in *Table 2.1*. Similarly, different montages can be used to highlight different features. It is necessary to use several montages, including at the least a bipolar and a referential method, in any routine recording. Typical examples of popular bipolar and reference montages are shown in *Figs 2.19* and *2.20*.

ARTEFACTS AND INTERFERENCE

The EEG is of extremely low amplitude, ranging from one or two microvolts, at the level of noise due to random movements of electrons, to some hundreds of microvolts. Consequently, it is difficult and important to minimise contamination of the trace from other sources of electrical activity. Artefacts may arise from electrochemical events at the surfaces of electrodes, equipment faults, bioelectric phenomena from the subject (muscle potentials, eye movements, etc.) and as interference from mains operated devices or electrically charged objects in the recording area. Here a brief account is given of how artefacts affect EEG interpretation. For a detailed consideration of the causation, recognition and elimination of artefacts, see Cooper et al (2005).

highlighted most conveniently by 'bipolar derivation', an extension of the method introduced by Grey Walter (see *Fig. 2.15(b)*). This involves connecting rows of electrodes to consecutive channels such that an electrode is attached to lead 2 of one channel and lead 1 of the next (*Fig. 2.18(c)*). In diagrams of bipolar connections, leads 1 are conventionally shown as solid lines, and leads 2 as broken lines (as in *Fig. 2.18(c)*). More simply, each channel may be represented by an arrow directed from the electrode on lead 1 to that on lead 2. An EEG event that causes an electrode in the middle of a chain to assume potential differences of similar polarity with respect to both its neighbours will cause two channels to deflect in opposite directions, giving a striking appearance that is traditionally, if perhaps incorrectly, called a 'phase reversal'. *Figure 2.18* shows a bipolar recording of the same focal event at electrode G used in some previous illustrations. From the bipolar recording of a single row of electrodes one can deduce the profile of the electrical field along the chain, as shown in *Fig. 2.18(b)*. According to the customary jargon of EEG there is a 'phase reversal

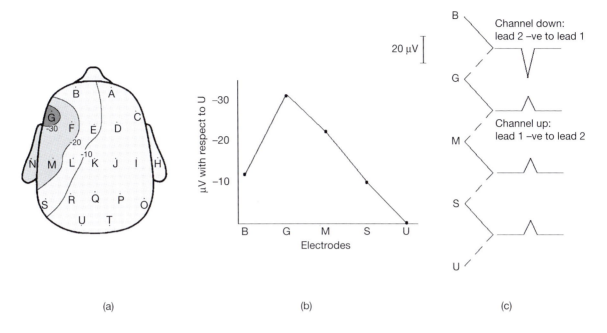

Fig. 2.18 Principle of bipolar derivation. Same event as in *Fig. 2.14*, also shown as (a) a contour map and (b) a potential profile along (c) the bipolar chain of electrodes. (From Binnie et al (2003), by permission.)

The presence of artefacts is generally considered evidence of poor technical standards. However, it may be fruitful to consider another point of view before detailing some common types. Although most 'physical' artefacts are highly undesirable and should be prevented by good technique, 'biological artefacts' arise from events within or relating directly to the patient and may reveal much about the state of the subject. They may reflect anxiety or tension, somnolence or biological reactivity to external stimuli in an apparently comatose individual. Other biological artefacts may indicate conditions such as cardiac dysrhythmia or nystagmus, possibly relevant to the clinical problem. In infants and children, one has to accept a much greater amount of movement and muscle artefacts.

Three particular instances may be cited where biological artefacts are of value. First, the electro-oculogram (EOG) provides a built-in calibration signal. Both the technologist and the neurophysiologist should be able to recognise unhesitatingly the topography and typical appearance of normal eye movement artefacts. Any anomaly should alert the observer to the presence of either a technical problem or a biological abnormality. Secondly, in association with apparently subclinical epileptiform EEG discharges, artefacts, electromyogram (EMG), oculogram or movement, for instance, may

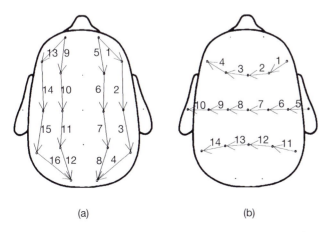

(a)   (b)

Fig. 2.19 Application of the IFSECN guidelines to the design of bipolar montages: chains go from anterior to posterior (a) and from right to left (b). (From Binnie et al (2003), by permission.)

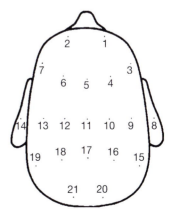

Fig. 2.20 Application of the IFSECN guidelines to a reference montage. The same principles would apply whatever the reference – which is not specified in the figure. (From Binnie et al (2003), by permission.)

indicate the occurrence of clinical ictal events – i.e. that the 'subclinical discharges' are in fact seizures. This is useful, particularly often in children. Thirdly, biological features considered 'artefacts' in the routine diagnostic EEG are essential for determining depth of sleep. For these reasons we can regard the biological artefacts with rather more enthusiasm than the physical ones and hope that the reader will enjoy the experience of using them in understanding more about the patient behind the EEG.

Physical artefacts
Electrodes and input leads
The electrodes in use for routine and special EEG recordings give rise to a variety of artefacts, from sudden gross deflections to the occasional spike produced by a pad electrode. All originate as a transient electrochemical change at the electrode–tissue interface due to imperfect contact, which creates a change in voltage across the input of the amplifier, which is amplified along with the genuine

Table 2.1   Properties of derivation methods

Methods	Bipolar	Common reference	Common average reference	Source reference
Principle	Displays local potential gradients between adjacent electrodes; localisation by 'phase reversal'	Displays potential difference between each electrode and common reference; localisation by amplitude	Displays potential difference between each electrode and average of entire array; localisation by amplitude and phase	Displays potential difference between each electrode and local weighted average; localisation by amplitude
Recording characteristics				
Discrete foci	Good detection and localisation; special montage may be needed	Good detection and localisation; may be difficult to select appropriate reference	Good detection unless amplitude of ongoing activity is high; easy localisation in montages, false localisation if too few channels	Good detection and localisation, exaggerates focal character; requires at least 16 channels; not applicable to non-standard electrode sites
Diffuse low amplitude activity	Poor detection	Good detection and localisation, if suitable reference	Adequate detection and localisation	Poor detection
General topography of ongoing activity	Good display; extent of ongoing activities exaggerated; slower components attenuated	Good display	Fair display; difficult for inexperienced interpreter	Good display; attenuates diffuse components
Local reduction of activity	Poor detection	Good detection	Fair detection	Poor detection
Asymmetries	Unreliable; must be checked against a reference derivation	Good display, except with ear references	Good display, unless asymmetry gross – then misleading	Good display
Waveform	Adequate for discrete stable focus, otherwise poor	Good display	Usually good; bad for widespread activities and travelling waves	Fair to good
Electrode and lead artefacts	Easily identified and located; affect few channels	Easily identified and located; affect all channels if at reference electrode	May be impossible to locate if number of channels is small; always affect all channels	Fairly easy to locate; affect several channels
Common mode rejection	Good	Fair	Poor – unequal impedances on leads I and II	Fair
Main uses	General display of background activity; locating stable foci	Displaying specific features – background asymmetries, local reduction of amplitude, diffuse activities, travelling waves; accurate display of localisation and waveform of focal activity	As for common reference, less reliable but more convenient for use; excellent for preliminary screening of EEG	As for average reference, better display of focal features but less satisfactory than other reference methods for ongoing activity

EEG signal. Artefacts are also induced if electrode leads swing in the Earth's magnetic field. Some examples of electrode artefacts are shown in *Fig. 2.21*.

Machine faults

Modern, digital EEG machines are more reliable than their predecessors, but faults, for instance incorrect calibration, may be more difficult to detect. In older machines, mechanical switches are a common source of artefact, giving rise to excessive noise, spiky transients or loss of sensitivity. Damage to or maladjustment of pen recorders can result in loss of high-frequencies, reduction of dynamic range or total loss of signal. More insidious, because less easily detected, are variations in chart speed, usually due to wear or mechanical friction in the paper drive mechanism, producing a spike whenever the movement of the paper is interrupted.

Electrical interference

The ubiquitous presence of electric and magnetic fields alternating at the frequency of the mains supply (50 Hz in Europe and 60 Hz in North America) in all hospital buildings has forced the designers of electrophysiological apparatus to take stringent precautions against interference from these sources. Consequently, no serious problems should arise in a properly designed laboratory. Nevertheless, the ever-increasing use of such apparatus in electrically hostile surroundings, such as an intensive therapy unit (ITU) or operating theatre, can present difficulties. Although a notch filter tuned to the frequency of the mains supply will greatly reduce interference from this source, its use should be regarded as a last resort rather than as a first expedient.

Mains frequency interference may arise by electrostatic inductance from an unscreened conductor such as the live lead of the mains supply fluctuating in potential with respect to the subject and to earth. Another electrostatic source is the high voltage that can occur on fabrics, for instance, clothing or bedding, particularly if these are made from synthetic fibre. Movement of such a source near the patient induces a variable voltage at the electrodes, which can give rise to dramatic deflections in the EEG tracing. Intravenous drips, some ventilators when moisture drops are moving in plastic tubing, and special ripple-type anti-bedsore beds can produce similar problems in the ITU or operating theatre, and may also induce interference at mains frequency.

Mains interference also arises when a conductor carrying an alternating current creates an electromagnetic field in its vicinity, which induces a voltage in conducting loops, such as those formed by the electrode leads and the patient's head.

Such sources are most likely to be encountered when recording in the operating theatre, ITU or neonatal unit (*Fig. 2.22*). Susceptibility to interference is increased by poor application of electrodes, sometimes inevitable in a restless child or a neonate to whom access is limited. Con-

sequently, interference presents particular difficulties in emergency paediatric work and in uncooperative children.

Biological artefacts

It has already been pointed out that biological signals can provide useful information about a patient and that it is only in certain circumstances that these should be regarded as undesirable artefacts during EEG recording. Nevertheless, in so far as they tend to obscure the EEG, they can hinder interpretation.

Well-illustrated reviews of the commoner biological artefacts are given by Beaussart and Guieu (1977), Saunders (1979), Barlow (1986), Brittenham (1990) and Blume and Kaibara (1995), and some rarer ones are given by Klass (1995).

Oculogenic potentials

There is a potential difference of a few millivolts between the cornea and retina, the former being positive with respect to the latter. A movement of the eyeballs or eyelids causes a change in the electrical field that will be picked up by electrodes in their vicinity. When the subject looks upwards, an electrode above the eye becomes electropositive with respect to one situated more posteriorly, causing downward deflections in the bipolar derivations shown in *Fig. 2.23(a)*. Conversely, looking down produces bilateral upward deflections. Eye opening also produces bilateral upward deflections. Vertical eye movements should produce equal deflections in channels connected to symmetrically placed frontal electrodes and provide a useful check on the symmetry of their placement. The artefacts produced by eye closure, upward gaze and blinking are similar but not identical, either morphologically or topographically, a fact that presents problems for automatic artefact removal procedures. Conjugate lateral eye movements cause approximately equal potential changes of opposite polarity at symmetrically placed frontotemporal electrodes, giving rise to the deflections shown in *Fig. 2.23(b)*.

The experienced observer should be able immediately to recognise any anomaly, for instance, a topography that is simply incompatible with an oculographic origin in a subject with normal eyes and eye muscles. The cause, whether technical or biological, then needs to be determined. Frontal delta activity in particular is all too easily discounted as eye movement. Contrary to the general principle that all apparent abnormalities should in the first instance be considered artefactual, frontal slow activity that does not show the topography of any known oculographic artefact should be regarded as being of cerebral origin, until proven otherwise. The experienced interpreter rarely has difficulty in discriminating eye movement from cerebral activity, but sometimes there are problems and the distinction may be critical, for instance when eye movement is wrongly reported as

(a)

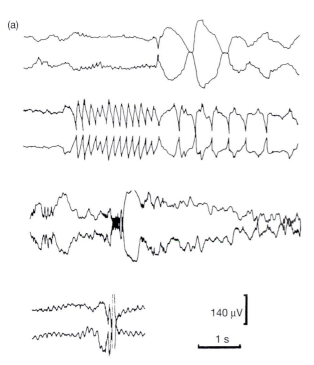

Fig. 2.21 (a) Two-channel recordings of electrode artefacts produced by movement and/or poor contact. (b) To show confirmation of an electrode artefact at C4 by use of bipolar and average reference montages. Same sample of digital EEG 're-montaged'. (From Binnie et al (2003), by permission.)

140 μV

1 s

(b)

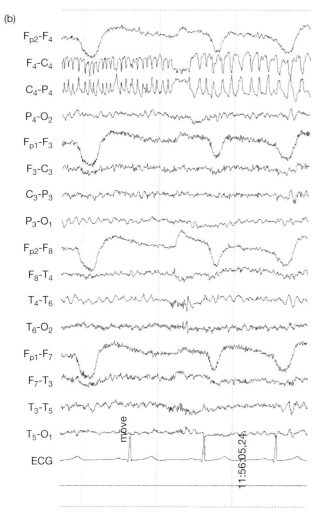

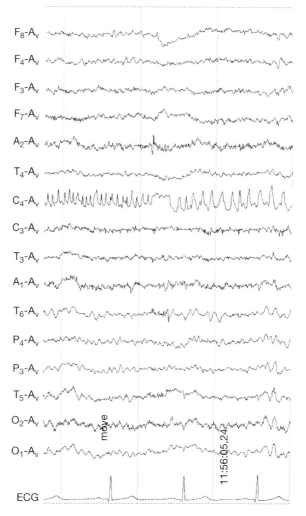

70 μV   HF 70 Hz

1 sec

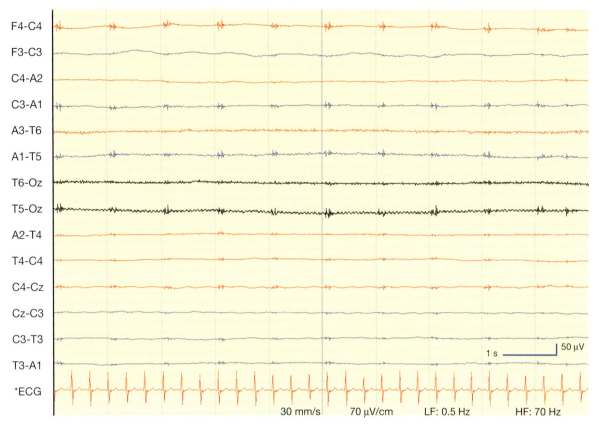

Fig. 2.22 Bed heater in an ITU cot producing an artefact in a 2-month-old infant after prolonged cardiac arrest and electrocerebral silence in the EEG.

spike-wave activity, or frontal delta during a seizure is mistaken for artefact.

Myogenic potentials

Muscle potentials on the scalp, which can be localised or widespread, occur in a variety of forms (*Fig. 2.24*) and may be almost continuous in an apprehensive or uncomfortable subject. These artefacts may be reduced by relax-

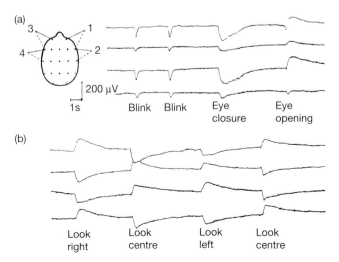

Fig. 2.23 (a) Vertical eye movement potentials. (b) Lateral eye movement potentials. (From Binnie et al (2003), by permission.)

ation, by a change of posture, or by slightly opening the mouth. If these manoeuvres are ineffective, the high-frequency filters may have to be used, but it should be realised that this may transform the artefact into a signal indistinguishable from genuine beta activity or spikes. In younger children and those with behavioural problems or learning difficulties there may be only a few pages of undisturbed EEG during a 20-min routine recording. It may be necessary to repeat the EEG as a sleep recording (see p. 72).

Potentials related to cardiac activity

Cardiac action is associated with an electrical field, the ECG, which is an order of magnitude larger than the EEG, and also produces a pulse pressure wave in the arterial system. The latter not only causes small movements of electrodes in the vicinity of arteries, but also gives a mechanical impulse to the body as a whole, the ballistocardiogram. These phenomena give rise to three kinds of repetitive artefact.

In most subjects, the electrical field of the heart is almost equipotential over the scalp so that it is not readily picked up, particularly with bipolar derivations. Anomalies of the cardiac axis commonly increase susceptibility to ECG artefact (Dirlich et al 1997), not only in cardiac disease, but also due to pregnancy, ascites, emphysema or obesity, and in subjects of markedly pyknic or sthenic habitus (*Figs 2.25* and *2.26*). The R-wave of the ECG complex, which is usually the largest component seen, is electronegative on the right side of the head and neck with respect to the left. It is unaffected by small alterations of electrode position, but changing the patient's posture, and thus the orientation of the cardiac axis, may help to reduce it. It should rarely be difficult to distinguish ECG artefact from spikes because of its regular occurrence. However, cardiac dysrhythmias may give rise to artefacts that are irregular, intermittent and of atypical waveform and can resemble spikes.

When an electrode is placed over or near an artery, a saw-toothed waveform synchronous with the pulse is sometimes recorded. The source of the artefact can be confirmed by simultaneous recording of the ECG. Pulse artefact can usually be eliminated by a slight change of electrode position.

The ballistocardiogram takes the form of a repetitive oscillation in time with the pulse. It is most likely to be due to slight movement of the electrode leads in synchrony with the heart beat and is most often seen when recording at high sensitivity in the ITU.

Artefacts due to changes of skin potential or resistance

Perspiration and psychogalvanic responses are the common causes of long-duration slow potentials or baseline sways in EEG recordings from patients who are hot or worried. Good planning of a comfortable room and

proper explanation about the procedure help eliminate such problems. Simple fans or removing clothes may help.

Artefacts due to movement and tremor

Few children are capable of remaining still throughout a recording of 30 min or more. Consequently, artefacts associated with movement are to be expected in every EEG record. Most arise from the electrodes and their connections, as discussed above. Special difficulty arises in patients with tremor at a frequency within the EEG range. Even the twitching of a leg is likely to manifest in the record by virtue of its ballistic effect.

Careful observation of the patient or, for example, the mother or nurse holding a small child, will often indicate the cause of artefacts from rocking or patting movements, head scratching, hiccups, scalp wiggling, sucking (*Fig. 2.27(a)*) or chewing. Intermittent rhythmical movements, such as rocking (*Fig. 2.27(b)*) or a mother gently rubbing her baby's back to help it bring up wind, will produce alarming periodic phenomena in the record that may closely resemble genuine EEG events. When there is doubt about the authenticity of an apparent movement

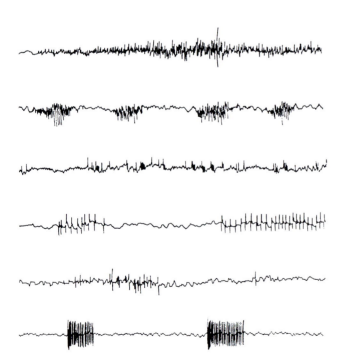

Fig. 2.24 A variety of myogenic potentials recorded from electrodes on the scalp. The fourth example shows repetitive unit potentials; the last is due to facial myokymia. (From Binnie et al (2003), by permission.)

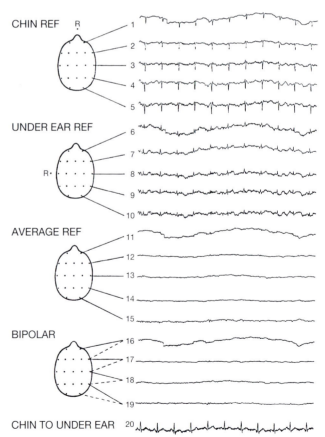

Fig. 2.25 ECG contamination of the EEG. Simultaneous common reference, average reference and bipolar derivations. Maximum ECG gradient between chin and scalp electrodes. Average reference and bipolar derivations are unaffected. (From Binnie et al (2003), by permission.)

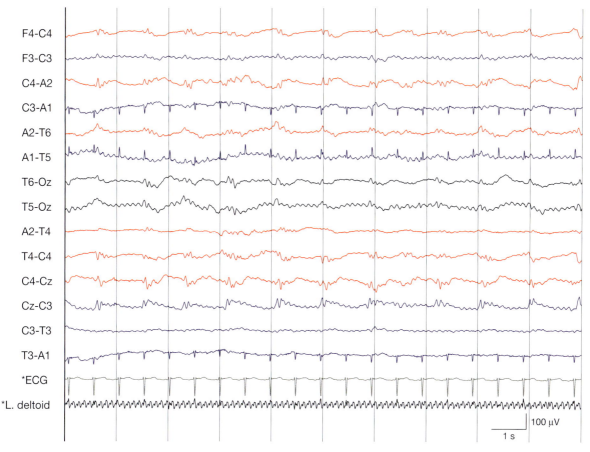

Fig. 2.26   Artefacts in a neonatal EEG recorded in the ITU. Term infant with meconium aspiration and poor APGAR scores; on oscillatory ventilator and sedation. Note that the ECG artefact is maximal over A1, and the oscillator artefact over the left posterior region (also seen in the EMG left deltoid) and electrographic seizure right central region.

artefact, recording the output of an accelerometer applied to the suspected source will often clarify the matter.

Movement artefacts during seizures (whether epileptic or otherwise) may resemble generalised epileptiform discharges and are an important source of misinterpretation. Generalised SW or slow wave discharges commonly exhibit bifrontal, and less typically occipital, maxima. Any other topography should be viewed with suspicion. Apparent discharges with more than one maximum on the same side of the head (producing double phase reversals in bipolar montages) are almost always artefactual.

Hospital beds often display a resonant frequency in the theta range, and personnel knocking the bed may produce bursts of rhythmic activity during recording at high sensitivity in the ITU.

Methods for artefact detection and rejection

As will be evident from the above, constant vigilance is necessary to detect and identify artefacts. Some are easily recognised by characteristic morphology and topography (e.g. normal eye movements, ECG in sinus rhythm, myogenic potentials); others are more subtle, for instance:

- sharp transients of unusual waveform, polarity or topography (electrode, lead or plug artefacts)
- isolated sharp waves (extrasystoles) or bursts of rhythmic theta activity (vibration of the bed) in patients where no other cerebral electrical activity is recordable
- periodic transients (drip and ventilator artefacts, and bubbles in ventilator tubing)
- generalised apparently epileptiform discharges of unusual topography during 'seizures' that are not necessarily epileptic (movement)
- sharply focal rhythmic spikes of uncertain origin, slowly declining in frequency and resembling a localised epileptiform discharge (electrode artefacts can show this time course)
- slow posterior delta activity (head movement on the pillow)
- sharp occipital theta activity in a seated patient (EMG from neck muscles – a common finding during telemetry).

Conversely, it is important to recognise cerebral activities that may be mistaken for artefacts. Examples include:

- apparent eye movement potentials with anomalous topography (frontal delta activity – liable to misinterpretation in hypermotor seizures of frontal origin)
- steep, sharply localised, positive spikes resembling electrode artefacts (can be true epileptiform discharges over a skull defect with underlying cortical lesion)
- short bursts of fast activity at about 50 Hz (may be of cerebral origin, for instance in cortical dysplasia).

The main approaches are five-fold:

1. Try to eliminate the cause by good technique and observation, for instance by continuous video monitoring of all patients so that movements and possible artefacts can be compared at the time of interpretation.
2. Routinely combine polygraphic recording from potential sources of biological artefacts with the EEG (e.g. Irrgang and Höller 1981).
3. Where a suspected artefact cannot otherwise reliably be identified, attempt either to elicit it deliberately (tongue movement, tapping the bed, voluntary eye movement), or temporarily to remove the source (disconnect infusion pump, overhead heater or ripple bed – with agreement of responsible personnel).
4. Use judicious and clearly annotated temporary adjustment of controls such as low- and high-

(a)

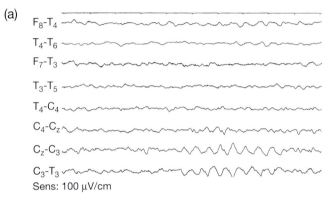

Sens: 100 µV/cm

Fig. 2.27 (a) Rhythmical slow waves produced by dummy-sucking in a 3-month-old infant. (b) Rhythmical delta activity maximal over left temporoposterior region produced by rocking. (From Binnie et al (2003), by permission.)

(b)

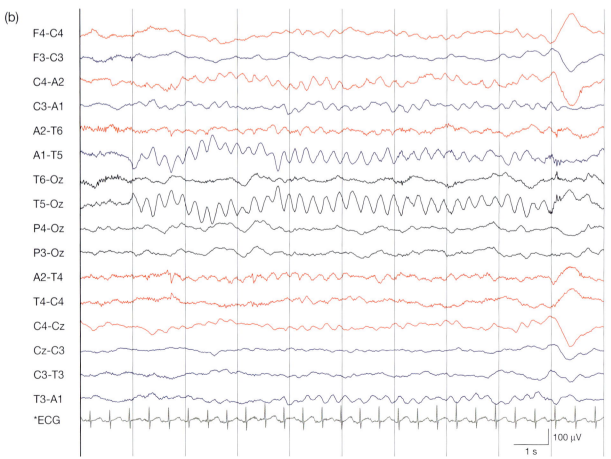

100 µV

1 s

frequency filters on the EEG machine when it proves impossible to reduce overwhelming artefact.

5. Use some of the automatic artefact-rejection systems developed for computer assessment of EEG (Barlow 1985, 1986).

EFFECTS OF AGE

A comprehensive account of the effects on the EEG of age and of arousal and sleep is outside the scope of the present book. However, a summary is provided to clarify subsequent references to age- or sleep-related phenomena. For a fuller account the reader is referred to standard texts, such as Binnie et al (2003).

The neonatal EEG

The earliest recordings from very premature infants at 22–23 weeks after conception show brief bursts of low amplitude delta and theta activity separated by long periods of electrical silence (tracé discontinu, *Fig. 2.28*),

lasting 10–40 s and sometimes up to 60 s. The bursts become longer and the silent periods briefer, rarely more than 20 s by 30 weeks postconceptual age (PCA), and disappear entirely towards term. They are replaced by episodes of low amplitude activity, alternating with higher amplitude bursts (tracé alternant). The low amplitude episodes in turn become less frequent and briefer, being seen only in sleep at term and disappearing by 4 weeks after term. The bursts by term are composed of mixed slow and faster activities, often with episodic sharp waves, particularly towards the front of the head.

The foregoing is a very brief summary of a complex evolution, and determining whether the pattern is appropriate to the PCA requires considerable skill and, being crucial to interpretation of the neonatal EEG, is described more fully in the clinical neonatal section on page 182.

Early childhood

The infant's EEG is dominated by theta and delta activities but shows clear differentiation between sleep

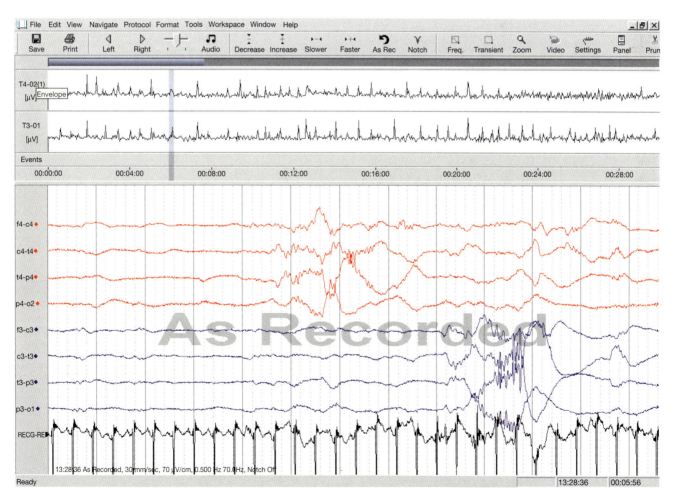

Fig. 2.28 Digital EEG recording showing typical tracé discontinu in an infant of 33 weeks gestational age with a good outcome. Trend window 30 min (upper two traces), EEG window 12.5 s (corresponds to vertical grey bar across trend display). Trend of log amplitude of EEG envelope on T4-O2 and T3-O1 shows tracé discontinu. (From Binnie et al (2003), by permission.)

and waking. Rhythmic activity in the mid-theta range appears in the central regions towards the end of the first year and then migrates posteriorly, increasing in frequency so that by 3 years an immature alpha rhythm at around 8 Hz is established.

The normal development of the EEG is rapid during the first year of life and has been described by Samson-Dollfus et al (1964). Characteristic developmental features are:

1. Irregular rhythmic activity with an amplitude of 50–100 µV in the posterior regions appears clearly between 3–5 months. This activity attenuates on eye opening. At 2 months of age the frequency is 2–3 Hz and at 4 months around 4 Hz (*Fig. 2.29*). After 5 months a progressive increase in frequency up to 7 Hz at the end of the first year of life is the main feature (*Fig. 2.30*).
2. A central rhythm of 6–8 Hz at 25–50 µV (*Fig. 2.31*) is seen in infants and toddlers, starting to emerge at around 4–8 months, peaking at 12–24 months. The frequency of this rhythm remains relatively stable over time, increasing from 6–7 Hz at 5 months to 8 Hz at 24 months.
3. From 6–8 months bilaterally synchronous runs of high amplitude slow activity (central or hypnagogic hypersynchrony) appear during drowsiness (*Fig. 2.32*).
4. Sleep spindles appear at 4–6 weeks and 'Rolandic humps' (possibly rudimentary vertex transients or K-complexes) from 3–4 months.

Central or 'hypnagogic' hypersynchrony comprises runs of frontocentral theta activity. This occurs in bursts of abrupt onset, which may be mistaken for epileptiform discharges. Organisation of sleep spindles occurs rapidly from about 6 weeks of age, with stability of measures such as density, duration, frequency, amplitude, asymmetry and asynchrony well established by 3 months (Louis et al 1992). Unlike those of older children and adults, infants' spindles tend to have a sharp, arcuate waveform, resembling a mu rhythm, and are open to misinterpretation as spikes. Particularly in the age group 3–6 months spindles can be prolonged for up to 10–15 s.

Visual, broadband frequency analysis and spectral analysis of the EEG in the awake state from birth to 1 year were examined in 29 healthy infants who were neurologically and developmentally normal on follow-up to 3.5–6 years (Hagne et al 1973). Considerable inter-

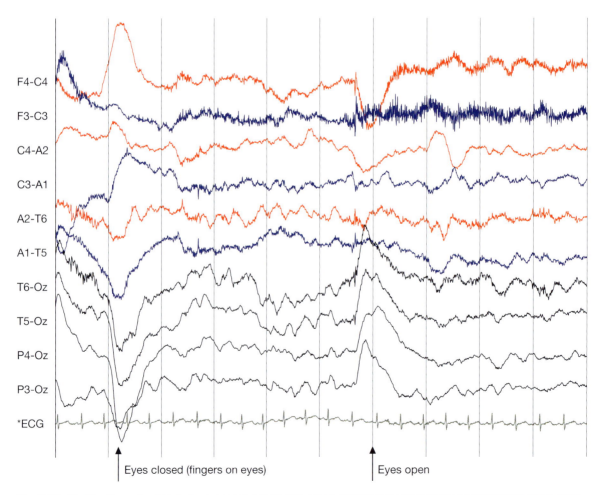

Fig. 2.29 Healthy 3-month-old infant: on passive eye closure a 3 Hz rhythm appears over the posterior third of the head and blocks on eye opening.

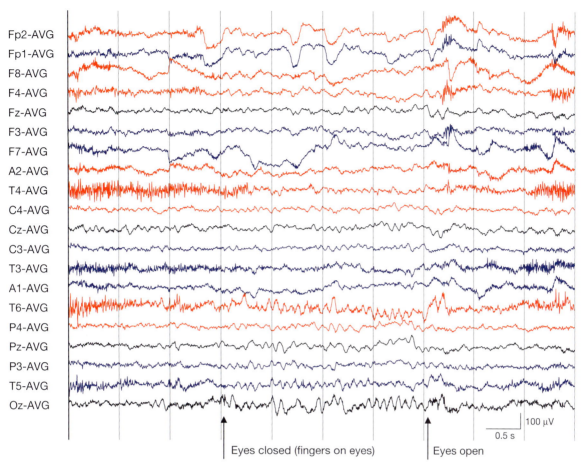

Fig. 2.30 Healthy 15-month-old toddler: on passive eye closure a 7 Hz rhythm is seen maximal over the posterior regions and blocks on eye opening.

and intrasubject variation was noted and the difficulties in obtaining scrupulously artefact-free traces in small infants tended to favour the visual assessment by an experienced observer over spectral analysis. Nonetheless, there was clear development of significant changes in spectral composition with age, comprising the appearance of peaks and a subsequent increase in their frequency and amplitude, with spectral values reflecting an increase in faster and a decrease in slower components.

Preschool period

The developing alpha rhythm increases in frequency and becomes more localised. The rate of maturation is variable and, although the mean postcentral dominant frequency is 9 Hz by 7 years, it may remain below 8 Hz in some children. Theta activity is prominent, and mostly appears as a slow alpha variant. Delta activity is prominent over the posterior regions (*Fig. 2.33*).

Later childhood and adolescence

Through the late pre-school years and into adolescence the characteristic features of the adult EEG emerge. Infantile slow activity becomes increasingly confined to the back of the head with posterior temporal maxima, as 'posterior slow waves of youth'. Within the alpha territory, much of the theta activity is at a subharmonic of the alpha frequency and shows responsiveness similar to that of the alpha rhythm (slow alpha variant; see *Fig. 2.11*). These two types of posterior slow activity become most conspicuous typically at about 12 years of age and decline thereafter (*Fig. 2.34*). In some normal subjects they are hardly detectable by the age of 20 years; in others they may persist up to age 30 years.

The amount of slow activity changes with age in normal subjects but, as slow activity can also be a pathological phenomenon, determining whether the amount of delta and theta activity is appropriate to the age of the subject is important for clinical interpretation.

The adult EEG

The establishment of an adult EEG pattern is the eventual outcome of maturation but may be achieved by age 16 years or not until 30 years of age. The postcentral dominant activity with eyes closed is an alpha rhythm, typically at 9–10 Hz. Sporadic, irregular temporal theta waves occur but some rhythmic slow alpha variant may also persist. Fast activity appears diffusely but is more readily visible over the anterior regions. Delta activity may be detectable by quantitative analysis but is usually little in evidence in waking adults, with the exception of a few individuals with persistent posterior slow waves of youth.

LEVEL OF CONSCIOUSNESS

Effects of wakefulness, arousal and sleep

Alterations of functional state, associated with changes in the level of consciousness, usually have a marked effect on the morphology of the EEG. This applies not only for pathological states such as coma but also for physiological states during which consciousness is either lowered (drowsiness or sleep) or heightened (mental attention or stress).

With increasing depth of drowsiness and sleep the typical features of the waking EEG gradually or abruptly disappear and are replaced by a gamut of sleep patterns with distinctive morphology, usually associated with specific levels or 'stages' of sleep. These are more or less regularly interrupted by yet other patterns, representing arousal or reactivation of the cortex. These are provoked by internal or external stimuli, which may or may not

result in a return to a higher level of consciousness or even wakefulness. The absence, or an abnormal presentation, of these sleep or arousal phenomena may reflect underlying cerebral dysfunction and thus add to the diagnostic value of the EEG recording. Moreover, the states of drowsiness or sleep may provoke transient EEG abnormalities that were absent or recorded less frequently during wakefulness, notably epileptiform discharges. The recording of drowsiness and sleep is thus not something to be confined to specialised sleep laboratories but should be part of many, if not all, diagnostic EEG studies in the standard department of clinical neurophysiology. The recognition of, possibly unwanted, changes in vigilance is also crucial to the interpretation of clinical EEGs, where slowing due to drowsiness may be mistaken for an abnormality. Those seeking more detailed information on sleep work are referred to the following: Parmalee et al (1968), Dutertre (1977), Ellingson (1979), Ellingson and Peters (1980), Blume (1982), Erwin et al (1984), Westmoreland (1982, 1990), Cooper (1994), Sheldon (1996), Aldrich (1999) and Blume and Kaibara (1999).

The limitations of the busy general clinical neurophysiology laboratory generally preclude the recording of extended periods of natural, especially nocturnal, sleep.

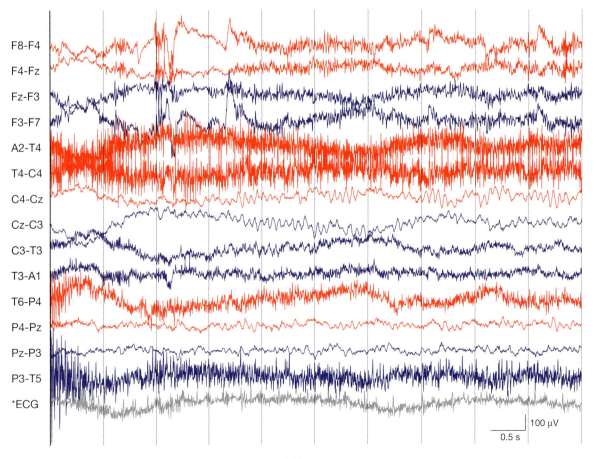

Fig. 2.31 Prominent 8 Hz central rhythm in an 18-month-old boy.

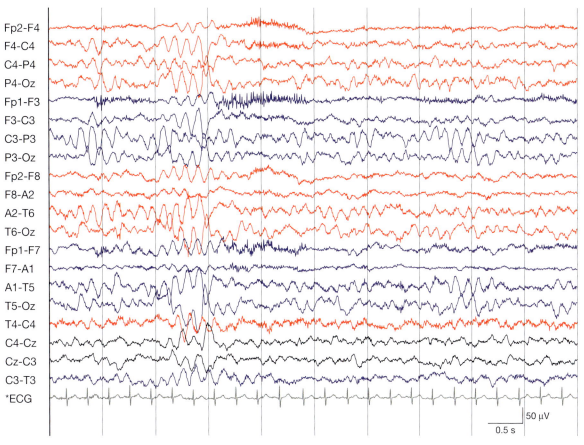

Fig. 2.32 Early drowsiness in a 12-month-old girl. Bursts of high amplitude 4–5 Hz activity are seen, at times with a notched appearance.

Under favourable conditions, however, some drowsiness or even short stretches of light sleep may be recorded during routine studies, and this can be further promoted by a work schedule that allows extra time for recording, beyond the regular 30–45 min, especially for investigations in the early afternoon, after lunch. Patients should be lying rather than sitting or reclining, should be covered with a blanket and the recording should be made in a quiet setting, preferably an isolated room with dimmed lighting. When necessary, sleep can be induced by short-acting hypnotics, which will, however, generally result in a sleep record with increased beta activity. Alternatively, sleep can be provoked by sleep deprivation during the night preceding an early morning EEG recording, but this is hardly practical in infants and younger children. Induction of sleep with melatonin is now the preferred method in many EEG departments (Wassmer et al 2001).

For most practical purposes, these transient states between wakefulness or drowsiness and light sleep, as recorded under usual laboratory conditions, will be sufficient to demonstrate any significant abnormalities of physiological sleep phenomena and to provoke a majority of the abnormalities expected to be seen during sleep in patients with seizure disorders. For more detailed studies concerning sleep disorders, sleep architecture or quantitative aspects of sleep, prolonged investigations are required; these are best performed in specialised laboratories and with special techniques beyond the scope of this book (see Cooper et al 2005, Chapter 12).

Arousal and mental activity during wakefulness

Even in the awake state, changes in the EEG may be seen as the result of sensory stimuli, mental activity, psychological stress, attention, etc. Well-known examples are the blocking or attenuation of the alpha rhythm by opening the eyes or concentrating on a mental task, the appearance of bi-occipital lambda waves when the eyes are scanning a patterned environment, and the suppression of the central mu rhythm by contralateral somatosensory stimuli.

A general attenuation of ongoing rhythmic activity and a diffuse low voltage pattern of fast irregular beta activity may be seen with increased arousal or in very tense subjects. Conversely, an increase of the alpha rhythm, sometimes of a slightly lower frequency than during regular eye closure in the same person, may be induced by relaxation.

From a clinical point of view, however, the most important changes in the EEG are those seen during lowered states of consciousness, associated with the progressive states of drowsiness and sleep.

Sleep stages

The progression from relaxed wakefulness, through drowsiness, to light and finally deep sleep is characterised by a specific sequence of patterns. These have been categorised in various ways but the classification now almost universally accepted is that defined by Dement and Kleitman (1957), and supplied with detailed scoring rules by Rechtschaffen and Kales (1968). According to this classification which was primarily intended for adults, sleep is grossly divided in two categories of non-REM and REM (or rapid eye movement) sleep; non-REM sleep is then further subdivided into stages I to IV.

In all healthy humans, with the exception of prematures, neonates and very young babies, these consecutive stages of sleep occur in a more or less regular and repetitive pattern over the night, a series of 'sleep cycles', the composition of which constitutes the basic 'sleep architecture' of the subject.

Changing states, corresponding to sleep and waking appear in the EEG of the premature neonate from 30 weeks, with differentiation between REM or 'active' sleep and 'quiet sleep' without eye movements (see p. 182). By term, three states of wakefulness, active and quiet sleep (and an indeterminate state) can be distinguished (see p. 188). Neonates sleep for some 16 h/day. Episodic reduction of amplitude (tracé alternant) is seen in quiet sleep up to 44 weeks postconceptual age. In the next few months there is a gradual evolution from the neonatal sleep pattern to a mature pattern. By 3–6 months a pattern of predominant diurnal wakefulness is established and nocturnal sleep occupies 9–12 h. Sleep spindles appear by 6 weeks and spontaneous K-complexes by 4–5 months. Stages II and III are recognisable at 2 months and stage IV by about 8 months. Thereafter, sleep structure resembles that of the adult but there is a progressive reduction, first of diurnal and later of nocturnal sleep, up to some 19 years of age.

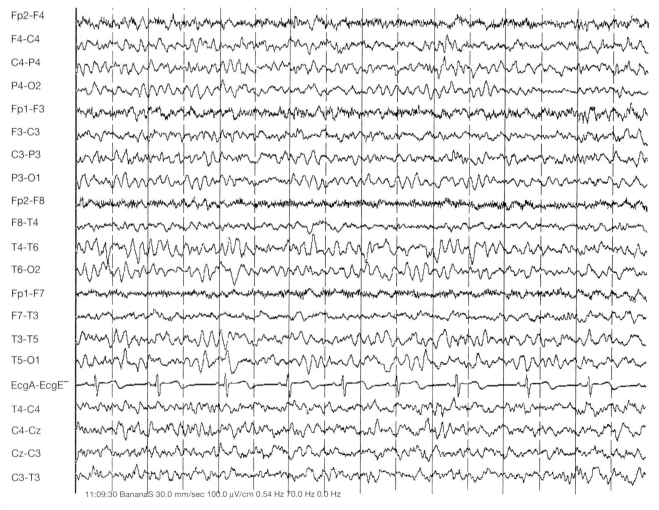

11:09:30 BananaS 30.0 mm/sec 100.0 µV/cm 0.54 Hz 70.0 Hz 0.0 Hz

Fig. 2.33 Four-year-old boy misdiagnosed as having epilepsy on the strength of prominent, normal posterior slow waves of youth. Note that when these are mixed with alpha activity a waveform resembling a spike wave may result. (From Cooper et al (2005), by permission.)

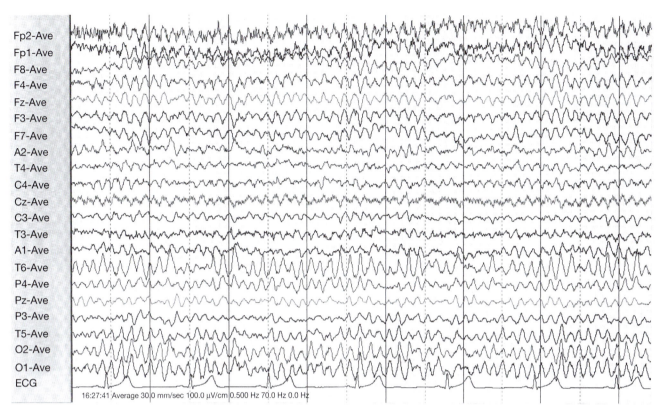

Fp2-Ave
Fp1-Ave
F8-Ave
F4-Ave
Fz-Ave
F3-Ave
F7-Ave
A2-Ave
T4-Ave
C4-Ave
Cz-Ave
C3-Ave
T3-Ave
A1-Ave
T6-Ave
P4-Ave
Pz-Ave
P3-Ave
T5-Ave
O2-Ave
O1-Ave
ECG

16:27:41 Average 30.0 mm/sec 100.0 μV/cm 0.500 Hz 70.0 Hz 0.0 Hz

Fig. 2.34 Adolescent with posterior slow waves of youth. Posterior slow waves in a 17-year-old volunteer. Note how an admixture of alpha and slower activity gives a spiky appearance (as in *Fig. 2.33*). (From Cooper et al (2005), by permission.)

All sleep stages are described in the following section for different ages (sleep stages in preterm and neonates are described on pp. 180 ff.). *Figure 2.35* illustrates the sleep stages in a healthy 1 year old and *Fig. 2.36* shows the sleep stages in a healthy 6 year old.

Non-REM sleep stage I

Infants and toddlers (1–36 months)   There is a slowing of ongoing activity into the delta range that is maximal in the occipital area. There is also hypnagogic hypersynchrony (see *Fig. 2.35(a)*) from 6 to 8 months of age. Bursts of 4–5/s activity may be seen.

Preschool children (3–5 years)   Hypnagogic hypersynchrony may be seen up to 6–8 years of age, and anterior theta activity is seen in early drowsiness. Positive bursts of 14 and 6 Hz are rarely seen.

School children (6–12 years)   There is increasing slow activity together with gradual fading of alpha rhythm (*Fig. 2.36(a, b)*). Positive bursts of 14 and 6 Hz are seen in up to 20% of this age group.

Adolescents and adults (> 12 years)   This is the transitional period, characterised by the gradual disappearance of the normal EEG patterns of wakefulness up to the appearance of the first specific sleep phenomena, vertex transients. The EEG is generally of low amplitude, but may be dominated by fast or theta activity. Slow, lateral eye movements generally occur, producing characteristic oculographic artefacts, and may give the first warning of incipient drowsiness. The incidence of 14 and 6 Hz positive bursts starts to decline after age 20 years.

Non-REM sleep stage II

Infants   Vertex transients (transition from stage I to II) start to appear at age 3–5 months, and are initially rudimentary and blunted, but often are quite prominent and of high amplitude. Spindles start to appear at age 1–2 months with an amplitude range of 12–15 Hz (typically 14 Hz in the first year of life, and 12–14 Hz later) and maximal over the central and parietal regions. Spindle trains may be long (up to 10 s), particularly in the first year of life (see *Fig. 2.35(b, c)*). K-complexes develop by 4–5 months and may be blunted and of rather low amplitude initially.

Preschool children   Vertex transients and K-complexes have a more sharp appearance. Spindles are maximum over the vertex.

School children   Vertex transients, which may be asymmetric, are of high amplitude and are frequently repetitive (see *Fig. 2.36(c)*).

Adolescents and adults   This is light sleep, which is predominantly characterised by theta and upper delta frequencies and the abundant presence of specific transient phenomena such as vertex transients, sleep spindles and K-complexes. Sleep spindles are spindle-shaped bursts of activity, typically at about 14 Hz and generally asynchronous but symmetrical in amount.

37

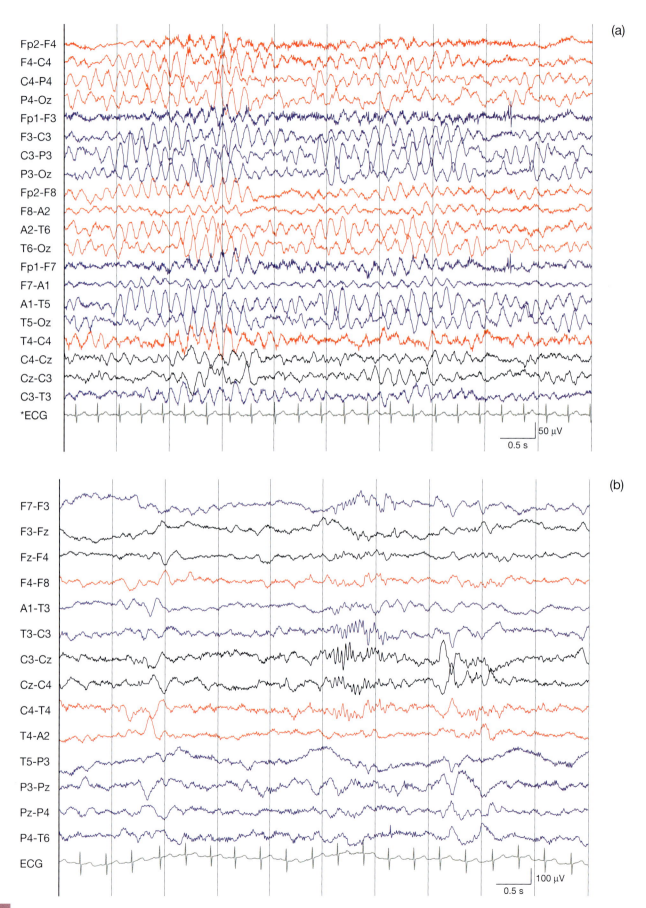

Fig. 2.35   Sleep stages in a 1-year-old girl. (a) Drowsiness: runs of bilateral, high amplitude, rhythmic slow activity (hypnagogic hypersynchrony). (b) Stage I: spindles and early K-complexes.

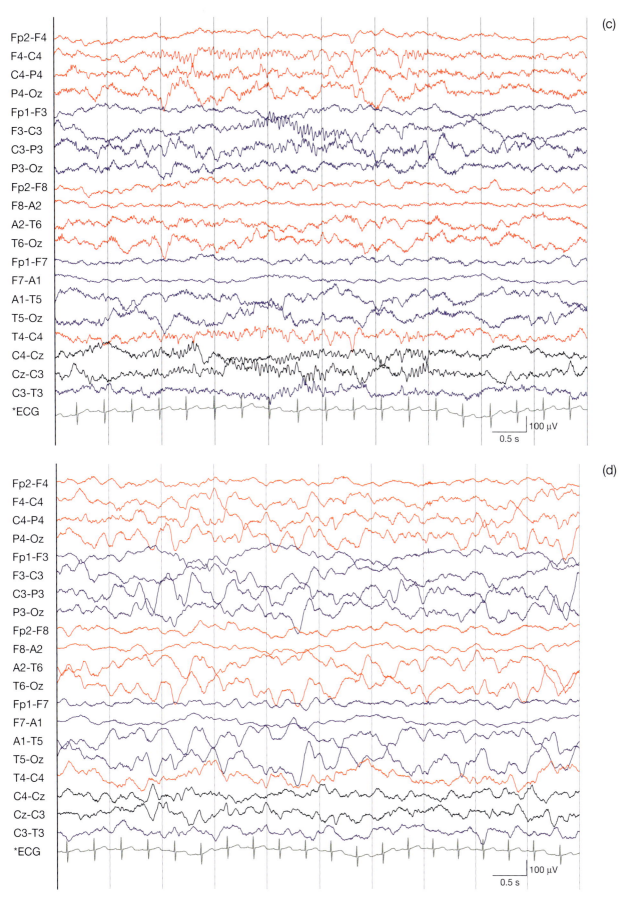

(c)

(d)

Fig. 2.35 *(cont.)* Sleep stages in a 1-year-old girl. (c) Stage II: asynchronous sleep spindles lasting several seconds. (d) Stage III: widespread irregular slow activity and occasional sleep spindles.

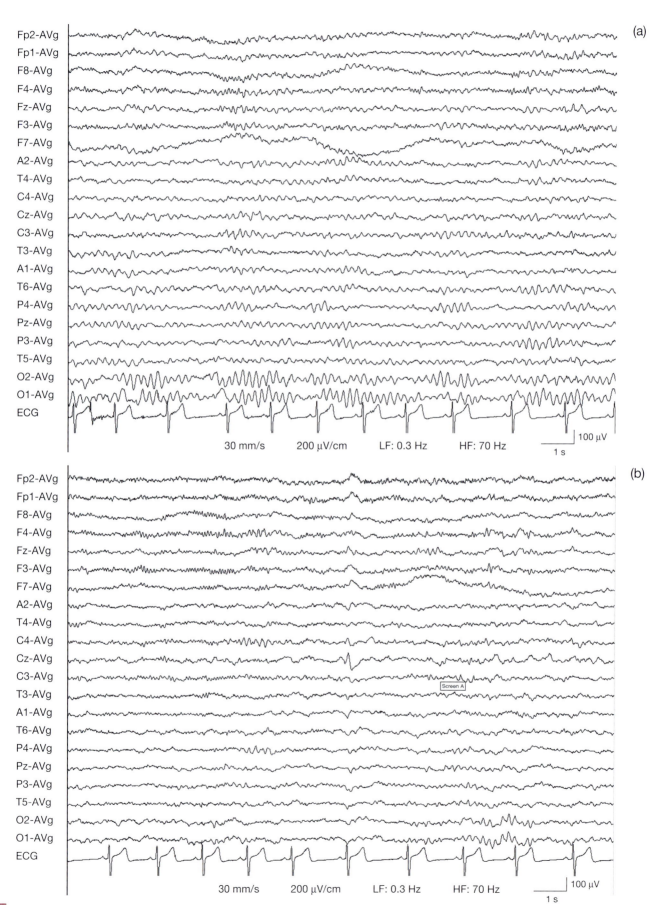

Fig. 2.36   Sleep stages in a 6-year-old boy. (a) Early drowsiness: some slow lateral eye movements, alpha rhythm beginning to become discontinuous. (b) Stage I: alpha has disappeared, vertex transient seen in middle of figure.

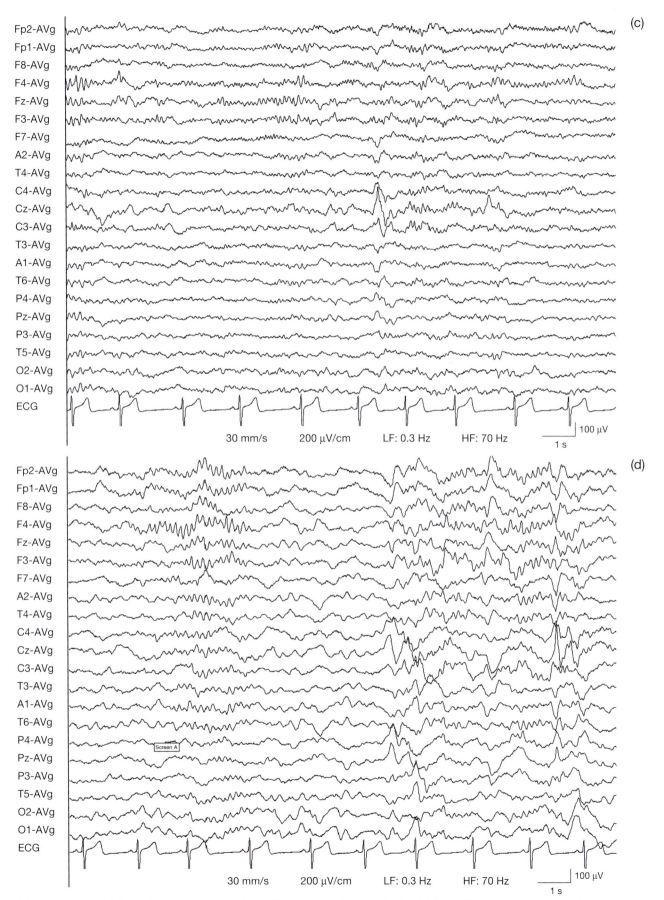

(c)

30 mm/s   200 µV/cm   LF: 0.3 Hz   HF: 70 Hz   100 µV | 1 s

(d)

30 mm/s   200 µV/cm   LF: 0.3 Hz   HF: 70 Hz   100 µV | 1 s

Fig. 2.36 *(cont.)*   (c) Stage II: sharp vertex transients, K-complex and sleep spindles. (d) Stage III: widespread irregular slow activity, K-complexes, and occasional sleep spindles.

41

K-complexes now appear and are often demonstrably related to arousal. They are of greatest amplitude at the midline electrodes, usually at the vertex but sometimes in the midfrontal or parietal regions, and typically comprise a sharp wave, a slow wave and a following sleep spindle.

Non-REM sleep stage III

Infants, children and adults (see *Figs 2.35(d)* and *2.36(d)*)   This is moderately deep sleep, with diffuse delta activity in addition to some sleep spindles and K-complexes. There should be not more than two waves per 10 s of more than 500 ms duration and exceeding 100 µV.

Non-REM sleep stage IV

Infants, children and adults (*Fig. 2.37*)   This is deep sleep characterised predominantly by diffuse delta activity and only occasional K-complexes. Spindles may persist or be absent. The record is dominated by activity of 2 Hz or slower, exceeding 100 µV in amplitude. For practical purposes stages III and IV are sometimes grouped together as deep, or slow wave sleep (SWS).

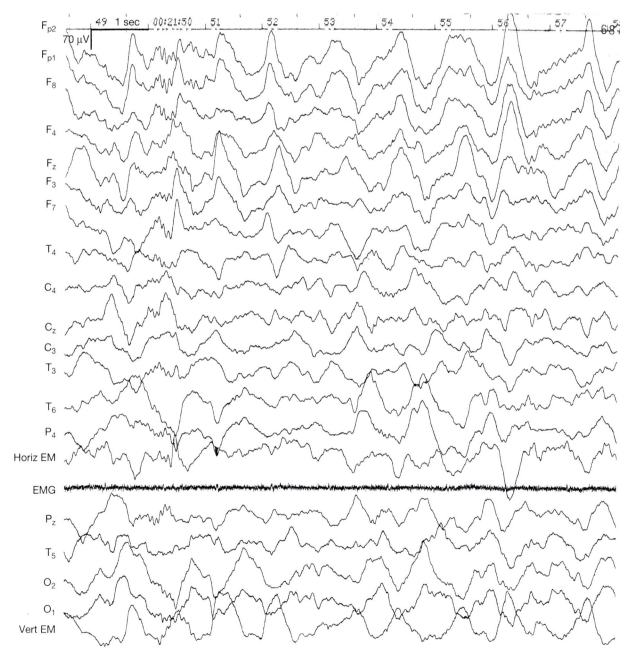

Fig. 2.37   Stage IV in a 15-year-old boy. Widespread high amplitude delta activity, sometimes frontally predominant, and residual sleep spindles. Reduced muscle tone. (From Binnie et al (2003), by permission.)

REM sleep (paradoxical sleep)

Infants The proportion of REM to non-REM sleep decreases from about 50% in the neonatal period to around 30% at the end of the second year of life. Lower amplitude delta and theta activity is seen, intermixed with some sharp-contoured transients in the occipital area.

Preschool children REM sleep is rarely seen in routine recordings. The EEG shows little desynchronisation.

School children The picture becomes more mature with increasing desynchronisation and mixed theta, delta and beta activity.

Adolescents and adults REM sleep is characterised by general desynchronisation, sometimes some fast rhythmic activity much as in the awake state, and by occasional runs of saw-tooth waves over the vertex, in addition to the typical rapid lateral eye-movement potentials (*Fig. 2.38*). Muscle tone is reduced, as may be demonstrated by monitoring the submental EMG. However, sudden movements may occur, including grinding of the teeth. Sometimes the eye movements are few or absent and this sleep stage is then recognisable only by the loss of muscle tone and the EEG appearance.

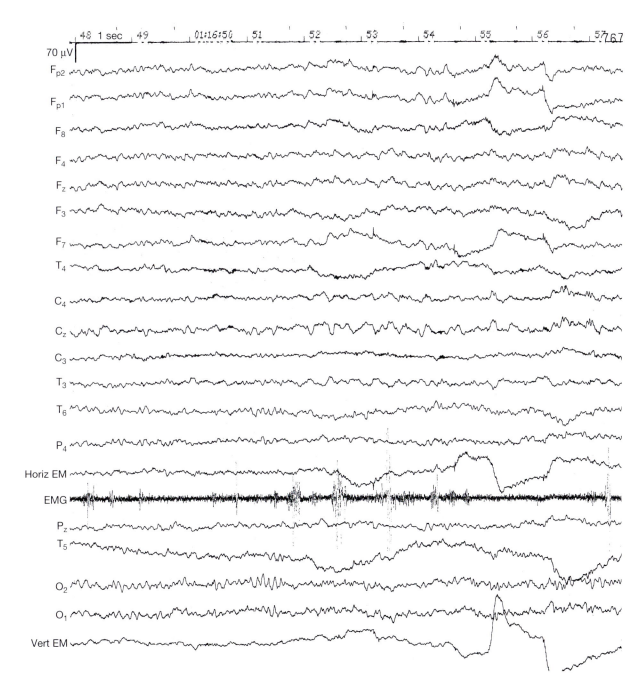

Fig. 2.38 Stage REM in a 22-year-old woman. Rapid eye movements recorded by EOG channels and frontal EEG derivations. Saw-toothed waves at vertex. Continuing residual EMG with muscle twitches. (From Binnie et al (2003), by permission.)

PATHOLOGICAL PHENOMENA

Introduction

An individual's EEG changes dramatically with age and state of awareness, and shows more subtle alterations during different cognitive activities and with various biorhythms. There are also marked differences between the EEGs of different people of the same age and in the same state of vigilance. Some differences are quantitative, for instance with respect to amplitude, frequency, symmetry and responsiveness of the alpha rhythm. Others are qualitative, as certain phenomena occur in some people but not in others; examples include 6 and 14/s positive spikes and various 'atypical' responses to visual stimulation, which are considered elsewhere. However, intersubject differences or intrasubject changes may result from pathological processes and are the basis of the clinical use of the EEG. Some pathological EEG findings are unequivocally abnormal and are not encountered in health. Others involve features within the range of normal variation but different from the individual's premorbid EEG. Some common abnormalities are described below.

Ongoing activities
Amplitude reduction

The most unequivocal sign of cerebral dysfunction is the reduction in amplitude of normal activities. Among other causes, this may result from neuronal loss, or suppression of neuronal activity by toxic agents including drug overdose, acute hypoxia and hypothermia. Extreme amplitude reduction (i.e. the absence of recordable activity, so-called 'electrocerebral silence') is therefore observed in brain death, after massive barbiturate overdose, during profound surgical hypothermia, and often for some tens of seconds after a convulsive seizure. Amplitude reduction in the surface EEG may also result from impaired conduction of signals from the cortex to the scalp, for instance due to the presence of an intervening subdural haematoma.

In neonates, hypoxia during birth may cause mild to extreme amplitude reduction lasting from hours to days, depending on the length of the of hypoxia and in relation to the severity of the subsequent hypoxic–ischaemic encephalopathy (*Fig. 2.39*).

A transitory reduction in EEG amplitude lasting 1–10 s sometimes occurs at the onset of an epileptic

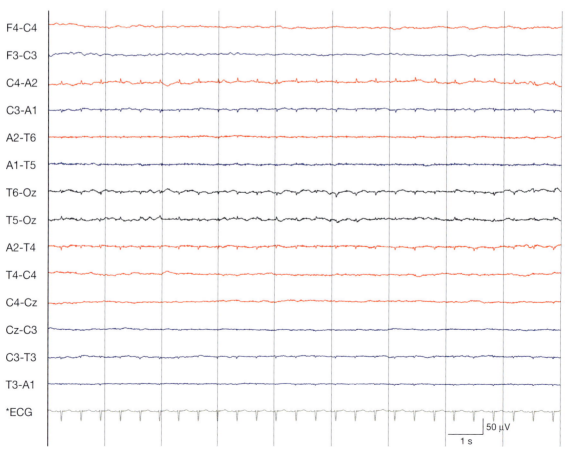

Fig. 2.39   Term infant with neonatal pneumonia, on extracorporeal membrane oxygenation (ECMO). The EEG shows low amplitude of < 5 µV. Over the posterior regions some low amplitude irregular 3–5/s activity can be seen with an amplitude of 5–10 µV.

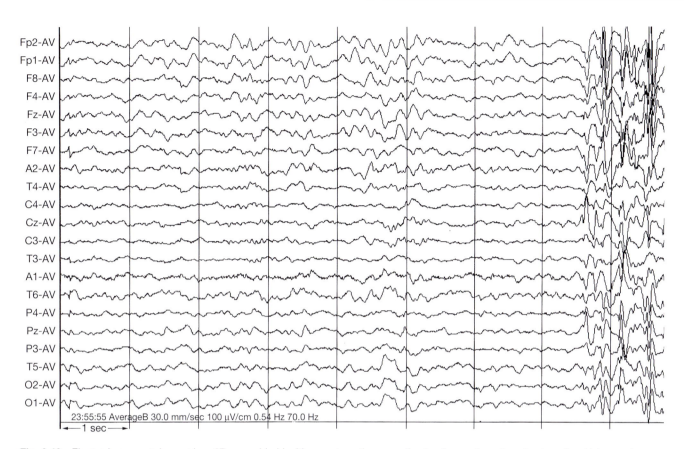

Fig. 2.40 Electrodecremental event in a 17-year-old girl with symptomatic generalised epilepsy. A nocturnal seizure in which ongoing activity is reduced in amplitude for 2.5 s at onset, followed by generalised spike wave discharge. The patient turns her head to the right and raises both arms. (From Binnie et al (2003), by permission.)

seizure (an 'electrodecremental event') (*Fig. 2.40*). This is seen particularly often in paediatric practice, specifically during infantile spasms and other dystonic seizures. Here a totally different mechanism is involved – desynchronisation of neuronal activity – and single unit recordings during a reduction of EEG or stereo EEG (SEEG) amplitude at seizure onset may show maintained or increased neuronal discharge (Babb et al 1987). Another rarer, or underreported, type of amplitude reduction occurs in patients with coma of various aetiologies and usually with fatal outcome. These 'episodic low amplitude events' (ELAEs) (Rae-Grant et al 1991) are generalised, last 0.5–4 s, and are often followed by a change in the pattern of ongoing activity. They are to be distinguished from the low amplitude periods of the burst-suppression pattern (which are usually of longer duration, interrupted by separate clear bursts of high amplitude activity) and from the effects of arousal by stimulation, to which they are not apparently related.

As many healthy subjects constitutionally have low amplitude EEGs (a genetically determined trait) and others exhibit a marked voltage reduction when anxious or hyperaroused, degrees of amplitude reduction short of electrical silence or near silence may not be recognisable as unequivocally abnormal unless they are also asymmetrical. An asymmetry greater than 50% in amplitude

of normal ongoing activities will generally reflect disease on the side where the amplitude is lower. An exception to this principle is seen in patients with a skull defect (e.g. a burr hole or craniectomy). EEG activity may be of greater amplitude over the defect, particularly with respect to fast components, which are ordinarily reduced by conduction through the skull. It should be apparent from the foregoing discussion that several pathologies could produce the picture shown in *Fig. 2.41*. In fact the teenager had suffered a left-sided middle cerebral artery (MCA) infarct, but a similar pattern might be seen, for instance, some months after a severe head injury, or in the presence of a left-sided subdural haematoma.

Slowing

An increase in lower frequency components of the EEG is a non-specific abnormality, which can reflect various pathological processes including cerebral hypoxia, oedema, raised intracranial pressure (which may be associated with both the foregoing), cerebral inflammatory or degenerative processes, the postictal state and various intoxications.

Sometimes the slow activity arises de novo, sometimes it appears to represent an increase in slow rhythms which were previously present, and on occasion it may appear to reflect a change in frequency of a pre-existing activity such as the alpha rhythm. Sometimes it may appear that more than one type of change occurs sequentially with increasing dysfunction. Thus serial EEGs in a patient with progressive metabolic disturbance may at first show an alpha rhythm slower than that previously exhibited by the same person in health, then activity with similar characteristics but in the upper theta range, then an abrupt transition to generalised delta waves of totally different topography.

Minor changes in ongoing frequencies may be recognised as pathological only if they are also asymmetrical. The normal EEG is much more symmetrical in terms of frequency than of amplitude. Thus asymmetries in the dominant alpha frequency of only 0.5 Hz may be clinically significant, and can be found not only within the territory of the alpha rhythm but also in the central areas. The slower component, which is likely to reflect the side of greater cerebral dysfunction, may be of either greater or lesser amplitude than the corresponding normal activity over the other hemisphere. The lateralising significance of an asymmetry of alpha amplitude may

therefore be misinterpreted unless frequency is also taken into account.

There is indeed a general problem in interpreting asymmetrical pathological slow activities. These may be expected to be of greater amplitude over the more disturbed hemisphere, but the reverse will apply if the underlying pathological process also leads to amplitude reduction. Thus, if a subdural haematoma produces bilateral slowing due to raised intracranial pressure and an asymmetry of amplitude, it may be impossible to lateralise from the EEG. The hypothesis that unilateral amplitude reduction is present may, however, be supported by the finding of a parallel asymmetry of beta activity. Because the alpha frequency of most alert older school children and adult subjects is somewhat higher than 8 Hz, essentially similar considerations apply to the interpretation of slowed alpha and of excess theta activity. It may be helpful in describing such pathological EEGs to employ the concept of the 'postcentral dominant rhythm' regardless of whether this is within or below the alpha frequency range. In younger children the maturational stage and normal variations have to be taken into account.

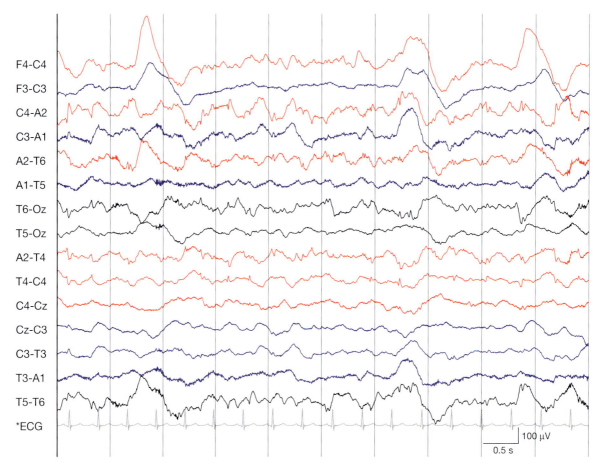

Fig. 2.41   A 14-year-old girl with MELAS (mitochondrial encephalopathy with lactic acidosis and stroke-like episodes). She had a previous left hemispheric infarct and presented now with an acute right MCA territory infarct. The EEG shows high amplitude slow activity over the right, especially over the frontocentral regions, intermixed with frequent sharp waves and spikes (acute infarct). Note the amplitude reduction and absence of fast activity over the left (old infarct).

Excess beta activity

The amount of beta activity is very variable, particularly in childhood. Beta activity is moreover increased in spontaneous drowsiness and by many sedative drugs, notably barbiturates and benzodiazepines. Because electroencephalographers' notions of normality are generally based on clinical recordings in which prominent beta activity is often associated with medication, there may be a tendency wrongly to assume that prominent beta activity is abnormal, and that patients showing this feature are taking drugs not mentioned by the referring physician. Excess beta activity is rarely a pathological phenomenon but may be seen briefly in the early stages of acute cerebral anoxia; short bursts of very high frequency beta activity (upwards of 40 Hz) occur in some patients with cerebral degenerative disorders and in some types of cortical dysplasia. Fast activity in the beta or gamma ranges may occur at onset of partial seizures, but is more readily detected in intracranial than scalp recordings. Drug-induced beta activity commonly has the same frontocentral emphasis as that occurring normally (*Fig. 2.42*). However, benzodiazepines sometimes induce a fast alpha variant with a topography similar to that of the alpha rhythm. Note, however, that a fast alpha variant also occurs in some normal subjects not taking medication. Prominent high amplitude fast activity can be seen in some patients with migration disorders, namely Type 1 lissencephalopathy, and is typical in children with infantile neuroaxonal dystrophy.

Altered reactivity

When the postcentral dominant rhythm is slowed it often shows a reduced reactivity to eye opening or alerting. However, some normal subjects exhibit only minimal alpha blocking, and most will show a reduced response (in terms of both EEG and behaviour) when drowsy. Thus poor reactivity of the postcentral dominant rhythm cannot reliably be regarded as pathological unless it is also asymmetrical (less responsive on the more abnormal side) or is seen in a subject whose EEG was previously known to be responsive in the same behavioural state.

Strikingly unresponsive activity of alpha frequency, often rather faster and more widespread than the alpha rhythm, may be seen in states of altered consciousness, usually due to midbrain lesions, following cerebral hypoxia or in some intoxications ('alpha coma').

Arousal in sleep, from stage II or deeper, produces various slow wave transients, usually maximal at the vertex, often followed by changes in ongoing activity reflecting a change of state (*Fig. 2.43(a)*). However, some patients with impaired vigilance show persistent or prolonged runs of bifrontal slow activity when brought by stimulation to a state of full behavioural arousal (*Fig. 2.43(b)*). This phenomenon of 'paradoxical slow wave arousal' has been described in some intoxications, but particularly with presumed brainstem or midbrain

lesions after head injury (Schwartz and Scott 1978). Surprisingly perhaps, the prognosis is generally favourable; the opportunity has not therefore arisen to establish the neuropathological basis.

Localised abnormalities

As indicated above, amplitude reduction and slowing may be bilateral or asymmetrical. Essentially similar disturbances can be more localised. Amplitude reduction over a small region of the scalp may involve all components of the EEG or selectively the higher frequencies. Global reduction of amplitude will generally reflect gross hypofunction or destruction of underlying cortical neurons. Less severe damage may produce loss only of faster activities and this may also be seen at sites where local slowing occurs.

Localised slow activities apart from so-called 'rhythms at a distance' (see p. 48) generally reflect structural abnormality in the cortex underlying the electrodes from which they are recorded. Localised slowing may result in abnormal activities in the lower alpha, theta or delta ranges or combinations of these. Local delta activity may show a variety of waveforms. Sometimes the waves are irregular in both shape and frequency (polymorphic delta activity) (*Fig. 2.44*) and are often associated with a reduction in amplitude of overlying faster components at the same site. Acute space-occupying or destructive lesions (e.g. rapidly growing tumours, abscesses, encephalitis and intracerebral haematomata) causing gross local cerebral dysfunction associated with localised destruction and oedema of white and grey matter may produce steep high voltage slow delta waves. The occurrence of polymorphic focal delta activity at the site of a lesion probably requires both cortical damage and partial deafferentation due to white matter involvement (Gloor et al 1977). Pathology confined to cortical grey matter produces only a reduction of amplitude, without delta activity either overlying or at the margins of the lesion; conversely, total cortical deafferentation does not produce a delta focus.

Theta activity and sporadic delta waves of lower amplitude and duration may be seen during recovery or in association with lesions less disruptive of local cerebral function; amplitude reduction may be the eventual outcome, when those damaged cells that have not died have recovered. Lesions that are static or only slowly progressive (e.g. porencephalic cysts and meningiomata) may produce little slowing, even if large. However, small chronic lesions giving rise to epilepsy may produce surprisingly large amounts of focal slow activity even when no seizure has occurred for several days.

Note that it is not cerebral lesions but dysfunctional neurons that generate abnormal EEG activity, and thus slowing may be seen around the periphery of a space-occupying lesion but not over its centre where, if it is sufficiently large, amplitude reduction may be detected. Another important category of abnormal phenomena, which may be localised, epileptiform activity, is considered on pages 51–57.

Rhythms at a distance

Structurally normal cortex may generate abnormal activities in response to altered afferent inputs from deep structures; these are termed 'rhythms at a distance'.

Frontal intermittent rhythmic delta activity (FIRDA)

Frontal intermittent rhythmic delta activity (FIRDA) is seen over the frontal regions, and is usually bilateral and synchronous. The frequency is in the range 1.5–2.5 Hz, the slower type tending towards an anterior temporal distribution. The waveform is typically sinusoidal (hence

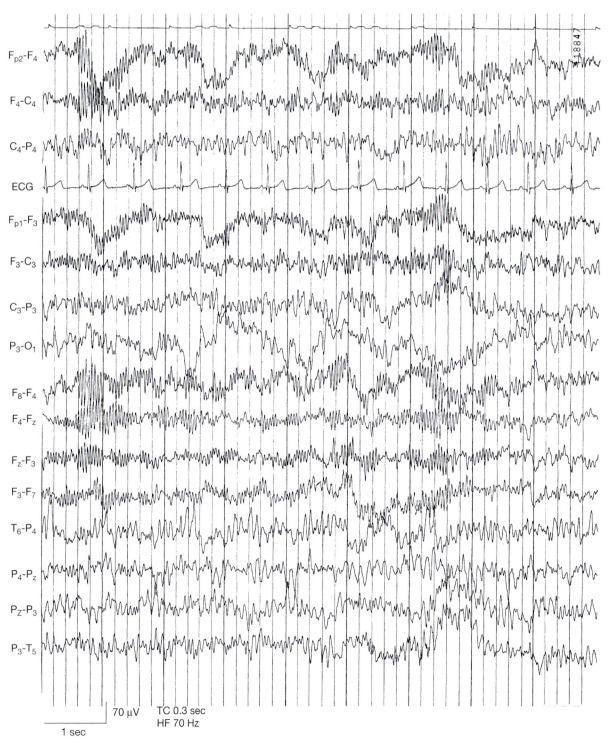

70 µV   TC 0.3 sec
        HF 70 Hz
1 sec

Fig. 2.42 Excessive beta activity in a 6-year-old girl who was receiving clonazepam and sodium valproate to control absence seizures. Note the distribution and spindled pattern of the beta activity. (From Binnie et al (2003), by permission.)

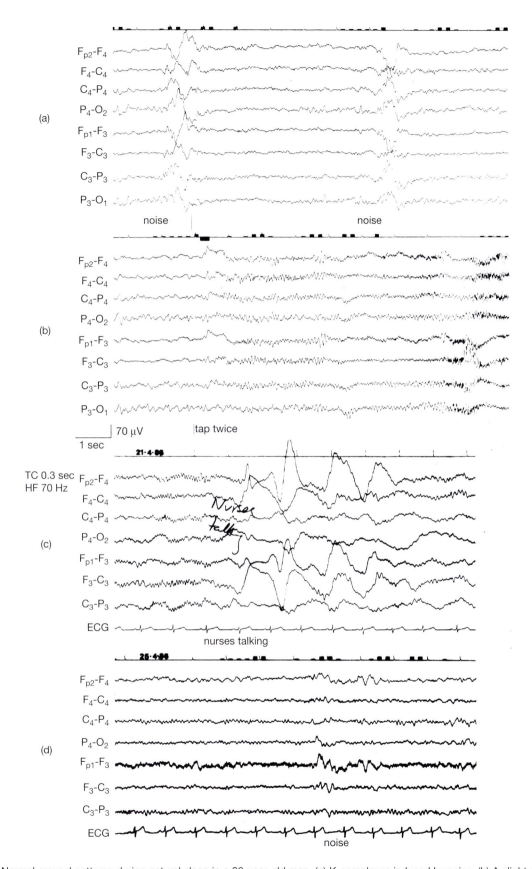

Fig. 2.43 Normal arousal patterns during natural sleep in a 26-year-old man. (a) K-complexes induced by noise. (b) A slight sound induces a change towards a more alert state with the appearance of alpha rhythm. Contrast this with abnormal arousal patterns evoked in a 24-year-old man in a light coma from water intoxication following heat stroke. (c) Paradoxical slow wave arousal response to quiet talking near the patient. (d) A much less pronounced response to noise 4 days later when the patient was conscious but disorientated. (From Binnie et al (2003), by permission.)

the term formerly used, 'monorhythmic sinusoidal frontal delta activity'), or saw-toothed (*Fig. 2.45*). The activity occurs in rhythmic bursts ranging from single waves to an almost continuous rhythm but typically lasting 2–5 s. It is markedly dependent on arousal level, being seen neither in the fully alert subject nor below stage I of sleep.

Frontal intermittent rhythmic delta activity can occur in metabolic and toxic disorders, status epilepticus and occasionally postictally, but most often in association with cerebral pathology. It is seen particularly, but not exclusively, with abnormalities of the diencephalon. It is thus found, for instance, with thalamic tumours, metastases presumed to include deep lesions, with obstruction of the aqueduct, and in ischaemia of the anterior thalamus due to vascular spasm following rupture of an anterior communicating artery aneurysm. However, the most common pathological correlate of FIRDA is diffuse disease involving grey and white matter (Gloor et al 1968).

Bitemporal theta activity

In early drowsiness, theta activity may increase over the temporal regions, but in various states causing pathological drowsiness bilateral rhythmic temporal or frontotemporal theta activity occurs. It occurs under similar circumstances to FIRDA and has no more specific pathophysiological significance.

Posterior slow activity

Various types of slow activity occur at the back of the head and are distinguished on grounds of topography, morphology and clinical context.

As was noted on page 33, posterior temporal slow activity is a normal finding in the young which disappears in the early 20s. It does, however, persist into the 30s in some individuals, many of whom are normal, but including a disproportionate number of persons exhibiting psychopathic or antisocial traits, particularly aggression. Posterior temporal slow waves may also be seen as a nonspecific abnormality following a variety of cerebral insults including trauma, cerebrovascular accidents and after exposure to severe hypoglycaemia. The slow activity may then be of greater amplitude on the side of greater cerebral abnormality, but in general it shows the same tendency to predominate over the non-dominant hemisphere, as shown by maturational posterior temporal slow waves of youth which it resembles in all other respects, blocking with the alpha rhythm on eye opening and increasing on overbreathing.

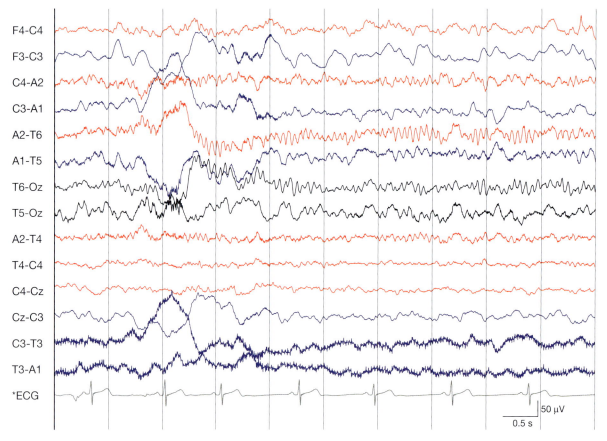

Fig. 2.44 A 15-year-old boy with dissection of the internal carotid artery. He presented acutely with drowsiness, aphasia and right-sided hemiplegia. Computerised tomography (CT) confirmed a left-sided MCA infarct. The EEG shows a loss of alpha rhythm and excess high amplitude slow wave activity on the side of the infarct.

Brainstem dysfunction can also produce posterior temporal slow waves, which occur in irregular bursts with vertebrobasilar insufficiency. Brainstem injury may indeed be the cause of posterior slow activity after cerebral trauma, as noted above.

Much slower delta waves (of more than 1 s duration) are sometimes seen over the occipital regions, in association with mass lesions in the posterior fossa and also with haemorrhage from the vertebrobasilar system. They occur most often in children where they may attain an amplitude of several hundred microvolts; in adults they are generally inconspicuous and easily mistaken for electrode artefacts from the head moving on the pillow. This slow activity too tends to show a right-sided preponderance and is therefore non-lateralising and, like FIRDA, it occurs most readily in a state of slight drowsiness.

Various types of rhythmic posterior slow activity are seen. Short runs of reactive, regular delta activity at 3–4 Hz occur, very rarely, in apparently healthy young adults and presumably represent an unusual slow alpha variant; they may be consistently or intermittently unilateral. Rhythmic, usually bilateral posterior slow activity at about 3 Hz is, however, also seen, less rarely, in children with absence epilepsy. The waveform is often notched (*Fig. 2.46(a)*), suggesting that this is a variant of SW activity and, at the start of an absence or on overbreathing, it appears to evolve into a typical generalised SW discharge (see pp. 51–57). Occipital intermittent rhythmic delta activity (OIRDA) resembles FIRDA but is located posteriorly (*Fig. 2.46(b)*); it is more likely to be associated with brainstem pathology.

Epileptiform activity

Cerebral electrical activity changes dramatically during epileptic seizures and characteristically spiky waveforms may be seen, some of which are illustrated in *Fig. 2.47*. Waves of sharp outline standing out from the ongoing background rhythms, and those lasting < 70 ms are conventionally called 'spikes'. Those lasting for 70–200 ms are termed 'sharp waves'. It will be noted that this definition begs the question of how obviously the waveform should stand out from the background and also implies that a wave that would be indistinguishable from ongoing beta or alpha activity may be designated as a spike or sharp wave when it arises from a background in which these rhythms are not prominent.

Official terminology recognises that spikes are often followed by delta waves either singly (spike-and-slow-wave complex) or in runs (spike-and-slow-wave activity). In fact most spikes are polyphasic but, where this is a conspicuous feature, the terms 'multiple spike complex', 'multiple spike-and-slow-wave complex', etc., may be used. Usage of the term 'polyspike' is discouraged. Close inspection also shows that very many spikes are followed by a slow wave; only if this is conspicuous is the term 'spike and slow wave' used. The most conspicuous

components of a focal spike discharge, and of both the spike and the wave of a spike-and-slow-wave complex, will usually be electronegative on the cortical surface and in the scalp EEG. Apart from '14 and 6 Hz positive bursts' (see below), positive-going focal spikes are rare. They are occasionally seen in status epilepticus in children, and over cortical defects that permit recording from the deep aspect of the cortex without intervening grey matter (e.g. after a penetrating wound). They sometimes clearly represent one pole of an unusually orientated dipole, the negative pole of which is seen as a synchronous focus nearby. The presumption that (with the exceptions just mentioned) spikes are negative at the cortical surface may be an important consideration in deciding whether a sharp transient is artefactual, and in locating dipolar sources. For instance, a spike that is positive below the sylvian fissure and negative above it probably arises from the superior aspect of the first temporal convolution.

Sharp or spiky waveforms occur in the interictal state, i.e. between overt seizures, in most people with epilepsy. Similar phenomena are also found in some patients with other cerebral disorders who are not thought to suffer from seizures. Normal subjects also exhibit a variety of sharp transients, the majority of which are generally distinguishable with little difficulty from those associated with epilepsy (see below).

There is no agreement as to a suitable collective term to describe this category of phenomena. 'Epileptic activity' is unacceptable as it wrongly attaches to these waveforms diagnostic significance that they do not possess. 'Seizure discharges', widely used in North America, appears even more inappropriate, as even in people with epilepsy the discharges may be interictal. 'EEG seizure pattern' is countenanced by the IFCN, but only to describe repetitive discharges with abrupt onset and termination, lasting several seconds – as seen during seizures. If not accompanied by a clinically overt seizure, they should be designated 'subclinical'. 'Paroxysmal activity', favoured in the UK, is unsuitable as many normal EEG phenomena have a paroxysmal quality, e.g. K-complexes or lambda waves. In continental Europe 'irritative activity' is a popular euphemism, which nevertheless implies epileptogenesis. 'Epileptiform pattern, discharge, or activity' is a compromise supported by the IFCN and favoured by many authorities on epilepsy in Europe and North America, which at least acknowledges the statistical association with epilepsy underlying the concept, while stressing that the term refers to a category of waveforms and not a diagnosis.

It will be clear from the foregoing that not only the terminology but also the definitions of epileptiform phenomena are in some disarray. Nevertheless, the IFCN has proposed a terminology and it serves the best interests of the discipline that we should use it, until something better can be devised. The recommendations concerning epileptiform phenomena (corrected for some

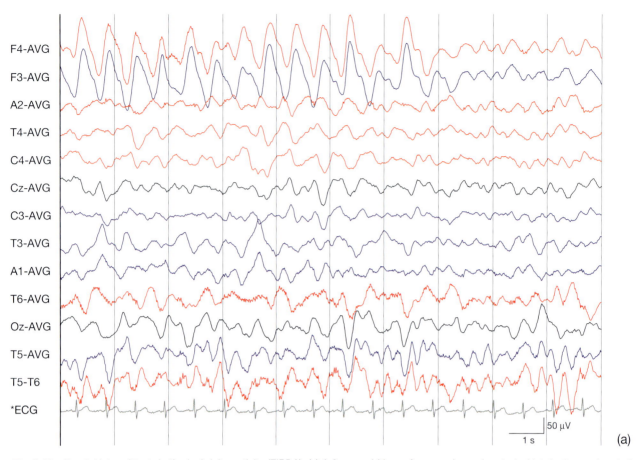

Fig. 2.45 Frontal intermittent rhythmical delta activity (FIRDA). (a) A 3-year-old boy after a prolonged period of intubation and sedation following repair of subglottic stenosis. He was awake, but subdued, during the recording. The EEG is dominated by irregular 4–6 Hz activities, with superimposed faster components. Intermittent high amplitude rhythmic delta activity is seen over the frontal regions.

obvious typographical errors) are as follows (Noachtar et al 1999):

Epileptiform pattern: Synonym: epileptiform discharge, epileptiform activity. Describes transients distinguishable from background activity, with a characteristically spiky morphology, typically, but neither exclusively nor invariably, found in interictal EEGs of people with epilepsy.

This recommendation ignores common usage, as many workers employ the term to include ictal phenomena also.

Seizure pattern, EEG: Phenomenon consisting of repetitive EEG discharges with relatively abrupt onset and termination and characteristic pattern of evolution, lasting at least several seconds. These EEG patterns are seen during epileptic seizures. Frequent interictal epileptiform discharges are usually not associated with clinical seizures and thus should be differentiated from electroencephalographic seizure patterns. The component waves or complexes vary in form, frequency and topography. They are generally rhythmic and frequently display increasing amplitude and decreasing

frequency during the same episode. When focal in onset, they tend to spread subsequently to other areas. Comment: EEG seizure patterns unaccompanied by clinical epileptic manifestations detected by the EEG technologist should be referred to as 'subclinical'.

Paroxysm: Phenomenon with abrupt onset, rapid attainment of a maximum and sudden termination; distinguished from background activity. Comment: commonly used to refer to epileptiform patterns and seizure patterns (cf. epileptiform pattern; seizure pattern).

Spike: A transient, clearly distinguished from background activity, with pointed peak at a conventional paper speed or time scale and a duration from 20 to under 70 ms, i.e. 1/50–1/15 s, approximately. Main component is generally negative relative to other areas. Amplitude varies. Comments: (1) term should be restricted to epileptiform discharges. EEG spikes should be differentiated from sharp waves, i.e. transients having similar characteristics but longer durations. However, it should be kept in mind that this distinction is largely arbitrary and primarily serves descriptive purposes. Generally, in ink-written EEG

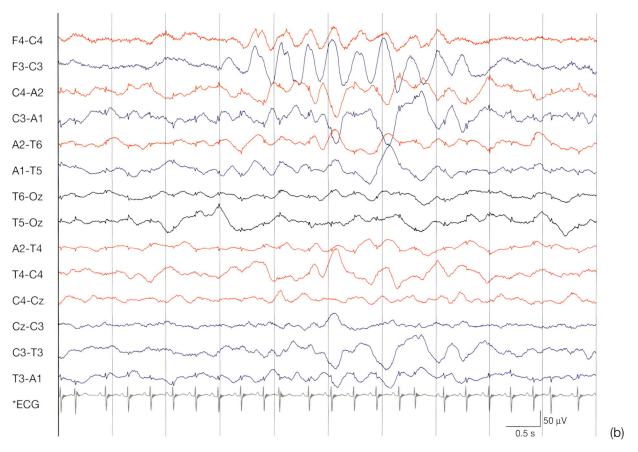

F4-C4
F3-C3
C4-A2
C3-A1
A2-T6
A1-T5
T6-Oz
T5-Oz
A2-T4
T4-C4
C4-Cz
Cz-C3
C3-T3
T3-A1
*ECG

50 µV

0.5 s

(b)

Fig. 2.45 *(cont.)* (b) Asymmetrical FIRDA in a 6-month-old boy with craniosynostosis. After cranial surgery he had several apnoeic episodes. The EEG shows runs of intermittent rhythmic delta activity (2–3 Hz) in the frontal regions. These were notched at times and of higher amplitude on the left. The ongoing activity is continuous with 1.5–3 Hz activity (mild excess of slowing for age).

records taken at 3 cm/s, spikes occupy 2 mm or less of paper width and sharp waves more than 2 mm. (2) EEG spikes should be clearly distinguished from the brief unit spikes recorded from single cells with microelectrode techniques (cf. sharp wave).

Sharp wave: A transient, clearly distinguished from background activity, with pointed peak at conventional paper speed or time scale and duration of 70–200 ms, i.e., 1/15–1/5 s approximately. Main component is generally negative relative to other areas. Amplitude varies. Comments: (1) term should be restricted to epileptiform discharges and does not apply to (a) distinctive physiologic events such as vertex transients, lambda waves and positive occipital sharp transients of sleep, (b) sharp transients poorly distinguished from background activity and sharp-appearing individual waves of EEG rhythms; (2) sharp waves should be differentiated from spikes, i.e. transients having similar characteristics but shorter duration. However, it should be kept in mind that this distinction is largely arbitrary and primarily serves descriptive purposes. As a rule, in ink-written EEG records taken at 3 cm/s, sharp waves occupy more than 2 mm of paper width and spikes 2 mm or less (cf. spike).

Spike-and-slow-wave complex: A pattern consisting of a spike followed by a slow wave. Comment: hyphenation supposedly facilitates use of term in plural form: 'spike-and-slow-wave complexes' or 'spike-and-slow-waves'.

Sharp-and-slow-wave complex: A sequence of a sharp wave and a slow wave. Comment: hyphenation facilitates use of term in plural form: 'sharp-and-slow-wave complexes'.

Three Hz spike-and-slow-wave complex: Characteristic paroxysm consisting of a regular sequence of spike-and-slow-wave complexes which: (1) repeat at 3–3.5 Hz (measured during the first few seconds of the paroxysm); (2) are bilateral in their onset and termination, generalised and usually of maximal amplitude over the frontal areas; and (3) are approximately synchronous and symmetrical on the two sides of the head throughout the paroxysm. Amplitude varies but can reach values of 1000 µV (1 mV) (see atypical spike-and-slow-wave complex).

Six Hz spike-and-slow-wave: Spike-and-slow-wave complexes at 4–7 Hz, but mostly at 6 Hz, occurring

53

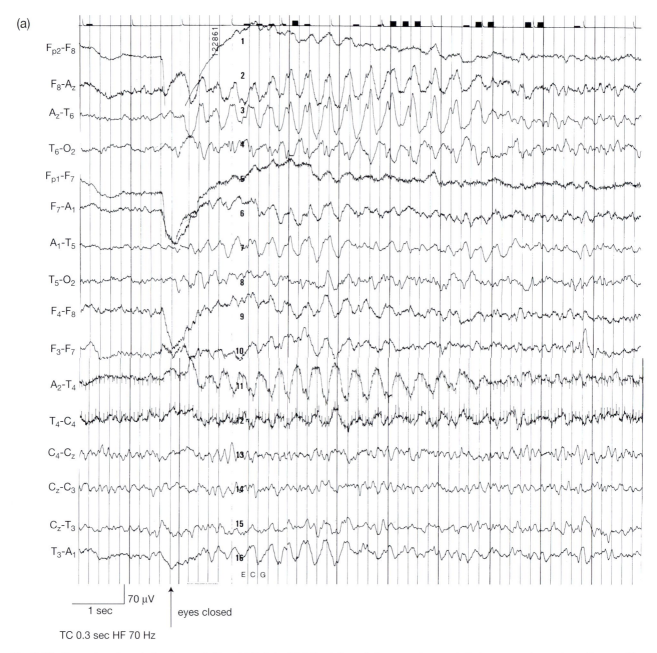

(a)

Fig. 2.46 Examples of posterior slow activities. (a) Notched 2.5 Hz delta activity appearing posteriorly on eye closure in an 8-year-old boy with childhood absence epilepsy. This feature of ongoing activity should not be mistaken for the 3 Hz SW complexes accompanying absence seizures. (From Binnie et al (2003), by permission.)

generally in brief bursts bilaterally and synchronously, symmetrically or asymmetrically, and either confined to or of larger amplitude over the posterior or anterior regions of the head. Amplitude varies but is generally smaller than that of spike-and-slow-wave complexes repeating at slower rates. Comment: This pattern is of little clinical significance and should be distinguished from epileptiform discharges.

Multiple spike complex: A sequence of two or more spikes. Preferred to synonym: polyspike complex.

Multiple spike-and-slow-wave complex: A sequence of two or more spikes associated with one or more slow waves. Preferred to synonym: polyspike-and-slow-wave complex.

Hypsarrhythmia: Pattern consisting of diffuse high voltage (> 300 μV) irregular slow waves interspersed with multiregional spikes and sharp waves over both hemispheres.

To add conceptual to linguistic confusion, there exist various sharp or frankly spiky waveforms that commonly occur in normal subjects and which are of little or no significance in relation to the diagnosis of epilepsy. Because of the importance of distinguishing them from

(b)

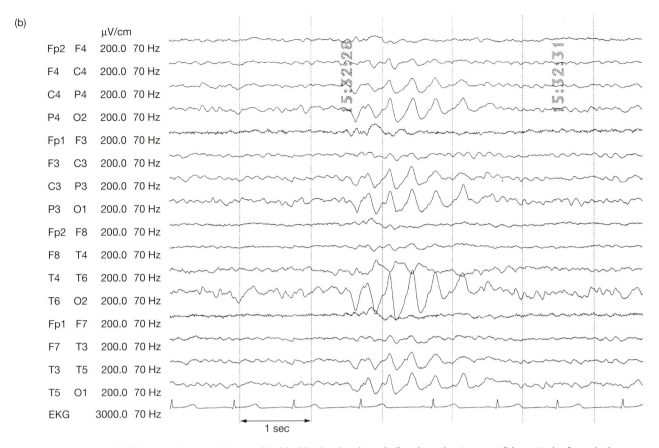

Fig. 2.46 *(contd.)* (b) OIRDA occurring in a 14-year-old girl with visual and cerebellar signs due to a possible posterior fossa lesion. (From Binnie et al (2003), by permission.)

pathological phenomena that are the topic of this section, they are described here.

> *Fourteen and 6-Hz positive burst: Burst of arch-shaped waves at 13–17 Hz and/or 5–7 Hz but most commonly at 14 and/or 6 Hz seen generally over the posterior temporal areas of one or both sides of the head during sleep. The sharp peaks of its component waves are positive with respect to other regions. Amplitude varies but is generally below 75 µV. Comments: (1) best demonstrated by referential recording using contralateral earlobe or other remote, reference electrodes; (2) this pattern is of no established clinical significance.*

An increased incidence has been claimed in a variety of conditions ranging from behaviour disorders to allergies. In any event they contribute nothing to the diagnosis of epilepsy.

> *Benign epileptiform transients of sleep (BETS): Small sharp spikes (SSS) of very short duration and low amplitude, often followed by a small theta wave, occurring in the temporal regions during drowsiness and light sleep. This pattern is of little clinical significance.*

Benign epileptiform transients of sleep can be demonstrated in many normal subjects, but in clinical material they may show a weak association with epilepsy (Hughes and Grunener 1984). Depth recording in patients with focal epileptiform discharges giving rise to seizures has shown that BETS arise from different sites and therefore probably play no part in the epileptic mechanism in these patients (Westmoreland et al 1979). Their presence in any event cannot be considered to offer any significant support to a diagnosis of epilepsy.

> *Rhythmic mid-temporal discharge (formerly termed 'psychomotor variant')* consists of rhythmic activity at about 6–7 Hz over one or other temporal region, often present in repeated records over many years in normal subjects and unaffected by eye opening or antiepileptic medication. It may be precipitated by hyperventilation or blocked by attention. The discharge starts abruptly, without any gradual build up or progressive change in frequency, and ends, often minutes later, equally suddenly. Despite a supposed relationship to temporal lobe epilepsy, as indicated by the earlier name, it appears that this phenomenon is without clinical significance (Hughes and Cayaffa 1977). This term does not appear in the recommended IFCN terminology, where its place seems to have been taken by 'rhythmic temporal theta burst of drowsiness', which the present authors consider to be a different phenomenon.

> *Subclinical rhythmical epileptiform discharges of adults (SREDA): This is a rhythmic pattern seen in the adult*

age group which consists of a mixture of frequencies, often predominant in the theta range. It may resemble a seizure discharge but is not accompanied by any clinical signs or symptoms. The significance of this pattern is uncertain, but it should be distinguished from an epileptic seizure pattern.

This is somewhat similar to rhythmic midtemporal discharge but often develops from a run of sharp waves, increasing in frequency from 0.5 to 6 Hz; it is located more posteriorly at the junction of the temporal, parietal and occipital lobes. The discharge may be unilateral or bilaterally synchronous and symmetrical. It is seen in normal adults above the age of 50 years and is not associated with epilepsy.

Midline spikes are spiky transients occurring at the vertex in full wakefulness. They are rare and are statisti-

cally associated with epilepsy, but the majority of subjects showing this phenomenon are apparently healthy. Where epilepsy is present, seizures arising from the supplementary motor area are occasionally found, but more usually the phenomenon is of no localising significance. There may be a subcategory, found in children, with a prominent surface positive phase, associated with partial seizures, not necessarily arising on the medial surface of a hemisphere.

Anomalous or marked responses to activation are readily misinterpreted. Both hyperventilation and photic stimulation are useful methods of eliciting epileptiform activity (see p. 61 ff.). However, it is important that neither the normal bifrontal slow wave response to overbreathing in young subjects, nor various spiky photic responses confined to the back of the head or time locked to the flicker, should be credited with clinical significance.

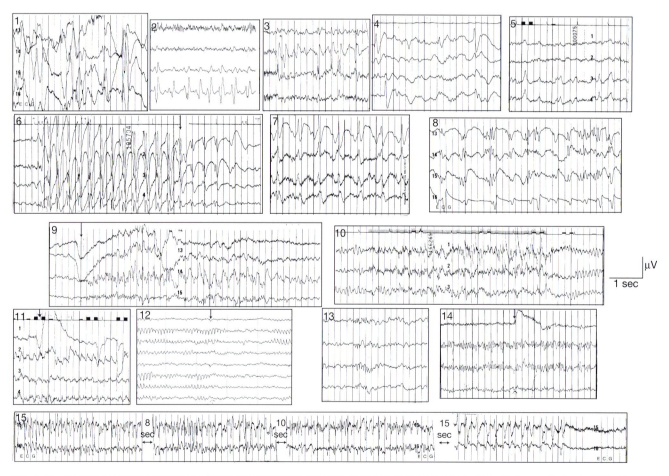

Fig. 2.47 Examples of spiky waveforms (amplitude calibration for each sample indicated in microvolts). (1) Hypsarrhythmia (100 µV); (2) occipital spikes (200 µV); (3) temporal–temporal spikes (100 µV); (4) independent temporal spike-and-slow-wave complexes (100 µV); (5) left temporal sharp waves (70 µV); (6) typical 3 Hz SW complexes (70 µV), changing to 200 µV at arrow; (7) slow SW complexes (150 µV); (8) multiple spike-and-slow-wave complexes (200 µV); (9) spiky discharge on eye closure (at arrow) (150 µV); (10) spiky discharge on photic stimulation at 30 f/s (70 µV); (11) spiky waveform produced by a combination of repetitive eye movement potentials and lateral rectus muscle spikes during nystagmus (70 µV); (12) spiky mu rhythm attenuated by imagined proprioceptive stimulation at arrow (150 µV); (13) 14 Hz positive spikes (70 µV); (14) subclinical rhythmic epileptiform discharge of adults (SREDA) seizure discharge, eyes open promptly to command at arrow (100 µV); (15) polymorphic spiky waveforms during partial seizure (70 µV). (From Binnie et al (2003), by permission.)

Breach rhythm is a spiky rhythmic waveform in the lower alpha or upper theta range seen over cranial defects, usually following a burr hole or craniotomy. It probably owes its appearance in part to enhanced transmission of beta activity through the opening in the bone; however, the often prominent theta components may suggest that there is also dysfunction of underlying brain, due either to the surgical intervention or the original pathology. Breach rhythm as such does not amount to evidence of epilepsy, although patients with cranial defects do, of course, have an increased risk of seizures.

It should be noted that these various spiky phenomena, which are normal or of little significance in relation to the diagnosis of epilepsy, are distinguishable by characteristic morphology, topography or circumstances of occurrence. They should not be confused with those spiky phenomena that are associated with epilepsy and are conventionally described as 'epileptiform'.

Genesis and significance of epileptiform activity

The cardinal feature of epileptiform activity is the occurrence of excessive and hypersynchronous neuronal activity. That is, a population of neurons not only becomes more than usually active, but also shows a greater than usual degree of synchronisation of activity between cells. These conditions are necessary to produce the abrupt changes in potential that appear in the EEG as spikes. The spikes seen on the scalp mostly arise from summated excitatory postsynaptic potentials on the dendrites of radially oriented neurons in the superficial layers of the immediately underlying cortex. They are often followed by a slow wave, which is generated by inhibitory processes causing inhibitory postsynaptic potentials, or hyperpolarisation on the cell bodies of the same or similar neurons, deeper in the cortex. As the excitation causes a negative-going change on the superficial dendrites, and the inhibition produces positive-going change deep in the cortex, both are expressed as negative potentials in the EEG. Spikes, sharp waves, and spike-and-slow-wave complexes are usually negative on the scalp immediately overlying the site where they are generated. Positive spikes (apart from the non-epileptic 14 and 6 Hz positive spike phenomenon) are rare and should raise a suspicion that they may be artefactual. They may, however, be recorded, presumably from the deep surface of the cortex, in the presence of a gross cortical defect, such as after a penetrating head injury.

Brief, epileptiform discharges involving neuronal populations that are small or not critical to maintenance of apparently normal behaviour may produce no clinical effects and are described as interictal (between seizures) or subclinical. Subclinical discharges are seen in up to 3% of healthy children without a history of clinical seizures (Eeg-Olofsson et al 1971, Cavazzuti et al 1980). Discharges accompanied by seizures are described as ictal. It should, however, be noted that seemingly interictal discharges may be accompanied by subtle clinical events detectable only by the closest observation (Binnie 2003), and that ictal discharges occurring deep in the brain or in small but critical neuronal aggregates may be undetectable in the scalp EEG.

Epileptiform discharges may be generated locally, and are then described as focal, or they may be over large, bilateral areas of cortex, in which case they are described as generalised. Focal and generalised discharges have importantly different clinical correlates. *Focal discharges* are associated with a localised disturbance of brain function, and when they lead to ictal symptoms these reflect reduced or disordered function of the affected brain area (*focal or 'partial' seizures*). Such disturbances may involve motor function (jerking of part of the body), sensory experiences (tingling in some region, flashing lights, etc.) or disturbances of more complex higher cerebral activities (dysphasia, a *déjà vu* sensation, fear, etc.). Except in the so-called 'benign partial childhood epilepsies', focal seizures arise in regions of cerebral pathology in the context of epilepsy syndromes also described as 'focal' or 'partial'.

Generalised discharges are generated by feedback loops involving the cortex, thalamic structures and neural pathways between cortex and thalamus, or between different regions of cortex, in the same or opposite hemispheres. Clinical ictal phenomena due to generalised discharges reflect simultaneous dysfunction of both hemispheres, ranging from brief immobility and loss of awareness (*absence*) to loss of muscle tone and collapse (*atonic seizure*) or generalised motor activity: stiffening (*tonus*), jerking (*clonus*), and particularly muscular rigidity followed by jerking (*tonic–clonic seizure*). Epilepsies characterised by these seizure types are called 'generalised'.

The distinction between focal and generalised events, both electrical and clinical, is less clear than the above simple account may suggest. In some partial seizures the discharges remain well localised and the clinical events reflect only dysfunction of correspondingly small brain areas (*simple partial seizures*). In others the discharges spread to involve brain regions critical to maintenance of consciousness. The patient ceases to be aware and may just lose contact with the surroundings and/or may engage in automatic activities (*complex partial seizures*). Eventually the discharges may become fully generalised, leading to a tonic–clonic convulsion. The evolution from clearly localised to generalised epileptic disturbance is commonly reflected by spread of the EEG discharges.

A challenge to the EEG technologist and electroencephalographer is to determine whether apparently generalised discharges have a local origin and are undergoing rapid secondary generalisation. Identification of ictal and interictal discharges, and the determination of their topography and morphological characteristics are fundamental to the use of the EEG in the care of people with epilepsy.

Periodicity

Various pathological EEG phenomena occur repetitively at more or less constant intervals. Some of these are considered in the following two sections. Perhaps the most dramatic example of periodicity is shown by stereotyped complexes of slow waves intermixed with spikes, sharp waves and other components that occur repeatedly at intervals of 10–20 s in subacute sclerosing panencephalitis. In various other contexts epileptiform discharges or bursts of slow waves occur repetitively at more or less regular intervals. If the conditions in which they occur have any common feature, it is probably diffuse dysfunction of both white and grey matter.

Other transients

Triphasic complexes (or waves)

Each complex consists of three waves of alternating polarity, one at least of which lies in the delta range. Typically the first and third components are electronegative and of greatest amplitude anteriorly, but the whole complex may spread across the head from front to back with a small time lag. The complexes occur in bilateral runs at 1.5–2.5 Hz, generally appearing as a continuous sequence rather than as isolated complexes or bursts. A variety of other waveforms comprising slow waves associated with some faster components (including periodic lateralised epileptiform discharges (PLEDs) – see below) may be misclassified as triphasic waves. This phenomenon, which is a sign of severe diffuse cerebral dysfunction, is almost always associated with a metabolic disorder and typically with hepatic encephalopathy. Triphasic waves appear only after the EEG has considerably slowed and the patient is in manifest coma or pre-coma.

Periodic lateralised epileptiform discharges (PLEDs)

These are stereotyped slow waves or complexes including sharp waves repeating at fairly regular intervals of 0.5–2 s, typically of about 1 s (*Fig. 2.48*). They generally show a localised maximum but appear widely over one hemisphere. Occasionally the phenomenon is bilateral (bi-PLEDs), or multifocal, although usually asynchronous; here the term 'lateralised' would not seem to apply. PLEDs are seen in a variety of conditions, both acute and chronic, and in adults are strongly associated with localised structural disease and possibly more generalised cerebral dysfunction. Examples include rapidly growing tumours, cerebrovascular accidents, cerebral abscess and

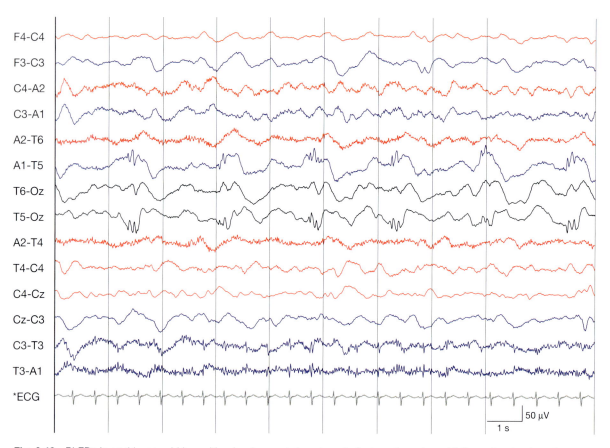

Fig. 2.48   PLEDs in an 11-year-old boy with relapsing acute lymphocytic leukaemia and renal failure who developed reversible posterior leukoencephalopathy. The EEG shows spikes and sharp waves over the left posterior region, occurring repetitively every 2 s. Note the generalised slowing of the ongoing activity.

following head injury. In children they are often associated with chronic, diffuse lesions of the central nervous system (CNS) (PeBenito and Cracco 1979). A particularly common correlate is herpes simplex encephalitis. PLEDs tend to be a self-limiting acute phenomenon, even with progressive pathology, typically becoming less frequent and then fading away over the course of a week or so, but may persist in children. They may also fluctuate within the course of a single recording, increasing during drowsiness and disappearing when the patient is alerted (Chatrian et al 1964, Markand and Daly 1971, Schwartz et al 1973). They may be associated with seizures, including convulsive or non-convulsive status epilepticus, and can often be suppressed by antiepileptic medication, such as intravenous diazepam. In adult patients with PLEDs acute seizures occur in about 70% of cases, with 20% having pre-existing epilepsy (Snodgrass et al 1989).

Other periodic phenomena

Generalised repetitive phenomena are encountered in the subacute encephalopathies, subacute sclerosing panencephalitis (SSPE) and Creutzfeldt–Jakob disease (CJD). The complexity of the discharges is greater, their morph-

ology more variable and the repetition rate slower in SSPE than in CJD. Early in the illness the SSPE complexes, often recurring at about one every 10–20 s, may be difficult to appreciate against the background of high amplitude ongoing activity in a child or teenager. Complexes may consist merely of a run of fast waves, and later with an associated slow wave. They tend to be bilaterally synchronous and similar in morphology, being stereotyped in any one area but often of different waveform and amplitude in different regions (*Fig. 2.49*).

Less elaborate transients with simple monophasic or diphasic morphology are encountered in comatose patients, for example with hypoxic encephalopathy after cardiac arrest (Binnie et al 1970, Prior 1973). Occasionally, in patients in a vegetative state after cardiac arrest, periodic transients may persist for many months, with a gradual increase in interdischarge intervals – a course not dissimilar from that of CJD (Takahashi et al 1993).

Burst-suppression pattern

The appearance of periods of amplitude suppression or total loss of activity (*Fig. 2.50*) in the EEG has long aroused interest. Early descriptions were in the context of circumstances as diverse as deep barbiturate anaes-

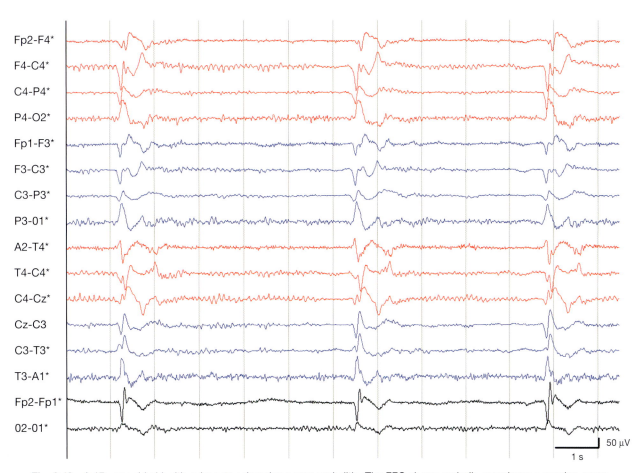

Fig. 2.49   A 17-year-old girl with subacute sclerosing panencephalitis. The EEG shows periodic complexes occurring every 5–7 s, each associated with small jerks of her right arm. Note the normal ongoing activity with mu rhythm in the central region. (Courtesy of Dr S Smith.)

thesia (Brazier and Finesinger 1945), physical isolation of the cerebral cortex from underlying white and deep grey matter by surgical undercutting (Henry and Scoville 1952) or by pathological processes such as tumours. In general terms, the phenomenon of burst suppression in the EEG may be considered as a potentially reversible expression of reduced cortical neuronal metabolic function.

Care needs to be used with terminology when describing the EEG of the newborn infant. Profound, episodic amplitude reduction (tracé discontinu) is a normal finding in the premature newborn of less than 34 weeks of conceptional age (CA), but may be pathological if the duration of the episodes of flattening is excessive for the gestational age (de Weerd et al 1999, Lamblin et al 1999, Noachtar et al 1999) (see p. 182). In the newborn of 34 weeks CA or above, tracé discontinu is replaced by tracé alternant, a less pronounced form of discontinuous pattern associated with non-REM (quiet) sleep. In strict phenomenological terms these patterns represent a form of burst-suppression pattern, but their significance as a normal developmental pattern justifies the preferred and distinctive nomenclature.

Electrocerebral inactivity – electrocerebral silence (ECS) – the isoelectric EEG

The IFCN recommendations on terminology discourage use of the term 'isoelectric EEG', preferring that of 'record of electrocerebral inactivity' (Noachtar et al 1999). An alternative term, 'electrocerebral silence' (ECS), is also commonly used, as it is here.

It is essential to pay great attention to recording technique. Considerable work from highly experienced experts has led to the American Electroencephalographic Society Guidelines (1994a, b) and the IFCN Recommendations (Chatrian et al 1996, Guérit et al 1999); they form essential reading. It is much more demanding to prove the absence of activity (*Fig. 2.51*) than to characterise active EEG phenomena.

The current IFCN Guidelines (Guérit et al 1999, embracing the earlier report of Chatrian et al 1996) give excellent guidance and specify appropriate technical requirements specific to neurophysiological recording in the intensive therapy unit (ITU). Their advice on EEG recording covers the need for polygraphic recordings, the number of channels and type of electrodes, and the

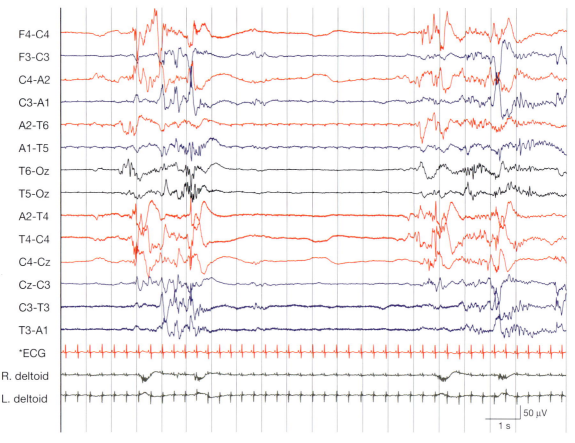

Fig. 2.50   Burst-suppression pattern; an ex-premature infant (gestational age 29/40, now 33 weeks) with seizures and apnoea since birth, treated with phenobarbitone. The EEG is discontinuous with variable interburst intervals 5–12 s. The bursts consist of high amplitude spikes, spike-and-slow waves, and sharp waves with no age-appropriate activity. Note the small myoclonia of the upper limbs in the EMG channels associated with each burst.

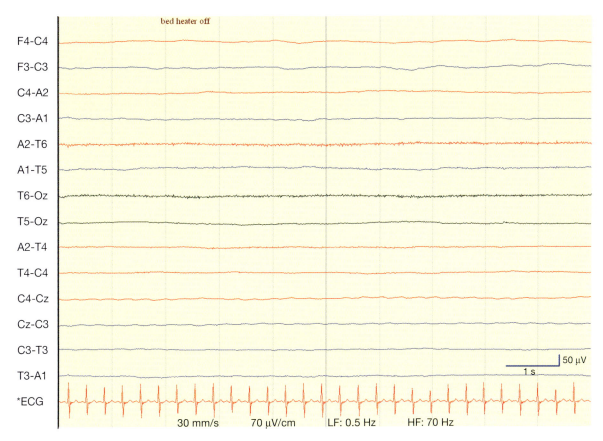

Fig. 2.51   A 2-month-old infant after prolonged cardiac arrest (same infant as in *Fig. 2.22*, after the bed heater was switched off). The EEG shows electrocerebral silence.

adaptation of technique to the clinical circumstances. The emphasis on matching technique to the clinical questions being asked carries the important rider that the clinical neurophysiologist should always be present to guide the investigation and discuss its interpretation with the ITU clinicians.

In the comatose patient, ECS is always a cause for concern, even though various reversible causes, such as high levels of CNS depressant drugs (due to overdose or therapeutic regimens in the ITU) or hypothermia, are not uncommon. This is particularly difficult in neonates, where antiepileptic drugs and sedatives can have a more pronounced effect on the EEG. Rigorous checks must be made to discover any reversible cause; reports should draw attention to this and to the need for further neurophysiological investigation, including repeat EEG examination and testing for auditory and somatosensory evoked potentials (SEPs). The demonstration of short latency evoked potential (EP) components, which are fairly resistant to major CNS depressant drugs sufficient to render the EEG isoelectric, is crucial evidence for the functional integrity of some central pathways, thereby refuting any suggestion of irreversible damage (Ganes and Lundar 1983).

It is also important not to equate ECS with brain or brainstem death. Sustained EEG silence not explicable by potentially reversible factors may be an important indication of irreversible loss of cortical function (at least in the areas accessible to recording with scalp electrodes). However, this provides only part of the diagnostic criteria of irrecoverable cerebral cortical damage and does not provide any indication of the functional state of the vital centres of the brainstem. Overwhelming, irrecoverable cerebral hemisphere damage may occur, for example, in neocortical death – a consequence of selective hypoxic brain damage after cardiac arrest in which function of the brainstem may be fully preserved with the possibility of prolonged survival in a vegetative state. It is also evident from frequent serial recordings that some transient return of activity may be seen even in patients sustaining severe hypoxic damage (Prior 1973).

ACTIVATION PROCEDURES

Clinically significant EEG phenomena may occur preferentially in specific states: arousal, drowsiness, sleep, eyes open or closed, etc. Manipulating the patient's state during the recording may therefore provide useful information. Routine recording procedures include asking the subject to open and shut his or her eyes, encouraging any spontaneous tendency to fall asleep but arousing patients who are already drowsy, and stimulating those in coma. However, more active measures may be adopted to 'activate' the EEG. Some of these form a part of the 'routine' EEG examination; others are used only to address specific questions, mostly concerning alleged precipitants of seizures.

It should be noted that, in late adolescence, the documentation of a clinical event, however subtle, during an epileptiform discharge elicited by overbreathing, establishes the occurrence of an epileptic seizure and may compromise eligibility for a driving licence.

Hyperventilation

Vigorous overbreathing induces slowing of the EEG in children and young adults and may elicit or increase various abnormalities.

Physiological effects

Various mechanisms have been postulated to explain EEG activation by overbreathing (for a review see Patel and Maulsby 1987; see also Kraaier et al 1988, Achenbach-Ng et al 1994). Hyperventilation leads to a fall in blood CO_2 level (hypocapnia), causing small cerebral arteries to constrict, whereas peripheral vessels dilate, lowering blood pressure. Cerebral blood flow is therefore reduced, in turn reducing the availability of oxygen and glucose to the brain. Other mechanisms are probably also involved in the EEG response to overbreathing, and remain the subject of ongoing controversy. EEG changes in normal subjects are in any event closely related to the magnitude of the fall in end-tidal pCO$_2$; frontal delta activity will generally appear when this drops by some 15 mmHg (2 kPa). Occasionally an exceptionally marked and prolonged response to overbreathing reveals clinically significant hypoglycaemia; more generally, responses are somewhat enhanced in patients who are fasting, and are reduced following administration of 50 g glucose by mouth.

Procedure

The degree of hypocapnia required to produce EEG changes, particularly in normal subjects, produces disagreeable symptoms, and consequently will be achieved only if the subject is well prepared for these, motivated and encouraged. With some ingenuity, experienced technologists may be surprisingly successful in persuading children as young as 2 years old to hyperventilate. They may be willing to blow on a tissue, streamers or a small windmill, and more sophisticated devices may be used, such as electronic toys activated by blowing on a microphone.

The technologist should be aware of the symptoms that may result from hyperventilation, and prepared to cope with patients who become distressed, shiver or develop tetany, and particularly those who continue to overbreathe when asked to stop. This last problem may usually be solved by re-breathing, achieved by placing a paper bag over the patient's nose and mouth. Some subjects may become distressed and overreact to the symptoms commonly elicited by overbreathing. Seizures, particularly absences, may be elicited by hyperventilation and, in susceptible people, psychogenic non-epileptic seizures may occur. Contraindications to overbreathing include: respiratory or cardiovascular disease, raised intracranial pressure, cerebrovascular disease including Moyamoya disease, or recent cerebrovascular accident. Caution must also be observed when these conditions are not established, but figure in the differential diagnosis of the symptoms being investigated.

EEG changes induced by hyperventilation in normal subjects

From the age of some 5–6 years, at which satisfactory overbreathing can reliably be obtained, there is a progressive decrease with age in the degree of EEG response. There are also age-related qualitative differences in overbreathing response: in children up to 10 years old the build up of slow activity begins at the back of the head and then generalises (Fig. 2.52(a)). In subjects up to the age of 30 years, 3 min of vigorous hyperventilation will generally produce bilaterally synchronous frontal delta activity (Fig. 2.52(b)).

This occurs at first in short runs of a few seconds' duration, then continuously if the subject persists. The instruction to stop overbreathing will generally be followed by a period of apnoea and the frontal delta activity may briefly continue to increase, and then disappears within some 30 s. A more prolonged response is rarely of clinical significance, except in hypoglycaemia. Posterior temporal slow waves, whether normal maturational phenomena or pathological, increase on overbreathing, maintaining any pre-existing asymmetry. There will in addition be a general increase in theta activity and a slight slowing of the alpha rhythm.

Abnormal EEG responses to hyperventilation

Many EEG abnormalities seen in the awake state can be elicited or increased by overbreathing. Most notable are the effects on generalised spike-and-slow-wave activity. In patients with uncontrolled absence seizures, generalised spike-and-slow-wave activity and indeed frank absence attacks are generally elicited by adequate hyperventilation for 3 min. This finding is so consistent that the absence of such a response in a patient who overbreathes sufficiently to produce generalised delta activity may be taken as evidence that the patient does not suffer from active absence epilepsy. Estimates of activation rate in absence epilepsies are in the range 90–100%, but the rate depends on the adequacy of overbreathing and the accuracy of the diagnosis. Curiously, overbreathing does not appear so consistently to activate spike-and-slow-wave activity in those patients who sometimes have typical absences in the context of other syndromes, such as the benign partial epilepsies of

(a)

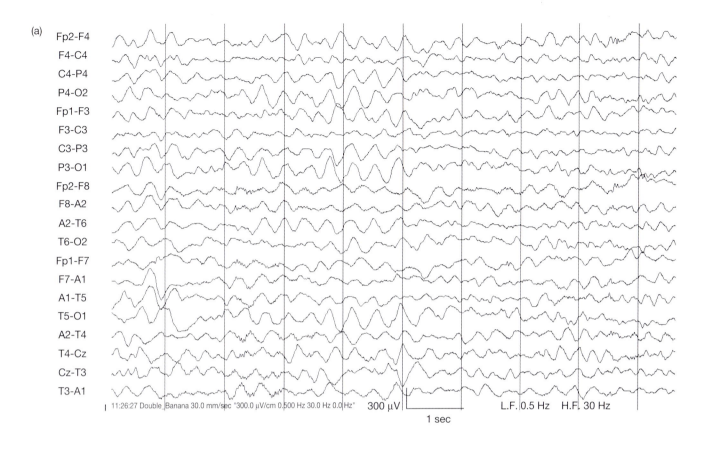

11:26:27 Double_Banana 30.0 mm/sec "300.0 µV/cm 0.500 Hz 30.0 Hz 0.0 Hz"   300 µV   L.F. 0.5 Hz   H.F. 30 Hz

1 sec

(b)

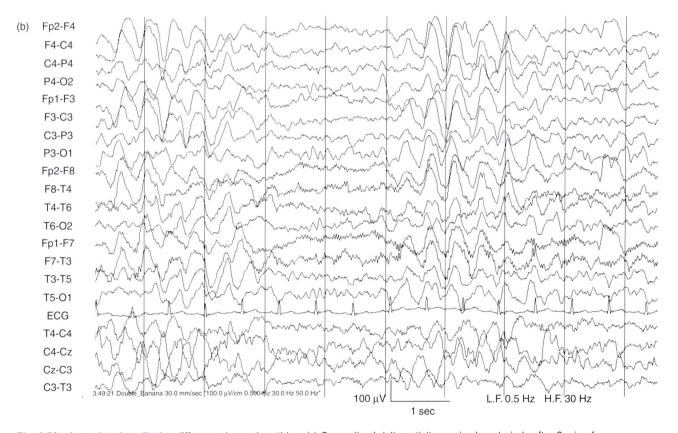

3:49:21 Double_Banana 30.0 mm/sec "100.0 µV/cm 0.500 Hz 30.0 Hz 50.0 Hz"   100 µV   L.F. 0.5 Hz   H.F. 30 Hz

1 sec

Fig. 2.52 Age-related qualitative difference in overbreathing. (a) Generalised delta activity maximal posteriorly after 2 min of hyperventilation in a 7-year-old girl (note gains). (b) Bilaterally synchronous frontal delta activity following 3 min of vigorous hyperventilation by a 29-year-old woman with a history of three nocturnal seizures. (From Binnie et al (2003), by permission.)

63

childhood. Other types of epileptiform discharge, both generalised and focal, are often but less consistently activated. The majority of epileptic seizures occurring during routine EEG recording are seen on hyperventilation. Epileptiform activity induced by hyperventilation may be blocked by acute administration of benzodiazepines, but the normal slowing is not.

The technologist should be prepared to observe and elicit signs of a possible epileptic seizure, if epileptiform activity appears during overbreathing. Breath-counting or counting on each finger in turn may be interrupted due to impaired consciousness. A phrase or number spoken to the patient during the episode may not be remembered. If the patient is asked to sit up with arms extended, momentary loss of muscle tone may be revealed by the hands falling away. Incidental findings of increased or diminished EMG activity, a sudden transitory change in heart rate, or a consistent eye movement artefact at the end of the discharge may provide further evidence of clinical events during the spike-and-slow-wave discharge. Rightly or wrongly, the clinical significance of epileptic activity during overbreathing is better appreciated by the referring physician if an overt seizure is also documented. Ideally, video monitoring should be available to document all clinical events during EEG recording, and this is becoming a standard facility in many departments.

A small number of children apparently lose consciousness when generalised delta activity appears during overbreathing. This is most probably due to the cerebral hypoxia noted above, but has been claimed as an atypical form of absence.

Enhancement of any abnormality of ongoing activity is sometimes difficult to distinguish from the normal effects of hyperventilation. However, patients whose baseline records contain excess slow activity will in general show a greater than usual increase in slow activity on overbreathing. In patients with asymmetries of ongoing activity, these are often enhanced, sometimes because slowing is more marked on the side that originally showed a greater amount of delta or theta activity. Localised or focal slow waves are generally increased on overbreathing, particularly so where cerebral perfusion is compromised by the underlying pathology. As noted above, pathological posterior temporal slow waves are increased. In general, hyperventilation should be terminated once any marked EEG abnormality appears.

Photic stimulation

Stimulation with single flashes elicits visual evoked responses (see Chapter 3). With regularly repeated intermittent photic stimulation (IPS) at increasing flash frequencies, the discrete EPs run together to produce a steady-state response, which in the present context of routine EEG activation is termed 'photic following' or 'driving'. Various anomalous responses may be obtained; some occur in normal subjects, others are associated with

epilepsy (see p. 66). Photic stimulation may also help to identify slow alpha variants, which are blocked, whereas other, pathological posterior slow activities are not (Zschoke 1995).

Methods of photic stimulation

Stimulators for visual evoked potential (VEP) studies are considered in Chapter 3. For purposes of EEG activation by IPS less stringent specifications may apply, but most photic stimulators supplied by manufacturers of EEG equipment are unfit for their intended purpose. The photic stimulator should be capable of delivering discrete manually triggered stimuli and stimulus trains at flash rates from 1/s to at least mains frequency. This last requirement, for a maximum flash rate of at least 50/s in Europe and 60/s in North America, is determined by the need to investigate seizures induced by common mains frequency sources of environmental flicker, such as television. A choice of flash intensities should be available and the highest should be not less than 100 nit-s/flash.

A consensus statement following discussions between experts from ten European countries proposed evidence-based recommendations for stimulators and IPS procedures (*Table 2.2*). The authors provide detailed justification for these. The main considerations are that the methods used should maximise the probability of eliciting abnormal responses in patients with photosensitive epilepsy, whilst minimising the chances of inducing such a response in others or of precipitating a seizure during testing. Users of EEG services should be aware that inadequate equipment or techniques will greatly reduce the effectiveness of IPS and result in failure to detect photosensitive epilepsy. For a detailed discussion of the instrumentation and procedures for IPS, see Cooper et al (2005).

Procedure

The probability of obtaining an abnormal response in epilepsy may be enhanced by the positioning of a grid or grating over the front of the lamp, but purists interested in photosensitive epilepsy will complain that this confounds the effects of flash and pattern stimulation. Others claim that more subjects are sensitive to diffuse than to patterned flicker (Harding 1996, Leijten et al 1998). The conclusion to be drawn from this debate is probably that some subjects are more sensitive to patterned, and others to diffused, light. In any event, a central fixation marker should be provided, as photosensitivity can rarely be demonstrated using IPS in peripheral vision.

IPS may be experienced as pleasant, producing vivid images of colour and linear patterns, or so disagreeable that the patient is unwilling to undergo the procedure. In any event, a full explanation is required before the lamp is positioned. In particular the possibility of inducing a seizure should be explained. If the technique described below is followed, with careful monitoring of the EEG,

Table 2.2 Summary of IPS methodology*

Category	Feature	Recommendation
Stimulus	Brightness	Up to 100 nit-s/flash
	Size of reflector; nature of light distribution	As Grass photic stimulator, with diffuser
	Frequencies used; sequence; duration of stimulation (per train)	Standard sequence of frequencies; separate trains; 10s per train
	Distance to the lamp	30 cm nasion to lamp front
	Angle of visual field stimulated	13°
Patient	Position	Seated, or reclining at angle which allows observation of fixation, behaviour and alertness
	Spectacles	Worn, if worn for reading
	Direction of gaze	Towards fixation point at centre of lamp
	Eye condition	Open, closed and eye closure
	Sleep deprivation	Note whether sleep deprived
	Medication	Note medication
Surroundings	Room lighting	Just sufficient to observe patient

*After Kasteleijn-Nolst Trenité et al (1999).

the risk of inducing a convulsive seizure is minimised. Nevertheless, there is an irreducible, small proportion of patients (perhaps 1/10,000 people with epilepsy) in whom a focal occipital ictal discharge appears after completion of IPS, evolving inexorably to a convulsive seizure. This fact cannot be disregarded when obtaining informed consent to the procedure. Again, the possible consequences for the patient of seizure induction must also be considered. Inducing a habitual attack in a child known to have frequent photogenic seizures is a matter of less consequence than precipitating the first seizure after many years in an adult who has a driving licence. It is questionable whether the use of photic stimulation as a component of the 'routine' EEG examination is justified. It should only be performed where the small risk of inducing a seizure is justified by its relevance to the clinical problem. This would imply chiefly the initial investigation of patients with possible epilepsy. It would generally not include investigation of delirium, headache, psychosis, head injury, etc. IPS would rarely be warranted in the long term follow-up of a patient with, for example, well-controlled juvenile myoclonic epilepsy, unless the persistence or otherwise of photosensitivity was specifically at issue.

Generally, photic stimulation is carried out with the patient recumbent as for a routine EEG recording, but if complex studies of IPS in photosensitive epilepsy are contemplated there may be some advantage in the patient being seated upright. IPS is effective only if the lamp is viewed in central vision. A fixation point should be placed on the centre of the lamp glass, which should be located at a standard distance (e.g. 30 cm) in front of

the nasion. It is important to ensure both correct positioning of the lamp and maintenance of fixation, when the eyes are open during the procedure.

The main clinical interest of photic stimulation lies in the possible induction of epileptiform discharges in persons with epilepsy, and this consideration will usually determine the procedure to be employed. Unless the subject is known to be photosensitive or a policy of extreme caution is being followed, stimulation should generally be carried out first at about 18 flashes per second (f/s), this being the frequency most likely to induce discharges in photosensitive subjects. There is an element of startle when stimulation is first carried out, which appears to increase the likelihood of a discharge (Topalkara et al 1998).

A range of flash rates should then be tested, including 1 or 2 f/s to demonstrate discrete VEPs, and other preset frequencies above and below 18 f/s, e.g. 6, 8, 10, 15, 20, 25, 30, 40, 50 and 60 f/s. These frequencies are equally spaced along the curve relating discharge probability to flash rate (*Fig. 2.53*). There is little purpose in testing more closely spaced frequencies as small changes in frequency make little difference to the EEG response. In neonates and infants, following may be limited to the lowest flash rates.

Both photic following and abnormal discharges are different for the eyes-open and eyes-closed states and are influenced by eye closure. A procedure should therefore be followed that allows the three conditions (eye closure, eyes closed and eyes open) to be tested at each flash rate. If an abnormal discharge is elicited at a particular flash rate, the stimulus train should be immediately terminated and stimulation at that frequency repeated only with great caution. Where photosensitivity is established or suspected, the lower and upper frequency thresholds for eliciting a discharge should be determined, approaching cautiously from the lowest and highest flash rates of the standard sequence. The photosensitivity range, the difference between these two thresholds, provides a quantitative measure of the degree of sensitivity. This can be used to monitor clinical progress, response to medication, etc. Once the upper and lower frequency thresholds have been found, stimulation should not be carried out at any flash rate within this range because of the risk of inducing a seizure. For further procedural recommendations see Binnie (1993).

Photosensitive patients can often protect themselves against visual stimuli by covering one eye. For purposes of subsequent counselling, it may be useful to determine the efficacy of this manoeuvre during photic stimulation.

Effects in normal subjects

In normal subjects discrete flashes at 1–3 f/s will produce VEPs, which may or may not be visible to the naked eye without the help of averaging. At higher flash rates, a rhythmic following response will generally be seen over the posterior regions, attaining a maximum amplitude at flash rates close to the alpha frequency (*Fig. 2.54*). Harmonic components at, for instance, half or twice the flash rate may be seen (*Fig. 2.55*). At higher flash frequencies the following response generally becomes smaller; marked following at flash rates above 15 f/s is rather uncommon and there have been unconfirmed claims that a high frequency (H) response is associated with migraine. Some normal subjects show clear fundamental photic following at flash rates as high as 60 f/s. Photic following is in general symmetrical, but there may be amplitude asymmetries of up to 50% in normal subjects. The symmetry of the photic response usually follows that of the alpha rhythm. No known clinical significance is attached to the amplitude of photic following, except in the rare instances of very high voltage responses of more than 100 μV peak-to-peak found in some types of neuronal ceroid lipofuscinosis. Neither the presence of harmonic components, nor indeed the absence of following detectable by naked eye inspection of the trace, is thought to be of clinical significance.

Atypical and abnormal responses

Many normal subjects, particularly children, exhibit photic responses that may be considered atypical. Some of these are unequivocally normal; others show a weak statistical association with cerebral disease, and some are strongly associated with epilepsy. Understanding of the differing clinical significance of the various types of atypical responses has been hindered by confusing terminologies, notably the use of the term 'photoparoxysmal' (now generally accepted) to describe both benign and pathological phenomena. Kasteleijn-Nolst

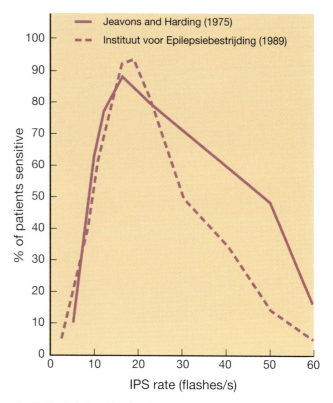

Fig. 2.53 Relationship of probability of photoparoxysmal response and flash rate in two populations of photosensitive patients. The series of patients of Jeavons and Harding (1975) (full line; n = 170) was investigated in the UK at a time when sodium valproate was not registered there. The series from the Instituut voor Epilepsiebestrijding in The Netherlands (Kasteleijn-Nolst 1989) (broken line; n = 138) comprised a group of patients of whom many were taking sodium valproate, and it may be for this reason that sensitivity at the higher flash rates was substantially less. (From Binnie et al (2003), by permission.)

Trenité et al (2001), again on the basis of a consensus statement by an international group of experts, have attempted to draw these into a single classification:

1. *Photic following at flash rate*. Confluent VEPs to successive stimuli producing a regular activity at the flash frequency, which ends as soon as the stimulus train is terminated. The commonest normal response (see *Fig. 2.54*).

   Photic following at harmonics. Regular activity at a sub- or supraharmonic of the flash frequency, which ends as soon as the stimulus train is terminated. Often intermixed with following at the flash rate; of no clinical significance if symmetrical (see *Fig. 2.55*).

2. *Orbitofrontal photomyoclonia*. This was described in detail and termed 'photomyoclonus' by Bickford et al (1952). It comprises bioelectrical signals elicited by successive stimuli producing a regular activity at the flash frequency, which end as soon as the stimulus train is terminated (*Fig. 2.56*). The signals are predominantly of electromyographic origin, arising in the orbicularis oculi and frontalis muscles,

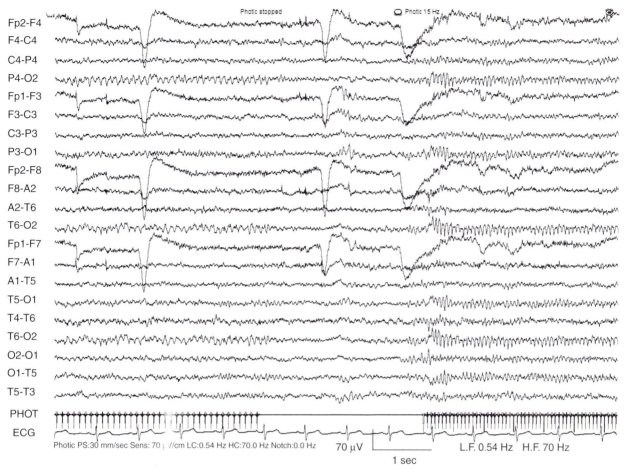

Fp2-F4
F4-C4
C4-P4
P4-O2
Fp1-F3
F3-C3
C3-P3
P3-O1
Fp2-F8
F8-A2
A2-T6
T6-O2
Fp1-F7
F7-A1
A1-T5
T5-O1
T4-T6
T6-O2
O2-O1
O1-T5
T5-T3
PHOT
ECG

Photic stopped   Photic 15 Hz

Photic PS:30 mm/sec Sens: 70 µV/cm LC:0.54 Hz HC:70.0 Hz Notch:0.0 Hz   70 µV   L.F. 0.54 Hz   H.F. 70 Hz

1 sec

Fig. 2.54   Photic following during stimulation at flash rates of 10 and 15 f/s. (From Binnie et al (2003), by permission.)

and are therefore maximal at the front of the head. Other muscles of head, neck and upper arms may also be involved. A cerebral component of frontal lobe origin may be present (Chatrian and Perez-Borja 1964, Obeso et al 1985, Artieda and Obeso 1993). While very rare in children, it is a normal finding in, particularly elderly, adults. Orbitofrontal photomyoclonia occurs more readily in the presence of high muscle tone, and this may explain a weak association with anxiety and some psychiatric disorders (Bickford et al 1952, Gastaut et al 1958). Consciousness is usually unimpaired. Photomyoclonia should be distinguished from the constant amplitude, flash-synchronous spikes, often of unusual distribution, which may be produced by electrical interference from the stimulator, and from the frontal spikes, which can (very rarely) result from photochemical effects at the more anterior electrodes.

Grades 4–6 are commonly known as 'photoparoxysmal responses', and have been classified into four grades by Waltz et al (1992).

4. *Posterior stimulus-dependent responses*. These are anomalous steady-state VEPs, of unusually sharp waveform or high amplitude. There are various clinical correlates, including occipital spikes after suppression of generalised photoparoxysmal response by medication and high amplitude VEPs in neuronal ceroid lipofuscinosis.

5. *Posterior stimulus-independent responses*. Activity is confined to or maximal at the back of the head, and not at the flash frequency or at an harmonic thereof. Such responses include delta and theta activity and frank epileptiform patterns (*Fig. 2.57*).

5a. *Self-sustaining posterior stimulus-independent responses*. These are rare, often last many seconds and may evolve to an overt seizure.

6. *Generalised photoparoxysmal response*. This comprises multiple spikes or SW activity, which are apparently generalised, but it may be of greater amplitude at the front or the back of the head. It was termed 'photoconvulsive response' by Bickford et al (1952); it corresponds to the type 4 response of Waltz et al (1992).

6a. *Self-sustaining generalised photoparoxysmal response (Fig. 2.58)*. Note: this may not be demonstrated unless the stimulus train is terminated as soon as generalised photoparoxysmal response (PPR) is identified. There is a strong association with

67

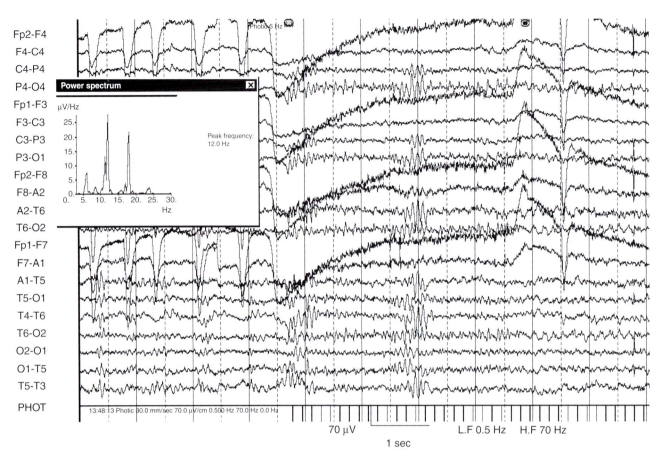

Fig. 2.55   Harmonic following to photic stimulation. Flicker of 6 f/s induces irregular waveforms, but the power spectrum of T_5-O_1 derivation shows these are made up of 6, 12, 18 and 24 Hz components. (From Binnie et al (2003), by permission.)

epilepsy. It was termed 'prolonged photoconvulsive response' by Reilly and Peters (1973), who reported a strong association with epilepsy in patients referred for clinical EEG examination. The prevalence in patients with epilepsy is approximately 5%; the prevalence in the general population is unknown, but it was found in 5/13,658 of apparently healthy aircrew (Gregory et al 1993).

Very rarely, photic stimulation may activate an existing focus, usually not posteriorly located, possibly eliciting a seizure (Seddigh et al 1999). Strictly speaking this is not a PPR and does not, therefore, figure in the classification above.

Three response characteristics are associated with epilepsy: generalisation of the response, asynchronous component waves independent of the individual flashes, and a self-sustaining response that persists after termination of stimulation. As most photosensitive epileptic patients suffer seizures precipitated by environmental visual stimuli, the finding of photosensitivity often has clinical significance. For the present purposes, the term 'photosensitive' is applied only to patients with a self-sustaining generalised photoparoxysmal response ((6a) above) and/or to epilepsy with photogenic seizures.

As always during the application of activation procedures, the technologist should be prepared for the possible occurrence of seizures and should observe and document all clinical events. Eyelid flutter synchronous with the spike-and-slow-wave discharges of absences should be distinguished from the faster eyelid movements of myoclonia, and from the slow, sustained upward deviation of the eyes employed by self-stimulating patients, sometimes even during IPS.

Close observation and questioning reveal that during a generalised PPR, over 70% of subjects exhibit or experience subtle clinical ictal phenomena, notably myoclonias or eyelid flutter. Overt convulsions are largely avoidable, if (1) the upper and lower frequency limits of photosensitivity are determined as described above, and stimulation between these limits is avoided; and (2) the stimulus train is terminated as soon as a generalised PPR appears. Sleep deprivation exacerbates photosensitivity, and seizures may result if photic stimulation is performed after sleep deprivation in a known photosensitive subject.

Photomyoclonus, although generally benign, may occur in acute encephalopathies and after withdrawal of alcohol or benzodiazepines, when it may be the prelude to a photogenic seizure, which can appear unexpectedly some seconds after the end of stimulation. In this clinical

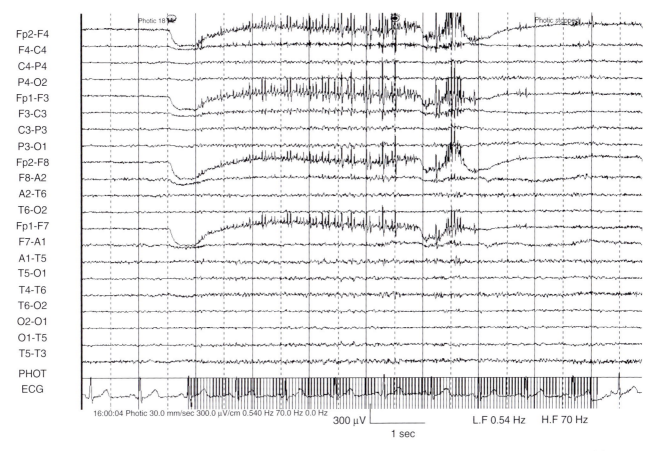

Fig. 2.56 Orbitofrontal myoclonias (photomyoclonic response). A crescendo of bilateral myogenic and eye movement potentials anteriorly synchronised with flash stimuli. (From Binnie et al (2003), by permission.)

context, IPS should be terminated as soon as photo-myoclonus appears. Psychogenic non-epileptic seizures are often precipitated by photic stimulation and are usually predictable; patients who display non-epileptic seizures during IPS usually express apprehension concerning the procedure, paradoxically more often than those with photosensitive epilepsy.

If these guidelines are strictly followed, virtually all convulsive seizures seen during photic stimulation in the EEG laboratory will be psychogenic. As noted above, a few patients will suffer occipital partial seizures with a very slow evolution, which remain unavoidable as the discharges develop too slowly to be recognised in time for IPS to be aborted.

To the above list of abnormal or atypical photic responses should be added the occurrence of unusually high amplitude VEPs at low flash rates, characteristic of, but not specific to, Batten's disease (neuronal ceroid lipofuscinosis II, discussed in more detail on page 303).

Other methods of visual stimulation: television, pattern, fixation-off

Because of the importance of external stimuli in the genesis of seizures in photosensitive patients with epilepsy, it may often be helpful to analyse these factors further. In Western Europe, 50% of photosensitive epileptic subjects suffer seizures induced by television, which may be due either to flicker of the screen or to the striped raster pattern from which the picture is built up. Television sensitivity may be investigated in much the same way as photosensitivity, asking the patient to view a television set during EEG recording. The main determinant of the epileptogenicity of television is viewing distance; thus the patient should first be exposed at a distance of at least 3 m and gradually brought closer until discharges appear. The set should display a stable, well-lit picture, not a random 'noise' pattern, as the line-synch is relevant to epileptogenesis. Malfunction is not an important factor in television epilepsy and does not need to be replicated.

Some 30% of photosensitive patients also exhibit epileptiform discharges when viewing static linear patterns and 70% do so if the patterns oscillate orthogonal to their line orientation. Seizures may thus be produced by environmental patterns, and this possibility can be investigated further by the presentation of suitable epileptogenic patterns under EEG control. The epileptogenicity of a pattern is critically determined by its physical characteristics. Photosensitive subjects with normal vision are most likely to exhibit epileptiform discharges whilst viewing a linear pattern of 2–4 cycles/degree (i.e. each black or white stripe subtends 7.5–15 min of arc). A high

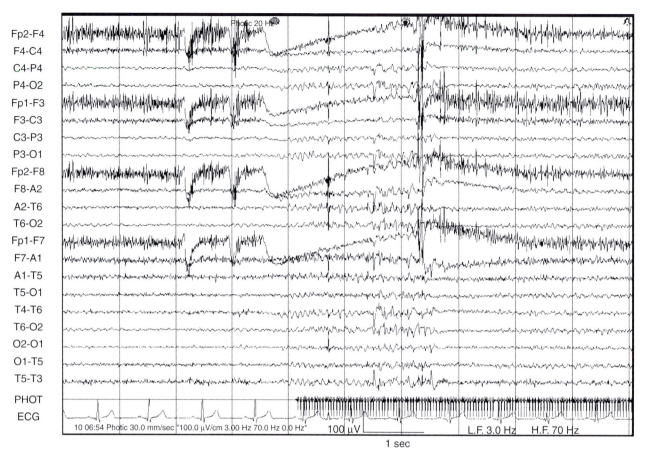

Fig. 2.57   Posterior stimulus-independent responses, irregular posterior theta activity in normal subject. (From Binnie et al (2003), by permission.)

level of illumination and contrast are required. Note that, in contrast to their efficacy for eliciting visual evoked responses, grids and checkerboards are far less effective in photosensitive subjects than are gratings. The total visual angle subtended by the pattern is also critical, but some subjects may exhibit discharges or even suffer seizures when fixating patterns at a visual angle as small as 2°.

Testing for pattern sensitivity may be performed using high-contrast black and white striped patterns pasted on circular cards of different sizes. Conventional 50 or 60 Hz television screens or visual display units (VDUs) should not be used as they introduce a flicker stimulus within the possible range of photosensitivity; a high-frequency (100 Hz), high resolution television or monitor is, however, an ideal solution. In calculating spatial frequency and total visual angle, it may be convenient to note that 1 cm subtends 1° at a distance of 57 cm. Thus stripes of 2.5 mm width will provide a suitable spatial frequency at this viewing distance, and a set of patterns up to 50 cm diameter will be required. A fixation point should be provided at the centre of each test pattern. The illumination of the pattern should be not less than 200 cd/m². If no response is obtained to a static pattern, the epileptogenicity of the display can be increased by

oscillation orthogonal to the stripe orientation. The optimal frequency is about 18 Hz, which is not achievable with a hand-held card, but can be generated with an electronically controlled display. For further details see Binnie (1993).

Pattern stimulation can provoke seizures and should be presented with as much care as IPS, starting with a small stimulus and increasing the size if no discharge is elicited (e.g. over the range 2–50° of visual angle). Normal subjects often display prominent lambda waves during pattern stimulation, which must not be mistaken for epileptiform discharges. In sensitive subjects, a stimulus only marginally above threshold often elicits discharges confined to the occipital regions; these rapidly generalise if a larger pattern is used.

Some patients, generally not photosensitive, exhibit continuous spike-and-slow-wave activity when deprived of patterned visual input. This, obviously, is achievable by eye closure or darkening the room, but true 'fixation-off sensitivity' can be proven by exposing the subject to an illuminated, unstructured visual field. This can conveniently be provided by applying ski or diving goggles with frosted lenses. Fixation-off sensitivity is typically seen in childhood epilepsy with occipital paroxysms.

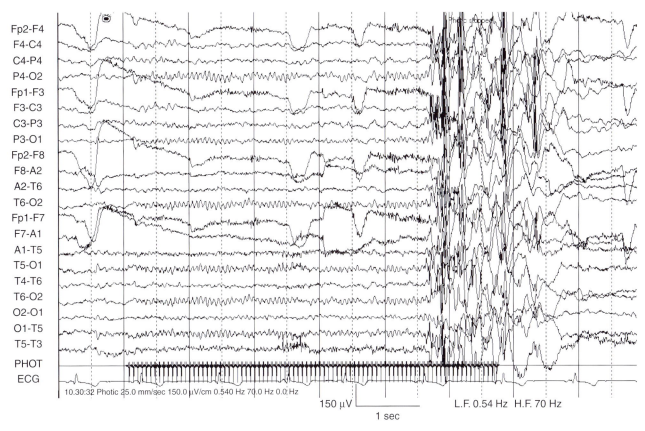

Fig. 2.58 Self-sustaining generalised photoparoxysmal response in a 16-year-old boy with juvenile myoclonic epilepsy. The characteristic features of generalised irregular spike wave discharge continue after the stimulus train (channel 22) has ended. (From Binnie et al (2003), by permission.)

Other types of stimulation

It has been claimed that repetitive auditory stimulation by trains of clicks may sometimes enhance abnormalities in temporal lobe epilepsy. Although some photic stimulators for EEG used to incorporate an auditory output, phonostimulation has never become a routine clinical procedure. However, it has been claimed that auditory stimulation may be almost as effective as photic stimulation for eliciting SW discharges in patients with generalised epilepsies (Hogan and Sundaram 1989).

Somatosensory stimuli, generally to some specific region of the body, may produce transients that may be regarded either as epileptiform discharges or as 'extreme evoked potentials'. These are seen in benign partial epilepsies and in some children with partial seizures arising from lesions of the peri-Rolandic region. Peripheral stimuli may also elicit discharges in many patients with myoclonic seizures, but also in some people who do not show any other evidence of epilepsy (De Marco and Negrin 1973). To investigate such subjects, peripheral nerves may be stimulated using an EMG stimulator or battery-operated vibrator. Taps may be administered to the appropriate part, ideally with a tendon hammer incorporating a microswitch so that the precise timing can be registered on the recording (*Fig. 2.59*).

Unstructured sensory stimuli, such as flashes, clicks and percussive taps, may have a non-specific activating effect on hyperexcitable cortex in the relevant primary projection areas. However, there are some patients who show more complex reflex epilepsies in which seizures and/or epileptiform EEG discharges are triggered by much more specific external stimuli or by cognitive activities. These triggers range in complexity from such fairly simple stimuli as applying water to a particular part of the body or hearing certain sounds, through more complex activities with sensory, cognitive, affective and motor components such as eating, reading or listening to music, to purely mental acts such as mathematical calculation. The stimulus may be highly specific; reported examples include: viewing a safety-pin, the voice of a particular radio announcer, Hebrew religious texts and Beethoven's symphonies. In general, the more complex the stimulus or activity, the longer it is necessary for it to be applied to produce an activating effect. Thus, photic stimulation at an optimal flash rate generally produces discharges within a second and responses to taps in susceptible subjects are immediate, whereas eating or reading may have to be carried out for many minutes before EEG activation is seen. The effects of a complex activity may also be dependent on situational and

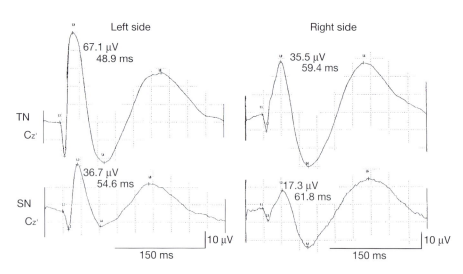

Fig. 2.59 High amplitude SEP elicited from tibial (TN) and sural (SN) nerves in a 12-year-old boy, with Rolandic spikes.

emotional factors. Thus a patient with eating epilepsy may have seizures only when dining at table, and not when food is placed in the mouth in an EEG laboratory; language-related attacks may occur only when speaking in public or conducting difficult negotiations by telephone. In many subjects with a clear history of seizures related to a specific stimulus, the stimulus itself has especial emotional significance to the patient; here too testing in an alien laboratory setting may fail to demonstrate activation.

The claim that a child's seizures are precipitated by external stimuli or specific activities tends to be dismissed too readily, and can often be confirmed by replicating the precipitating circumstances under EEG control. Where there is any possibility of reflex hypoxic attacks, the ECG must also be monitored. In the case of simple stimuli, conventional EEG recording methods may be appropriate but, where it is necessary to replicate a complex situation, EEG or polygraphic telemetry is required.

Drowsiness and sleep
Importance in clinical work
As indicated elsewhere (see p. 34) the EEGs of normal subjects change dramatically with alteration of state of awareness, and recording during sleep may be necessary to investigate sleep-related symptoms, parasomnias, sleep apnoea, etc. Drowsiness and sleep may also be used as a non-specific activating procedure in routine EEG investigation. Some pathological EEG phenomena, such as FIRDA or PLEDs, occur characteristically in a state of depressed consciousness and may thus not be detected unless the patient is allowed to become drowsy during the investigation. Various abnormal EEG patterns seen in diffuse encephalopathies, such as subacute sclerosing panencephalitis, may be markedly influenced by state of awareness, so that detection at some stages in the course of the illness may depend on obtaining a record during drowsiness or sleep.

Topographic EEG abnormalities produced by cerebral pathology may be shown not only by phenomena of the awake state but also by those confined to sleep; thus patients with lateralised pathology may show asymmetries of sleep spindles, K-complexes, etc.

However, the main application of sleep activation of the EEG is to be found in epilepsy, where it may enhance focal discharges or produce new foci not evident in the awake state (*Fig. 2.60*) and may produce marked changes in morphology of generalised epileptiform activity. In presurgical assessment of patients with normal waking EEGs, temporal foci emerging in sleep are of particular localising value (Adachi et al 1998). In large populations of people with epilepsy about 50% exhibit epileptiform activity in a single waking EEG (Ajmone Marsan and Zivin 1970). The yield increases to some 80% if this is combined with a sleep recording (Binnie and Stefan 1999). Expressed more compellingly, the number of false negatives is reduced from 50% to 20% by the addition of a single sleep recording to the initial diagnostic EEG protocol. However, others could not confirm that sleep or sleep deprivation increases the yield in paediatric EEGs (Gilbert et al 2004).

Methods of sleep induction
The circumstances of EEG recording are in general conducive to sleep and a relaxed approach on the part of the technologist. A comfortable recording environment and the avoidance of pad or cap electrodes (which become uncomfortable within half an hour) will often permit a period of recording during drowsiness or light sleep to be achieved in the course of a routine EEG examination. Daytime sleep, whether natural or drug-induced, can generally be obtained most easily after lunch, and sleep EEGs can often be obtained from infants and young children without sedation after they have been fed. These considerations should be taken into account in making appointments.

There is perhaps an excessive reluctance to employ hypnotic drugs to obtain sleep, on the grounds that they may alter EEG patterns. Drug effects are of course undesirable in studies of sleep as such. There is, however,

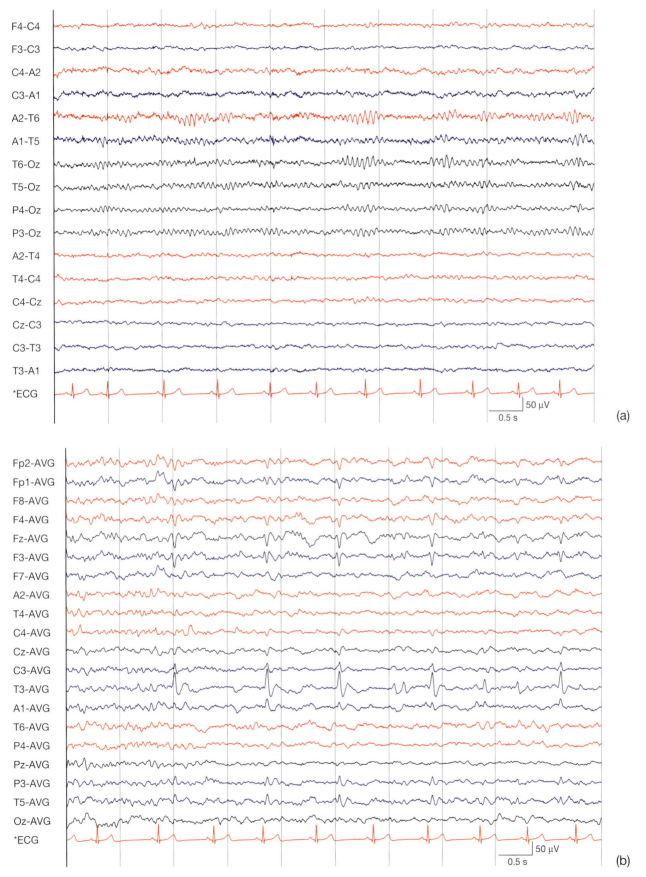

Fig. 2.60 (a) A 7-year-old boy who had two nocturnal generalised tonic–clonic seizures. His wake EEG is normal. (b) In sleep (stage II: some spindles and vertex transients) frequent sharp slow waves are seen over the left centrotemporal region with a positivity over the midfrontal regions, confirming the diagnosis of benign partial epilepsy with centrotemporal spikes.

no evidence that hypnotics materially affect those phenomena of diagnostic significance, which the sleep EEG is intended to demonstrate. Indeed the beta activity induced by barbiturates in particular may serve to highlight cerebral abnormalities that produce asymmetry. Thus some 80% of patients undergoing assessment for possible surgical treatment of epilepsy arising from the temporal lobes exhibit asymmetries of barbiturate-induced fast activity over the frontal regions or at sphenoidal electrodes. An apparent lack of fast activity on one side may help to lateralise the source of seizures.

The hypnotic drugs most often used were, until recently, short-acting barbiturates, such as oral secobarbital. Oral diazepam is also of value as a sleep-inducing agent and will generally induce sleep as efficiently as secobarbital in benzodiazepine-naive subjects; sleep may be less deep, and subsequent recovery is more rapid than after oral barbiturates. Diazepam has a selective suppressive action on generalised discharges as opposed to focal epileptiform discharges (Gotman et al 1982). This is an obvious disadvantage if it is hoped to demonstrate generalised discharges, but may be useful in patients with generalised epileptiform activity where an underlying focus is suspected. After oral administration, but particularly with slow intravenous administration carefully titrated against response, it is possible to produce suppression of the generalised discharges followed by activation of the focus as the patient falls asleep.

Infants and children unwilling to swallow tablets may conveniently be sedated, both to induce sleep and indeed to obtain an EEG, by means of alimemazine syrup 2 mg/kg. Rectal chloral is highly effective in infants, but the assault required for rectal administration of this, or of any sedative agent, in young children is rarely justified.

Melatonin is currently gaining favour as a benign means of inducing sleep. For children less than 5 years of age 2.5 mg, and for older subjects 5–10 mg, is as effective for inducing sleep as sleep deprivation, and produces less after-effects (Wassmer et al 2001); apparently the yield of clinically useful information is similar. Nevertheless, some concerns remain: melatonin may have antiepileptic action, which could suppress epileptiform activity; and in many countries it is unregistered as a pharmacological agent, and is seen as an unorthodox 'alternative therapy'.

Nocturnal sleep deprivation is an effective method of inducing sleep in the laboratory during working hours. A child below the age of 10 years may be allowed to go to sleep but woken again at about 1 a.m. Older children should be kept awake all night. The EEG investigation is commenced at about 8 a.m. the following morning and generally permits a natural sleep recording to be obtained, including one or two complete REM cycles if required. This technique may be of particular use in a research context where there is a requirement to study the effects of sleep stages on, for instance, epileptiform activity, without the complicating effects of drugs and

where all-night sleep recording is not feasible. Sleep deprivation facilitates the occurrence of seizures; indeed prolonged sleeplessness may produce seizures in persons not ordinarily regarded as having epilepsy. It might therefore be expected that sleep deprivation would have a specific activating effect on EEG phenomena relevant to the diagnosis of epilepsy. However, the only evidence from randomised controlled trials suggests that there is no such effect independent of induction of drowsiness and sleep (Veldhuisen et al 1983). The case for using sleep deprivation routinely as a more effective or more 'physiological' alternative to hypnotics is thus not compelling. One should also appreciate that its use in an outpatient generally inconveniences an entire household. Moreover, sleep deprivation has precipitated tonic– clonic convulsions in patients, with juvenile myoclonic epilepsy for instance, who had previously suffered only minor seizures. Sleep induction with medication is more convenient.

Evaluation studies on sleep recording in epilepsy mostly suffer from methodological weaknesses due to the constraints of clinical practice. Sleep recording, whether drug-induced or after sleep deprivation, will usually be carried out only in those patients in whom a conventional EEG in the awake state has failed to demonstrate epileptiform discharges. Because of the marked variability in occurrence of epileptiform activity in persons with epilepsy, a substantial yield of new information may be expected simply from repeating the EEG in those patients whose previous records were negative. Most evaluation studies therefore confound the effects of sleep and sleep deprivation with those of obtaining a repeat EEG which, in these circumstances, is likely to be of longer duration than the routine recording. The few investigations that have taken account of these factors show an undoubted increase in epileptiform discharge rate during sleep in most patients (see *Fig. 2.60*), but little effect of sleep deprivation as such, except in stage II after prolonged (36 h) wakefulness (Veldhuisen et al 1983, Molaie and Cruz 1988). The ideal is to adopt routine practices that facilitate spontaneous occurrence of drowsiness and sleep. Departments where this is achieved will find a smaller yield of additional information from formal sleep recordings.

Procedure and practical aspects

It should be self-evident that sleep will more readily be obtained in a relaxed, comfortable patient. Sleep recordings, except after sleep deprivation, should where possible be carried out in the early afternoon. A suitable recording environment is required: quiet with reduced lighting and an absence of extraneous noise. A temperature that is comfortable for active subjects will be too cold for sleep, and blankets should be provided. The patient should be fully recumbent unless dyspnoeic in this position. Pad electrodes are not suitable for sleep recordings: the pressure of the electrodes and rubber retaining straps becomes uncomfortable after 20–30 min and contact resistances increase as the pads dry out. Elasticised electrode caps may prove adequate but are not ideal, for much the same reasons. Relatively insecure types of self-retaining electrodes (discs retained with adhesive paste) may be acceptable, but fail if the patient becomes restless. The most reliable electrode type for general use in sleep studies is the stick-on disc electrode. These are also the most comfortable type, allow the use of an ordinary pillow, rather than a hard neck support, and permit the patient to lie on one side if preferred. Surprisingly, where a need arises to apply additional electrodes during sleep without waking the subject, needles prove most satisfactory.

Some children become anxious at the prospect of being required to sleep during the daytime and are kept awake by their very anxiety, particularly if pressurised by their parents. Here too, a relaxed approach is helpful and it should not be suggested that the patient will or should sleep, but rather that he or she is likely to feel sleepy and may doze.

Space does not permit a detailed discussion of the legal and practical considerations concerning the administration of hypnotic drugs, the avoidance and management of adverse reactions, etc., which are the responsibility of the prescribing physician. Local and national regulations regarding the storage, prescription and administration of controlled drugs will apply. If oral hypnotics are administered in the afternoon, the patient should have only a light lunch, as nausea is sometimes experienced when these drugs are taken in the daytime. The patient should be advised to continue all regular medication as usual.

Preparation for drug-induced sleep, or recording after sleep deprivation, must include arrangements for subsequent care of the patient. An inpatient will require at least an escort and possibly a wheelchair or stretcher to return to the ward. Drug-induced sleep recording should not be undertaken on outpatients unless adequate arrangements have been made for transport home and for supervision during the journey and for some hours thereafter. Obviously the patient must not ride a bicycle, and the use of public transport should be avoided if possible. These matters should be addressed when the appointment is made and checked on the arrival of the patient in the department. In any event, patients should not be permitted to leave the laboratory after taking sedative drugs until they have been assessed by a physician, and the escort instructed about the precautions required. In psychiatrically disturbed children, the administration of hypnotics in the daytime may produce frank abreaction or conversion symptoms. When a patient remains unresponsive on completion of the sleep EEG although the tracing shows a responsive waking record, or lies on the floor apparently unable to stand, the problem is most probably psychiatric in nature, although no less difficult to manage on that account; however, organic causes will need to be excluded.

Other uses of drugs
Activation tests
Various drug agents with more or less epileptogenic properties have been used for the investigation of epilepsy, including psychotropic drugs, methohexital and procaine. Formerly, frank convulsants such as Metrazol (pentalene tetrazole) were widely used in the investigation of epilepsy. A threshold test was employed whereby Metrazol was slowly injected during photic stimulation to determine the dose at which photosensitivity appeared (Bickford et al 1952). There is little evidence that this threshold was of any significance in determining a constitutional liability to seizures. Metrazol was also used to induce seizures and discharges, when ictal EEG or depth recordings were required for purposes of focus localisation prior to possible surgery. This technique became discredited when it was realised that both the clinical ictal manifestations and, more importantly, the EEG changes induced in this manner often bore little resemblance to the patient's habitual spontaneous seizure patterns (Wieser et al 1979).

Suppression tests
The use of intravenous benzodiazepines to suppress generalised discharges has been considered above. Similarly, intravenous barbiturates (as thiopental and, formerly, methohexital) can selectively suppress the less important of multiple foci. Antiepileptic drugs may also be administered during EEG recording to determine the relationship of clinical symptoms to a particular epileptiform EEG phenomenon. Thus, if the intravenous injection of a small amount of diazepam in a confused patient with frequent SW discharges suppresses the epileptiform activity and produces a rapid improvement in mental state, it may be concluded that the clinical picture was due to non-convulsive status epilepticus. Occasionally the effects are dramatic and unexpected, as when a patient who has been mute for years begins to speak when continuous focal discharges are suppressed by benzodiazepines. Intravenous injection of pyridoxine or pyridoxal phosphate is routinely used to establish the diagnosis of pyridoxine-dependent seizures in the newborn infant.

Drugs to identify areas of diseased cerebral cortex
Drugs such as barbiturates, which produce fast activity in the EEG over areas of intact cerebral cortex, have been used to outline areas of diseased tissue or damage. The technique generally involves a small intravenous injection of a bolus of thiopental (or formerly methohexital) under EEG control and careful evaluation of the amount and amplitude of fast activity at each electrode site (Pampiglione 1952, Fenton and Scotton 1967, Brazier 1969). Formal mapping techniques may be utilised to highlight cortical regions with more subtle but significant depression of drug-induced fast activity (Duffy et al

1984). The technique has been widely applied in presurgical evaluation of patients with epilepsy due to localised cortical lesions (e.g. Dasheiff and Kofke 1993).

Other methods of activation

Various tests of vasomotor instability have been employed, particularly in some European countries, for differential diagnosis of syncope and epileptic seizures. Some of these are based on reflex vagal stimulation and involve significant hazards, as for instance ocular compression (retinal detachment) and carotid massage (embolism). Others, such as exercise on an ergometer or the use of a tilt table, are harmless but require special equipment. In general the tests appear to show neither specificity (many healthy adolescents will display a bradycardia or asystole if subjected to carotid massage) nor sensitivity, except possibly in children with syncope. The use of tilt tables is of undoubted value in autonomic function testing and is free of the risks noted above.

EEG IN EPILEPSY

The EEG and the pathophysiology of epilepsy

At an international workshop 'What is Epilepsy?' (Trimble and Reynolds 1986) the many distinguished participants were unable to arrive at a comprehensive definition and we are no closer two decades later. Epilepsy is generally described as a liability to recurring, spontaneous epileptic seizures, but this begs important questions. What is a liability to recurrence; if a first seizure is associated with risk factors predictive of recurrence, must one wait for the second seizure before diagnosing epilepsy? Are seizures triggered by external factors such as sleep deprivation, alcohol withdrawal or flickering light to be included in the definition – indeed, as seizures are presumably determinate events, are they ever truly spontaneous? Of more immediate concern here, what is an epileptic seizure? As long ago as 1873, Hughlings Jackson offered a definition based, not on clinical phenomena, but on pathophysiology, namely: 'occasional sudden, excessive rapid and local discharges of grey matter'. Half a century later the EEG confirmed this speculative suggestion, and the central unifying concept of epilepsy remains a greater than usual liability to a characteristic type of episodic neuronal dysfunction, involving increased, hypersynchronous and autonomous activity.

Excessive activity due to an imbalance between excitation and inhibition is not the only abnormality; some models do indeed show deficient inhibitory activity, others offer evidence of increased GABAergic inhibitory innervation. Hypersynchrony is probably a crucial feature of epileptogenesis and may involve either excitatory or recurrent inhibitory innervation, or non-synaptic transmission.

Recordings from individual neurons in animals with experimental epilepsy and from epileptic foci in man often show a characteristic electrical abnormality. Spontaneous episodes of membrane depolarisation occur (paroxysmal depolarisation shifts (PDSs)) accompanied by bursts of action potentials. Electrical stimuli, which ordinarily would elicit a single action potential, may also produce a more sustained depolarisation and burst firing. More than one mechanism may be responsible: in the CA3 region of the hippocampus depolarisation is due to entry of calcium ions; in the CA1 region, sodium currents are involved. Recording with extracellular microelectrodes shows that burst firing occurs synchronously, involving groups of neurons. The propensity of individual units to fire in bursts is not peculiar to 'epileptic neurons'; some 5% of cells in the normal human hippocampus show such a discharge pattern. In epileptic foci an increased proportion of neurons exhibit burst firing; the individual units fire less regularly, at slightly greater mean intervals and with a much greater degree of synchronisation between neurons (Isokawa-Akesson et al 1987, 1989, Jensen et al 1994). There is a four-fold increase in the duration of inhibition following electrical stimulation, suggesting increased recurrent inhibition, which probably helps to synchronise the burst firing (Isokawa-Akesson et al 1989). However, non-synaptic mechanisms such as ephaptic transmission appear to contribute, as synchronous burst firing may continue even after blockade of both excitatory and inhibitory synapses. The relationship of burst firing to epileptogenesis, moreover, is not simple: Colder et al (1996) found burst firing to be more frequent in the temporal lobe contralateral to seizure onset.

Larger, more distant electrodes recording from the cortical surface or the scalp display the averaged activity of larger neuronal populations. They display spikes corresponding to the burst firing and often slow waves during periods of hyperpolarisation and cessation of action potentials – the familiar 'spike-and-wave' (SW) activity (*Fig. 2.61*). The cortical surface shows a negative-going change during both the excitatory spike and the inhibitory slow wave. This paradox may be explained by depolarisation of dendrites near the cortical surface during the spike, and hyperpolarisation of cell bodies deep in the cortex during the slow wave; both cause the surface to become negative with respect to the deeper layers.

Seizures are conventionally divided into two types, generalised and partial (or focal). Generalised seizures are characterised physiologically by a sudden widespread disturbance of cerebral activity, usually the onset of bilateral synchronous and fairly symmetrical spikes or SW discharges, but less commonly a desynchronisation leading to a reduction in EEG activity. The clinical features often include loss of consciousness, which is immediate or follows within a second of discharge onset; any motor phenomena that occur are essentially symmetrical. The discharges and clinical manifestations may

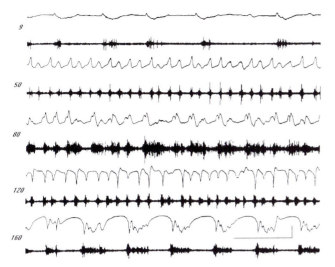

Fig. 2.61 Two-channel intracortical and epicortical recording of ictal activity elicited by electrical stimulation in the chronically isolated supra-sylvian gyrus of the cat. Upper trace: ECoG; lower trace: unit activity from intracortical extracellular electrode. Excerpts recorded at 9, 50, 80, 120 and 160 ms after stimulation. (From Scherrer and Calvet (1973), by kind permission of Springer Science and Business Media.)

cease as abruptly or there may be a gradual recovery accompanied by unconsciousness or confusion. In partial seizures, the physiological onset is localised and the clinical features reflect dysfunction of the brain region involved. The disturbance may remain well localised or spread to a greater or lesser degree, possibly producing a 'secondarily generalised' seizure. The clinical features reflect the degree of spread: disturbance of the corresponding elemental brain function with a sharply circumscribed discharge, alteration of consciousness if a larger region is involved, or a bilateral convulsion with generalisation of the abnormal activity.

Generalised seizures: the centrencephalic and corticoreticular models

The abrupt onset of bilaterally synchronous cortical activity invited the explanation of a pacemaker deep in the centre of the brain which both triggered and synchronised the discharges; the term 'centrencephalic' was introduced in the 1950s (Penfield and Jasper 1954) to describe this mechanism. This model has been rendered untenable by a large body of experimental and clinical evidence, largely amassed by Gloor and co-workers at the Montreal Neurological Institute, of which only a few examples can be cited here. In cats, administration of penicillin intravenously or by direct application to an area of cerebral cortex produces generalised SW discharges. Unilateral cortical application of penicillin results in bilateral discharges similar to those obtained with intravenous administration, but these become confined to the side of penicillin application if the corpus callosum is divided (callosotomy) (Marcus et al 1968). Thus, rendering the cortex of only one hemisphere hyperexcitable in an intact

animal with a normal thalamus is sufficient to produce bilaterally synchronous generalised discharges. On the other hand, thalamocortical connections are not sufficient to maintain such discharges if the cortico-cortical pathways of the corpus callosum are interrupted. Similarly, mesial frontal lesions in man can give rise to generalised seizures with bilaterally symmetrical SW activity.

Particularly compelling evidence from humans was obtained in the course of the carotid Amytal test, which is performed during assessment of patients being considered for surgical treatment of epilepsy. This test involves injecting Amytal (sodium amobarbital) and formerly also the pro-convulsant drug pentalene tetrazole (Metrazol) into one or other carotid artery. It occasionally happens that the injection is made inadvertently into the vertebro-basilar system which furnishes the main blood supply to the thalamus. According to the centrencephalic hypothesis, it might be expected that intravertebral Amytal (an anticonvulsant) would suppress the supposed thalamic trigger mechanism, abolishing generalised SW discharges, whereas intravertebral Metrazol should have an activating effect. Intracarotid amobarbital or Metrazol would be expected to produce, respectively, ipsilateral suppression or activation of the generalised SW discharges, or possibly to be without effect. In fact, what is found is the reverse of the above: unilateral intracarotid Amytal suppresses generalised SW activity bilaterally, and Metrazol activates it; intravertebral injections have little effect (Gloor 1968).

These and similar observations led Gloor (1968) to postulate a new model of the 'generalised corticoreticular epilepsies'. This places the primary functional abnormality in hyperexcitable cerebral cortex that responds abnormally with generalised discharges to essentially normal physiological afferents originating from the thalamus and reticular activating system. The thalamus does become secondarily entrained in the synchronous discharges through corticothalamic pathways, but is not their primary source. Once oscillatory firing of cortex and thalamus is established the activity of either structure may lead, but neither can be regarded as the pacemaker; both are necessary to maintain the discharge (Avoli and Kostopoulos 1982). Blumenfeld and McCormick (2000) offer an elegant validation of this model. In ferret lateral geniculate tissue slices, they simulated cortical feedback by an electronic circuit that stimulated the corticothalamic tract. The circuit could replicate normal cortical function by delivering single pulses in response to bursting, or cortical hyperexcitability by delivering a train of six pulses. The single stimuli gave 6–10 Hz oscillations resembling normal spindles, but the pulse trains set up 3–4 Hz paroxysmal oscillations. A γ-aminobutyric acid b ($GABA_B$) antagonist converted the slow oscillations back to 6–10 Hz spindles, whereas a $GABA_A$ antagonist had the opposite effect.

The oscillatory burst firing apparently represents a perturbation of a normal physiological mechanism, which produces spindle activity in the thalamocortical neurons during drowsiness (Avoli et al 1983). Following a burst of thalamocortical activity, recurrent inhibition leads to hyperpolarisation due to prolonged inhibitory postsynaptic potentials (IPSPs) mediated by GABA$_B$ receptors. These reactivate low-threshold T-calcium channels, leading to calcium influx, depolarisation and a burst of action potentials, and hence to a further period of recurrent inhibition, generating rhythmic generalised SW discharges (Deschênes et al 1982, Snead 1992). This mechanism again serves to emphasise the role of inhibition in the pathophysiology of epilepsy. A breakdown of the GABAergic transmission underlying the hyperpolarisation can allow evolution of the discharge to the continuous spiking of a tonic seizure (Kostopoulos et al 1983). For a review of the pathogenesis of SW and absence seizures, see Stefan and Snead (1997).

It may be a mistake to seek a single pathophysiological mechanism for all thalamocortical SW activity. Certainly genetic animal models of generalised epilepsy are heterogeneous, and similar electrophysiological phenomena may involve different mechanisms. Whether all human epilepsy accompanied by generalised SW activity involves a common pathophysiological mechanism is unknown. The discharges themselves show unexplained differences of morphology in different syndromes.

The corticoreticular model still leaves some important issues unresolved. A localised discharge may spread, eventually giving rise to 'secondarily generalised' SW activity. Similarly, a seizure may commence with symptoms reflecting local brain dysfunction and then evolve to a generalised convulsion. Gloor (1968) accommodated such 'secondary bilateral synchrony' due to 'a focal cortical pacemaker' within the corticoreticular model, but by so doing undermined the whole concept of primarily generalised epileptogenesis, which the centrencephalic hypothesis had sought to explain. In patients with generalised seizures and no evidence of secondary generalisation from a focus there is presumed to be generalised cortical hyperexcitability. (Such seizures are often colloquially termed 'primarily generalised', but curiously, although secondarily generalised seizures are explicitly recognised in the international classification of seizures (Dreifuss 1981), 'primarily generalised' seizures are not.) However, a localised abnormality may be sufficient to produce generalised seizures (as in cats after focal application of penicillin), and rapid generalisation after local epileptogenesis in a discrete neuronal aggregate may appear more plausible than simultaneous activation of the entire cortex, once the centrencephalic model is abandoned. It may then be argued that the distinction between generalised and focal epileptogenesis, fundamental to present classifications of both seizures and epileptic syndromes, is simply a question of the rate and extent of spread of a physiological disturbance of local onset. There is clinical evidence, admittedly somewhat arcane, to support this heretical view (*Box 2.1*):

1. On close inspection, the ictal phenomenology of that archetypal generalised seizure, the absence, includes features that can be explained only by cerebral dysfunction that is asymmetrical or indeed focal. One of the first discoveries arising from the use of simultaneous video-EEG investigations was that automatisms occur in most patients with absence seizures (Penry et al 1975). Versive movements have been reported in otherwise clinically and electrographically typical absences, which responded selectively to valproate therapy (So et al 1984). Stefan (1981), on the basis of careful scrutiny of video recordings, describes an asymmetric craniocaudal march of motor signs in some absences: eyelid flutter is followed by oral movements, then twitching of the arms and fiddling with the hands and, finally, leg movements. Similarly, juvenile myoclonic epilepsy, another supposedly idiopathic generalised syndrome, is often associated with generalised and focal EEG discharges, and rarely by seizures in which the patient rotates in one direction. In some subjects, seizures are precipitated by voluntary movements of the limbs on one side.

2. A useful method for studying cortical hyperexcitability is provided by patients with pattern-sensitive epilepsy; hemifield pattern stimulation in susceptible subjects with idiopathic generalised epilepsy demonstrates an asymmetrical threshold for initiating discharges in some 50% (Wilkins et al 1981). In over 60% of photosensitive patients, the clinical features of some or all photogenic seizures indicate a focal onset rapidly followed by generalisation (Hennessy and Binnie, 2000). These considerations apply equally to patients with generalised and partial epilepsies.

3. Other types of reflex seizures involving highly specific cognitive triggers, presumably localised to particular regions or neuronal systems, are typically found in generalised idiopathic epilepsy, once again calling into question the concept of generalised cortical epileptogenesis.

4. Callosotomy in patients with apparently generalised discharges and seizures often unmasks focal discharges and partial seizures.

Paradoxically, it may not yet be possible entirely to dismiss the centrencephalic model, as it is clearly possible for experimentally generalised seizures to be of subcortical onset. Brainstem-driven seizures typically produce tonic motor phenomena and can be elicited in various animal models after transcollicular transaction (Burnham 1987, Browning 1994). However, in the intact animal the discharges can spread to invade the forebrain. Examples include the finding by Rodin et al (1971, 1977) and Velasco et al (1980) of high-frequency multiunit brainstem activity preceding cortical involvement in biculline and pentalene tetrazole induced seizures. Lateral geniculate kindling, by repeated subthreshold electrical

---

Box 2.1   Evidence of local epileptogenesis in IGE

---

1. Local structural abnormality
Focal pathology
- Cortical dysgenesis (Meencke and Janz 1984, Woermann et al 1999)

---

2. Local abnormalities on functional imaging

---

3. Postulated diffuse cortical hyperexcitability is not uniform
Spontaneous epileptogenesis
- EEG semiology indicates asymmetric or focal cortical activation
- Focal interictal discharges in IGE, e.g. 34% in JME (Panayiotopoulos et al 1991)
- High prevalence of asymmetrical discharge onsets (32%) and foci (39%) in IGE (Binnie 1996)
- Seizure semiology indicates asymmetric or focal cortical activation
- Versive or rotary seizures and focal motor activation in JME
- Versive absences (So et al 1984)
- Craniocaudal march in absences (Stefan 1981)

Reflex epileptogenesis
- Visual – *Papio papio*
- Genetic or pharmacologically induced generalised epilepsy, not associated with any lesion. Discharges are initiated in frontocentral cortex, before other cortical or subcortical structures, generalisation is secondary (Fischer-Williams et al 1968). Single unit spike bursts are synchronous with spike-wave in frontal cortex but with flashes in occipital region (Menini et al 1981)

Visual – *Homo sapiens*
- Associated with IGE. Discharges are initiated in parieto-occipital cortex, generalisation is secondary (Wilkins et al 1980)
- Epileptogenesis depends on stimulus characteristics that determine patterns of neuronal activity in visual cortex:
  - level of activity within specific neuronal population
  - volume of cortex activated in either hemisphere
  - epileptogenesis originates from parieto-occipital cortex
  - when VPA has selectively abolished generalised EA, posterior focal spikes persist and critical pattern size is unchanged (Darby et al 1985)
  - marginally supraluminal stimuli elicit posterior discharges, these may then be seen to generalise
  - hemifield pattern stimulation activates contralateral posterior quadrant (Wilkins et al 1981)
  - threshold pattern size is asymmetrical in 50% of patients – as commonly in IGE as in other epilepsies
  - IPS responses are larger on side with lower pattern threshold (Binnie et al 1981)
- Subliminal 8 Hz flicker elicits occipital spikes, resistant to VPA (Harding et al 1978)
- 60% of patients have photogenic partial seizures, majority with visual aura (Hennessy and Binnie 2000)

Other reflex epilepsies
- Somatosensory, touch, startle, eating: often lesional and partial, focal trigger zone
- Idiopathic generalised reflex myoclonic epilepsies of infants (Deonna and Despland 1989)
- Cognitive triggers: reading: generalised/partial, idiopathic/ symptomatic, music: partial, thinking, complex percepts, etc.; mostly IGE
- Motor programming: IGE, especially JME (Matsuoka et al 2000)

EA, epileptiform activity; IGE, idiopathic generalised epilepsies; IPS, intermittent photic stimulation; JME, juvenile myoclonic epilepsy; VPA, valproate

---

stimulation, eventually produces spontaneous tonic–clonic seizures, which have been claimed as a model of idiopathic generalised epilepsy (Shouse and Ryan 1984). The role of the substantia nigra has also caused some confusion. Both lesions and micro-injection of anticonvulsants in the substantia nigra elevate seizure thresholds, but stimulation does not induce seizures and ictal activity cannot be recorded there. It appears that the substantia nigra has a tonic inhibitory action on a GABAergic inhibitory system of the superior colliculus (Gale 1985, Garant and Gale 1987).

Some of these problems may be resolved by the current view of the Montreal school: classifications provide practical aids to communication and prognosis, but necessarily involve oversimplification; a neurobiological description of the aetiology and pathophysiology of the seizure disorder in a particular patient allows a more individualised approach (Berkovic et al 1987). Surprisingly perhaps, this apparently rather complex, academic approach has found favour with the Task Force on Classification and Terminology of the International League Against Epilepsy (ILAE) (Engel 2001), which has produced a discussion document proposing a description of each patient's condition in terms of several dimensions, rather than a prescriptive classification.

Partial seizures

The approved term, 'partial seizures', is unfortunate and has an historical origin in the notion that such attacks were not complete or 'genuine' epilepsy. At seizure onset, an extracellular recording from the site of origin typically shows sharply localised rhythmic fast activity at frequencies as high as 50 Hz or more, probably a spatiotemporal sum of excitatory and inhibitory postsynaptic potentials. Individual neurons may or may not be involved; some fire rapidly for up to 2 s but do not continue throughout the seizure, only a small proportion, some 7–14%, of neurons contribute at any time to the population discharge (Babb et al 1987). Typically, after a few seconds, or sometimes after as much as a minute, the discharge spreads to involve a larger volume of tissue and changes in character, producing either SW activity or, more usually, rhythmic, sharp waveforms, which gradually increase in amplitude and diminish in frequency – 'rhythmic ictal transform-ation' (Geiger and Harner 1978) (*Fig. 2.62*). This activity typically becomes slower and irregular and may give way to SW discharges, which in their turn slow to about 1/s, become irregular and abruptly cease. Sometimes SW activity is seen from the start of the seizure; however, this is less sharply localised than the fast activity described

above, suggesting either that a significantly different 'regional' pathophysiological process is involved, or possibly just that there is no electrode positioned appropriately to detect seizure onset. In the postictal state, there may be a reduction in amplitude of all activity, then slow waves appear gradually increasing in frequency until the normal rhythms are restored. These events may be confined to a small part of the brain, or spread more widely in the same hemisphere or to a localised, usually homotopic, area of the opposite hemisphere; they may eventually become generalised.

The relationship of these intracranial electrical phenomena to clinical events and to the scalp EEG is variable. Clinical changes may be obvious or more subtle, possibly detectable only by repeated close observation with the help of video recording, or by the use of psychological tests to detect brief disturbances of cognitive function (Aarts et al 1984). Indeed, many patients with partial epilepsy display electrical events typical of those seen at onset of their overt attacks without evident clinical change. In intracranial recordings, such 'electrographic seizures' (if one may use such a term without risk of confusing an EEG phenomenon with epilepsy) may outnumber clinical events by a factor of ten or more (Binnie et al 1994). Clinical ictal signs, when present, appear after the electrical seizure onset, at an interval from some tens of milliseconds to several minutes. Their nature will depend upon the site, duration and extent of spread of the discharges. If these remain confined to an area close to the original focus, the clinical changes will reflect a disturbance of the elemental functions of that part of the brain – for instance, jerking of a limb if the motor cortex is affected, or dysphasia with involvement of the dominant temporal lobe. Such attacks are termed 'simple partial seizures'. It is postulated that more widespread involvement, particularly with bilateral discharges in the limbic system, will produce an alteration of consciousness and possibly involuntary behaviour, or automatisms. Such attacks are called 'complex partial seizures'. The meaning of the term 'complex' has changed over the years. In the current ILAE seizure classification, it does not relate to the complexity of the symptoms, but to a disruption of the total integrative function of the brain; disturbances of elemental higher functions, such as speech and memory, with intact consciousness occur in simple partial seizures. Where there is extensive bilateral spread a secondarily generalised tonic–clonic convulsion may result.

The distinction between simple and complex symptomatology was based on physiological speculation that now appears questionable. Some patients investigated with bilateral intracerebral electrodes show ictal loss of consciousness with discharges apparently confined to one temporal lobe. Conversely, rare but well-documented instances are known of ictal automatisms, or even of tonic–clonic convulsions with retained consciousness (Alarcón et al 1997).

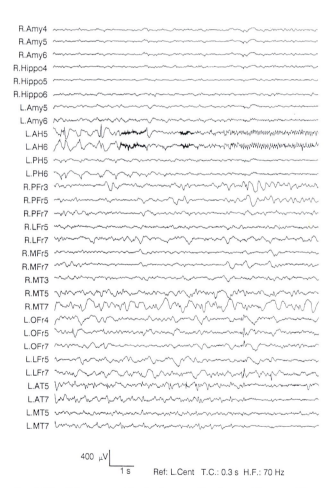

400 μV

1 s   Ref: L.Cent   T.C.: 0.3 s H.F.: 70 Hz

Fig. 2.62 Seizure onset in depth recording from a 23-year-old woman with intractable complex partial seizures. Seizure onset is focal at the left anterior hippocampal depth electrodes LAH5 and LAH6 and is characterised first by some irregular spikes and slow waves followed by two short bursts of very high frequency fast activity and then a transition to rhythmic activity at about 12 Hz (rhythmic ictal transformation). (From Binnie et al (2003), by permission.)

Depending on the volume, location and orientation of the tissue involved, the events described above may be clearly or poorly represented in the EEG. During a simple partial seizure involving a small, possibly deep, focus no visible change may occur in the EEG. Sufficient spread of discharge to produce a complex partial seizure will usually be reflected in the EEG, but possibly with a delay, so that clinical onset precedes EEG changes.

Both generalised and partial seizures are often preceded by an electrodecremental event – a brief attenuation of ongoing activity, rarely lasting much more than a second, but sometimes of up to 15 s duration (*Fig. 2.63*). In partial seizures this may be localised or generalised. When focal it does not necessarily correspond to the site of seizure onset even in depth recordings, as determined by subsequent focal discharges and successful surgical treatment (Alarcón et al 1994a). The occurrence of an event that may be generalised or falsely localising at the start of a partial seizure is perplexing. One explanation is that it reflects an afferent event in the reticular activating system which triggers the seizure; another is that the cortical desynchronisation is itself due to activity propagated from an earlier focal ictal discharge, which is itself undetectable. In any case, the phenomenon suggests some serious deficiency in our understanding of the underlying pathophysiology.

The interictal EEG

About 90% of patients with epilepsy sometimes exhibit abnormal EEG discharges in the intervals between attacks. Typically these take the form of spikes, sharp waves or SW complexes, which may be isolated or repetitive but generally are much briefer than the ictal discharges. In patients who suffer only generalised seizures, these discharges are themselves bilateral, symmetrical and synchronous, generally of greatest amplitude over the frontal regions but sometimes located posteriorly. In patients with partial seizures their topography usually corresponds more or less closely to the focus from which seizures arise. Occasionally in patients with partial seizures, bursts of high-frequency activity can be seen similar to, but much briefer than, those occurring at seizure onset, but in general interictal focal discharges resemble those seen late in the seizure and lack the rhythmicity of the early ictal phenomena. Focal interictal discharges may spread, producing 'secondarily generalised' spike-and-slow-wave activity. If the generalisation is rapid, the focal onset may be difficult to detect (*Fig. 2.64*).

Serial EEG recordings often show changing patterns of interictal discharge. In particular, new foci appear, and the question arises whether these have always been present intermittently, but not previously detected, or

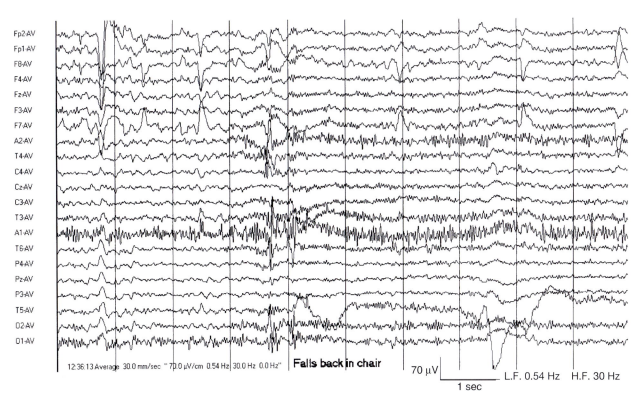

Fig. 2.63 Electrodecremental event at seizure onset in a 13-year-old boy with learning difficulties and symptomatic generalised epilepsy. The EEG at rest was diffusely slowed but more so on the left. At the beginning of the excerpt the patient had been hyperventilating for 2 min, increasing the amount of slow activity. The onset of the electrodecremental event corresponded to the first clinical manifestations: he raised his left hand, turned his head to the right and stared. As generalised epileptiform activity appeared he began rocking his trunk backwards and forwards. (From Binnie et al (2003), by permission.)

whether they have arisen de novo, possibly as a consequence of repeated interictal discharges, ictal anoxia, progression of underlying pathology, or head injuries. Similarly, an established focus may appear to migrate; in children it has long been claimed that occipital foci shift to the temporal regions (Gibbs et al 1954). Again, it is uncertain whether this is a sampling problem or a true evolution of the pathophysiology (Andermann and Oguni 1990, Blume 1990).

In experimental epilepsy, the development of a focus may be followed by the appearance of a 'mirror focus' in the homotopic area of the contralateral hemisphere. At first this is time-locked to the discharges of the primary focus and will disappear if the latter is excised. Later the mirror focus becomes autonomous, fires independently and will persist after removal of the primary. Ascending the evolutionary scale, the speed with which mirror foci appear decreases; moreover, antiepileptic medication discourages their development. It is disputed whether or not mirror foci occur in man. Certainly many patients have bilateral temporal foci, but the causative pathology may also be bilateral. Morrell and Whisler (1980) reported

reversible contralateral foci in a series of patients with cerebral tumours, and Hughes (1985) found that 40% of foci eventually become bilateral. However, many authors contend that it has not yet been reliably established whether or not mirror foci actually occur in humans (Goldensohn 1984, Blume 1990). Another experimental model of doubtful relevance to man is 'kindling'. This is produced by electrical stimulation, typically at daily intervals, of susceptible brain structures, notably the amygdala, at an intensity sufficient to cause a local after-discharge but not a seizure. After repeated presentations the same stimulus eventually gives rise to overt seizures, and finally spontaneous seizures develop. Analogies, of uncertain validity, may be claimed between kindling and the occurrence of repeated seizures or of subclinical discharges.

Both the mirror focus and the kindling model, if applicable to man, predict that epilepsy should be a progressive process: seizure frequency should increase, multiple types of attack should appear in place of a single type of partial seizure, an increasing proportion of partial seizures should generalise, deterioration should be particularly rapid in patients with many interictal

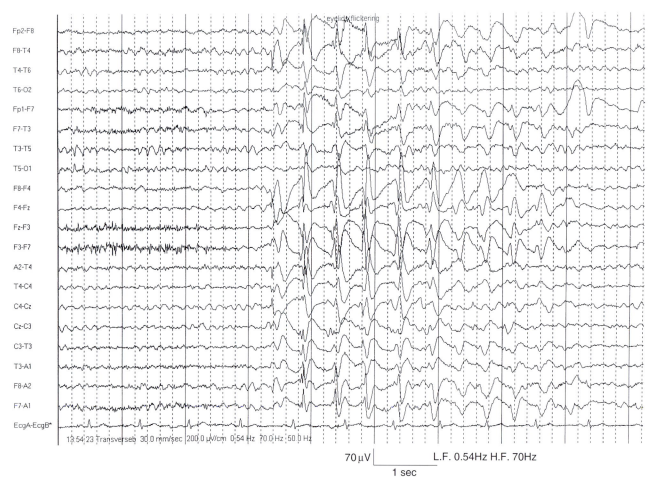

Fig. 2.64   A 33-year-old man with intractable epilepsy with multiple seizure types. Focal discharge in the right frontocentral region (F4, C4) undergoing rapid secondary generalisation to produce an attack with brief loss of awareness and flickering of the eyelids – apparently an absence but in fact a complex partial seizure. (From Binnie et al (2003), by permission.)

discharges, and the effectiveness of surgical treatment should diminish with time as mirror foci become independent. These predictions are not generally true (Bengzon et al 1968, Engel 1989). Whatever the physiological basis of interdependence of bilateral foci, the second focus often disappears gradually after removal of an epileptogenic lesion (Falconer and Kennedy 1961). In Morrell's terms, such foci must presumably have reached the intermediate phase of secondary epileptogenesis.

The distinction between ictal and interictal discharges has been confused by the realisation that, with suitable testing, deficits of cerebral function (transitory cognitive impairment (TCI)) can be detected during apparently subclinical discharges in at least 50% of patients tested. In the case of focal discharges the nature of the disturbances may reflect the side and possibly the region of the brain involved. Thus left-sided discharges may be accompanied by momentary lapses of verbal memory, and right-sided discharges interfere with performance of spatial tasks (Aarts et al 1984). Reaction times may also be increased when stimuli are presented in the visual field contralateral to the focus or responses made with the contralateral hand (Shewmon and Erwin 1988). Nevertheless, to abandon the generally accepted concept of interictal discharges at this time would be unnecessarily confusing and the term is therefore in this text despite the uncertainties concerning its validity. Operationally, an event is subclinical if it is not detectable by routine clinical observation, and the concept of TCI is justified as it draws attention to the occurrence of cognitive changes not detectable by such means.

The EEG and seizure type

As indicated above, the pathophysiological distinction between partial and generalised seizures and between simple and complex partial seizures are both open to question. Nevertheless, the seizure classification of the ILAE (Dreifuss 1981) (Table 2.3), which incorporates all these concepts, does provide a pragmatic basis for communication and a structure within which a particular patient's disorder can be categorised. It is described as a 'clinical and electroencephalographic classification'. However, rigorous application of the EEG criteria is often impossible, sometimes because certain clinical seizure types are frequently associated with EEG findings materially different from the 'typical' patterns described, or more usually because an ictal EEG is not available.

EEG findings in patients with partial seizures

The three basic categories of partial seizure (simple, complex or evolving to become secondarily generalised tonic–clonic seizures) were noted in the previous section.

The classification gives some indication of the range of clinical phenomena that may occur in simple partial seizures, but these different features do not have any very distinctive EEG signatures. Simple partial seizures are associated with ictal and interictal discharges in the affected region, which can readily be detected in intracranial recordings, but are often not apparent in the scalp EEG. As focal discharges spread they are more likely to appear in the EEG, but may produce ictal symptoms reflecting dysfunction of structures at some distance from the site of onset. Specifically, the concept of complex partial seizures is based on the assumption that impairment of consciousness occurs in partial seizures only when there is bilateral invasion of neocortical and/or limbic structures. The ictal discharge can be initiated by a focus in the temporal, frontal, temporo-parietal or occipital region. For this reason, 'temporal lobe epilepsy' may have become an unhelpful clinical term, as it is often used indiscriminately to describe both partial epilepsy with simple ictal symptoms of temporal dysfunction, and complex partial seizures. It must also be realised that, once propagation of discharges has occurred, the most intense seizure activity may be remote from the site of onset; neither the distribution of the epileptiform activity nor indeed the clinical events necessarily indicate the origin of the seizure (Gloor 1975). This is also reflected in the postictal picture: although postictal slowing or amplitude reduction most commonly occurs over the region from which the seizure arose (Kaibara and Blume 1988, Hufnagel et al 1995) they are of limited localising value, and may occur contralateral to the seizure onset as established by depth recording (Binnie et al 1994). Ictal EEG patterns and dynamics of ictal behavioural changes can be analysed clinically with a nucleus shell structure model (Stefan 1999). The initial ictal EEG changes or seizure signs can be analysed in order to get a first hypothesis where the region of seizure onset (nucleus) can be expected. Following the changes in the temporal course of the seizure in EEG or ictal signs may indicate propagation to distant brain regions (or shells) (Stefan 1999).

More generally, the relationship of focal discharges in the scalp EEG to the primary site of epileptogenesis is open to misinterpretation. In patients with seizures arising from deep structures, such as the mesial temporal region, interictal recordings with deep (intracerebral or foramen ovale) and superficial (subdural or scalp) electrodes show discharges occurring more or less simultaneously at deep sites and on the surface. However, the spikes detectable on the surface commonly show a latency of some tens of milliseconds with respect to the deep sources, greater temporal dispersion, and an amplitude two orders of magnitude too great to be explained by simple volume conduction (Alarcón et al 1994b). Thus the focal scalp EEG activity arises from superficial neocortical generators activated by propagation from the deep source and is not directly recorded from the ictal pacemaker in, for instance, the mesial temporal region. The fact that the surface topography of the discharges can be modelled by deep dipole generators does not indicate either that such generators exist, or that

Table 2.3   International Seizure Classification*

I.   Partial (focal, local) seizures

Partial seizures are those in which, in general, the first clinical and electroencephalographic changes indicate activation of a system of neurons limited to part of one cerebral hemisphere. A partial seizure is classified primarily on the basis of whether or not consciousness is impaired during the attack. When consciousness is not impaired, the seizure is classified as a simple partial seizure. When consciousness is impaired, the seizure is classified as a complex partial seizure. Impairment of consciousness may be the first clinical sign, or simple partial seizures may evolve into complex partial seizures. In patients with impaired consciousness, aberrations of behaviour (automatisms) may occur. A partial seizure may not terminate, but instead progress to a generalised motor seizure. Impaired consciousness is defined as the inability to respond normally to exogenous stimuli by virtue of altered awareness and/or responsiveness.

There is considerable evidence that simple partial seizures usually have unilateral hemispheric involvement and only rarely have bilateral hemispheric involvement; complex partial seizures, however, frequently have bilateral hemispheric involvement.

Partial seizures can be classified into one of the following three fundamental groups:
A. Simple partial seizures
B. Complex partial seizures:
   1. with impairment of consciousness at onset
   2. simple partial onset followed by impairment of consciousness
C. Partial seizures evolving to generalised tonic–clonic convulsions (GTC):
   1. simple evolving to GTC
   2. complex evolving to GTC (including those with simple partial onset)

Clinical seizure type	EEG seizure type	EEG interictal expression

A. Simple partial seizures (consciousness not impaired)
1. With motor signs:
  (a) focal motor without march
  (b) focal motor with march (Jacksonian)
  (c) versive
  (d) postural
  (e) phonatory (vocalisation or arrest of speech)
 	
2. With somatosensory or special-sensory symptoms (simple hallucinations, e.g. tingling, light flashes, buzzing):
  (a) somatosensory
  (b) visual
  (c) auditory
  (d) olfactory
  (e) gustatory
  (f) vertiginous
 	
3. With autonomic symptoms or signs (including epigastric sensation, pallor, sweating, flushing, piloerection and pupillary dilatation)
 	
4. With psychic symptoms (disturbance of higher cerebral function). These symptoms rarely occur without impairment of consciousness and are much more commonly experienced as complex partial seizures:
  (a) dysphasic
  (b) dysmnesic (e.g. *déjà vu*)
  (c) cognitive (e.g. dreamy states, distortions of time sense)
  (d) affective (fear, anger, etc.)
  (e) illusions (e.g. macropsia)
  (f) structured hallucinations (e.g. music, scenes)
	Local contralateral discharge starting over the corresponding area of cortical representation (not always recorded on the scalp)	Local contralateral discharge

B. Complex partial seizures (with impairment of consciousness; may sometimes begin with simple symptomatology)
1. Simple partial onset followed by impairment of consciousness:
  (a) with simple partial features (A.1–A.4) followed by impaired consciousness
  (b) with automatisms
	Unilateral or frequently bilateral discharge; diffuse or focal in temporal or frontal regions	Unilateral or bilateral; generally asynchronous focus; usually in the temporal or frontal regions

Continued

Table 2.3 *Continued*

2. With impairment of consciousness at onset:
   (a) with impairment of consciousness only
   (b) with automatisms

C. *Partial seizures evolving to secondarily generalised seizures (this may be generalised tonic, tonic–clonic, tonic or clonic)*
1. Simple partial seizures:
   (a) evolving to generalised seizures

   Above discharges become
   secondarily and rapidly generalised

2. Complex partial seizures:
   (a) evolving to generalised seizures

3. Simple partial seizures evolving to complex partial seizures evolving to generalised seizures

II. Generalised seizures (convulsive or non-convulsive)

Generalised seizures are those in which the first clinical changes indicate initial involvement of both hemispheres. Consciousness may be impaired and this impairment may be the initial manifestation. Motor manifestations are bilateral. The ictal electroencephalographic patterns initially are bilateral, and presumably reflect neuronal discharge which is widespread in both hemispheres.

Clinical seizure type	EEG seizure type	EEG interictal expression
A.1. *Absence seizures:* (a) impairment of consciousness only (b) with mild clonic components (c) with atonic components (d) with tonic components (e) with automatisms (f) with autonomic components ((b)–(f) may be used alone or in combination)	Usually regular and symmetrical 3 Hz, but may be 2–4 Hz, spike-and-slow-wave complexes and may have multiple spike-and-slow-wave complexes. Abnormalities are bilateral	Background activity usually normal, although paroxysmal activity (such as spikes or spike-and-wave complexes) may occur. This activity is usually regular and symmetrical
A.2. *Atypical absence seizures* May have: (a) changes in tone that are more pronounced than in A.1 (b) onset and/or cessation that is not abrupt	EEG more heterogeneous; may include irregular spike-and-slow-wave complexes, fast activity or other paroxysmal activity. Abnormalities are bilateral but often irregular and asymmetrical	Paroxysmal activity (such as spikes or spike-and-slow-wave complexes) frequently irregular and asymmetrical
B. *Myoclonic seizures:* Myoclonic jerks (single or multiple)	Polyspike and wave, or sometimes spike-and-wave or sharp and slow waves	Same as ictal
C. *Clonic seizures*	Fast activity (l0 c/s or more) and slow waves: occasional spike-and-wave patterns	Spike-and-wave or polyspike-and-wave discharges
D. *Tonic seizures*	Low voltage, fast activity or a fast rhythm of 9–10 c/s or more decreasing in frequency and increasing in amplitude	More or less rhythmic discharges of sharp and slow waves, sometimes asymmetrical. Background is often abnormal for age
E. *Tonic--clonic seizures*	Rhythm at 10 c/s or more, decreasing in frequency and increasing in amplitude during tonic phase; interrupted by slow waves during clonic phase	Polyspike and waves or spike and wave, or sometimes sharp and slow wave discharges
F. *Atonic seizures (astatic)* (combinations of the above may occur, e.g. B and F, B and D)	Polyspikes and wave or flattening or low voltage fast activity	Polyspikes and slow wave

III. Unclassified epileptic seizures

Includes all seizures that cannot be classified because of inadequate or incomplete data and some that defy classification in hitherto described categories. This includes some neonatal seizures, e.g. rhythmic eye movements, chewing and swimming movements.

Continued

Table 2.3 *Continued*

IV. Addendum

Repeated epileptic seizures occur under a variety of circumstances:

1. (1) as fortuitous attacks, coming unexpectedly and without any apparent provocation; (2) as cyclic attacks, at more or less regular intervals (e.g. in relation to the menstrual cycle, or the sleep-waking cycle); (3) as attacks provoked by (a) non-sensory factors (fatigue, alcohol, emotion, etc.) or (b) sensory factors, sometimes referred to as 'reflex seizures'.

2. Prolonged or repetitive seizures (status epilepticus). The term 'status epilepticus' is used whenever a seizure lasts for a sufficient length of time or is repeated frequently enough that recovery between attacks does not occur.

   Status epilepticus may be divided into partial (e.g. Jacksonian) or generalised (e.g. absence status or tonic–clonic status). When very localised motor status occurs, it is referred to as 'epilepsia partialis continua'.

*Reprinted by permission from Dreifuss (1981).

their activity is detectable on the scalp. As focal discharges in partial epilepsies arising from sites other than the cerebral convexity represent propagated activity, their frequently anomalous location is unsurprising. *Table 2.4* summarises seizure semiology and ictal EEG changes in partial seizures.

Seizures arising in the temporal lobes   Seizures of temporal lobe origin are typically associated with interictal temporal spikes, sharp waves or isolated SW complexes. In some 20–50% of patients these occur over both hemispheres, albeit usually independently (King and Ajmone Marsan 1977, Delgado-Escueta et al 1982, So et al 1989). A four-to-one preponderance on one side may be regarded as evidence of lateralisation (Polkey 1983). Some authors claim that in patients undergoing assessment for possible surgery, interictal anterior temporal discharges usually correctly indicate the side of seizure onset as determined by intracranial recording (Kanner et al 1993). Where the focus is located in mesial temporal structures (amygdala and hippocampus) the spikes are usually of greatest amplitude (on extracranial recording) at the sphenoidal or anterior temporal electrodes. The focus on the scalp will typically be some 30% less in amplitude than at the sphenoidal site and located about 1 cm posterior to the external canthus of the eye (*Fig. 2.65*). A temporal focus not maximal at the sphenoidal electrode is more likely to arise from a lesion involving lateral temporal neocortex (Binnie et al 1989). However, the typical anterior temporal topography is generally found irrespective of the distribution of temporal spikes in intracranial recordings. It seems that it is the combination of spatiotemporal averaging of discharges propagating over the surface of the temporal lobe, and preferential conduction through various low-resistance foramina in the skull (e.g. the optic foramen and inferior orbital fissure) that determine this characteristic topography of temporal discharges (Fernandez Torre et al 1999a,b).

In patients with mesial temporal epilepsy the discharges typically exhibit a bipolar field, with a sharply

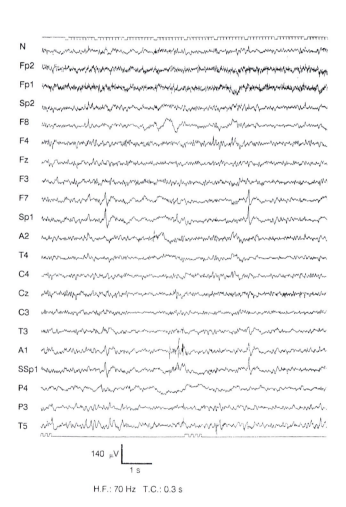

Fig. 2.65   Sphenoidal recording of interictal spikes in a 25-year-old woman with complex partial seizures of left temporal origin. This common average reference recording was performed with multipolar sphenoidal electrodes. Non-standard electrode designations: N, nasion; Sp2, right sphenoidal; Sp1, left sphenoidal; SSp1, additional contact on left sphenoidal bundle some 5 mm below the skin surface. Spikes are recorded at sphenoidal and scalp electrodes on the left. Note that there is a gradient from the deep sphenoidal, through the more superficial sphenoidal contact to F7 and A1, but the sphenoidal spikes are clearly detectable on the surface. (From Binnie et al (2003), by permission.)

Table 2.4   Seizure semiology, interictal and ictal EEG changes in partial seizures*

Focus	Seizure semiology	Interictal EEG	Ictal onset pattern
Temporal	Aura: epigastric, olfactory, autonomic, psychic, visual and auditory, simple in younger children	May be normal	Rhythmic theta in anterior temporal or midtemporal region
	Behavioural arrest	Temporal theta activity	
	Dystonic posturing (contralateral)	Spikes or sharp waves	
	Automatism (ipsilateral)		
	Ictal or postictal dysphasia (dominant)		
	Relative preservation of awareness (non-dominant)		
	Late forceful head turning (contralateral)		
	Asymmetric ending of secondary generalised seizure (ipsilateral)		
	Motor manifestations in infants and younger children		
	Postictal (Todd's) paresis (contralateral)		
Frontal	Frequent, brief, often nocturnal	Often normal	Generalised or widespread slow activity, fast activity or attenuation
	Automatism, often violent, involving legs: cycling, kicking, posturing, turning	Theta unilateral or bilateral	
	Motor restlessness	Interictal spikes or sharp waves may be seen in the frontal or temporal regions	Movement artefact may obscure EEG
	Supplementary motor area: asymmetric tonic limb posturing, speech arrest, minimal loss of awareness	Generalised spike-and-wave may be seen	
	Early forceful head turning and eye deviation (contralateral)		
	Postictal (Todd's) paresis (contralateral)		
Parietal	Aura: somatosensory	May be normal or non-specific	Often normal in simple partial seizures, particularly with sensory symptoms
	Somatic illusions (metamorphosia, rotatory sensation)	Epileptiform discharges often non-localised or in other areas	
	Visual illusions or hallucinations, vertigo		
	Features depend on propagation		
Occipital	Visual hallucinations (elementary, colours, forms, complex illusions or hallucinations, visuospatial distortion)	Lateralised slow waves in posterior regions	Localised/lateralised ictal onset may be seen
	Amaurosis	Occipital spikes or sharp waves	May be normal
	Blinking, nystagmus		Fast activity or spikes in occipital region or more widespread
	Ictal or postictal headache		
	Ictal vomiting		
	Forced head or eye deviation (not lateralising)		
	Features depend on propagation		

*Adapted from Chee et al (1993), Foldvary et al (2001), Leutmezer et al (2002), Gallmetzer et al (2004) and Ray and Kotagal (2005).

localised negative spike in the anterior temporal region and a more diffuse low amplitude positivity in the contralateral central area. This characteristic topography is not seen in patients with seizures arising in lateral temporal neocortex (Ebersole 1992). As noted above, the fact that a bipolar field can be demonstrated does not necessarily indicate that the activity is in fact recorded from a single dipole generator.

Often associated non-epileptiform abnormalities will be found: reduction of fast activity over the frontotemporal region ipsilateral to the focus, or possibly localised theta or delta activity, occurring either at the site of the spike focus or more widely. It might be expected that abnormalities of ongoing activity would reflect underlying pathology. In fact temporal delta activity is closely related to lateral temporal hypometabolism, as demonstrated by fluorodeoxyglucose positron emission tomography (FDG-PET) scans, but not to gross pathology, nor (except perhaps in early postictal recordings) to the recent occurrence of seizures (Koutroumanidis et al 1998).

The initial ictal EEG change may consist of clear temporal spikes, but often only theta or slower activity is seen, which may be bitemporal or generalised (*Fig. 2.66*).

Ictal scalp EEG abnormalities are lateralising in only 50% of patients and even then indicate the side of origin of temporal lobe seizures with only 80% reliability (Lieb et al 1976, Spencer et al 1985). Some seizures of temporal lobe origin are preceded by a generalised electrodecremental event and sometimes this is the only EEG change. Others, notably simple partial seizures with viscerosensory or psychic symptoms, may produce no visible alteration of the scalp EEG. In view of the postulated bilateral limbic involvement in complex partial seizures it might be expected that these would always produce bilateral changes in the EEG. Although this is generally the case, complex partial seizures can occur with unilateral discharges and even with no apparent change in the scalp EEG (*Fig. 2.67*).

Seizures of frontal origin   In patients with partial seizures of frontal origin, appropriately localised interictal or ictal foci in the scalp EEG are uncommon (Williamson and Spencer 1986). A frontal interictal focus may be found in as few as 9% of patients, and some 59% exhibit regional discharges over the frontocentral or frontotemporal areas (Quesney et al 1992). The difficulty of EEG localisation may reflect: the inaccessibility of foci on the extensive mesial and orbital aspects of the frontal lobes, rapid spread of discharges within the frontal lobes and a marked tendency to secondary generalisation, and sometimes an extensive epileptogenic zone (Fegersten and Roger 1961, Quesney 1986).

In patients with seizures of mesial or orbital frontal origin, ictal and interictal records may show secondarily generalised slow waves or SW discharges (Tükel and Jasper 1952, Ralston 1961, Tharp 1972). An irregular waveform and a possible asymmetry distinguish these from the typical 3/s SW activity of idiopathic generalised epilepsy. Ictal discharges consisting only of frontal slow waves are easily overlooked or mistaken for oculographic artefact in patients actively moving during a complex partial seizure. As seizures of mesiobasal frontal origin are often bizarre and readily supposed to be psychogenic, the EEG findings if not carefully interpreted may serve only to compound the diagnostic confusion. Seizures arising from the cingulate gyrus may resemble absences, both clinically and electrographically. Those from the supplementary motor area typically produce characteristic ictal posturing and are often associated with ictal and interictal sharp waves at the vertex. It should be noted that the montages most often used in many departments both for routine EEGs and telemetry omit either the

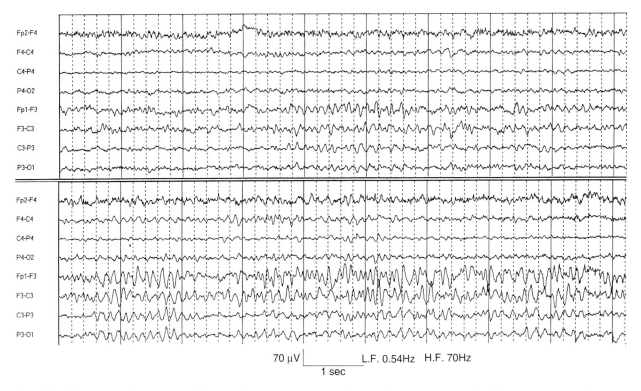

Fig. 2.66   Two consecutive excerpts (above and below) during onset of a complex partial seizure in a 32-year-old man, without frank spikes. Left-sided rhythmic activity at about 8.5 Hz appears widely in the parasagittal leads (above), then becomes slightly slower and appears intermittently on the right (below). Patient was staring and inaccessible. (From Binnie et al (2003), by permission.)

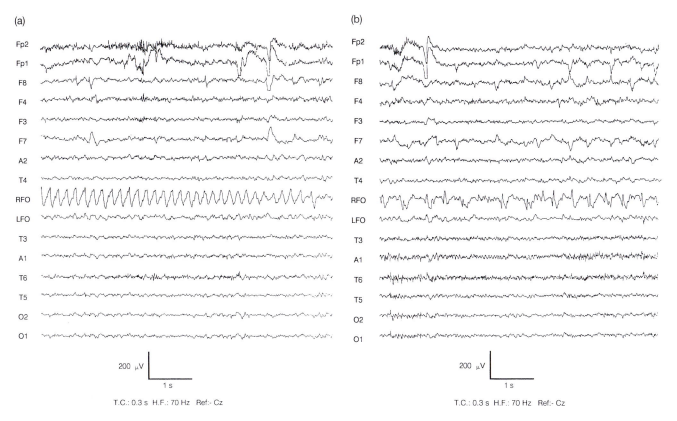

Fig. 2.67 Unilateral intracranial discharge during a complex partial seizure. Recording using foramen ovale electrodes (RFO and LFO) from a 25-year-old man with left mesial temporal sclerosis. (a) During this excerpt, taken some 10 s after seizure onset, he is manifestly confused and ignoring attempts by his mother to attract his attention. (b) 30 s later he is still confused and is showing version to the right. The abnormal electrical activity seen in this recording is virtually confined to the right foramen ovale contact. Bilateral spread is not seen despite impairment of consciousness and the substantial duration of the seizure. (From Binnie et al (2003), by permission.)

midfrontal or the vertex electrode and may fail to demonstrate these discharges.

Simple partial seizures with motor symptoms may be associated with central foci of ictal and interictal discharges (most strikingly in the syndrome of benign epilepsy of childhood with centro-temporal spikes; see p. 253). Interictal temporal spikes or generalised spike and slow waves may be present. However, a lack of interictal abnormality is more usual in symptomatic peri-Rolandic epilepsies, and even ictal records may show no epileptiform activity if the motor phenomena are brief and of restricted extent.

Seizures of parieto-occipital origin Occipital foci are uncommon and do not necessarily reflect the site of origin of seizures. Symptomatic epilepsies arising from the occipital lobes are yet rarer. Blume et al (1991) reported that 80% of subjects showed appropriately localised ictal and interictal focal discharges; however, Williamson et al (1992a) found that interictal epileptiform activity maximal in the occipital region was uncommon; temporal foci or generalised discharges were more often seen. Where complex partial seizures supervene, ictal discharges spread widely over the corresponding hemisphere. Posterior temporal and occipital epileptogenic lesions commonly produce abnormalities of ongoing EEG activity, particularly alpha asymmetry. The occipital SW discharges of 'benign epilepsy of childhood with occipital paroxysms' (see p. 254) are usually bilateral but may be focal in one or other posterior–temporal–occipital region.

Parietal epileptogenic lesions are rarely associated with interictal EEG abnormality in the parietal area, beyond a possible local reduction in amplitude of ongoing activity. Interictal discharges tend to be located in the mid- to posterior temporal region. Ictal spikes are often generalised but may be lateralised to the side of seizure origin; a parietal focal onset in the scalp EEG is unusual (Williamson et al 1992b).

Partial seizures evolving to secondarily generalised The ILAE seizure classification includes a category of those partial seizures that evolve into secondarily generalised attacks, usually tonic–clonic convulsions. A corresponding EEG evolution is seen with the focal or bilateral discharges of the simple or complex partial seizure, as described above, giving way to generalised spikes and later SW activity. Once generalisation has occurred, seizure activity and postictal disturbances may be more apparent contralateral to the site of onset; asymmetries during or after secondarily generalised seizures are therefore of no lateralising value. Note that the occurrence in the interictal EEG of secondary generalisation of focal discharges (see *Fig. 2.65*) is not necessarily predictive of secondarily generalised seizures.

EEG findings in patients with generalised seizures

Absences   Of all seizures, the 'typical' absence has the most characteristic EEG signature. Generalised SW activity at about 3/s commences abruptly with the onset of the attack and ceases at the end (*Fig. 2.68*). The patient, usually a child, exhibits an impairment of consciousness, does not fall but may show changes in muscle tone and usually displays some abnormal movements, eyelid fluttering, mouthing or small rhythmic movements of the hands. The paradox of a dramatic EEG change with minimal clinical symptoms may be explained by the nature of the SW discharge itself. Burst firing of cortical neurons occurs during the spike, but the slow wave, which is of much longer duration, is

accompanied by hyperpolarisation due to recurrent inhibition; the clinical manifestations of the absence seizure appear to be mainly inhibitory. The spike and the slow wave are topographically distinct, although there remains some disagreement about the details of their distribution (Rodin and Ancheta 1987). The discharges usually show bifrontal maxima, and a single midline maximum is claimed to be associated with an atypical clinical picture and a poor response to medication (Dondey 1983). Clinical seizures are not necessarily observed during every SW discharge but, it is claimed, can always be detected given sufficiently close observation; thus 'in absence seizures interictal discharges probably do not occur' (Delgado-Escueta 1979). The importance of adopting appropriate measures during recording to detect clinical concomitants of SW discharges is noted repeatedly elsewhere. The interictal EEG is usually normal but ongoing rhythms may be slightly slowed. Some patients exhibit striking rhythmic posterior slow activity at about 3 Hz, which builds up to

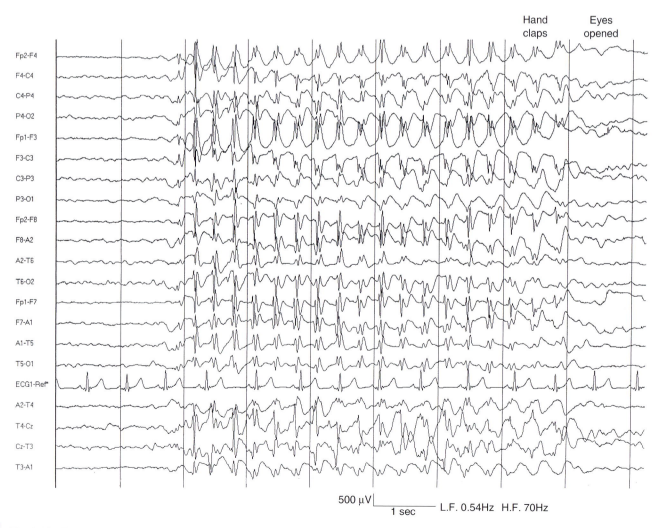

Fig. 2.68   Typical absence seizure in a 9-year-old girl. Note abrupt and symmetrical onset and termination of discharge with opening of the eyes when the technologist clapped her hands. (From Binnie et al (2003), by permission.)

a generalised SW discharge on overbreathing. This feature is reported to be a favourable prognostic sign and is more common in the childhood form of absence epilepsy. Overbreathing elicits SW activity in patients with absence seizures so consistently that the lack of this finding in an untreated patient who hyperventilates efficiently for no less than 2 min must cast doubt on the diagnosis of an active absence epilepsy.

The principal differential diagnosis of absences is complex partial seizures. The clinical phenomenology of the absence can be identical to that of a brief complex partial seizure and, if focal ictal discharges rapidly generalise, their origin may not be apparent. Minor asymmetries of SW activity are common in absences, particularly over the frontal regions at seizure onset. There appear to be no established criteria concerning what degree of asymmetry of generalised SW activity should be regarded as evidence of a focal onset. The risk of misdiagnosing complex partial seizures in a patient with absences is probably no less than the opposite. A good case can be made for the heretical view that absences and brief complex partial seizures with SW activity are physiologically similar (Aird et al 1989) but this is not explored further here.

'Typical absences' occur in several epilepsy syndromes and the discharges have slightly different characteristics. Note that 'atypical absence seizures' are a separate category in the ILAE classification. The term does not simply signify absences that are in some way unusual, but specifically refers to those with a more than usually marked change of muscle tone and lacking the usual sudden onset and termination. Atypical absences are generally accompanied by more irregular discharges in which the spikes are often multiple, a pattern also seen in those complex partial seizures that clinically most resemble absences. A specific type of atypical absence is seen in the Lennox–Gastaut syndrome, characterised by SW activity at not more than 2 Hz, and loss of muscle tone.

Myoclonic seizures In juvenile myoclonic epilepsy generalised ictal discharges of multiple spikes and slow waves are seen, and brief, frequent bursts also occur in the interictal EEG. About half of the patients with this syndrome also exhibit photosensitivity. Similar findings may be present in other epilepsies associated with myoclonus, although many of these syndromes are associated with diffuse cerebral disease, the effects of which may dominate the EEG picture. Often routine EEG shows no obvious changes in cortical myoclonus, but back averaging (Brown et al 1999) can be helpful to differentiate cortical from non-cortical myoclonus.

Tonic seizures Tonic seizures are typically accompanied by fast low voltage activity, starting in the beta or mid-alpha range, slowing and increasing in amplitude during the attack. These too often occur in the context of the various syndromes of symptomatic generalised epilepsy, with other associated EEG abnormalities.

Tonic–clonic seizures The tonic–clonic seizure is characterised by abrupt onset, possibly a cry as air is expelled through the closed glottis. The patient becomes rigid (tonic phase), falls if standing and may suffer injury. After some 20–60 s during which increasing cyanosis develops, rhythmic jerking appears, affecting all limbs. The jerks slow, become irregular and cease, and the patient may then remain unconscious or confused for a period ranging from seconds to upwards of an hour. EEGs recorded during tonic–clonic seizures are generally obscured by large amounts of muscle artefact. However, characteristically the tonic phase is accompanied by generalised spikes and the clonic phase by SW activity that is synchronous with the jerks and slows and becomes irregular as they do (*Fig. 2.69*). In the postictal state a generalised profound reduction in amplitude of the EEG may be seen, followed by the gradual appearance of slow activity, which may at first be episodic but becomes

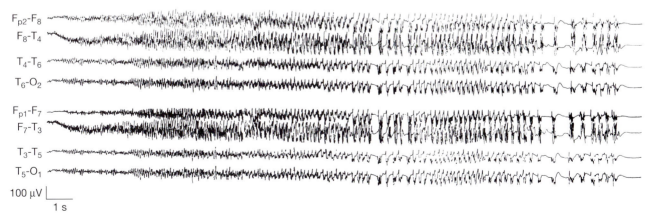

Fig. 2.69 Spontaneous tonic–clonic seizure in a 28-year-old man. Initial crescendo of generalised 10 Hz activity obscured by increasing muscle action potentials of the tonic phase. The clonic phase is associated with bilateral SW discharges and synchronised bursts of muscle potentials. (From Binnie et al (2003), by permission.)

continuous and faster as the normal interictal pattern is restored. Interictal recordings from patients who suffer only tonic–clonic seizures may be normal or characterised by occasional bursts of generalised, multiple spikes and slow waves, particularly during sleep.

Atonic seizures Atonic seizures produce a sudden loss of muscle tone, often without loss of consciousness. The patient may fall and may suffer injury. These seizures occur mainly in the context of syndromes of symptomatic generalised epilepsy, which are characterised by a variety of interictal EEG abnormalities. In pure atonic seizures, without other clinical features such as loss of consciousness, the ictal EEG often shows only an electrodecremental event (Egli et al 1985).

The EEG and epileptic syndromes

Box 2.2 sets out the most recent ILAE classification of epilepsies and epileptic syndromes (Commission on Classification and Terminology of the International League Against Epilepsy 1989). The distinction between seizures, epilepsies and syndromes may appear artificial but serves a useful purpose. The seizure classification provides a convenient means of describing the attacks to which a patient is subject. It involves certain questionable assumptions about pathophysiology, but none concerning aetiology, pathogenesis or prognosis, and is of some value in selecting treatment. The classification of the epilepsies, by contrast, provides a framework for describing the overall disease from which the patient suffers. It involves two dichotomies, between a generalised and localised (partial/focal) disorder of brain function, and between idiopathic (formerly 'primary') epilepsies without apparent structural cause and symptomatic (or 'secondary') epilepsies due to cerebral pathology. Unfortunately, in the 1989 Classification of the Epilepsies, localised dysfunction is not termed 'partial', in conformity with the seizure classification, nor even 'focal', which might be more appropriate to emphasise the pathophysiological basis, but rather as 'localisation related' (this term is not used in this text). The latest classification allows an intermediate category on the idiopathic–symptomatic dimension, 'cryptogenic', to accommodate presumed symptomatic epilepsies of unknown aetiology, thus avoiding the absurdity of a Lennox–Gastaut syndrome in a multiply handicapped patient being classified as 'idiopathic' if the cause is unknown. Since the 1989 classification further genetic epilepsies have been described and subsequently validated by gene localisation and sequencing, originally, and now paradoxically many of these syndromes would have been described as idiopathic. A further development of this classification will have to take this into account. The classification of syndromes is a useful addition, because within the very broad structure provided by the classification of the epilepsies there is a need to distinguish many distinctive seizure disorders with character-

istic aetiology, age of onset and prognosis, with or without associated clinical features such as dementia or learning difficulties. Most of the distinctive epilepsy syndromes are age dependent and are described in detail on pages 230–270.

Both seizure and syndrome classifications have attracted criticism; some of the objections are set out above. A task force set up by the ILAE (Engel 2001) has, in effect, proposed replacing rigorous classification with a multidimensional description of each patient's condition. Whether this will prove feasible is uncertain; it could create a new, important role for the EEG. Considering the EEG in relation to the traditional, general classification of the epilepsies, idiopathic generalised epilepsy (formerly 'primary generalised epilepsy'), occurring in a structurally normal brain, will not generally be associated with abnormalities of ongoing activity. The interictal and ictal EEG findings are those characteristic of the main generalised seizure types, absences, tonic–clonic and myoclonic seizures, i.e. generalised SW activity of various types.

By contrast, symptomatic generalised epilepsies (formerly 'secondary generalised epilepsies') involve, by definition, diffuse cerebral pathology or multiple lesions, which are likely to be reflected in abnormalities of the ongoing EEG. Interictal discharges commonly occur at multiple foci together with generalised discharges, some of which may be secondarily generalised after a focal onset (*Fig. 2.70*). The similarity of the terms, 'secondarily generalised discharges' and 'secondary generalised epilepsy' is fortuitous and caused much confusion. The finding of secondary generalisation was supposed to support a diagnosis of secondary generalised epilepsy – wrongly so, as secondarily generalised discharges and seizures often occur in partial epilepsy. The substitution in the 1989 ILAE classification of the term 'symptomatic' for 'secondary' should have overcome this problem. Cryptogenic generalised epilepsy, being largely a classification by default, is applied mainly to patients with symptomatic generalised epilepsy in which the underlying pathology has not been determined and the EEG findings are those of the symptomatic group. However, the use of this category has also served to highlight the existence of idiopathic forms of some conditions, such as West's syndrome, which were presumed to be invariably symptomatic. Thus some patients with epilepsies provisionally classified as cryptogenic are found to show an unexpected lack of EEG abnormality, particularly with respect to ongoing activity, and have a relatively good prognosis. In time EEG criteria may emerge for reclassifying such epilepsies as idiopathic. The idiopathic type of partial epilepsy is represented by the benign epilepsies of childhood, the most common form of which (benign childhood epilepsy with centrotemporal spikes) is characterised by high amplitude, sharply focal spikes or SW complexes in the central or centrotemporal region of one or both hemispheres ('Rolandic spikes') (*Fig. 2.71*). The

Box 2.2 International Classification of Epilepsies and Epileptic Syndromes*

1. Localisation-related (focal, local, partial) epilepsies and syndromes

1.1 Idiopathic (with age-related onset) – at present, the following syndromes are established but more may be identified in the future:
- Benign childhood epilepsy with centro-temporal spike
- Childhood epilepsy with occipital paroxysms
- Primary reading epilepsy

1.2 Symptomatic:
- Chronic progressive epilepsia partialis continua of childhood (Kojewnikow's syndrome)
- Syndromes characterised by seizures with specific modes of precipitation
- Temporal lobe epilepsies
- Frontal lobe epilepsies:
  - supplementary motor seizures
  - cingulate
  - anterior frontopolar region
  - orbitofrontal
  - dorsolateral
  - opercular
  - motor cortex
  - Kojewnikow's syndrome
- Parietal lobe epilepsies
- Occipital lobe epilepsies

1.3 Cryptogenic
Cryptogenic epilepsies are presumed to be symptomatic and the aetiology is unknown. This category thus differs from the previous one by the lack of aetiological evidence

2. Generalised epilepsies and syndromes

2.1 Idiopathic (with age-related onset – listed in order of age):
- Benign neonatal familial convulsions
- Benign neonatal convulsions
- Benign myoclonic epilepsy in infancy
- Childhood absence epilepsy (pyknolepsy)
- Juvenile absence epilepsy
- Juvenile myoclonic epilepsy (impulsive petit mal)
- Epilepsy with grand mal (GTCS) seizures on awakening
- Other generalised idiopathic epilepsies not defined above
- Epilepsies with seizures precipitated by specific modes of activation

2.2 Cryptogenic or symptomatic (in order of age):
- West syndrome (infantile spasms, Blitz–Nick–Salaam Krämpfe)
- Lennox–Gastaut syndrome
- Epilepsy with myoclonic–astatic seizures
- Epilepsy with myoclonic absences

2.3 Symptomatic
2.3.1 Non-specific aetiology:
- Early myoclonic encephalopathy
- Early infantile epileptic encephalopathy with suppression burst
- Other symptomatic generalised epilepsies not defined above
2.3.2 Specific syndromes:
- Epileptic seizure may complicate many disease states. Under this heading are included diseases in which seizures are a presenting or predominant feature

3. Epilepsies and syndromes undetermined whether focal or generalised

3.1 With both generalised and focal seizures:
- Neonatal seizures
- Severe myoclonic epilepsy in infancy
- Epilepsy with continuous spike waves during slow wave sleep
- Acquired epileptic aphasia (Landau–Kleffner syndrome)
- Other undetermined epilepsies not defined above

3.2 Without unequivocal generalised or focal features. All cases with generalised tonic–clonic seizures in which clinical and EEG findings do not permit classification as clearly generalised or localisation related, such as in many cases of sleep–grand mal (GTCS), are considered not to have unequivocal generalised or focal features

4. Special syndromes

4.1 Situation-related seizures (Gelegenheitsanfälle):
- Febrile convulsions
- Isolated seizures or isolated status epilepticus
- Seizures occurring only when there is an acute metabolic or toxic event due to factors such as alcohol, drugs, eclampsia or non-ketotic hyperglycaemia

*Reprinted by permission of the Commission on Classification and Terminology of the International League Against Epilepsy (1989). GTCS, generalised tonic–clonic seizures.

children typically suffer, usually infrequent, nocturnal partial seizures starting in the mouth or face. Less common variants include childhood epilepsy with occipital paroxysms, characterised as the name suggests by posterior spike wave discharges when the eyes are closed. These and other variants are described on page 245.

Finally, partial epilepsy or, in the terminology of the ILAE's current classification, symptomatic localisation-related epilepsy, is caused by a localised structural abnormality of the brain producing partial seizures, the interictal and ictal EEG features of which have been described at some length above. In addition to abnormal discharges, there may be abnormalities of ongoing activity reflecting the underlying pathology, ipsilateral reduction of spontaneous and drug-induced beta activity, or focal slowing in the theta or delta ranges. Many of the epileptic syndromes have characteristic EEG findings that often relate to the total cerebral disorder, of which epilepsy is not necessarily the most important feature. Most are age dependent, appearing in infancy or childhood, and are considered on pages 230–270.

Diagnostic EEG investigation of epilepsy

EEG investigation of patients with suspected epilepsy serves two main purposes: to support or otherwise the clinical diagnosis of epilepsy and to aid syndromic classification. The EEG may provide a range of information relating to cerebral pathology, anticonvulsant intoxication, etc., but the primary concern is to demonstrate epileptiform activity supportive of epilepsy. Spontaneous seizures do not often occur during a 30 min recording in the laboratory, and the contribution of the EEG to the investigation of epilepsy must normally depend upon interictal records. This is a subject of some misunderstanding. Discussions of the sensitivity of diagnostic EEG investigation tend to focus on the yield of epileptiform activity from a single EEG in the awake state, commonly estimated at 50%, but ignore the intrasubject variability of the findings and the effects of sleep. These together demand a rather more sophisticated diagnostic strategy. Conversely, prevailing views about specificity are also uninformed, as they fail to distinguish the prevalence of spiky waveforms in healthy subjects from that of

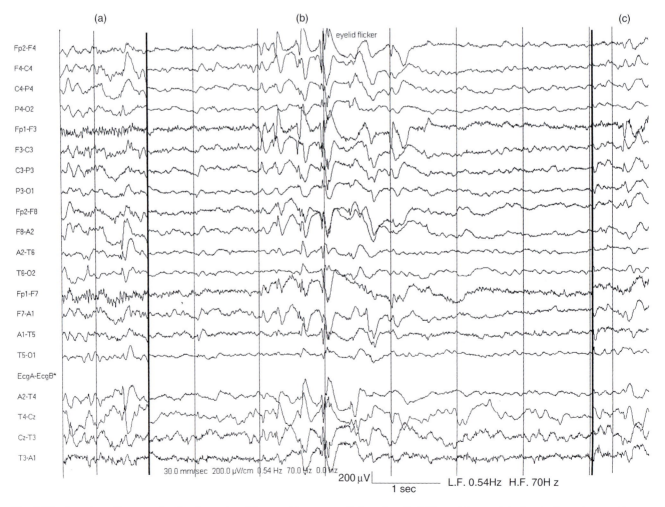

Fig. 2.70   Symptomatic generalised epilepsy in a 34-year-old woman with learning disability and frequent tonic–clonic and several types of complex partial seizures. Note slowed background and irregular generalised discharges (b); also multifocal discharges in the right midtemporal to posterior temporal region (a) and left frontal region (c). (From Binnie et al (2003), by permission.)

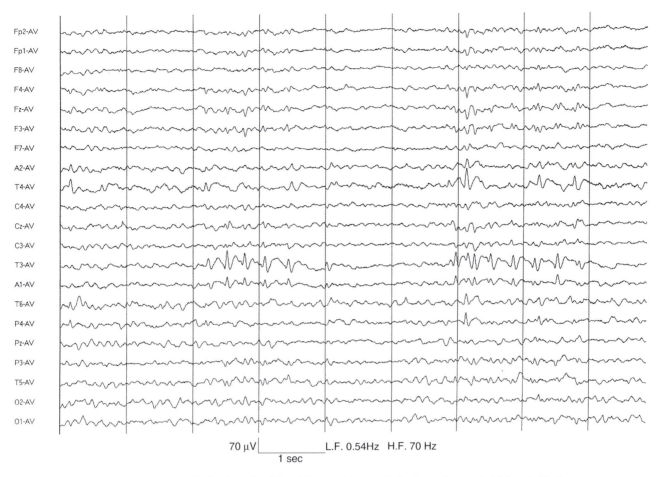

| Fp2-AV |
| Fp1-AV |
| F8-AV |
| F4-AV |
| Fz-AV |
| F3-AV |
| F7-AV |
| A2-AV |
| T4-AV |
| C4-AV |
| Cz-AV |
| C3-AV |
| T3-AV |
| A1-AV |
| T6-AV |
| P4-AV |
| Pz-AV |
| P3-AV |
| T5-AV |
| O2-AV |
| O1-AV |

70 µV   L.F. 0.54Hz   H.F. 70 Hz

1 sec

Fig. 2.71 Centrotemporal sharp waves ('Rolandic spikes') in a 10-year-old boy with benign epilepsy of childhood. Discharges occur independently over both hemispheres, maximally at the sylvian electrodes. Note that these are usually, but not invariably, accompanied by a midfrontal positive component. (From Binnie et al (2003), by permission.)

various more specific phenomena associated with epilepsy. In this discussion there is also often a failure to distinguish between the findings in neurologically screened volunteers and patients with cerebral disease who do not have seizures.

Diagnostic sensitivity

In a broad cross-section of patients with epilepsy, a routine EEG in the awake state of at least 30 min duration (i.e. actual recording time) and with periods of hyperventilation and photic stimulation, demonstrates interictal epileptiform activity in about 50% (*Table 2.5*). However, on repeated examination it emerges that a third of patients consistently exhibit discharges in every record, some never exhibit epileptiform activity on repeated EEG recording in the awake interictal state, and about half do so on some occasions but not others (Ajmone Marsan and Zivin 1970). Thus, simply repeating the awake EEG increases the yield of positive findings until eventually epileptiform activity will have been demonstrated in some 85% of subjects. With serial EEGs, 92% of the patients with interictal discharges in the awake state will be identified within four recordings (Salinsky et

al 1987). There are, of course, various clinical factors that influence the probability of interictal discharges being found; Sundaram et al (1990) reported the yield of epileptiform activity from a single record to be substantially increased (to 77%) in patients investigated within 2 days after a seizure, and to be much greater (68%) in patients with monthly seizures than in those who had been seizure-free for a year (39%). The yield can be substantially increased, and new abnormalities found, if the EEG is repeated during sleep (see p. 72).

Specificity

The incidence of false-positive EEG findings leading to a possibly incorrect diagnosis of epilepsy is also a subject of widespread misunderstanding. As mentioned on page 16, the EEG phenomena that occur in normal subjects include various spiky waveforms, some common, some rare, which are of little or no significance in relation to the diagnosis of epilepsy. These include 14 and 6/s positive spikes, 6/s SW 'phantom', benign epileptiform transients of sleep, rhythmic midtemporal discharges, subclinical rhythmic epileptiform discharges of adults, and many instances of midline spikes. All are character-

Table 2.5   Diagnostic EEG investigation of epilepsy: percentage yield of epileptiform activity in serial EEGs from 3,000 subjects*

Procedure	Interictal epileptiform activity in waking EEG			Cumulative yield of epileptiform activity
	Always	**Sometimes**	**Never**	
Record wake EEG	33	53	14	
EA present	33	16	0	49
EA absent	0	37	14	
Record sleep EEG		37	14	
EA present	–	27	5	81
EA absent	–	10	9	
Repeat wake and sleep EEG	–	10	9	
EA found	–	10	1	92
Still no EA found	0	0	8	

*Reprinted with permission from Binnie (1996).

ised by a distinctive morphology and occur in particular types of subjects or under particular circumstances. Other normal phenomena that may be wrongly credited with diagnostic significance include various atypical responses to photic stimulation (photoparoxysmal responses grades 1–3 of Waltz et al (1992)) and the normal phenomenon of bifrontal slow activity on hyperventilation in children and young adults. Probably the most common single cause of misreading EEGs is a failure to identify correctly posterior slow waves of youth (see *Fig. 2.33*).

When such phenomena are excluded from the category of epileptiform activity, the incidence of epileptiform discharges in non-patients is found to be low. Moreover the overwhelming majority of adult volunteers exhibiting epileptiform activity, on further enquiry turn out to have a history either of epilepsy or of other significant cerebral disorder. The best evidence concerning the occurrence of epileptiform activity in large series of adults, who were not self-selected and were screened and followed up to exclude current or incipient neurological disorder, is probably that obtained from studies of candidates for pilots' licences. These data suggest the prevalence of epileptiform activity in the narrow sense indicated above in normal adult males to be not more than 4/1,000 (Robin et al 1978, Gregory et al 1993). Specificity is probably lower, and sensitivity higher, in children but a reliable estimate is not available due to a lack of comparable large series that used rigorous definitions of epileptiform activity and included adequate neurological screening and follow-up.

Although epileptiform activity in the EEGs of awake healthy subjects is rare, a different picture emerges in patients requiring diagnostic investigation, as a variety of systemic or cerebral disorders are associated with epileptiform EEG activity in the apparent absence of seizures. In adult patients without previous history of epilepsy, the prevalence of unequivocal epileptiform activity is typically 2–3%. In a series of over 6,000 patients, Zivin and Ajmone Marsan (1968) found that about one-third of those with EEG discharges had primary cerebral disease. The prevalence of epileptiform activity was much greater in certain groups: 6% in those with mental handicap, 10% with congenital or perinatal brain damage, 11% after cranial operations and 8% in patients with cerebral tumours; 14% of those with discharges subsequently developed epilepsy.

Eeg-Olofsson et al (1971) performed EEGs in 743 normal children aged 1–15 years. They found paroxysmal activity in 1.9% of children during wakefulness and in 8.1% during sleep. However, these included normal non-epileptiform paroxysmal activity, such as hypersynchrony during drowsiness. Epileptiform activity during hyperventilation was seen in 0.3% of children.

The diagnostic value of detecting epileptiform EEG activity therefore depends on the clinical context. If a neurologically intact child or teenager with no past history of cerebral disease suffers episodes that could be either panic attacks or partial seizures, the finding of temporal epileptiform activity strongly supports the latter diagnosis. By contrast, if a child with learning disability displays episodes of aggression, the presence of interictal spikes does little to resolve the question of whether or not these attacks are epileptic in nature.

The available evidence on sensitivity and specificity is open to a more positive interpretation than that which is usually presented. Goodin and Aminoff (1984) suggest that in the general population about 4.3% will have epileptiform activity in a single waking EEG (a generous estimate), which includes half of the 0.5% who have active epilepsy. However, in a clinical population of patients with episodic symptoms, 50% of whom actually have epilepsy, 28% exhibit epileptiform activity, of whom 93% have epilepsy. Thus the finding of epileptiform activity increases the a priori probability of a seizure disorder from 0.5 to 0.93, whereas negative findings reduce it only slightly to 0.33. One may speculate about the consequences of performing both wake and sleep recording, assuming that one interprets the results conservatively in accordance with the findings of Beun et al (1998) and that the yield in persons with epilepsy is 80%. This leads to a probability of epilepsy of 0.95 in those who show epileptiform activity in a combined wake and sleep recording, and of only 0.17 in those who do not. This is an outcome greatly at variance with the received wisdom that the interictal EEG is of marginal diagnostic value in epilepsy.

Role of special procedures
long term monitoring

As indicated above, the routine use of the EEG in epilepsy is of limited value except where there is a defined problem, such as uncertainty concerning the diagnosis or difficulties of classification. The fact that in nearly 10% of patients with epilepsy repeated interictal EEG recordings in wakefulness and sleep fail to demonstrate epileptiform activity is generally a matter of no great concern, except where the clinical diagnosis is in doubt.

There is, however, an important group of patients requiring investigation of episodic symptoms of uncertain, possibly epileptic origin. They comprise perhaps 5% of all people with epilepsy, and an equal number of persons whose attacks eventually prove to be non-epileptic. Difficulties arise chiefly from the intermittent nature of the clinical and electrophysiological manifestations of epilepsy, with variable and sometimes inconclusive interictal EEG findings and the fact that most valuable clinical or electrophysiological information can be obtained only during seizures. The solution of these problems lies in prolonged observation of both behaviour and the EEG until sufficient salient events, usually seizures, have been captured. The development of low-cost video equipment has made it possible accurately to document behaviour over long periods. Prolonged EEG recording can, in principle, be carried out using conventional technology. However, the requirement that the patient remains seated or lying down whilst attached to an EEG machine is burdensome to the subject and limits the range of behaviours and situations that can be investigated. For this reason, prolonged EEG recording is generally carried out either by means of telemetry systems, which allow signals to be transmitted from a moving subject over distances of some tens of metres by a radio link or cable, or by a fully portable (ambulatory) recorder carried by the patient.

In addition to its principal application of diagnostic assessment, intensive monitoring is used to identify and determine the frequency of subtle seizures, which may ordinarily be difficult to recognise, to study suspected seizure precipitation by cognitive activities or environmental factors, and in preoperative assessment. Monitor-

ing techniques and their applications are the subject of the companion book (Smith et al 2006).

Preoperative assessment

Contrary to popular belief and indeed the impressions of many physicians, epilepsy is in general a disease with an excellent prognosis. The proportion of newly diagnosed patients becoming seizure-free with optimal medical treatment is some 70%, and many of these will subsequently not need antiepileptic drug treatment in the longer term. The remainder, however, develop chronic epilepsy, and indeed comprise a large proportion of the 5/1,000 population who at any time have more or less uncontrolled seizures. Of these, probably a quarter are potentially treatable by surgery and have a sufficiently severe seizure disorder to justify surgical intervention. There is at this time no country in the world with facilities to meet this demand; consequently, only a small proportion of patients with drug-resistant epilepsy receive surgical treatment.

There are various surgical approaches to different types of epilepsy, but the most widely employed depends on removing tissue that is both structurally and functionally abnormal. The identification of structural abnormality requires sophisticated neuroimaging techniques, whereas the investigation of dysfunction depends on clinical observation, neuropsychological assessment, and particularly electrophysiological data. Beyond the basic clinical and EEG assessment required to establish that a patient has intractable partial epilepsy, preoperative assessment is a complex interdisciplinary activity that should be undertaken only in specialist centres. Well-intentioned attempts to carry out part of the work-up before neurosurgical referral are unhelpful, as the investigations generally need to be repeated in accordance with the established practices of the surgical team. Indeed, negative findings may, wrongly, discourage further referral. Preoperative electrophysiological assessment frequently involves specialised techniques, including telemetry and intracranial recording.

Special problems
Febrile convulsions

Whether febrile convulsions should be regarded as a form of epilepsy is largely a matter of semantics, and it is all too easy, particularly in a research context, to define the condition in such a way that it becomes a diagnosis by hindsight, when it is established at follow-up that no subsequent afebrile seizures have occurred. This approach ignores the very real problem of differential diagnosis in a child with fever who suffers a seizure. In the acute phase the EEG may be of value for detecting acute cerebral disease underlying both fever and convulsion, for instance encephalitis. In the first few days after recovery, epileptiform or other EEG abnormalities are not uncommon and are of no prognostic significance in relation to subsequent development of epilepsy. Later

EEG follow-up is generally unhelpful: in the latent period between the first febrile convulsion and onset of epilepsy, neurologically intact children do not usually show any epileptiform activity. Most clinicians therefore believe that an EEG for a neurologically intact child, who had a febrile seizure and made a rapid recovery, is not indicated.

Psychiatric disorder and epilepsy

EEG investigation is often requested to help determine whether episodic, abnormal (sometimes criminal) behaviour has an epileptic basis. More often than not such referrals reflect a lack of familiarity with the clinical phenomenology of epilepsy. Aggressive behaviour during a seizure is uncommon and directed aggression exceedingly rare. Most aggressive behaviour associated with epilepsy occurs in response to restraint during postictal confusion (and is usually undirected), or during related psychoses, which are more prolonged and recognisable by other features. Similarly, impulsiveness in children rarely has an epileptic basis. On the other hand, the bizarre behaviours in seizures of mesiobasal frontal origin are readily misinterpreted as having a psychiatric basis. In such cases assessment of the interictal EEG may be complicated by the use of psychotropic drugs producing EEG abnormalities, including temporal sharp waves. The issue can often be resolved by capturing an episode of the behaviour in question during video telemetry. By contrast, an abnormal mental state in which the possibility of an epileptic basis is often overlooked is non-convulsive status epilepticus, presenting as mild confusion and slight clumsiness. Here the EEG will be conclusive.

Non-epileptic attacks

A major diagnostic problem is presented by attacks that appear to be epileptic in origin but are not. These may have some other physical basis, for instance systemic disorders, ranging from syncope to pheochromocytoma, primary neurological conditions such as movement disorders, and parasomnias. In many instances, the causes are psychogenic; these include conversion disorders, malingering (rarely) and other episodic psychiatric phenomena such as abreaction or panic attacks. The term 'pseudoseizures' has been used loosely to cover various attacks that may be mistaken for epilepsy. This name can be seen as pejorative or trivialising the problem; moreover, it denies the fact that the diagnosis may be iatrogenic – it is usually the physician, not the patient who misdiagnoses syncope, panic attacks or abreaction as epilepsy. A further source of terminological and diagnostic confusion is that both epileptic and non-epileptic seizures may be 'psychogenic', as psychological factors can precipitate epileptic seizures, non-epileptic organic attacks such as syncope, and acute psychiatric symptoms such as anxiety or abreaction. The term 'non-epileptic seizures' (NES) is to be preferred to pseudoseizures for 'psychogenic' non-epileptic seizures caused by an action of the mind, and 'psychogenic

epileptic seizures' should be clearly specified as such (for a review see Binnie 1994). 'Non-epileptic attack disorder' (NEAD) has gained some currency, but in psychiatric taxonomy 'disorder' implies a syndrome; NEAD is not a unitary syndrome and does not represent a full diagnosis unless the underlying psychiatric condition is specified. NES are more common in adolescence, but are also seen in younger children. People with NES may have abnormal EEGs, and indeed many of them suffer from epilepsy. Comorbid epilepsy diagnosis in children is 20–30%. This last, perhaps surprising, fact often reflects the difficulty that many people with chronic epilepsy experience in coping with a changed lifestyle if their seizures are controlled. The interictal EEG is rarely of value in the diagnostic assessment of suspected NES. In 8% of people with epilepsy, repeated awake and sleep EEG examinations fail to demonstrate epileptiform activity, which may lead to the diagnosis of epilepsy being mistakenly doubted. Conversely, the presence of epileptiform activity or other EEG abnormalities may encourage a diagnosis of epilepsy, which would not otherwise have been entertained. This may be a chance finding unrelated to the episodic symptoms in question, may be due to some other organic condition such as hypoxia or hypoglycaemia, or indeed may sometimes be wrongly reported due to misinterpretation of normal episodic EEG phenomena. It should particularly be noted that some 2% of psychiatric patients and at least 10% of those with learning disorders exhibit EEG discharges in the absence of epilepsy; some of these patients may have episodic psychiatric symptoms, which are liable to be misinterpreted on account of the EEG findings. Generally the differential diagnosis requires the capture of some of the attacks by telemetric and video EEG monitoring, but even this does not always provide a solution.

Status epilepticus

Where obvious repeated seizures occur, the diagnosis of status epilepticus rarely presents a problem. However, non-convulsive status can be present in many forms, ranging from loss of speech to a confused or apparently psychotic state possibly associated with slight clumsiness (see *Fig. 2.72*). The EEG shows continuous epileptiform activity, either focal or generalised depending upon the type of status, and may be crucial to establishing the diagnosis.

Where uncertainty remains as to whether the epileptiform discharges are an incidental finding or indicate that status is the cause of the patient's clinical state, the response to intravenous antiepileptic medication (usually a benzodiazepine) may be conclusive. Rapid clinical improvement following suppression of the discharges confirms the diagnosis. However, it should be noted that, in addition to producing the other complications of benzodiazepines, diazepam can convert non-convulsive to convulsive status (Prior et al 1972). In convulsive status, whether generalised or partial, the EEG will show

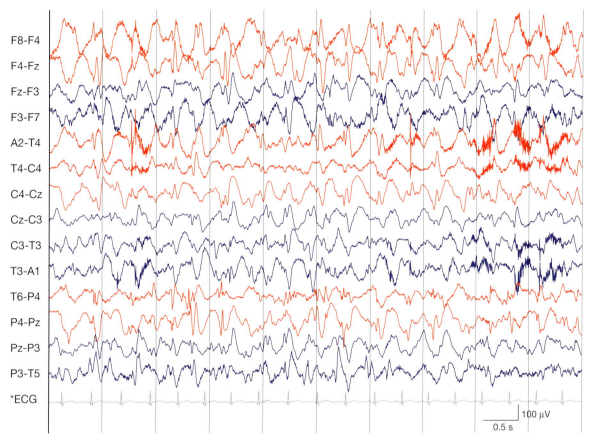

Fig. 2.72 A 2-year-old boy with multiple seizures and a recent increase of seizure frequency. For the preceding few days he was less aware and less active, but otherwise there was no change. The EEG shows generalised, continuous, irregular spike-and-wave discharges (non-convulsive status).

episodic changes relating to the seizures. In the early stages these may be easy enough to recognise and correspond to the ictal patterns appropriate to the seizure type in question. Between attacks the EEG will generally remain slowed or of low amplitude, reflecting the incomplete recovery of consciousness between attacks. However, during prolonged generalised convulsive status epilepticus in man, Treiman et al (1990) describe a progressive evolution in which five stages may be distinguished:

1. discrete seizures with interictal slowing
2. confluent seizures reflected in continuous but waxing and waning ictal discharges
3. continuous rhythmic sharp or SW discharges
4. continuous discharges interrupted by periods of flattening lasting 0.5–8 s
5. periodic stereotyped sharp waves against a relatively flat background.

Stage IV may present problems of interpretation in a patient with seizures in the course of an acute cerebral illness, as there may be uncertainty whether the picture represents continuous discharges with episodic flattening, or burst suppression related to the underlying condition, or is due to medication. These studies were necessarily limited by clinical constraints, but a closely similar progression was found in untreated rats. During this evolution the clinical ictal events become less florid and may eventually be unrecognisable without the help of EEG monitoring. Asymmetries during partial status epilepticus are potentially misleading: ictal discharges may occur only over the side contralateral to the original causative focus, which has presumably become incapable, due to local anoxia or other acute changes, of producing epileptiform discharges recognisable in the surface EEG. Patients with epilepsia partialis continua may show appropriately localised epileptiform discharges, but often show only focal slowing or no localised abnormality. An acute presentation of unconsciousness with the EEG pattern described as stage 4 above may present a difficult problem of differential diagnosis, as a similar picture may be seen in encephalitis. Continuous EEG recording in status epilepticus provides a valuable means of monitoring the effects of intravenous antiepileptic drugs. Indeed, when it is necessary to anaesthetise the patient and administer a muscle relaxant during artificial ventilation, the EEG provides the only reliable guide as to whether seizures have been controlled or are continuing.

EFFECTS OF DRUGS ON THE EEG

Types of drug effect

Drug-induced EEG changes are usually diffuse, and symmetrical in the normal brain, but epileptiform activity in particular may sometimes be focal. The range of common types of drug effect distinguishable by visual assessment is small, and they are generally non-specific (*Table 2.6*).

Increased beta activity is often seen with sedative drugs, hypnotics (barbiturates in particular) and antiepileptics. The fast activity is typically in the 18–25 Hz range with a bifrontal emphasis. Benzodiazepines, however, often induce slow beta activity with the characteristics of a fast alpha variant. Slowing of ongoing activity with increased theta components is common with overt sedation, and delta activity appears with frank intoxication. A bitemporal increase of theta activity is seen in particular with various psychotropic agents, notably antipsychotics and thymoleptics. More marked and generalised slowing may be seen with therapeutic doses of clozapine. Among the antiepileptics, carbamazepine is noted for producing slowing of ongoing activity into the theta range, even with good therapeutic response and without obvious sedation.

Epileptiform activity may be induced, ranging from equivocal sharp bitemporal theta activity (phenothiazines, butyrophenones) to frank generalised SW discharges (lithium, clozapine). Triphasic transients, burst-suppression patterns and, eventually, an isoelectric EEG may be seen in gross intoxication.

Interpretative problems due to drugs

Drug effects may complicate EEG interpretation in various ways. Abnormalities of diagnostic value may be abolished, notably in epilepsy. Valproate, benzodiazepines, lamotrigine, levetiracetam and ethosuximide suppress generalised SW discharges and photosensitivity; lamotrigine reduces epileptiform activity in most forms of epilepsy. Few other established antiepileptics greatly reduce EEG abnormality, and any apparent suppression of epileptiform activity due to effective treatment is more probably related to an increased interval between the EEG recording and the most recent seizure. Effective treatment of systemic disorders producing EEG abnormalities, e.g. hormone replacement therapy, will generally have a normalising effect. Drug withdrawal, again most notably of antiepileptics or sedatives with an incidental anticonvulsant action, can induce abnormalities, both spontaneous epileptiform discharges and photosensitivity in normal subjects. The foregoing are important considerations when the EEG proves negative in a patient in whom antiepileptic treatment has been started before the diagnosis was established. If it is decided for purposes of diagnostic EEG investigation to withdraw a drug which it is thought may be suppressing discharges, it may be necessary to delay repeating the EEG until some weeks after withdrawal, to avoid the possible misinterpretation of acute withdrawal effects as spontaneous epileptiform activity. Particular problems arise if abruptly stopping antiepileptic drugs induces clear discharges or an undoubted seizure, in a patient with suspected non-epileptic seizures. Once this has happened

Table 2.6   Summary of EEG effects of drugs in doses short of gross intoxication

Drug class	Increased beta activity	Slowed alpha rhythm	Excess theta activity	Temporal sharp waves	Epileptiform discharges
Barbiturates	+++ (25–35 Hz)	+			On withdrawal
Benzodiazepines	+++ (15–20 Hz)				On withdrawal
Neuroleptics		+	++	+	
Lithium		++	+++	++	+
Thymoleptics	++	++	++	+++	+/–
Phenytoin	+	+	+		
Carbamazepine		+	++		Possibly +
Bromides		++	+++		
Opiates		+			
Cocaine	++				+/–
Ethanol		+			On withdrawal
Solvents		++	+++		

it becomes difficult convincingly to sustain the claim that the patient does not have epilepsy. Routinely stopping antiepileptic drugs a few days before EEG investigation of possible epilepsy cannot therefore be recommended. An exception, however, is the use of drug withdrawal to assist the capture of seizures during EEG telemetry where the diagnosis of epilepsy is not in doubt.

By the same token, withdrawing any substance with an anticonvulsant action may induce spontaneous epileptiform activity or photosensitivity in normal people. Epileptiform discharges may for instance be seen, and possibly misinterpreted, in barbiturate abusers whose supplies are interrupted on admission to hospital, and during recovery after acute benzodiazepine overdose or extreme alcoholic intoxication. The slowing and particularly the equivocal, or sometimes frank, epileptiform activity induced by many psychotropic drugs are a recurring source of diagnostic difficulty. Typically, in a patient on treatment for a psychosis, the suggestion is made either that some form of episodic abnormal behaviour or experience may be epileptic in nature, or that atypical symptomatology suggests a possible, and unspecified, organic condition. Minor abnormalities may be considered to support these possibilities, and it is then usually necessary to repeat the investigation 2 or 3 weeks after drug withdrawal to determine whether the abnormalities persist.

Clinical significance of drug-induced EEG changes

The presence of drug-induced EEG changes is not invariably unhelpful. The use of drug activation and suppression tests in epilepsy was noted on page 75. Intoxication with various drugs may be first suspected on account of EEG changes. Examples include recognition of oversedation due to the synergy of two antiepileptics neither of which is individually at a level considered 'toxic', notably the combination of valproate with phenobarbital. EEG findings may raise a suspicion of substance abuse, or of the Munchausen syndrome by proxy. The potential epileptogenicity of many psychotropic drugs is often given more emphasis in the manufacturer's information inserts than is warranted in practice. Nevertheless, while epileptiform EEG activity may be misleading in psychiatric patients who do not have epilepsy, its presence may be important in identifying patients who do suffer drug-induced seizures, a striking recent example being clozapine, which often induces generalised SW activity, and indeed seizures.

GENERAL PHILOSOPHY OF CLINICAL EEG INVESTIGATION AND INTERPRETATION

Uses and abuses of the EEG

Amongst both users and providers of EEG services, there is little agreement concerning the clinical utility and scientific validity of this method of investigating brain function. The EEG is undoubtedly generated by brain activity but, within certain limits of frequency, amplitude, etc., appears random. It may be regarded by optimists as an encoded signal of great complexity that will eventually be deciphered, or by sceptics as noise from which there is little information to be extracted. These differing viewpoints are reflected in the ways the EEG is used.

Few laboratory investigations rely to the same extent as does the EEG on human interpretation to derive a clinical conclusion from data, which bear no simple relationship to anatomy or known pathophysiological mechanisms. This encourages subjectivity, the use of jargon and overinterpretation by those providing the service and, on the part of the users, attitudes ranging from uncritical trust, through scepticism, to frank hostility. The clinical misuse of the EEG can take two extreme forms.

Some users overvalue the investigation, give insufficient weight to the total clinical context, and attach excessive importance to negative results. This leads to requests for an EEG 'to exclude epilepsy', to psychiatric disorders being classified as 'organic' on the basis of minimal, possibly drug-induced EEG anomalies, and to medication being prescribed with the intention of correcting EEG abnormalities with scant regard to the clinical picture. Such misuse is particularly prevalent in paediatric practice because of the pressure on the clinician to 'do something', however hopeless.

The opposite extreme is to view the EEG as a rarely useful, but harmless and inexpensive test, to be used routinely for screening and follow-up of patients with possible cerebral disease. This attitude is self-perpetuating: as inappropriate EEG investigations elicit reports of minimal clinical value, the user sees no reason to make more discriminating use of the service.

Electroencephalographers are no less culpable than their clients. They may try to protect the reputation of their discipline by avoiding the acknowledgement of its limitations. In an attempt to appear helpful or to expand their departments, they may encourage unnecessary investigations and excessive workload at the expense of quality.

A balanced view is that the EEG demonstrably makes a unique contribution to the solution of certain specific clinical problems, and should be used selectively for those applications in which it is of proven value. It is then worth recording and interpreting to the limits of the current state of the art. Just as the attitudes leading to misuse are self-validating, the converse also applies. In centres where the users and providers of the service share an understanding of the role of the EEG in problem-solving, thoughtful, selective referrals, often for complex and technically exacting investigations, prove rewarding to all concerned and benefit patients. There is a huge difference in clinical practice and job satisfaction between the electroencephalographer who is a key player in a neonatal ITU, or the leader of an epilepsy surgery programme, and his unfortunate colleague reporting non-specifically abnormal records for applications in which they are unlikely to be useful.

The point is well illustrated by the use of the EEG in epilepsy. The cost of 48 h telemetry is equivalent to that of some eight standard EEGs performed for 'routine' follow-up. Evaluation studies have shown the yield of findings resulting in changes of management to be 65–80% from telemetry addressing specific questions, and about 3% from routine follow-up (Binnie 1990). Similarly, one session of intraoperative EEG monitoring, which may help to prevent death or catastrophic disability, uses no more resources than three routine EEGs in patients with behaviour disorders, with a negligible chance of clinically relevant findings.

Normality

In considering any phenomenon that is as variable as the EEG and generated by mechanisms that are so poorly understood, concepts of normality must be statistical and pragmatic. Normality is largely a statistical concept: findings which are not uncommon in health (such as 14 and 6/s positive spikes), or which are rare but show no association with disease (e.g. rhythmic midtemporal discharges) must be regarded as 'normal'. Conversely, changes seen in serial recordings in the course of a cerebral disease may be pathological for the individual concerned but not outside the range of common findings in a normal population. The issue of normality is further complicated by a lack of rigorously screened non-patient populations for normative studies. The best available data on the EEG in health are from studies of aircrew and astronauts, both of which are highly selected adult populations. There have been many studies of particular quantitative EEG features in children, but for data on the visual analysis used in clinical practice the most often cited source appears to be the work by Eeg-Olofsson et al (1971a, b), over 30 years ago.

The concepts of 'normality' and 'abnormality' can be used in two ways. The first is to describe as normal that which is common, i.e. of high incidence, in a healthy population. That which is unusual is then viewed as 'abnormal'. The second, more practical approach is to identify those findings that, in a particular context, are or are not of clinical significance. From this standpoint, a feature that is rare of no known significance may be designated normal. When observations are readily quantifiable, such as in EMG and EPs, the former, statistical approach can be supported by observational data. An EP component with a latency 3 standard deviations (SD) above the mean of a control population may in this sense properly be called 'abnormal'. This does not necessarily address the issue of clinical significance: a value more than 3 SD from the mean will be found in a small, quantifiable proportion of a healthy population, and if extreme values have no known clinical correlates, it may be unhelpful to call them 'abnormal'.

Greater difficulty arises with features that are not readily quantified. There are few reliable, quantitative normative data on visual EEG analysis. The electro-encephalographer relies on experience and subjective judgements to recognise, for instance, an excess of posterior slow activity at a particular age, or to distinguish normal temporal theta components in drowsiness from pathological sharp waves. However, the variability of the EEG within and between subjects dictates that such experience can be gained only by the assessment of many normal records. Few clinical neurophysiologists or technologists have had the opportunity to review large series of records from neurologically screened non-patients. Most base their notions of normality on an oral tradition handed down by their teachers, supplemented by the necessarily inadequate descriptions and scaled down single-page illustrations in textbooks. Those features that, in a clinical practice based entirely on a patient population, are reported as normal then become self-validating criteria of normality. The problems are especially serious in paediatric practice, as the 'normal' findings to be learnt change rapidly over the first few years of life.

It is also important to appreciate that the EEG is continually changing and a given individual does not have a single constant EEG picture. There is an important difference between such statements as '50% of routine, waking interictal EEGs from people with epilepsy are normal', which is true but reflects a sampling problem, and '50% of people with epilepsy have normal EEGs', which is proved false by repeated, more prolonged recording. A further general consideration is that the EEG reflects cerebral function; structural abnormality is manifest only in so far as it causes functional changes. The ways in which the EEG is known to change in disease are limited and only rudimentary knowledge exists of their pathological correlates.

It should also be recognised that the detection of abnormality is not the only objective of EEG interpretation. In a particular clinical context, the finding of a normal EEG may be of considerable diagnostic value. For instance, a normal EEG in the presence of marked papilloedema lends support to a diagnosis of benign intracranial hypertension, whereas a normal record in a disorientated patient excludes a toxic confusional state or non-convulsive status epilepticus as the underlying cause. In serial records of an individual, changes that remain within the normal limits for a healthy population may be clinically significant.

Abnormality

EEG abnormalities show little specificity with respect to basic pathological phenomena, such as cerebral oedema, raised intracranial pressure or neuronal loss, and even less in relation to diagnostic categories. Clinical context may, however, so narrow the possible explanations of the findings as to permit diagnostically specific interpretation.

For instance, in a confused patient the EEG may establish the diagnosis of non-convulsive status epilepticus, offer a high probability of a metabolic disorder and some suggestion concerning its nature, or suggest encephalitis. Similarly, in a patient with an acute presentation of hemiparesis, a focal slow wave abnormality in one mid-temporal region, showing rapid resolution over 10 days before the appearance of computerised tomography (CT) scan density changes suggests an infarct rather than other types of lesion. In subacute encephalopathies, two almost specific EEG patterns are known, characteristic of subacute sclerosing panencephalitis and Creutzfeldt–Jakob disease, in children and adults respectively. However, such specific EEG findings are very few.

Evolution

On grounds of cost and safety, the EEG may be more suited to repeated use than some other investigations of the brain. Serial EEGs may be valuable for following the progression or recovery of cerebral disease, and often permit early prediction of eventual outcome, for instance after stroke, head injury or cerebral hypoxia. This type of application is often not appreciated by users: thus a diagnostic record may (to no very useful purpose) be requested in the acute phase of an obvious cerebrovascular accident, but the use of serial records to predict outcome may be neglected.

Referral policy

The need for a coherent referral policy should be apparent from the preceding sections. Because of the considerable interpretative element in clinical EEG assessment there is a close relationship between the utility of the investigation and the quality of the referral. A large proportion of the workload of many departments concerns questions that cannot be answered by EEG or which are posed in such a form as to be unlikely to elicit a clinically relevant response. Failure then of the investigation to yield relevant results tends to maintain the pattern, promoting trivial referrals and squandering resources that could have been used for the investigation of clearly formulated problems to which EEG can make a valuable contribution.

The first step towards implementing an effective EEG referral policy is for the providers or the users of the service to ensure that the reason for every request is explicitly stated. Such remarks as: 'Developmental delay – EEG please' do not meet this requirement. Nor does: 'Two afebrile convulsions 3 years ago, none since on Tegretol but poor school performance, previous EEG normal, please repeat'. Although more informative, and suggesting several pertinent questions that could be asked, it does not clearly indicate the purpose of the request. By contrast, 'Probable benign childhood epilepsy, seizure-free but educational problems – sleep record to exclude ESES' is clear and ensures problem-orientated investigation and interpretation. If the reasons

for referrals can be established, there is a possibility of dialogue, creating an awareness of which questions can be answered and which cannot, and explaining the indications for further investigations, sleep recording, serial records, telemetry, etc. If the most valuable applications of EEG investigation are demonstrated in practice, it should be possible to persuade any reasonable user that resources will be better applied to addressing properly those questions that can be answered, than to performing indiscriminate routine investigations.

Clinical neurophysiologists with a major commitment to a hospital should attend ward rounds or case conferences, hold regular meetings to review EEGs with the principal users of the service, and participate in any similar meetings organised by such departments as neuroradiology or neuropathology. Technologists should also have the opportunity to attend such meetings. This helps to provide feedback concerning diagnostic successes and failures, which is essential for quality control and improving the service. Demonstrating actual records to users gives them a better understanding and, it is to be hoped, greater confidence in the evidence on which EEG reports are based, and encourages more thoughtful use of the service.

REFERENCES

Aarts JHP, Binnie CD, Smith AM, et al 1984 Selective cognitive impairment during focal and generalised epileptiform EEG activity. Brain 107: 293–308.

Achenbach-Ng J, Siao TC, Mavroudakis N, et al 1994 Effects of routine hyperventilation on PCO_2 and PO_2 in normal subjects: implications for EEG interpretations. J Clin Neurophysiol 11: 220–225.

Adachi N, Alarcón G, Binnie CD, et al 1998 Predictive value of interictal epileptiform discharges during non-REM sleep on scalp EEG recordings for the lateralization of epileptogenesis. Epilepsia 39: 628–632.

Adams A 1959 Studies on the flat electroencephalogram in man. Electroencephalogr Clin Neurophysiol (Suppl 11): 35–41.

Aird RB, Masland RL, Woodbury DM 1989 Hypothesis: the classification of seizures according to systems of the CNS. Epilepsy Res 3: 77–81.

Ajmone Marsan C, Zivin LS 1970 Factors related to the occurrence of typical paroxysmal abnormalities in the EEG records of epileptic patients. Epilepsia 11: 361–381.

Alarcón G, Binnie CD, Elwes RDC, et al 1994a Power spectrum and seizure onset in partial epilepsy. Electroencephalogr Clin Neurophysiol 94: 326–337.

Alarcón G, Guy CN, Binnie CD, et al 1994b Intracerebral propagation of interictal activity in partial epilepsy: implication for source localisation. J Neurol Neurosurg Psychiatry 57: 435–449.

Alarcón G, Elwes RDC, Polkey CE, et al 1997 Ictal oroalimentary automatisms with preserved consciousness. J Neurol Neurosurg Psychiatry 62: 205–206.

Aldrich MS 1999 Sleep Medicine. Oxford University Press, New York.

American Electroencephalographic Society 1994a Guidelines in electroencephalography, evoked potentials and polysomnography. Report of the Committee on Infectious Diseases. J Clin Neurophysiol 11: 128–132.

American Electroencephalographic Society 1994b Guideline thirteen: guidelines for standard electrode position nomenclature. J Clin Neurophysiol 11: 111–113.

Andermann F, Oguni H 1990 Do epileptic foci in children migrate? The pros. Electroencephalogr Clin Neurophysiol 76: 96–99.

Artieda J, Obeso JA 1993 The pathophysiology and pharmacology of photic cortical reflex myoclonus. Ann Neurol 34: 175–184.

Avoli M, Kostopoulos G 1982 Participation of corticothalamic cells in penicillin induced spike and wave discharges. Brain Res 247: 159–163.

Avoli M, Gloor P, Kostopoulos G, et al 1983 An analysis of penicillin-induced generalised spike-and-wave discharges using simultaneous recording of cortical and thalamic single units. J Neurophysiol 50: 819–837.

Babb TL, Wilson CL, Isokawa-Akesson M 1987 Firing patterns of human limbic neurons during stereoencephalography (SEEG) and clinical temporal lobe seizures. Electroencephalogr Clin Neurophysiol 66: 467–482.

Barlow JS 1985 A general-purpose automatic multichannel electronic switch for EEG artefact elimination. Electroencephalogr Clin Neurophysiol 60: 174–176.

Barlow JS 1986 Artefact processing (rejection and minimization) in EEG data processing. In: Clinical Applications of Computer Analysis of EEG and Other Neurophysiological Signals, revised series, Vol. 2: Handbook of Electroencephalography and Clinical Neurophysiology (eds FH Lopes da Silva, W Storm van Leeuwen, A Rémond). Elsevier, Amsterdam, pp. 15–62.

Barry W, Jones BM 1965 Influence of eye lid movement upon electro-oculographic recording of vertical eye movements. Aerospace Med 36: 855–858.

Beaussart M, Guieu JD 1977 Artefacts. In: Clinical EEG, I. Part A, Handbook of Electroencephalography and Clinical Neurophysiology, Vol. 11 (ed WA Cobb). Elsevier, Amsterdam, pp. 80–96.

Bengzon ARA, Rasmussen T, Gloor P, et al 1968 Prognostic factors in the surgical treatment of temporal lobe epileptics. Neurology 18: 717–731.

Berkovic SF, Andermann F, Andermann E, et al 1987 Concepts of absence epilepsies: discrete syndromes or biological continuum? Neurology 37: 993–1000.

Beun AM, Van Emde Boas W, Dekker E 1998 Sharp transients in the sleep EEG of healthy adults: a possible pitfall in the diagnostic assessment of seizure disorders. Electroencephalogr Clin Neurophysiol 106: 44–51.

Bickford RG, Sem-Jacobsen CW, White PT, et al 1952 Some observations on the mechanism of photic and photo-metrazol activation. Electroencephalogr Clin Neurophysiol 4: 275–282.

Binnie CD 1990 EEG audit: increasing cost efficiency of EEG investigations in epilepsy. Electroencephalogr Clin Neurophysiol 76: 29P.

Binnie CD 1993 Techniques of visual stimulation in the EEG laboratory. In: Textbook of Epilepsy, 4th edn (eds J Laidlaw, A Richens, D Chadwick). Churchill Livingstone, Edinburgh, pp. 277–278.

Binnie CD 1994 Nonepileptic attack disorder. Postgrad Med J 70: 1–4.

Binnie CD 1996 Epilepsy in adults: diagnostic EEG investigation. In: Recent Advances in Clinical Neurophysiology (eds J Kimura, H Shibasaki). Elsevier, Amsterdam, pp. 217–222.

Binnie CD 2003 Cognitive impairment during epileptiform discharges: is it ever justifiable to treat the EEG? Lancet Neurol 2: 725–730.

Binnie CD, Stefan H 1999 Modern electroencephalography: its role in epilepsy management. Clin Neurophysiol 110: 1671–1697.

Binnie CD, Prior PF, Lloyd DSL, et al 1970 Electroencephalographic prediction of fatal anoxic brain damage after resuscitation from cardiac arrest. BMJ 4: 265–268.

Binnie CD, Wilkins AJ, De Korte RA 1981 Interhemispheric differences in photosensitive epilepsy. II. Intermittent photic stimulation. Electroencephalogr Clin Neurophysiol 52: 469–472.

Binnie CD, Marston D, Polkey CE, et al 1989 Distribution of temporal spikes in relation to the sphenoidal electrode. Electroencephalogr Clin Neurophysiol 73: 403–409.

Binnie CD, Elwes RDC, Polkey CE et al 1994 Utility of stereoelectroencephalography in preoperative assessment of temporal lobe epilepsy. J Neurol Neurosurg Psychiatry 57: 58–65.

Binnie CD, Cooper R, Mauguière F, et al 2003 Clinical Neurophysiology, Vol. 2. Elsevier, Amsterdam.

Birbaumer N, Elbert T, Canavan AGM, et al 1990 Slow potentials of the cerebral cortex and behaviour. Physiol Rev 70: 1–41.

Blume WT 1982 Atlas of Pediatric Electroencephalography. Raven Press, New York.

Blume WT 1990 Do epileptic foci in children migrate? The cons. Electroencephalogr Clin Neurophysiol 76: 100–105.

Blume WT, Kaibara M 1995 Atlas of Pediatric Electroencephalography. Raven, Philadelphia, pp. 5–38.

Blume WT, Whiting SE, Girvin JP 1991 Epilepsy surgery in the posterior cortex. Ann Neurol 29: 638–645.

Blumenfeld H, McCormick DA 2000 Corticothalamic inputs control the pattern of activity generated in thalamocortical networks. J Neurosci 20: 5153–5162.

Brazier MAB 1969 Prenarcotic doses of barbiturates as an aid in localizing diseased brain tissue. Anesthesiology 31: 78–83.

Brazier MAB, Finesinger JE 1945 Action of barbiturates on the cerebral cortex. Arch Neurol Psychiatr 53: 51–58.

Bressler SL 1990 The gamma wave: a cortical information carrier? Trends Neurosci 13: 161–162.

Brittenham DM 1990 Artefacts. Activities not arising from the brain. In: Current Practice of Clinical Electroencephalography, 2nd edn (eds DD Daly, TA Pedley). Raven Press, New York, pp. 85–105.

Brown P, Farmer SF, Halliday DM, et al 1999 Coherent cortical and muscle discharge in cortical myoclonus. Brain 122: 461–472.

Browning RA 1994 Anatomy of generalized convulsive seizures. In: Idiopathic Generalized Epilepsies: Clinical, Experimental and Genetic Aspects (eds A Malafosse, P Genton, C Maresaux, et al). John Libbey, London, pp. 399–413.

Burnham WM 1987 Electrical stimulation studies: generalized convulsions triggered from the brain-stem. In: Epilepsy and the Reticular Formation: the Role of the Reticular Core in Convulsive Seizures (eds GH Fromm, CL Faingold, RA Browning, et al). Alan Liss, New York, pp. 25–38.

Butler S 1988 Alpha asymmetry, hemispheric specialization and the problem of cognitive dynamics. In: The EEG of Mental Activities (eds D Giannitrapani, L Murri). Karger, Basel, pp. 79–93.

Cavazzuti GB, Cappella L, Nalin A 1980 Longitudinal study of epileptiform EEG patterns in normal children. Epilepsia 21: 43–55.

Chatrian GE, Perez-Borja C 1964 Depth electrographic observations in two cases of photo-oculoclonic response. Electroencephalogr Clin Neurophysiol 17: 71–75.

Chatrian GE, White LE Jr, Shaw C-M 1964 EEG pattern resembling wakefulness in unresponsive decerebrate state following traumatic brain-stem infarct. Electroencephalogr Clin Neurophysiol 16: 285–289.

Chatrian GE, Bergamasco B, Bricolo A, et al 1996 IFCN recommended standards for electrophysiologic monitoring in comatose and other unresponsive states. Report of an IFCN committee. Electroencephalogr Clin Neurophysiol 99: 103–122.

Chee MW, Kotagal P, Van Ness PC, et al 1993 Lateralizing signs in intractable partial epilepsy: blinded multiple-observer analysis. Neurology 43: 2519–2525.

Colder BW, Frysinger RC, Wiulson CL, et al 1996 Decreased neuronal burst discharge near site of seizure onset in epileptic human temporal lobes. Epilepsia 37: 113–121.

Commission on Classification and Terminology of the International League Against Epilepsy 1989 Proposal for revised classification of epilepsies and epileptic syndromes. Epilepsia 30: 389–399.

Cooper RA 1994 Normal sleep. In: *Sleep* (ed RA Cooper). Chapman and Hall, London, pp. 3–46.

Cooper R, Winter AL, Crow HJ, et al 1965 Comparison of subcortical, cortical and scalp activity using chronically indwelling electrodes in man. Electroencephalogr Clin Neurophysiol 18: 217–228.

Cooper R, Binnie C, Billings R 2005 *Techniques in Clinical Neurophysiology*. Elsevier, Oxford.

Daniel RS 1966 Electroencephalographic pattern quantification and the arousal continuum. Psychophysiology 2: 146–160.

Darby CE, Park DM, Smith AT, et al 1985 Electroencephalographic characteristics of epileptic pattern sensitivity and their relation to the nature of pattern stimulation and the effect of sodium valproate. Electroencephalogr Clin Neurophysiol 63: 517.

Dasheiff RM, Kofke WA 1993 Evaluation of the thiopental test in epilepsy surgery patients. Epilepsy Res 15: 253–258.

Davidson RJ 1988 EEG measures of cerebral asymmetry: conceptual and methodological issues. Int J Neurosci 39: 71–89.

Delgado-Escueta AV 1979 Epileptogenic paroxysms: modern approaches and clinical correlations. Neurology 29: 1014–1022.

Delgado-Escueta AV, Bascal FE, Treiman DM 1982 Complex partial seizures on closed-circuit television and EEG. A study of 691 attacks in 79 patients. Ann Neurol 11: 292–300.

Dement W, Kleitman N 1957 Cyclic variations in EEG during sleep and their relation to eye movements, body motility and dreaming. Electroencephalogr Clin Neurophysiol 9: 673–690.

De Marco P, Negrin P 1973 Parietal focal spikes evoked by contralateral tactile somatotopic stimulation in four non-epileptic subjects. Electroencephalogr Clin Neurophysiol 34: 308–312.

Deonna T, Despland PA 1989 Sensory evoked (touch) idiopathic myoclonic epilepsy of infancy. In: *Reflex Seizures and Reflex Epilepsies* (eds A Beaumanoire, H Gastaut, R Naquet). Éditions Médicine & Hygiène, Geneva, pp. 99–102.

Deschênes M, Roy JP, Steriade M 1982 Thalamic bursting mechanism: an inward slow current revealed by membrane hyperpolarization. Brain Res 239: 289–293.

Deuschl G, Eisen A (eds) 1999 *Recommendations for the Practice of Clinical Neurophysiology*. Clinical Neurophysiol (Suppl 52): 304.

de Weerd AW, Despland PA, Plouin P 1999 Neonatal EEG. In: *Recommendations for the Practice of Clinical Neurophysiology* (eds G Deuschl, A Eisen). Clin Neurophysiol (Suppl 52): 149–157.

Dirlich G, Vogl L, Plasohke M, et al 1997 Cardiac field effects on the EEG. Electroencephalogr Clin Neurophysiol 102: 307–315.

Dondey M 1983 Transverse topographical analysis of petit mal discharges: diagnostical and pathogenic implications. Electroencephalogr Clin Neurophysiol 55: 361–371.

Dreifuss FE (Chairman) 1981 Commission on Classification and Terminology of the International League Against Epilepsy. Proposal for revised clinical and electroencephalographic classification of epileptic seizures. Epilepsia 22: 489–501.

Duffy FH, Jensen F, Erba G, et al 1984 Extraction of clinical information from electroence-phalographic background activity: the combined use of brain electrical activity mapping and intravenous sodium thiopental. Ann Neurol 15: 22–30.

Dutertre F 1977 Catalogue of the main EEG patterns. In: *Semiology in Clinical EEG. Handbook of Electroencephalography and Clinical Neurophysiology*, Vol. IIa (eds M Dondey, J Gaches). Elsevier, Amsterdam, pp. 40–79.

Ebersole JS 1992 Equivalent dipole modelling – a new EEG method for localisation of epileptogenic foci. In: *Current Problems in Epilepsy* (eds TA Pedley, BS Meldrum). John Libbey, London, pp. 51–71.

Eeg-Olofsson O 1971a The development of the electroencephalogram in normal children and adolescents from the age of 1 through 21 years. Acta Paediatr Scand (Suppl 208).

Eeg-Olofsson O 1971b The development of the electroencephalogram in normal children from the age of 1 through 15 years. 14 and 6 Hz positive spike phenomenon. Neuropädiatrie 2: 405–427.

Eeg-Olofsson O, Petersen I, Sellden U 1971 The development of the electroencephalogram in normal children from the age of 1 through 15 years. Paroxysmal activity. Neuropädiatrie 2: 375–404.

Egli M, Mothersill I, O'Kane M, et al 1985 The axial spasm: the predominant type of drop seizure in patients with secondary generalized epilepsy. Epilepsia 5: 401–415.

Ellingson RJ 1979 EEGs of premature and full-term newborns. In: *Current Practice of Clinical Electroencephalography* (eds DW Klass, DD Daly). Raven Press, New York, pp. 149–177.

Ellingson RJ, Peters JF 1980 Development of EEG and daytime sleep patterns in low risk premature infants during the first year of life: longitudinal observations. Electroencephalogr Clin Neurophysiol 50: 165–171.

Engel J 1989 *Seizures and Epilepsy*. Contemporary Neurology Series, No. 31. Davis, Philadelphia.

Engel J 2001 A proposed diagnostic scheme for people with epileptic seizures and epilepsy: report of the ILAE task force on classification and terminology. Epilepsia 42: 796–803.

Erwin CW, Somerville ER, Radtke RA 1984 A review of the electroencephalographic features of normal sleep. J Clin Neurophysiol 1: 253–274.

Falconer M, Kennedy WA 1961 Epilepsy due to small focal temporal lesions with bilateral spike discharging foci: a study of seven cases relieved by operation. J Neurol Neurosurg Psychiatry 24: 205–212.

Fegersten L, Roger A 1961 Frontal epileptogenic foci and their clinical correlations. Electroencephalogr Clin Neurophysiol 13: 905–913.

Fenton GW, Scotton L 1967 The use of methohexitone in sleep electroencephalography. Electroencephalogr Clin Neurophysiol 23: 273–276.

Fernandez Torre JL, Alarcón G, Binnie CD, et al 1999a Comparison of sphenoidal, foramen ovale and anterior temporal placements for detecting interictal epileptiform discharges in temporal lobe epilepsy. Clin Neurophysiol 110: 895–904.

Fernandez Torre JL, Alarcón G, Binnie CD, et al 1999b Mechanisms involved in the propagation of interictal epileptiform discharges in partial epilepsy. J Neurol Neurosurg Psychiatry 67: 51–58.

Fischer-Williams M, Ponset M, Richie D, et al 1968 Light induced epilepsy in the baboon *Papio papio*: cortical and depth recordings. Electroencephalogr Clin Neurophysiol 25: 557–569.

Foldvary N, Klem G, Hammel J, et al 2001 The localizing value of ictal EEG in focal epilepsy. Neurology 57: 2022–2028.

Gale K 1985 Mechanisms of seizure control mediated by p-aminobutyric acid: role of the substantia nigra. Fed Proc 44: 2414–2424.

Gallmetzer P, Leutmezer F, Serles W, et al 2004 Postictal paresis in focal epilepsies – incidence, duration, and causes: a video-EEG monitoring study. Neurology 62: 2160–2164.

Ganes T, Lundar T 1983 The effect of thiopentone on somatosensory evoked responses and EEGs in comatose patients. J Neurol Neurosurg Psychiatry 46: 509–514.

Garant DS, Gale K 1987 Substantia nigra-mediated anticonvulsant actions: role of nigral output pathways. Exp Neurol 97: 143–159.

Gastaut H 1952 Etude électrocorticographique de la réactivité des rythmes rolandiques. Rev Neurol 87: 176–182.

Gastaut H, Terzian H, Gastaut Y 1952 Etude d'une activité électro-encéphalographique méconnue: 'Le rythme rolandique en arceau'. Marseille Méd 89: 296–310.

Gastaut H, Trevisan C, Naquet R 1958 Diagnostic value of electroencephalographic abnormalities provoked by intermittent photic stimulation. Electroencephalogr Clin Neurophysiol 10: 176–196.

Geiger LR, Harner RN 1978 EEG patterns at the time of focal seizure onset. Arch Neurol 35: 276–286.

Gengerelli JA, Parker CE 1966 Spectrographic analysis of electro-encephalograms under conditions of alertness and relaxation. J Psychol 63: 67–72.

Gibbs FA, Gibbs EL 1950 *Atlas of Electroencephalography*, Vol. 1: *Methodology and Controls*, 2nd edn. Addison-Wesley, Cambridge, MA.

Gibbs EL, Gillen HW, Gibbs FA 1954 Disappearance and migration of epileptic foci in childhood. Am J Dis Child 88: 596–603.

Gilbert DL, DeRoos S, Bare MA 2004 Does sleep or sleep deprivation increase epileptiform discharges in pediatric electroencephalograms? Pediatrics 114: 658–662.

Gloor P 1968 Generalised corticoreticular epilepsies: some considerations on the pathophysiology of generalised bilaterally synchronous spike and wave discharge. Epilepsia 9: 249–263.

Gloor P 1975 Contributions of electroencephalography and electrocorticography to the neurosurgical treatment of the epilepsies. In: *Neurosurgical Management of the Epilepsies. Advances in Neurology*, Vol. 8 (eds DP Purpura, JK Penry, RD Walter). Raven Press, New York, pp. 59–105.

Gloor P, Kalabay O, Giard N 1968 The electroencephalogram in diffuse encephalopathies: electroencephalographic correlates of grey and white matter lesions. Brain 91: 779–802.

Gloor P, Ball G, Schaul N 1977 Brain lesions that produce delta activity in the EEG. *Neurology* 27: 326–333.

Goldensohn ES 1984 The relevance of secondary epileptogenesis to the treatment of epilepsy: kindling and the mirror focus. Epilepsia 25(Suppl 2): 156–168.

Goldman D 1950 The clinical use of the 'average' reference electrode in monopolar recording. Electroencephalogr Clin Neurophysiol 2: 209–212.

Goncharova II, Barlow JS 1990 Changes in mean frequency and spectral purity during spontaneous alpha blocking. Electroencephalogr Clin Neurophysiol 76: 197–204.

Goodin DJ, Aminoff MJ 1984 Does the interictal EEG have a role in the diagnosis of epilepsy? Lancet i: 837–839.

Goodwin JE 1947 The significance of alpha variants in the EEG, and their relationship to an epileptiform syndrome. Am J Psychiatry 104: 369–379.

Gotman J, Gloor P, Quesney LF, et al 1982 Correlations between EEG changes induced by diazepam and the localization of epileptic spikes and seizures. Electroencephalogr Clin Neurophysiol 54: 614–621.

Gregory RP, Oates T, Merry RTG 1993 Electroencephalogram epileptiform abnormalities in candidates for aircrew training. Electroencephalogr Clin Neurophysiol 86: 75–77.

Guérit J-M, Fischer C, Facco E, et al 1999 Standards of clinical practice of EEG and EPs in comatose and other responsive states. In: *Recommendations for the Practice of Clinical Neurophysiology: Guidelines of the International Federation of Clinical Neurophysiology* (eds G Deuschl, A Eisen). EEG (Suppl 52): 117–131.

Hagne I, Persson J, Magnusson R, et al 1973 Spectral analysis via fast Fourier transform of waking EEG in normal infants. In: *Automation of Clinical Electroencephalography* (eds P Kellaway, I Petersén). Raven Press, New York, pp. 103–143.

Harding GFA 1996 Eye closure and EEG abnormalities, darkness, fixation-off and photosensitivity. In: *Eyelid Myoclonia with Absences* (eds JS Duncan, C Panayiotopoulos). John Libbey, London, pp. 69–76.

Harding GFA, Herrick CE, Jeavons PM 1978 A controlled study of the effect of sodium valproate on photosensitive epilepsy and its prognosis. Epilepsia 19: 555.

Hennessy M, Binnie CD 2000 Photogenic partial seizures. Epilepsia 41: 59–64.

Henry CE, Scoville WB 1952 Suppression-burst activity from isolated cerebral cortex in man. Electroencephalogr Clin Neurophysiol 4: 1–22.

Hogan T, Sundaram M 1989 Rhythmic auditory stimulation in generalized epilepsy. Electroencephalogr Clin Neurophysiol 72: 455–458.

Hoovey ZB, Heineman U, Creutzfeldt OD 1972 Inter-hemispheric 'synchrony' and of alpha waves. Electroencephalogr Clin Neurophysiol 32: 337–347.

Hoptman MJ, Davidson RJ 1998 Baseline EEG asymmetries and performance on neuropsychological tasks. Neuropsychologia 36: 1343–1353.

Hufnagel A, Pörsch M, Elger CE, et al 1995 The clinical and prognostic significance of the postictal slow focus in the electroencephalogram. Electroencephalogr Clin Neurophysiol 94: 12–18.

Hughes JR 1985 Long-term clinical and EEG changes in patients with epilepsy. Arch Neurol 42: 213–223.

Hughes JR, Cayaffa JJ 1977 The EEG in patients at different ages without organic cerebral disease. Electroencephalogr Clin Neurophysiol 42: 776–784.

Hughes JR, Grunener G 1984 Small sharp spikes revisited: further data on this controversial pattern. Clin Electroencephalogr 15: 208–213.

Hughlings Jackson J 1873 On the anatomical, physiological and pathological investigation of epilepsies. West Riding Lunatic Asylum Med Rep 3: 315–339.

IFSECN 1958 Report of the Committee on Methods of Clinical Examination in Electroencephalography. Electroencephalogr Clin Neurophysiol 10: 370–375.

IFSECN 1974 Report of the Committee on EEG Instrumentation Standards. Electroencephalogr Clin Neurophysiol 37: 549–553.

IFSECN (ed) 1983 *Recommendations for the Practice of Clinical Neurophysiology*. Elsevier, Amsterdam.

Irrgang U, Höller L 1981 Polygraphic recording in the EEG laboratory. J Electrophysiol Tech 7: 98–111.

Isokawa-Akesson M, Wilson CL, Babb TL 1987 Structurally stable burst and synchronized firing in human amygdala neurones: auto- and cross-correlation analyses in temporal lobe epilepsy. Epilepsy Res 1: 17–34.

Isokawa-Akesson M, Wilson CL, Babb TL 1989 Inhibition in synchronously firing human hippocampal neurons. Epilepsy Res 3: 236–247.

Jasper HH, Solomon P, Bradley C 1938 Electroencephalographic analyses of behaviour problem children. Am J Psychiatr 95: 641–658.

Jeavons PM, Harding GFA 1975 *Photosensitive Epilepsy*. Heinemann, London.

Jensen MS, Azouz R, Yaari Y 1994 Variant firing patterns in the rat hippocampal pyramidal cells modulated by extracellular potassium. J Neurophysiol 71: 831–839.

Kaibara M, Blume WT 1988 The postictal electroencephalogram. Electroencephalogr Clin Neurophysiol 70: 99–104.

Kanner A, Morris H, Lüders H, et al 1993 Usefulness of unilateral interictal sharp waves of temporal origin in prolonged video-EEG monitoring studies. Epilepsia 34: 884–889.

Kasteleijn-Nolst Trenité DGA 1989 Photosensitivity in epilepsy. MD Thesis, University of Utrecht, The Netherlands.

Kasteleijn-Nolst Trenité DGA, Binnie CD, Harding GFA, et al 1999 Photic stimulation: standardization of screening methods. Epilepsia 40(Suppl 4): 75–79.

Kasteleijn-Nolst Trenité DGA, Guerrini R, Binnie CD, et al 2001 Classification of responses to photic stimulation. Epilepsia 42: 692–701.

Kiloh LG, McComas AJ, Osselton JW, et al 1980 Clinical Electroencephalography, 4th edn. Butterworths, London.

King DW, Ajmone Marsan C 1977 Clinical features and ictal patterns in epileptic patients with temporal lobe foci. Ann Neurol 2: 138–147.

Klass DW 1995 The continuing challenge of artefacts in the EEG. Am J EEG Tech 35: 239–269.

Kostopoulos G, Avoli M, Gloor P 1983 Participation of recurrent inhibition in the genesis of spike and wave discharges in feline generalized epilepsy. Brain Res 267: 101–112.

Koutroumanidis M, Binnie CD, Elwes RDC, et al 1998 Interictal regional slow activity in temporal lobe epilepsy correlates with lateral temporal hypometabolism as imaged with 18FDG PET: neurophysiological and metabolic implications. J Neurol Neurosurg Psychiatry 65: 170–176.

Kraaier V, Van Huffelen AC, Wieneke GH 1988 Changes in quantitative EEG and blood flow velocity due to standardized hyperventilation; a model of transient ischaemia in young human subjects. Electroencephalogr Clin Neurophysiol 70: 377–387.

Kulman WH 1980 The mu rhythm: functional topography and neural origin. In: Rhythmic EEG Activities and Cortical Functioning (eds G Pfurtscheller, P Buser, FA Lopes da Silva, et al). Elsevier, Amsterdam, pp. 105–120.

Lamblin MD, d'Allest AM, Andre M, et al 1999 EEG in premature and full-term infants: developmental features and glossary. Neurophysiol Clin 29: 123–219.

Leijten FSS, Dekker E, Spekreijse H, et al 1998 Light diffusion in photosensitive epilepsy. Electroencephalogr Clin Neurophysiol 106: 387–391.

Leutmezer F, Woginger S, Antoni E, et al 2002 Asymmetric ending of secondarily generalized seizures: a lateralizing sign in TLE. Neurology 62: 1252–1254.

Lieb J, Walsh GO, Babb TL, et al 1976 A comparison of EEG seizure patterns recorded with surface and depth electrodes in patients with temporal lobe epilepsy. Epilepsia 17: 137–160.

Lindsley DB, Cutts KK 1940 Electroencephalograms of 'constitutionally inferior' and behaviour problem children: comparison with those of normal children and adults. Arch Neurol Psychiatry 44: 1199–1212.

Louis J, Zhang JX, Revol M, et al 1992 Ontogenesis of nocturnal organization of sleep spindles: a longitudinal study during the first 6 months of life. Electroencephalogr Clin Neurophysiol 83: 289–296.

Marcus EM, Watson CW, Simon SA 1968 An experimental model of some varieties of petit mal epilepsy: electrical–behavioral correlations of acute bilateral epileptogenic foci in cerebral cortex. Epilepsia 9: 233–248.

Markand ON, Daly DD 1971 Pseudoperiodic lateralized paroxysmal discharges in electroencephalogram. Neurology 21: 975–981.

Matsuoka K, Takahashi T, Sasaki M, et al 2000 Neurophysiological EEG activation in patients with epilepsy. Brain 123: 318–330.

McCallum WC, Curry SH (eds) 1993 Slow Potential Changes in the Human Brain. Plenum Press, London.

Medvedev AV 2001 Temporal binding at gamma frequencies in the brain: paving the way to epilepsy? Australas Phys Eng Sci Med 24: 37–48.

Meencke HJ, Janz D 1984 Neuropathological findings in primary generalised epilepsy. Epilepsia 25: 8–21.

Ménini C, Stutzmann JM, Laurent H, et al 1981 Cortical unit discharges during photic stimulation in the Papio papio. Relationships with paroxysmal fronto-Rolandic activity. Electroencephalogr Clin Neurophysiol 52: 42–49.

Molaie M, Cruz A 1988 The effect of sleep deprivation on the rate of focal interictal epileptiform discharges. Electroencephalogr Clin Neurophysiol 70: 288–292.

Morrell F, Whisler WW 1980 Secondary epileptogenic lesions in man: prediction of results of excision of the primary focus. In: Advances in Epileptology: the XIth Epilepsy International Symposium (eds R Canger, F Angelieri, JK Penry). Raven Press, New York, pp. 123–128.

Noachtar S, Binnie CD, Ebersole J, et al 1999 A glossary of terms most commonly used by clinical electroencephalographers and proposal for the report form for the EEG findings. In: Recommendations for the Practice of Clinical Neurophysiology (eds G Deuschl, A Eisen). Clin Neurophysiol (Suppl 52): 21–40.

Obeso JA, Rothwell JC, Marsden CD 1985 The spectrum of cortical myoclonus: from cortical reflex myoclonus to epilepsy. Brain 108: 463–483.

Offner FF 1950 The EEG as potential mapping: the value of the average monopolar reference. Electroencephalogr Clin Neurophysiol 2: 213–214.

Pampiglione G 1952 Induced fast activity in the EEG as an aid in the location of cerebral lesions. Electroencephalogr Clin Neurophysiol 4: 79–82.

Panayiotopoulos CP, Tahan R, Obeid T 1991 Juvenile myoclonic epilepsy: factors of error involved in the diagnosis and treatment. Epilepsia 32: 672–676.

Parmalee AH, Schulte FJ, Akiyama Y, et al 1968 Maturation of EEG activity during sleep in premature infants. Electroencephalogr Clin Neurophysiol 24: 319–329.

Patel VM, Maulsby RL 1987 How hyperventilation alters the electroencephalogram: a review of controversial viewpoints emphasizing neurophysiological mechanisms. J Clin Neurophysiol 4: 1010–1020.

PeBenito R, Cracco JB 1979 Periodic lateralized epileptiform discharges in infants and children. Ann Neurol 6: 47–50.

Penfield W, Jasper H 1954 Epilepsy and the Functional Anatomy of the Human Brain. Churchill, London.

Penry JK, Porter RJ, Dreifuss FE 1975 Simultaneous recording of absence seizures with video tape and electroencephalography. A study of 374 seizures in 48 patients. Brain 98: 427–440.

Polkey CE 1983 Prognostic factors in selecting patients with drug-resistant epilepsy for temporal lobectomy. In: Research Progress in Epilepsy (ed FC Rose). Pitman, London, pp. 500–506.

Prior PF 1973 The EEG in Acute Cerebral Anoxia. Excerpta Medica, Amsterdam.

Prior PF 1987 The EEG and detection of responsiveness during anaesthesia and coma. In: Consciousness, Awareness and Pain in General Anaesthesia (eds M Rosen, J Lunn). Butterworths, London, pp. 34–45.

Prior PF, Maclaine GN, Scott DF, et al 1972 Tonic status epilepticus precipitated by intravenous diazepam in a child with petit mal status. Epilepsia 13: 467–472.

Quesney LF 1986 Seizures of frontal lobe origin. In: *Recent Advances in Epilepsy*. Electroencephalogr Clin Neurophysiol (Suppl 37): 81–110.

Quesney LF, Constain M, Rasmussen T, et al 1992 Presurgical EEG investigation in frontal lobe epilepsy. In: *Surgical Treatment of Epilepsy* (ed WH Theodore). Epilepsy Res (Suppl 5): 55–69.

Rae-Grant AD, Strapple C, Barbour PJ 1991 Episodic low-amplitude events: an under-recognized phenomenon in clinical electroencephalography. J Clin Neurophysiol 8: 203–211.

Ralston BL 1961 Cingulate epilepsy and secondary bilateral synchrony. Electroencephalogr Clin Neurophysiol 13: 591–598.

Ray A, Kotagal P 2005 Temporal lobe epilepsy in children: overview of clinical semiology. Epileptic Disord 7: 299–307.

Rechtschaffen A, Kales A (eds) 1968 *A Manual of Standardized Terminology, Techniques and Scoring System for Sleep Stages of Human Subjects*. Brain Information Service, UCLA, Los Angeles, CA.

Reilly EL, Peters JF 1973 Relationship of some varieties of electroencephalographic photosensitivity to clinical convulsive disorders. Neurology 23: 1050–1057.

Robin JJ, Tolan GD, Arnold JW 1978 Ten-year experience with abnormal EEGs in asymptomatic adult males. Aviat Space Environ Med 49: 732–736.

Rockstroh B, Elbert T, Canavan AGM, et al 1989 *Slow Cortical Potentials and Behaviour*. Urben and Schartzenberg, Munich.

Rodin E, Ancheta O 1987 Cerebral electrical fields during petit mal absences. Electroencephalogr Clin Neurophysiol 66: 457–466.

Rodin E, Onuma T, Wasson S, et al 1971 Neurophysiological mechanisms involved in grand mal seizures induced by metrazol and megimide. Electroencephalogr Clin Neurophysiol 30: 62–72.

Rodin E, Kitano H, Nagao B, et al 1977 The results of penicillin G administration on chronic unrestrained cats: electrographic and behavioral observations. Electroencephalogr Clin Neurophysiol 42: 518–527.

Salinsky M, Kanter R, Dasheiff RM 1987 Effectiveness of multiple EEGs in supporting the diagnosis of epilepsy: an operational curve. Epilepsia 28: 331–334.

Samson-Dollfus D, Forthomme J, Capron E 1964 EEG of the human infant during sleep and wakefulness during the first year of life. Normal patterns and their maturational changes; abnormal patterns and their prognostic significance. In: *Neurological and Electroencephalographic Correlative Studies in Infancy* (eds P Kellaway, I Petersén). Grune and Stratton, New York, pp. 208–229.

Saunders MG 1979 Artefacts: activity of noncerebral origin in the EEG. In: *Current Practice of Clinical Electroencephalography*, 1st edn (eds DW Klass, DD Daly). Raven Press, New York, pp. 37–67.

Scherrer J, Calvet J 1973 Normal and epileptic synchronisation at the cortical level in the animal. In: *Synchronisation of EEG Activity in the Epilepsies* (eds H Petsche, MAB Brazier). Springer, New York, pp. 112–132.

Schwartz MS, Scott DF 1978 Pathological stimulus-related slow wave arousal responses in the EEG. Acta Neurol Scand 57: 300–304.

Schwartz MS, Prior PF, Scott DF 1973 The occurrence and evolution in the EEG of a lateralized periodic phenomenon. Brain 96: 613–622.

Seddigh S, Thömke F, Vogt TH 1999 Complex partial seizures provoked by photic stimulation. J Neurol Neurosurg Psychiatry 66: 801–802.

Sheldon SH 1996 *Evaluating Sleep in Infants and Children*. Lippincott-Raven, Philadelphia.

Shewmon DA, Erwin RJ 1988 The effect of focal interictal spikes on perception and reaction time. II: Neuroanatomic specificity. Electroencephalogr Clin Neurophysiol 69: 338–352.

Shinomiya S, Urakami Y, Nagata K, et al 1994 Frontal midline theta rhythm: differentiating the physiological theta rhythm from the abnormal discharge. Clin Electroencephalogr 25: 30–35.

Shouse MN, Ryan W 1984 Thalamic kindling: electrical stimulation of the lateral geniculate nucleus produces photosensitive grand mal seizures. Exp Neurol 86: 18–32.

Smith NJ, van Gils M, Prior P 2006 *Neurophysiological Monitoring During Intensive Care and Surgery*. Elsevier, Oxford.

Snead OC 1992 $GABA_B$ receptor mediated mechanisms in experimental absence seizures in rat. Pharmacol Commun 2: 63–69.

Snodgrass SM, Tsuburaya K, Ajmone-Marsan C 1989 Clinical significance of periodic lateralized epileptiform discharges: relationship with status epilepticus. J Clin Neurophysiol 6: 159–172.

So EL, King DW, Murvin AJ 1984 Misdiagnosis of complex absence seizures. Arch Neurol 41: 640–641.

So N, Gloor P, Quesney LF, et al 1989 Depth electrode investigations in patients with bitemporal epileptiform abnormalities. Ann Neurol 25: 423–431.

Spencer SS, Williamson PD, Bridgers SL, et al 1985 Reliability and accuracy of localization by scalp ictal EEG. Neurology 35: 1567–1575.

Stefan H 1981 Pseudospontanbewegungen bei Patienten mit Petitmal-Anfallen. Arch Psychiatr Nervenkrankheiten 229: 277–290.

Stefan H 1999 *Epilepsien: Diagnose und Behandlung*, 3 Auflage. Thieme, Stuttgart.

Stefan H, Snead OC 1997 Absence seizures. In: *Epilepsy: A Comprehensive Textbook* (eds J Engel, TT Pedley). Lippincott-Raven, Philadelphia, pp. 579–590.

Sundaram M, Hogan T, Hiscock M, et al 1990 Factors affecting interictal spike discharges in adults with epilepsy. Electroencephalogr Clin Neurophysiol 75: 358–360.

Takahashi M, Kubota F, Nishi Y, et al 1993 Persistent synchronous periodic discharges caused by anoxic encephalopathy due to cardiopulmonary arrest. Clin Electroencephalogr 24: 166–172.

Tharp BR 1972 Orbital frontal seizures: an unique electroencephalographic and clinical syndrome. Epilepsia 13: 627–642.

Topalkara K, Alarcón G, Binnie CD 1998 Effects of flash frequency and repetition of intermittent photic stimulation on photoparoxysmal responses. Seizure 7: 249–255.

Treiman DM, Walton NY, Kendrick C 1990 A progressive sequence of electroencephalographic changes during generalized convulsive status epilepticus. Epilepsy Res 5: 49–60.

Trimble MR, Reynolds EH (eds) 1986 *What is Epilepsy?* Churchill Livingstone, Edinburgh.

Tükel K, Jasper H 1952 The electroencephalogram in parasagittal lesions. Electroencephalogr Clin Neurophysiol 4: 481–494.

Vanhatalo S, Tallgren P, Andersson S, et al 2002 DC-EEG discloses prominent, very slow activity patterns during sleep in preterm infants. Clin Neurophysiol 113: 1822–1825.

Velasco F, Velasco M, Romo R 1980 Specific and non-specific multiple unit activities during pentylenetetrazol seizures: I. Animals with 'encéphale isolé'. Electroencephalogr Clin Neurophysiol 46: 600–607.

Veldhuisen R, Binnie CD, Beintema DJ 1983 The effect of sleep deprivation on the EEG in epilepsy. Electroencephalogr Clin Neurophysiol 55: 505–512.

Vitouch O, Bauer H, Gittler G, et al 1997 Cortical activity of good and poor spatial test performers during spatial and verbal processing studied with slow potential topography. Int J Psychophysiol 27: 183–199.

Walter WG 1936 The location of cerebral tumours by electroencephalography. Lancet 2: 305–308.

Walter WG, Dovey VJ 1944 Electroencephalography in cases of sub-cortical tumour. J Neurol Neurosurg Psychiatry 7: 57–65.

Waltz S, Christen H-J, Doose H 1992 The different patterns of the photoparoxysmal response – a genetic study. Electroencephalogr Clin Neurophysiol 83: 138–145.

Wassmer E, Quinn E, Whitehouse W, et al 2001 Melatonin as a sleep inductor for electroencephalogram recordings in children. Clin Neurophysiol 112: 683–685.

Westmoreland BF 1975 Posterior slow wave transients associated with eye blinks. Am J EEG Technol 15: 14–19.

Westmoreland B 1982 Normal and benign EEG patterns. Am J EEG Tech 22: 3–31.

Westmoreland B 1990 Benign EEG variants and patterns of uncertain clinical significance. In: *Current Practice of Clinical Electroencephalography*, 2nd edn (eds DD Daly, TA Pedley). Raven Press, New York, pp. 243–252.

Westmoreland BF, Reiher J, Klass DW 1979 Recording small sharp spikes with depth electroencephalography. Epilepsia 20: 599–606.

Wheless JW, Kim HL 2002 Adolescent seizures and epilepsy syndromes. Epilepsia 43: Suppl 3, 33–52.

Wieser HG, Bancaud J, Talairach J, et al 1979 Comparative value of spontaneous and chemically and electrically induced seizures in establishing the lateralization of temporal lobe seizures. Epilepsia 20: 46–59.

Wilkins AJ, Binnie CD, Darby CE 1980 Visually-induced seizures. Prog Neurobiol 15: 163–171.

Wilkins AJ, Binnie CD, Darby CE 1981 Interhemispheric differences in photosensitive epilepsy: I. Pattern sensitivity thresholds. Electroencephalogr Clin Neurophysiol 52: 461–468.

Williamson PD, Spencer DD 1986 Clinical and EEG features of complex partial seizures of extratemporal origin. Epilepsia 27(Suppl 2): 546–563.

Williamson PD, Thadani VM, Darcey TM, et al 1992a Occipital lobe epilepsy: clinical characteristics, seizure spread patterns, and results of surgery. Ann Neurol 31: 3–13.

Williamson PD, Boon PA, Thadani VM, et al 1992b Parietal lobe epilepsy: diagnostic considerations and results of surgery. Ann Neurol 31: 193–201.

Wilson NF, Johnston FD, MacLeod AG, et al 1934 Electrocardiograms that represent the potential variations of a single electrode. Am Heart J 9: 447–458.

Woermann FG, Free SL, Koepp MJ, et al 1999 Abnormal cerebral stricture in juvenile myoclonic epilepsy demonstrated with voxel-based analysis of MRI. Brain 122: 2101–2107.

Zivin L, Ajmone Marsan C 1968 Incidence and prognostic significance of 'epileptiform' activity in the EEG of non-epileptic subjects. Brain 91: 751–778.

Zschoke S 1995 *Klinische Elektroencephalographie*. Springer, Heidelberg, pp. 145–149.

Chapter 3

General Characteristics of Evoked Potentials

PRINCIPLES OF EVOKED POTENTIALS

General considerations

Evoked potentials (EPs) are electrical potential changes of sensory receptors, neural pathways and the brain following external or endogenous stimuli. This broad definition includes the many types of EPs available to the clinician as diagnostic tests in the fields of neurology, ophthalmology, audiology, psychiatry, as well as paediatrics.

Although various novel methods of EP analysis are theoretically applicable and of research interest, the usual clinical approach is the display of potential changes as a function of time. EPs recorded from electrodes located on the surface of the scalp are of small amplitude but can be distinguished from other electrical signals of physio-

logical or environmental origin ('noise') by averaging, i.e. summation of a sufficient number of stimulus-locked responses followed by division by the number of responses to give the 'average EP' (*Fig. 3.1*).

The EP usually follows a constant time course after the stimulus, and if sections of time (epochs) are averaged from a fixed point with respect to the stimulus the average of many responses should be typical of any one of them. By contrast, unwanted noise occurs randomly and is reduced in amplitude in the average. Averaging is a mathematical operation and the electrical signals are first sampled and converted to numerical values (analogue to digital conversion). For technical details the reader is referred to standard texts such as Cooper et al (2005) or Binnie et al (2004).

The number of responses (trials) required (usually between 30 and 5,000) and the stimulus rate (50/s to 6/min) depend on the type of EP under study and determine the duration of the investigation. Noise is reduced in the average by a factor equal to the square root of the number of epochs. Thus, to improve signal-to-noise ratio by 2 it is necessary to quadruple the number of trials and duration of the investigation. Moreover, trials contaminated by artefacts, such as eye movement, blink and muscle potentials, must be rejected during data acquisition, further prolonging the procedure. This is a critical consideration in paediatric practice, as it may be difficult to persuade the child to sit still or concentrate on a particular stimulus for more than a few seconds.

An EP waveform can be viewed as a sequence of waves of a specific latency, amplitude and polarity (*Fig. 3.2*). The latency is measured as the time elapsed between stimulus and onset or peak of the wave. The amplitude is often related to the strength of the signal and can be measured from the baseline or from the onset of the preceding or following peak.

Polarity of EP components

Conventionally, it is usually assumed that an EP is a localised event, which can be picked up by a suitably placed 'active' electrode that is referred to a 'neutral' reference electrode. For those EPs that produce widespread electrical fields, this is an oversimplification. As explained in Chapter 2 (see pp. 18 ff.), the way the electrodes are connected to the two input leads of a

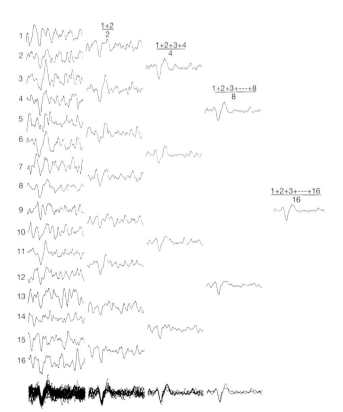

Fig. 3.1 Improvement of the signal-to-noise ratio by averaging. The noise is progressively reduced as the number of trials is increased. Superimposition of the trials is shown at the base of the figure. (From Cooper et al (1980), by permission.)

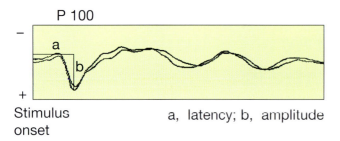

Fig. 3.2 Example of a visual evoked potential (VEP) waveform. Note the sequence of negative/positive waves: the P100 component is defined by polarity and expected normal latency. Amplitude (b) is measured from the preceding peak and the latency (a) from the onset of the stimulus.

differential amplifier determines the direction of deflection of the final display. For EEG recording, the input leads are designated 1 and 2, and are wired such that if lead 1 is connected to the negative pole and lead 2 to the positive pole of a source of potential difference, the display will deflect upwards. If standard EEG practice were followed for EPs, the electrode regarded as 'active' would be connected to lead 1, and the reference to lead 2. The 'negative-up' EEG convention is used for most EPs, except for early brainstem and middle latency auditory evoked potentials (BAEPs and MLAEPs), which are traditionally recorded with the active electrode on lead 2 (with some justification, as both electrode sites may be considered 'active'). The 'positive-up' convention is still in use, however, by workers for other EP modalities such as retinal potentials (electroretinogram (ERG)), which are recorded with the upward deflection corresponding to a positive-going change at the cornea. Whichever convention is followed, it is essential that it is clearly shown on any figures.

Nomenclature
Several different systems of naming components have been used:
1. the label N or P indicating the polarity, followed by latency (e.g. N20 for a negative wave appearing at a latency of 20 ms)
2. label N or P indicating the polarity with numbering based on order of appearance (N1, P1, N2, P2, etc.)
3. Roman numerals numbering the peaks sequentially (I, II, III, etc.)
4. lettering indicating the sequence of the negative and positive components (Na, Pa, etc.).

A single convention of designating polarity and latency has yet to be agreed. The International Federation for Clinical Neurophysiology (IFCN) (Deuschl and Eisen 1999) has proposed a nomenclature for each EP type, which is generally followed in adult practice.

Rapid changes in the morphology of EPs during maturation make standardisation of nomenclature in neonatal and paediatric age groups particularly difficult,

and the lack of a standard nomenclature for these age groups has led to difficulties in comparing normal, and especially pathological, results from different laboratories. An example of how nomenclature can generate confusion, especially in the field of neonatal and paediatric EPs, is given on page 123, where the nomenclature and maturation of flash visual evoked potentials (FVEPs) are described.

The concept of the EP generator
An important issue of clinical relevance to EP testing is the relationship of the EP waves to underlying brain physiology; i.e. what activity in what structures can be inferred from the EP waveform?

The concept of an EP generator is highly ambiguous, since it refers to any mechanism capable of producing a potential difference between two recording sites regardless of the physiological or anatomical basis. We need to envisage biophysical models of generators, but these can never be other than oversimplified views of the complex neuronal networks of the brain.

Like the electroencephalogram (EEG), EPs are generated by the synchronous activity of neuronal populations aligned in parallel. Such a favourable configuration exists when a population of cells activated by a common input is made up of a row of units with their long axes orientated in parallel (the 'open field' of Lorente de Nò (1947)) (*Fig. 3.3*). An afferent volley will then trigger synchronised postsynaptic potentials (PSPs), generating extracellular potentials orientated in the same direction, since currents will flow mainly along the long axis of each cell. The resulting compound potential can be modelled as a unique dipole, the orientation being along the long axis of the whole cell population. This model fits, for instance, the organisation of the neocortical pyramidal cells, the apical dendrites of which are perpendicular to the cortical surface and the synaptic contacts of which are concentrated on dendritic arborisations close to the surface and on the cell body in the depth of the cortical grey matter.

Conversely, if the cytoarchitecture of the cell population activated by a synchronous volley is such that potential fields generated by individual cells are multidirectional, then a cancellation of potential fields can result. This occurrence is illustrated by the 'closed-field' model of Lorente de Nò (1947) (see *Fig. 3.3*). In this model the neurons are multipolar with multidirectional dendrites spreading radially away from cell bodies clustered in the centre of the structure. It can be demonstrated that in this situation synchronised PSPs occurring either in the centre at the level of cell bodies, or at the periphery in the dendritic arborisation, will produce extracellular spherical concentric fields with a reversal of polarity situated midway between the external dendritic layer and the central core of cell bodies. Consequently, only an electrode situated in this structure itself will record a potential, whereas at a distance from it the potential will be zero.

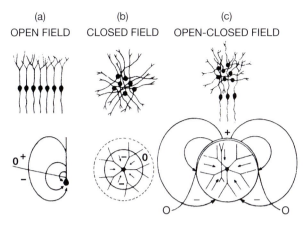

(a) OPEN FIELD (b) CLOSED FIELD (c) OPEN-CLOSED FIELD

Fig. 3.3 Lorente de Nò's models of potential field generators. Predicted current flows and potential fields generated by the synchronous depolarisation of the cell bodies. (A) In the open-field model, which is a schematic representation of the arrangement of pyramidal cortical neurons, the cell bodies, axons and dendritic tree are supposed to be aligned in a row. Depolarisation of the soma produces an inward current flow of the soma (sink) and an outward current flow from the dendrites (source), whereas depolarisation of the dendrites would produce the reverse. The field generated by the activation of such a structure can be picked up at a distance from the source and sink, provided that the two electrodes are not located on the same isopotential line. Note that the current flow from the axon to the cell bodies is not represented; this flow is considered to be negligible compared with that arising from apical dendrites, which have a broader surface. (B) In the closed-field model, which corresponds to a nuclear structure, the somas are gathered in the centre of the structure and dendrites extend radially outwards. Somatic depolarisation produces an inward current and maximal negative potential at the centre of the nucleus. This radial inward current causes the potential gradient to be zero anywhere outside the nucleus. Consequently, the electrical activity cannot be recorded using electrodes situated at the periphery of the nucleus. (From Lorente de Nò (1947), by permission.**)**

This model fits with the anatomical organisation of some nuclear structures of the brain in which synaptic transmission will not produce an external potential field and thus will be missed by surface electrodes.

Each category of afferents will usually make its synapses onto a specific region of the target cell(s) (either dendrites or somas) but not to both equally, and will produce either excitatory or inhibitory PSPs, but not both kinds of PSPs. Thus, the underlying unitary responses can be indirectly inferred from EP fields recorded on the surface only if the type and the anatomy of the synapses activated by the stimulus are known.

Equipment
Electrodes
Electrodes for surface EP recording should be non-polarising and are similar or identical to those used for EEG, typically discs made of gold or of silver coated with silver chloride. The electrodes are filled with saline gel

and attached to the skin or scalp with collodion and gauze, or with an electrically conducting adhesive paste.

Disposable electrodes with a water-soluble gel that adhere very strongly when dry are a good choice when long-duration recordings in uncooperative children are planned. Electrodes, also disposable, made of a solid gel where the recording wire is embedded are a good, but expensive, item for surface recordings. Neonates and infants have very delicate skin and the best method for electrode application is with adhesive paste and gauze, as shown in *Fig. 3.4*. Recordings of retinal visual EPs (the ERG) require special electrodes (see p. 121).

Amplifiers
Equipment for recording EPs must be suitable for amplifying electrical signals with amplitudes down to 0.1 μV and bandwidth from below 1 Hz up to 10 kHz. Most investigations require multichannel recordings. These considerations make exacting demands for specification and design of amplifiers. Fortunately, modern amplifiers are very stable and trouble free. For technical details the reader is referred to standard texts of EEG and EP technology (e.g. Cooper et al 2005).

Recordings are usually made in laboratories in which there is a high level of mains frequency (50 or 60 Hz) electrical interference. The effect of this can be minimised by suitable amplifier design and careful electrode preparation and placement. Susceptibility to interference is reduced if the impedance of the electrodes is stable and low (< 10 and preferably < 5 kΩ). Low electrode impedance requires vigorous skin preparation, which may not be well tolerated by children and, indeed, may damage the fragile skin of neonates (see p. 3).

Calibration
While it should be mandatory for a calibration signal to be recorded on all channels of every EP record, the amplitude calibration on averaging systems is rarely checked; indeed, regrettably, calibration facilities are not provided on many modern averagers.

Montages
The minimum number of channels varies with the type of investigation, the clinical problem, the patient's degree of collaboration and the need for special forms of analysis or display of results (standard voltage/time plots versus brain maps, etc.). The patterns of connections between electrodes and amplifiers (montages) determine the presentation of results and are considered in Chapter 2 (see pp. 18 ff.) (see also Cooper et al 2005). The IFCN (Deuschl and Eisen 1999) has set the minimum montages required for the different EPs.

Stimulators
Practically all EP investigations require sensory stimulation – visual (flashes, changing patterns), somatosensory (electrical pulses), auditory (tone pulses, clicks), etc.

Stimulators in modern machines are an integral part of the equipment. The stimuli are synchronised to the sampling epoch and the trigger point advanced or delayed from the start of the epoch to facilitate viewing the baseline data on the screen. Various types of stimulation for visual, somatosensory and auditory EPs (VEPs, SEPs and AEPs) are described in the appropriate sections of this chapter.

Displays

The EPs are displayed on a computer screen, or VDU, after averaging. Amplitude and time scales can be manipulated to optimise the display and traces can be written out on a printer. Latencies and amplitudes can be obtained from the displayed data, either automatically or by cursor measurement. The spatial distribution of the EP at particular times in the epoch can be displayed, usually as coloured maps (Binnie et al 2003).

Normative data

Local normative data need to be collected in every laboratory, as recording conditions, the types of stimulators, etc., influence the results. Methods for collection of normative data in adults are considered in the IFCN Recommendations for the Practice of Clinical Neurophysiology (1999). In paediatric and neonatal populations, difficulties arise on account of the ethical implications of performing an EP test in a normal child and because different norms apply at different ages.

Maturational changes in the waveform, latency and amplitude of the EP components take place during development, at a rate inverse to age. Therefore, during the neonatal period and the first months of life data collection needs to be done by weeks, then by months and only after at least 2 years of age at yearly intervals. An enormous effort is involved for each laboratory to collect its own normal data.

The Italian study group for Neurophysiology in Paediatrics has been working since 2002 on a multicentre normal data collection project. It has been building up a reference database for BAEPs across various ages from birth to adulthood: each centre can compare their own normal values with the multicentre database. The system is particularly useful to check the appropriateness of practices in new laboratories or changes in the recording systems, and gives correction factors for identification of abnormal records independently of age (Scaioli et al 2003). Work is now in progress to create a similar system for SEPs and VEPs.

Clinical role

EPs represent an inexpensive and excellent means for assessing the brain at work. Compared to neuroimaging techniques (ultrasound, computerised tomography (CT) and magnetic resonance imaging (MRI)), they give unique information about function from sensory receptors to the brain. With respect to other available functional investigations, such as functional MRI (fMRI), single proton emission computerised tomography (SPECT) and positron emission tomography (PET), EPs have a far better time resolution, are less invasive and are more cost effective. They may, however, be non-specific, highly variable or susceptible to artefacts; some (e.g. pattern VEPs) require cooperation by the subject.

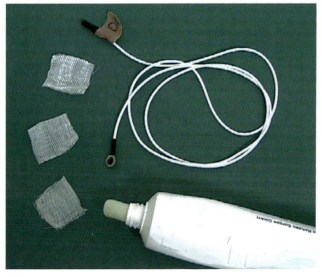

(a)

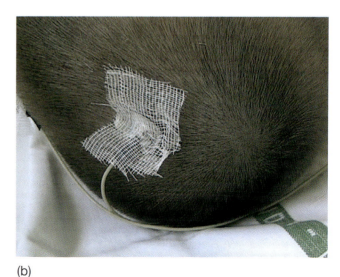

(b)

Fig. 3.4 Electrode application in neonates and children. The electrode is filled with paste, then gently pushed onto the scalp; the gauze is superimposed and attached to the small amount of paste around the edge of the electrode. The procedure is very quick and not invasive. If the child moves and the electrode becomes detached it can easily be replaced.

The relative utility of the different types of EPs in the specific clinical conditions in the field of paediatric neurology, psychiatry, ophthalmology and audiology are outlined in the specific sections on each modality. EP testing often provides clinical information that cannot be obtained by other means. Examples include: discrimination between organic and 'functional' symptoms, when clinically or neuroimaging silent neurological abnormalities need to be proved in a patient at risk; in critically ill patients in intensive therapy units (ITUs); during surgery when monitoring of spinal cord or cerebral function is needed; or when sensory functions have to be tested in an uncooperative patient.

In young patients or in children with behavioural or mental problems, EPs are an integral part of the clinical examination. It has to be realised, however, that the techniques often have to be adjusted by adopting a very simple, basic methodology capable of producing robust and reproducible results in this population. The laboratory has to be ready to deal with infants and anxious parents and gently to introduce older children to the procedure. Flexibility is needed because the test duration, in relation to cooperation and to the need for a sleep recording, cannot be predicted (Chiozza et al 1997).

Of the many types of EPs, only a few are routinely used as a diagnostic tool in paediatrics and neonatology. These are the so-called 'sensory EPs' in the visual, auditory and somatosensory modalities. These EPs are easy to record even in a young child, have a low inter- and intrasubject variability and, therefore, are clinically reliable. Furthermore, they provide diagnostic information that is particularly useful in subjects unable to describe sensory symptoms and where clinical examination is not reliable.

This chapter describes, in ways practical for the beginner, the standard methodology of EP testing, the normal findings and maturation of VEPs, AEPs and SEPs.

RETINAL AND CORTICAL VISUAL EVOKED POTENTIALS

Introduction
The electroretinogram

The electroretinogram (ERG) may be regarded as a special type of EP, which does not arise in the brain per se. It is the electrical potential evoked in the retina by a flash of light or changing pattern and is usually recorded near to or at the cornea. It is only relatively recently that the ERG has entered routine clinical practice. This probably reflects the introduction of ERG recording electrodes (Arden et al 1979, Dawson et al 1979, Hawlina and Konec 1992) that are easily applied with minimal risk of corneal abrasion, do not usually require the use of corneal anaesthesia, are well tolerated by the patient, and

do not interfere with the optics of the eye. This last property allows the recording of the retinal potentials evoked by structured pattern stimuli (PERG).

Few neurophysiologists offer the range of expertise available in dedicated ophthalmic electrodiagnostic clinics; the aspects of ERG most likely to be met in routine neurophysiological practice are described below. The reader is referred elsewhere for more complete reviews of the ERG in ophthalmic practice (Heckenlively and Arden 1991, Fishman et al 2001). Standards for both ERG and PERG have been published by the International Society for Clinical Electrophysiology of Vision (ISCEV) (Marmor and Zrenner 1999, Bach et al 2001) and the IFCN (Deuschl and Eisen 1999). Compliance with these recommendations will facilitate interlaboratory comparisons and communication, and also increase the likelihood of correct diagnosis by facilitating understanding of published data.

Cortical VEPs

VEPs can be evoked by brief changes either in the luminance (flash VEPs (FVEPs)) or in the contrast (pattern VEPs (PVEPs)) within the field of vision; although both types of change can be combined in a single stimulus, this simple distinction between flash and pattern stimulation guides the strategy of the clinical neurophysiologist when testing the function of the visual pathways and cortex in patients. Until about 1970, only the transient response to a diffuse stroboscopic flash was routinely used in the clinical setting (Cigánek 1961, Gastaut and Régis 1965); this type of VEP proved to be useful mainly in testing the visual function in patients with poor visual acuity and/or fixation and also revealed some specific abnormal waveforms in encephalopathies (Pampiglione and Harden 1977). The clinical application of VEPs in neurological disease remained limited until the introduction of the pattern reversal technique and the first demonstration by Halliday et al (1973) that it could reveal clinically silent demyelination of the visual pathways, a finding that represented a major breakthrough in the history of clinical neurophysiology.

However, with pattern stimulation it is difficult and time consuming to obtain reliable results in neonates and young or uncooperative children. In paediatric departments FVEPs are most often used (Harden et al 1989). The complex FVEPs of adults are highly variable and of limited clinical value. In young children the FVEP has a simpler and more reproducible waveform. When special care is devoted to standardisation of the setting, vigilance carefully monitored and a standard protocol adopted in patients and controls, FVEPs can give reliable diagnostic results in paediatric and neonatal practice. However, whenever the child is sufficiently cooperative, pattern stimulation should also be used. The PVEP has, in fact, a greater sensitivity to optic pathway lesions and is complementary to the FVEP, since it tests different (more central) retinocortical pathways.

Normal VEP waveforms in adults and their generators
The ERG

EP morphology evolves progressively from the neonatal period to adulthood. The normal waveforms of the mature EPs are described here; an account of their maturation is given on pages 122 ff. The typical flash ERG (FERG) waveform consists of a negative wave 'a' generated by the photoreceptor activation, followed by a positive 'b' wave representing the complex cell interactions in the retinal inner nuclear layer.

Since the rods and cones have different temporal and spectral sensitivities, manipulation of parameters such as flash intensity and colour, frequency of stimulation, and the state of light and dark adaptation of the eye can be used to achieve separate evaluation of rod and cone function. Additionally, use of both flashes and pattern stimuli (the latter with no overall luminance change) allows separate assessment of both peripheral and central (macular) retinal function.

The response to dim blue flashes obtained from the dark-adapted eye is a rod-specific response consisting only of a b-wave (scotopic, dark-adapted ERG) of high amplitude, broad morphology and long peak latency; when the intensity of the stimulus is below the cone threshold there is insufficient photoactivation to record an a-wave (*Fig. 3.5(a)*).

The response to high intensity white flashes in the light-adapted eye is a cone-specific response (rod suppressing background illumination) and consists of an a- and b-wave complex of short duration (photopic, light-adapted ERG; *Fig. 3.5(b)*). The 30 Hz flicker response is also a cone-specific, steady-state response (the rods have poor temporal resolution and saturate at rates > 24/s) (*Fig. 3.5(c)*).

The oscillatory potentials, small wavelets seen on the ascending limb of the b-wave, are considered to be generated by complex inhibitory circuits in the inner retinal layers, probably originating in amacrine cells. They can be recorded under scotopic or photopic conditions. Many laboratories use off-line filtering of the maximal or photopic single flash ERGs (*Fig. 3.6*).

The PERG waveform is composed of a P50–N95 complex and is thought to express the ganglion cell activity.

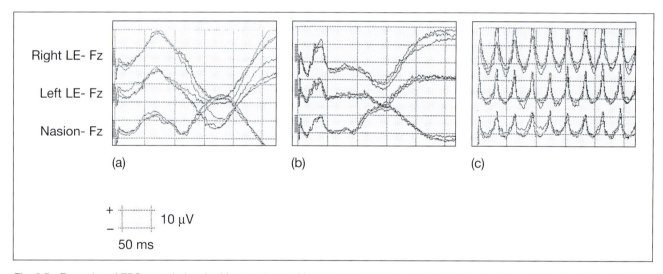

Fig. 3.5   Examples of ERG recorded under (a) scotopic and (b) photopic conditions. (c) The ERG to repetitive photic stimulation at 30 Hz. LE, lower eyelid.

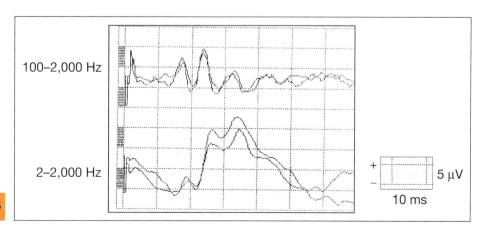

Fig. 3.6   Oscillatory potentials recorded with different filters under photopic conditions with skin electrodes in a 4-year-old normal child. Recording montage: left lower eyelid to Fz.

The VEP

In adults the PVEP consists of three main waves labelled by polarity and mean latency N70, P100 and N145 (*Fig. 3.7(a)*); they reflect activation of cortical areas related to perception of patterned visual stimuli. The morphology of the FVEP in adults, though quite variable, is composed of six major peaks. The nomenclature most widely used is that proposed by Cigánek (1961). The most reliable components are peaks III and IV, corresponding to latency and polarity, respectively, to the N70 and P100 of the PVEP (*Fig. 3.7(b)*). The FVEP generators are probably cortical, but a contribution of subcortical generators has been claimed.

Retinal and cortical VEPs: methodology
Recording environment

In adults FERGs are performed following protocols recommended by the International Society for Clinical Electrophysiology of Vision (ISCEV) (Marmor and Zrenner 1999, Bach et al 2001) and the IFCN (Deuschl and Eisen 1999). The procedure implies corneal recording electrodes, fully dilated pupils and a Ganzfeld stimulator (a diffusely illuminated sphere in which the child's head is inserted, enabling uniform whole-field illumination and strict control of stimulus and background intensity). The IFCN guidelines include specifications for paediatric ERG recordings: special sized corneal electrodes and, when there is non-cooperation, the use of restraint for small infants and sedation or general anaesthesia for older children (especially for those aged 2–6 years in whom restraint can be difficult) (Celesia et al 1999). Even with care to minimise corneal and psychological trauma, this approach is quite invasive and poses ethical problems both in normative data collection and in clinical application. Furthermore, when sedation is needed, problems in standardised application of the technique arise from possible interference of sedatives (Jaffe et al 1989) and anaesthetics on retinal function. For these reasons, laboratories involved mainly with children have adopted non-invasive surface electrodes for routine assessment of gross retinal function (Harden 1974, 1982, Harden et al 1989, Kriss 1994). Kriss and colleagues (1992) describe a clinical protocol for separate evaluation of rod and cone function in which a stroboscope is hand-held close to the patient's eyes by a skilled technologist so as to follow the patient's head and eye movements, giving a 'pseudo-Ganzfeld' stimulus. Dilation of the pupil carries a small risk of inducing an attack of acute closed-angle glaucoma, but this can be avoided as it has been shown that less than maximal pupillary dilatation does not significantly affect ERG recordings in children (Kriss et al 1992). Unless the child blinks a lot or is restless, the recordings can be done in the awake state, but when muscle artefacts are prominent it is better to record with the child asleep, gently holding the eyelids open and monitoring the eyeball position. If recordings are performed during sleep the interpretation of abnormal results has to be viewed with caution because of the possible effects sleep on the ERG (Peña et al 1999).

High-quality ERG recordings can be reliably achieved, even in the incubator, with techniques that are only minimally invasive and which are well tolerated by preterm and term neonates (Mactier et al 1988, Harding et al 1989). Mactier's group recorded in the incubator with a stroboscope and a specially constructed light-proof covering with a hole to allow for video monitoring of the neonate's state. They found that the technique compared well with formal Ganzfeld recording when applied to adults.

FVEPs are best obtained from patients and controls in a quiet, semi-dark room, the illumination of which is kept constant. In the author's laboratory, a small light source of known intensity is used in an otherwise fully darkened room. The patient has to be fully awake throughout the recording, since sleep, and especially drowsiness, affects the waveform in an unpredictable way; an effect that decreases with age.

Changes of response with the state of arousal is especially important in the newborn. Whyte et al (1987) compared VEPs from binocular stimulation with light-

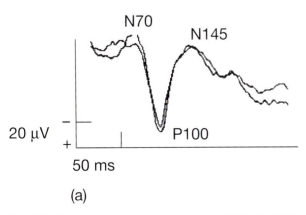

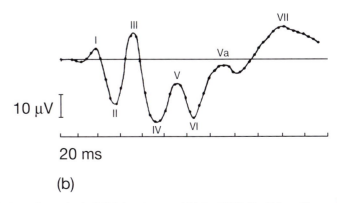

(a) (b)

Fig. 3.7 Normal adult waveform and nomenclature of: (a) the PVEP, from the author's (AS) laboratory; and (b) the FVEP (Oz-Pz) positive down, from Cigánek (1961).

emitting diode (LED) goggles in seven preterm neonates in four awake–sleep stages: awake, transitional or atypical, quiet sleep (QS) and active sleep (AS) (see p. 36). They demonstrated no difference in the N300 between the awake, atypical or AS stages; however, there was a significant amplitude decrease in QS. The P200, present in waking newborns of 40 weeks gestational age (GA), disappeared in both active and QS, and a later positive component, P400, was variable but reliably present in the awake or atypical states (*Fig. 3.8*). In longitudinal or follow-up studies valid comparisons of latency and amplitude can only be made in comparable states of arousal.

Apkarian et al (1991) extended this work with a study of 45 full term newborns between 2 and 7 days of age and assessed the state across the sleep–wake cycle by clinical scoring and polygraphic recordings of EEG, electro-oculogram (EOG), electrocardiogram (ECG), chin electromyogram (EMG) and respiration. They showed that state affects the amplitude, latency and waveform of the VEP, with the most significant differences being between sleep and wakefulness (*Fig. 3.9* and *Table 3.1*). They consider that control of state will reduce intra- and intersubject variability. More recently, Mercuri et al (1995) showed an increase in VEP amplitude, with no changes in latency, in preterm and full term neonates when changing from sleep to the awake state. Their distinction between behavioural states was made by clinical observation. The effect of sleep in the newborn and children with encephalopathy is not known.

If the patient is very restless, as often happens in children with behaviour disorders, a recording may be obtained in sleep but only normal results should be considered as clinically valid. *Figure 3.10* shows FVEP traces obtained in a neurologically normal 10-month-old infant in an awake state and after he fell asleep during the same recording session.

PVEP studies require active cooperation of the subject, who is usually instructed to gaze at a dot, placed either in the centre of the pattern for full-field stimulation, or at the edge of the pattern for partial-field stimulation. The most elaborate means of controlling fixation is to record eye movements and automatically to reject all sweeps contaminated by eye-movement potentials. It may also be useful to monitor ongoing EEG activity during the test and to stop the averaging when the EEG is contaminated by excessive artefact. EEG monitoring is also helpful by detecting the presence of an alpha rhythm, which signals eye closure or drowsiness with open eyes. If automatic artefact rejection or raw EEG monitoring are not available, the stability of the gaze can be checked visually by the technologist throughout the examination; and in practice usually ensures fixation, even in young infants, averaging being suspended when the gaze is not directed at the pattern.

The setting for PVEP in children above 5–6 years of age is the same as it is in adults. To encourage the child to concentrate on the screen the ambient light is kept low and environmental distractions are minimised. The

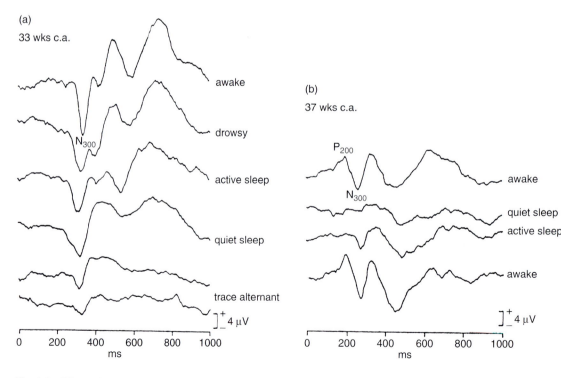

Fig. 3.8   Effect of awake–sleep stages on the VEP elicited by binocular stimulation with LED goggles in two preterm neonates of 33 and 37 weeks' postconceptional age. (a) Marked reduction of N300 during tracé alternant. (b) Loss of P200 in quiet sleep. Note that the second trace in (b) shows no components during quiet sleep in this neonate, emphasising the need to record VEPs in preterm neonates only when awake or in active sleep. (From Binnie et al (2003), by permission.)

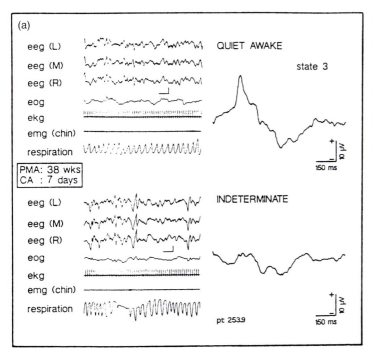

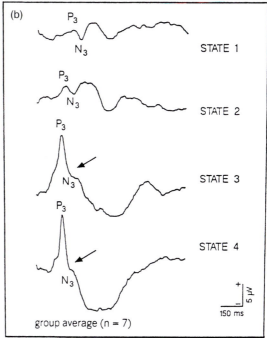

Fig. 3.9 Effects of behavioural state on the FVEP. (a) Polygraphic recording and corresponding VEP in a 1-week-old term neonate during quiet wakefulness (upper traces) and while in transition to sleep (lower traces). EEG aids discrimination between state 3 and the indeterminate condition reflecting drowsiness. The prominent positive P3 deflection with latency of about 200 ms is reliably present only during wakefulness. Horizontal EEG calibration bar equals 2 s. Vertical EEG and EMG calibration bar equals 90 μV. (b) Averaged luminance flash VEPs as a function of state in 7 healthy term neonates 2–7 days of age. Latency of the positive P3 deflection for states 3 and 4 is about 200 ms and is much more clearly defined than the comparable deflections (also labelled P3) of states 1 and 2. (From Apkarian et al (1991), with permission.)

author uses alternate monocular stimuli to the left and right eyes, in each trial, in order to reduce fatigue and, by ensuring a similar level of attention, for a more reliable comparison of responses from the two eyes. In younger children a more sophisticated system to ensure gaze fixa-

Table 3.1 Effect of state on the characteristics of the neonatal transient luminance VEP (group mean ± SD) in 45 healthy term infants between the 2nd and 7th days of life*			
State†	Latency P3 (ms)	Latency N3 (ms)	Amplitude difference P3–N3 (μV)
1	255 ± 38	330 ± 26	6.2 ± 2.7
2	249 ± 51	309 ± 41	5.7 ± 3.5
3	189 ± 16	242 ± 15	12.8 ± 5.0
4	191 ± 13	238 ± 8	15.1 ± 6.0

*From Apkarian et al (1991).

†States are defined according to criteria of Prechtl and Beintema (1964) as: 1 = quiet sleep; 2 = active (REM) sleep; 3 = quiet, awake; 4 = active, awake.

SD, standard deviation.

tion has to be adopted; some television stimulators offer superimposition of a cartoon film on the pattern in order to attract attention. It should be noted here that the time necessary to average responses to a set of 100 pattern-reversals at a rate of 2/s in a subject maintaining stable fixation is less than 1 min. Even when such optimal conditions are not obtainable, causing rejection of some sweeps or transient interruption of averaging, the whole duration of the test usually remains reasonably short. The above system works quite well in young children affected mainly by ophthalmological disorders, but in the presence of neurological disorders it may be difficult if not impossible to obtain reliable fixation. In these circumstances, only normal results are reliable for diagnostic purposes.

Observance of stringent recording standards is essential to obtain clinically usable tracings of PERG, e.g. the avoidance of muscle artefact from jaw clenching or head movement, careful monitoring of fixation and careful positioning of the corneal electrodes. The use of a headrest is recommended to ensure absolute head stability, and therefore constant relative vertical position of the corneal electrodes and the eyes. Variability, especially of the N95 component, may be reduced by regularly suspending averaging to let the patient blink. Binocular PERG registration has the great advantage that, in patients with unilateral visual acuity loss, fixation can be maintained by the eye in which vision is unimpaired. There may, however, be problems with an eye with a strabismus, amblyopia or a latent squint; such an eye may show lateral movements or a drift in fixation. Monocular recording is indicated under such circumstances. Any spectacles used by the patient to correct refractive error should be worn, as appropriate. Pupillary diameter should be examined. It is very difficult, if not impossible, to achieve the recording conditions described above, even in a very cooperative child, and therefore PERG is not routinely used in children but is reserved for investigation of special clinical issues, and it is not discussed further here.

In paediatric recordings the need of obtaining the minimum clinically relevant information as quickly as possible will guide the choice of the protocol. It is often convenient to record retinal and cortical VEPs simultaneously from electrodes positioned respectively on the skin near the eye or from the conjunctiva and on the occipital areas of the scalp. The setting is that used for VEP testing and has the advantage of giving additional information on gross retinal function (see p. 121).

Stimulators and stimulus parameters

A range of stimulus types, wider than for EPs of any other modality, has been used to produce visual EPs, the characteristics of which are highly dependent upon the nature of the stimuli.

The most used type of visual stimulation in neonates and infants is the stroboscopic flash. A disadvantage of the stroboscope is that often there is a click, which can evoke an auditory potential at the vertex in the same latency range as the visual response. White noise through earphones helps to mask this potentially confusing stimulus. Flash stimulation can also be presented via goggles fitted with a mosaic of LEDs (Taylor et al 1987). These are usually red and less intense than a stroboscope, although closer to the eye. The red light is less attenuated by closed eyelids than is white light. A disadvantage is that the eyes cannot be monitored. An awake child may

refuse to wear the goggles, but in other settings, such as the operating theatre, they may be more convenient than a stroboscope (Keenan et al 1987).

Ganzfeld stimulators can be used in paediatrics only if provided with a facility for video monitoring the eye position; many young or restless awake children will not accept this procedure, but a stroboscope with neutral density or coloured filters may be placed very close to the eyes giving a 'pseudo-Ganzfeld' stimulus, suitable for paediatric ERG.

The flash duration, colour, intensity, rate of the stimuli and background illumination should follow the IFCN guidelines (Celesia and Brigell 1999) and should be kept constant, since they influence the VEP waveform (Fichsel 1969, Pryds et al 1988). Commonly, high intensity, white light flashes at 1–2 Hz in a dimly illuminated room are used for FVEP recordings. For neonates a lower rate of stimulation should be used (≤ 0.5 Hz), since repetition tends to reduce the amplitude and increase the latency with a possible absence of the response in the preterm neonate at 1/s or more (Pryds 1992).

Moving or changing patterns, such as checkerboard reversal, can be generated on a VDU and switched electronically at the required rate (usually about 2/s). When checkerboard patterns are used, the check size is important. Check sizes of < 15' (minutes of arc) stimulate the spatial frequency receptors of the fovea; check sizes < 30' stimulate all spatial frequency detectors; check sizes > 40' stimulate luminance channels, as do light flashes. This means that the VEP to large checkerboard patterns is very similar to the FVEP.

During the first 6 months of life the optimal check size to elicit the VEP varies. *Table 3.2* summarises the optimal check size at different ages as recommended by the IFCN following the work by Sokol and Moskowitz (1981) and Sokol et al (1983). If no responses are obtained with the check size appropriate to the infant's age, the check size should be increased by a factor of 2.

The PERG can be elicited with the same checkerboard stimulator as for PVEP recording. It is recom-

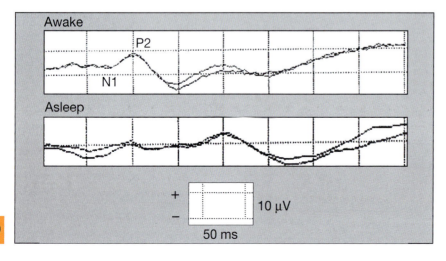

Fig. 3.10 FVEPs obtained in a 10-month-old infant in an awake state and after he fell asleep during the same recording session. Recordings from Oz–Fz and binocular stimulation.

Table 3.2 Optimal check size to elicit PVEPs in infants up to 6 months of age	
Age (months)	Check size (minutes of arc)
1	120–140
2	60–120
3–5	30–60
≥ 6	15–20

mended that the maximum available contrast be used for routine clinical work.

Binocular registration, i.e. simultaneous recording from both eyes, is recommended for both FERG and PERG.

In most patients the PVEP is elicited only by monocular stimulation; however, for FVEP monocular and binocular stimulation help to distinguish between pre- and postchiasmatic abnormalities. Monocular stimulation is obtained by covering the non-stimulated eye with a patch; for FVEPs the patch must be made of densely opaque material and light should not leak around its edges. In the neonatal ITU it is most practical to limit the examination to binocular stimulation using a hand-held device to follow the eyes.

Recording parameters

The standard clinical averager used in neurophysiology departments for VEP recordings, is also suitable for FERG with skin electrodes and for PERG. For technical details about recording parameters for FERG, PERG and VEP the reader is referred to standard texts such as that by Binnie et al (2004).

In neonates and infants, a longer analysis time, up to 1000 ms, is recommended because of the immaturity of the VEP waveform. It is also necessary to reduce the average to less than 100 sweeps, to avoid effects of habituation. Filters are usually set between 1 and 100 Hz. For ERG recordings a bandwidth of 1–1,500 Hz is preferred; the high-pass filter needs to be set at 100 Hz for optimal visualisation of the oscillatory potentials of the FERG.

Electrodes and montages

The electrodes for ERG recordings in adults include the Burian–Allen contact lens, the Arden gold foil (Arden et al 1979), the DTL fibre (Dawson et al 1979) and the H-K loop (Hawlina and Konec 1992), as described in standard texts (IFCN 1999, Binnie et al 2004). Disposable skin electrodes attached close to the child's eyes are the preferred option in paediatrics. The recording is well tolerated and can be performed repeatedly in follow-up studies. The application of skin electrodes for ERG recordings has been recently validated (Meredith et al 2003, Bradshaw et al 2004). The simplest technique is a surface electrode on the nasion and is recommended for simultaneous recording of FERG and FVEP in infants and small children (Harden 1982). In the author's department, disposable electrodes with incorporated solid gel adhesive paste are applied to the bridge of the nose and (in a slightly modified form to allow a close fit) just under each lower eyelid, very close to the lower eyelashes (Fig. 3.11). The common reference is Fz. This procedure has the advantage of being little affected by eye movements in the horizontal plane and permits monitoring of movements in the vertical plane by comparison of recordings from lower eyelids with those from the nasion (or upper eyelids) (Fig. 3.12).

The use of this non-invasive methodology for ERG testing based on skin electrode recordings has greatly expanded the opportunity of detecting and categorising retinopathies associated with paediatric genetic syndromes and neurodegenerative disorders. However, it is important to be aware of the limitations of using this non-standard methodology:

1. The morphology of the ERG is the same as that recorded with standard corneal electrodes, but the amplitude is reduced about ten-fold. Therefore, these electrodes can be reliably used only above 6 months of age when most of the maturational increment of the amplitude has already taken place.

2. Careful consideration of the position of the reference electrode(s) is essential if the recordings are to be free of contamination from the cortically generated VEP or from the ERG of the other eye. This is more of a problem for PERG recording than for FERG because of the much smaller amplitude.

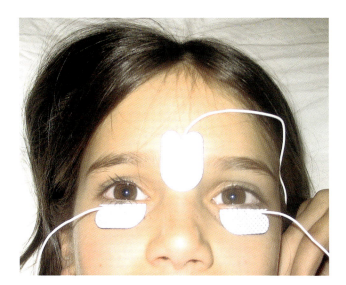

Fig. 3.11 Position of right and left lower eyelid and nasion electrodes in the author's laboratory for recording the FERG, all referred to Fz. With the eyes in the primary position the ERG amplitude is bigger at the lower eyelid electrodes. If the gaze is upwards the ERG at the nasion electrode is larger.

Neonatal ERG has been recorded by Mactier et al (1988) and Harding et al (1989) using the lightweight silver-impregnated thread (DTL) electrode developed by Dawson et al (1979). This electrode is placed in the lower conjunctival sac and supported at the outer canthus. The technique has been refined by Thompson and Drasdo (1987). Although initially used with local anaesthesia and pupillary dilatation, it is now common experience that DTL electrodes can be used in preterm and term neonates without pharmacological aids – the neonate typically remains asleep throughout.

In older children, as in normal adult subjects, components of FVEP peaking before 200 ms as well as the main (positive) P100 potential to pattern reversal are maximal over the occipital region at the midline of the head. Thus an electrode placed at Oz, or 5 cm above the inion on the midline, picks up VEPs with maximal amplitude. The main P100 component of the PVEP to full-field stimulation spreads widely over the occipital region and is maximal at the midline. It is recommended that at least two symmetrical electrodes are placed at 5 cm on each side of the midoccipital electrode across the back of the head or, better, that a five-channel montage be used, with two lateral electrodes placed 5 and 10 cm lateral to the midline on each side. This five-channel montage with a midfrontal reference has proved to be the most suitable to assess the PVEP waveform to half-field stimulation (Halliday et al 1982). The ground electrode is usually placed in the frontal region.

For FVEP recordings in younger children and infants it is useful to place at least three recording electrodes on the posterior scalp: O1, O2 and Oz locations of the International 10–20 System. These locations are identified with proportional measurement (i.e. Oz corresponds to 10% of the nasion–inion distance), and are therefore preferable for work in children to the absolute measurements suggested for adults (Halliday et al 1982), which do not take account of changes in size and shape of the child's head with somatic and brain growth.

Following the principles outlined above, in simultaneous recordings of FERG and FVEP, the suggested montage is:
- channels 1 and 2: right and left lower eyelids
- channels 3, 4 and 5: Oz, O1 and O2
- a common reference at Fz.

If the recording system does not provide five-channel recordings, only one channel should be devoted to the ERG, with the recording electrode placed at the nasion.

In the neonatal ITU electrode placement may be limited to three electrodes for the minimal examination of a single-channel VEP. The 'active' electrode is placed on the midline, just above the level of the inion, and is referred to a midfrontal electrode; the ground electrode is placed either on the mastoid process or on an earlobe. Electrode placement is important because of variation in distribution around the occiput of the maturing VEP (Stanley et al 1987, Lupton et al 1990). With multichannel equipment it is perfectly feasible, even in the preterm neonate, to record from a transverse row of three occipital electrodes each referred to the midfrontal electrode, thus providing additional information about the function of left and right postchiasmal visual pathways. Computer-based topographic brain mapping technique permits responses to be recorded over large areas of brain, even in the newborn.

Retinal and cortical VEPs: maturation and normative data
The ERG

Detailed studies concerning retinal morphology (Hendrickson and Drucker 1992) and other aspects of normal and abnormal visual development (Fulton 1988, Van Hof-van Duin 1992, Simons 1993) show that retinal photoreceptors are structurally mature at birth, except for those in the foveal region, which only achieve maturation between 15 and 45 months of age (Hendrickson and Yuodelis 1984).

Electrophysiological studies in neonates born at term and preterm indicate major postnatal retinal changes. The FERG has been recorded as early as 26 weeks GA with DTL electrodes. Both longitudinal and cross-sectional data show a decrease in a-wave latency and an increase in a-b amplitude with age. ERG measures do not correlate with postconceptional age, suggesting that duration of light exposure after birth may be an important factor in

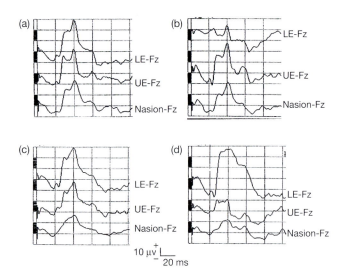

Fig. 3.12 ERG recording with skin electrodes placed at different periorbital locations. Note that with the eyes in the primary position (a) the amplitude is slightly higher when recording from lower eyelids; the situation is not very different with temporal deviation of the gaze (c); with the gaze directed upwards, the amplitude drops and the morphology is affected in recordings from the lower eyelids (b), while with the gaze directed downwards this happens in recordings from the upper eyelids (d). LE, lower eyelid; UE, upper eyelid; negativity is displayed downwards in all traces.

the development of the ERG in preterm infants (Mactier et al 1988, Leaf et al 1995) (*Fig. 3.13*).

More recently, electrophysiological studies that selectively evaluate cone and rod systems development have suggested that in young infants peripheral cone function is relatively more mature than rod function, and that the kinetics of deactivation of the rod photoresponse are slower (Hansen and Fulton 2005a, b).

The postnatal maturation of the ERG is very rapid during the first few months. By 6 months of age the amplitude is already 90% of the corresponding adult ERG, and thereafter gradually increases until adulthood. The differences between different age groups are not statistically significant after 6 months of age (Kriss et al 1992, Flores-Guevara et al 1996). Normal parameters for the ERG at different ages are reported in the literature for corneal (Flores-Guevara et al 1996, Fulton et al 2003) and skin electrode recordings (Harden et al 1989, Kriss et al 1992).

Normal FERG values recorded with skin electrodes at different ages from 1 to 24 months are summarised in *Table 3.3*.

The VEP

The nomenclature of the components of the FVEP during maturation can be confusing as different authors label the same peaks differently, as is evident in the comparison of published data reported by Harden (1982) and that of adult nomenclature by Cigánek (1961) (*Table 3.4*).

The adoption of a common nomenclature is desirable, and recently the IFCN has suggested labelling the occipitally recorded oscillations from the first positive–negative complex (i.e. P1, N1, P2, N2, ...) (Celesia et al 1999). From 6 months of age the basic components of FVEPs corresponding to components III and IV of Cigánek are N1 and P2.

Maturation of the neonatal FVEP has been studied and normative data collected for many years since the classical work of Ellingson (1958, 1960), Engel and Butler (1963) and Hrbek and Mares (1964). Norms are well established for various GAs for flash stimulation, although less extensively for LEDs and pattern stimulation. From 28 to 32 weeks' GA, the FVEP comprises a small occipital positive wave with a latency of about 200 ms, followed by a more prominent negative potential at about 300 ms, called N300 by Parmelee et al (1967, 1968), which may be bifid in some preterm infants (Taylor et al 1987). During this period the earlier positive component becomes increasingly evident. Between 35 and 40 weeks' GA the positive wave increases in amplitude and a preceding negativity appears, giving a negative–positive–negative complex with some lesser late components.

This sequence of development of the VEP, and in particular the increasing prominence of the positive wave, appears to be a more reliable indicator of maturation than either latency or amplitude measurement

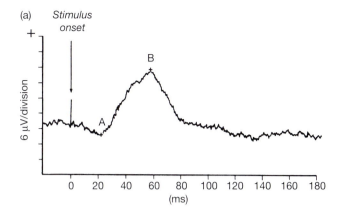

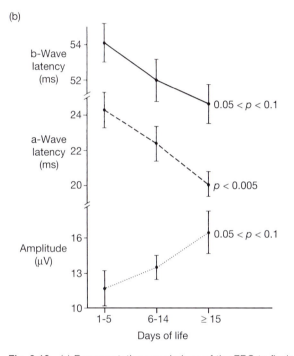

Fig. 3.13 (a) Representative morphology of the ERG to flash stimulation in a 32-day-old infant born after 32 weeks' gestation with birth weight of 1800 g. (b) Latency and amplitude values (group mean ± 1 standard error of the mean) of sequential weekly ERGs in 15 preterm infants; Friedman's two-step non-parametric analysis of variance was used to estimate the significance of the slopes. (From Mactier et al (1988), by permission.)

(*Fig. 3.14*) (Watanabe et al 1972, Mushin 1988); the described changes have been shown to parallel the anatomical maturation of the human visual cortex (Takashima et al 1980, Volpe 2000).

The origin of the VEP in premature infants is unclear. VEPs, even though with an abnormal positive–negative waveform, have been recorded in the complete absence of visual cortex in a child that was able to focus and track briefly, raising the possibility that some aspect of visual function may not be cortically mediated (Dubowitz et al 1986). However, Mushin (1988) suggests that the characteristic negative VEP, which occurs before 30 weeks' GA, probably reflects the activity of grossly immature pyramidal and non-pyramidal neurons. The appearance of the positive wave in the VEP at around 32 weeks' GA is associated with extensive dendrite differentiation and the development of dendritic spines in pyramidal neurons. After 40 weeks' GA there is still a rapid maturation of the FVEP; the waveform becomes more complex and the latency of components decreases.

Morphological changes are completed soon after birth. The initial large positive wave at around 200 ms (P200) becomes a bifid component with an earlier positivity (P100) usually emerging before 4 weeks of age. The P100 component may be present at birth in 88% of term infants (Taylor and McCulloch 1992). The two positive components remain quite prominent and distinct for several months, despite gradual shortening of the latencies of these components (*Fig. 3.15*). They usually merge into a single large positive component (P100), with a near-adult appearance by 6 months of age; there is also a very prominent P400 component after the age of 3 months (Mushin 1988, Taylor 1992). The morphological modifi-

Table 3.3   Skin ERG amplitude and latency (all values mean ± SD) in normal children: first 24 months of life*

Age (months)	No. of children	a-Wave latency (ms)	b-Wave latency (ms)	a–b Amplitude (µV)
1[†]	9	-	-	9.6 ± 1.8
2[†]	6	-	-	13.3 ± 4.6
3[†]	5	-	-	18.0 ± 2.0
4[†]	3	-	-	29.7 ± 7.2
5[†]	3	-	-	24.6 ± 7.1
6[†]	5	-	-	27.8 ± 5.5
0–6	10	17.5 ± 2.9	41.3 ± 3.6	7.5 ± 3.9
7–12	10	16.6 ± 1.7	41.1 ± 3.3	10.3 ± 4.1
13–24	10	12.2 ± 3.0	41.8 ± 4.0	11.9 ± 7.2

*Data from author's laboratory, mean from left and right eyes.
[†]Modified from Kriss et al (1992).
SD, standard deviation.

Table 3.4   Summary of FVEP nomenclature adopted in studies on the maturation of VEPs (modified from Harden et al (1982)) compared with corresponding peaks of the adult waveform reported by Ciganek (1961)

Ciganek (1961)	Dustman and Beck (1969) (1 month–16 years)	Dustman et al (1977) (4–16 years)	Laget et al (1977) (4 months–15 years)	Blom et al (1980) (4 months–5 years)	Barnet et al (1980) (3 months–3 years)	All studies
I 39.12 ± 4.18	A 34.8–41.1	P1 27.8–28.5 N1 39.0–44.8	N0 38.8–40.9	-	N1 26–32 P1 21–35 N2 36–66	N 35–66
II 53.40 ± 4.42	B 48.1–53.6	P2 49.7–55.3	P1 49.8–50.8	I 42–75	P2 51–73	P 42–75
III 73.33 ± 6.36	C 63.4–74.0	N2 66.1–71.4	N1 66.1–73.3	II 63–100	N3 71–86	N 63–100
IV 94.19 ± 7.13	D 103.0–119.8	P3 100.7–111.4	P2 94.6–99.7	III 102–154	P3 85–135	P 85–154
V 114.00 ± 7.41	E 136.0–141.8	N3 133.6–145.9	–	IV 165–220	N4 103–159	N 103–220
VI 134.55 ± 9.95	F 168.4–191.6	P4 168.4–175.3	–	V 127–298	P4 153–195	P 127–298

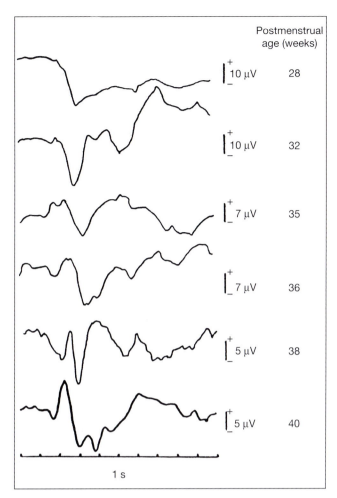

Fig. 3.14 Maturation of the FVEP in preterm infants (LED stimulation, see text). Representative morphology of the responses at different postmenstrual ages from 28 weeks to term. Positivity at the midline occipital electrode is recorded as an upward deflection. Before 30 weeks there is only a negative wave; from 32–35 weeks a preceding positivity appears and increases in amplitude; by 40 weeks an early negative wave precedes the positive component. (From Mushin (1988), by permission.)

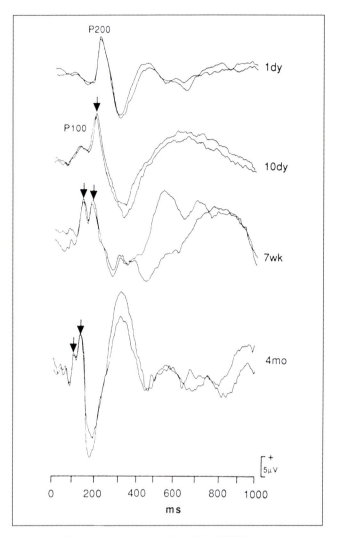

Fig. 3.15 Early postnatal maturation of the FVEP in term infants (LED stimulation). The P100 emerges soon after birth, becoming two clearly separated peaks in the early months of life. These rapidly decrease in latency with increasing age. (From Taylor (1992), by permission.)

cations of the FVEP during maturation are shown in *Fig. 3.17*; by 6 months the more consistent waves N1 and P2 have a near-adult appearance, subsequent maturation being associated with slight reduction of latency and increased morphological complexity.

PVEPs can be elicited in the awake neonate with eyes open by reversal of black and white checks. The responses are dominated by a major positive component (P1) around 280 ms in preterm neonates from 33–37 weeks' conceptional age (CA) (Harding et al 1989). The P1 latency shows a negative correlation with CA (*Fig. 3.16*). The morphology of the PVEP is dependent on the check size in young neonates as well as in older children (Moskowitz and Sokol 1983). There are advantages to using PVEPs in the newborn as they reflect the neuronal mechanisms for the treatment of spatial information, and

therefore determine the functional integrity of the visual pathway. However, they are highly influenced by state of alertness. Roy et al (2004) have recently examined the influence of state on the PVEP in the newborn and have proposed a method to improve testing particularly in preterm newborns.

Normal values at different ages for the N1 and P2 components of the FVEP, based on potentials with normal morphology, are summarised in *Fig. 3.17*.

For PVEP after 5–6 years of age, adult normal values can be used in assessment of normality. For children under 5 years old, see Moskowitz and Sokol (1983).

Maturation of the P1 latency of the PVEP occurs more quickly in VEPs elicited by large than by small checks (*Fig. 3.18*). The adult value is reached in VEPs elicited by checks larger than 400' at 1 year of age and by

125

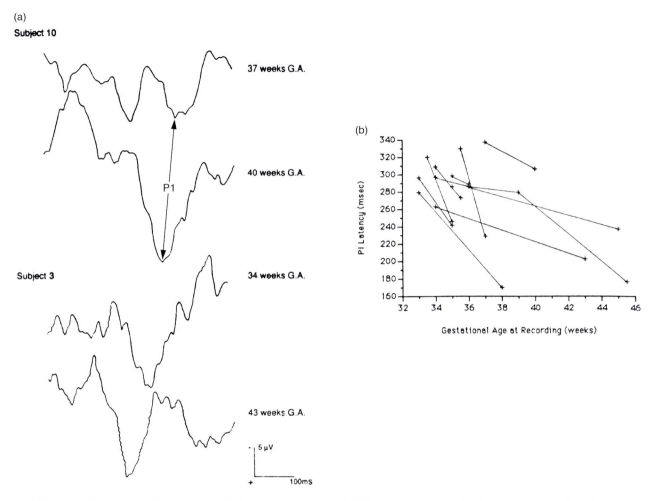

(a)

Subject 10

37 weeks G.A.

40 weeks G.A.

P1

Subject 3

34 weeks G.A.

43 weeks G.A.

5 µV

100ms

(b)

Gestational Age at Recording (weeks)

Fig. 3.16 Maturation of the pattern reversal VEP in preterm neonates. (a) VEPs to pattern reversal in two preterm neonates, showing changes with maturation. Recordings are from Oz to C3 and are the average of 20 responses. In neonate 10 the main positive component (P1) at 37 weeks' GA was at 340 ms, which reduced to 300 ms at 40 weeks' GA. In neonate 3, the comparable latency decrease was from 260 to 200 ms between 34 and 43 weeks' GA. (b) Latencies of P1 in 10 preterm neonates at various GAs. Latency decrease occurs at different rates but is significantly related to GA at recording ($p = 0.004$). (From Harding et al (1989), by permission.)

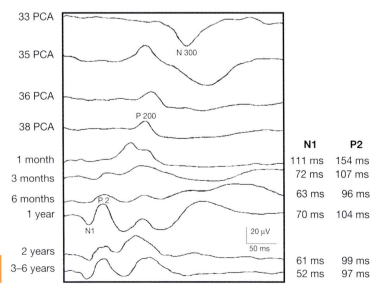

33 PCA

35 PCA — N 300

36 PCA

38 PCA — P 200

1 month

3 months

6 months

1 year — P 2

N 1

20 µV
50 ms

2 years

3–6 years

	N1	P2
1 month	111 ms	154 ms
3 months	72 ms	107 ms
6 months	63 ms	96 ms
1 year	70 ms	104 ms
2 years	61 ms	99 ms
3–6 years	52 ms	97 ms

Fig. 3.17 Representative traces of morphological modifications of the FVEP during maturation: from the author's normative database. Note the dramatic changes in morphology before 6 months of age; by 6 months the more consistent waves NI and P2 have a near-adult appearance, subsequent maturation being associated with slight reduction of latency and increased morphological complexity. Age given on the left (PCA, postconceptional age); NI and P2 latencies given on the right. Recordings from Oz–Fz derivations; binocular flash stimulation. (From Binnie et al (2003), by permission.)

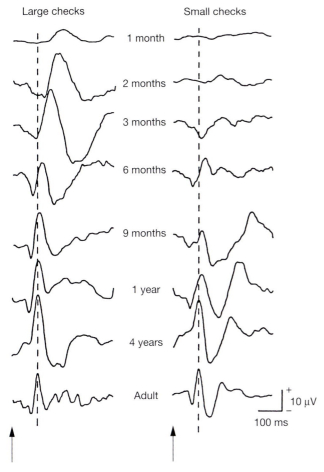

Large checks   Small checks

1 month

2 months

3 months

6 months

9 months

1 year

4 years

Adult

+10 μV

100 ms

Fig. 3.18 Age dependency of checkerboard VEP in response to large (60′) and small (15′) checks reversed at 1 Hz (changeover points at arrows). (From Moskowitz and Sokol (1983), by permission.)

checks of less than 150′ at 3 years of age (Moskowitz and Sokol 1983, Sokol 1986a, b, Taylor and McCulloch 1992).

Retinal and cortical VEPs: processing

The amplitude and latency of the a- and b-waves of the photopic ERG, the amplitude and latency of the first component of the 30 Hz ERG, and the amplitude and latency of the scotopic ERG b-wave should be measured for each eye and the interocular differences calculated.

The amplitude and peak latency of the main waves N70 and P100 of the PVEP should be measured, at locations Oz, O1 and O2. When the PERG is recorded simultaneously to the PVEP, the interpeak latency between P50 and P100 should be measured.

The amplitude and peak latency of the main waves N1 and P2 of the FVEP should be measured at location Oz in recordings following monocular stimulation, and the results from the left and right eyes compared. Waveforms obtained at locations O1 and O2 following binocular stimulation should be compared.

Retinal and cortical VEPs: interpretation

Cone system ERGs are obtained with both single-flash and 30 Hz flicker stimuli superimposed upon a rod-saturating background. The rods have poor temporal resolution, and 30 Hz stimulation combined with the presence of a rod-suppressing background gives a cone-specific response. This is perhaps the more sensitive measure of cone dysfunction and, being generated at an inner retinal level (Bush and Sieving 1996), may be affected in both photoreceptor dysfunction and inner retinal dysfunction. The cone b-wave reflects post-photo-transduction activity. When the intensity of the stimulus is below the cone threshold a rod-specific response is obtained which consists only of a b-wave; this is a measure of 'rod system' dysfunction.

The oscillatory potentials reflect inhibitory activity in the inner retinal layers supplied by the central retinal artery. The ERG, being a mass retinal response, is normal when dysfunction is confined to small retinal areas. For example, chorioretinitis from congenital cytomegalovirus or *Toxoplasma* infections shows usually a normal FERG. Visual dysfunction in focal retinal lesions localised to the central retinal areas which give the major retinocortical input can give rise to an abnormal FVEP. This also applies to macular dysfunction; an eye with disease confined to the macula has a normal full-field ERG but an abnormal VEP.

The clinical (flash) ERG contains no significant retinal ganglion cell contribution; ganglion cell function is explored with PERG.

The simultaneous recording of VEPs is also a useful adjunct to ERG in retinal dystrophies. The major retinal input to VEPs is from the central retina, allowing information to be obtained about extension of dysfunction to this region.

The VEP is a mass response of cortical and possibly subcortical visual areas. Although the cortical generators of FVEPs and PVEPs are not yet fully understood, this is not essential for their clinical use on an empirical basis. With PVEPs, it is possible to localise lesions within the visual pathways this is done by separate stimulation of different parts of the visual field (i.e. hemifield stimulation).

With FVEP, localisation is not possible but the comparison of responses to binocular stimulation recorded in the left and right occipital derivations and of responses recorded at Oz to right and left monocular stimulation, help in distinguishing between pre- and postchiasmal lesions. Caution is necessary and results should always be evaluated within the context of clinical and neuroimaging findings. A simplified approach to localisation is summarised in *Table 3.5*.

Retinal and cortical VEPs: clinical role

An integrated approach to the use of clinical neurophysiology, rather than the rigid separation of the different techniques in specialist departments that stemmed from their early development, is recommended.

Integrated investigation of specific groups of disorders of paediatric neurology is described in Chapter 6 (see pp. 275–281), while clinical applications peculiar to the neonatal age are discussed in Chapter 5 (see p. 209–217). Some neuro-ophthalmological applications specific to the visual system are described in Chapter 6 (see p. 321).

A suggested clinical protocol for ERG and VEP testing is given in *Box 3.1*.

AUDITORY EVOKED POTENTIALS

Introduction

A variety of potentials can be recorded following acoustic stimulation that reflect sequential activation of the structures of the auditory system from the cochlear receptor through to the primary and secondary projection areas in the auditory cortex. Historically, the major components have been classified according to their time of appearance after the stimulus as early (short), middle and long latency

Table 3.5　Pattern of dysfunction of VEP localisation

Location of dysfunction	ERG	Monocular in Oz VEP	Binocular O1/O2 VEP	EEG
Peripheral retina	Abnormal	Normal	Normal	Normal
Central retina	Normal	Abnormal	Normal	Normal
Prechiasmatic visual pathway	Normal	Abnormal	Normal or abnormal	Normal
Optic chiasm	Normal	Abnormal	Abnormal	Normal
Postchiasmatic visual pathway	Normal	Abnormal	Abnormal	Normal
Occipital cortex	Normal	Abnormal	Abnormal	Normal or abnormal
Cerebral cortex	Normal	Normal or abnormal	Normal or abnormal	Abnormal

Box 3.1　Suggested clinical protocol for ERG and VEP testing

Recording commences with dim stimuli and proceeds to brighter stimuli. With skin electrodes averaging is necessary. The routine ERG clinical protocol should start with scotopic (dark adapted) ERG obtained (after at least 20 min of adaptation in a fully darkened room) with dim blue light flashes at a slow rate. An interstimulus interval (ISI) of 20 s is sufficient to prevent significant changes in adaptive state in a dark-adapted patient, but a 2 s ISI can be used with blue light at low intensity. Darkness should be total, apart from a dim red light, which should not significantly affect adaptation. After the scotopic response has been recorded, the background light is switched on. Photopic responses are recorded after 10 min light adaptation with the background light on, the stimulation being superimposed upon the rod-suppressing background. Stimulation should be presented for a few seconds before data collection is begun. Caution should be exercised with regard to the use of flicker if there is any possibility of photosensitive epilepsy. Monocular stimulation may be used in such cases. Pupillary diameter should be checked at the end of the recording session. The protocol for the ERG is summarised in *Table 3.6*.

FVEP usually follows the ERG in the awake child, starting with binocular stimulation, since some children are upset by monocular occlusion. Then follows monocular stimulation with alternating right and left eyes if the child will tolerate it. If a retinal or prechiasmatic lesion is suspected, every effort should be made to record with monocular stimuli. In cooperative children the PVEP can be performed first.

Table 3.6　Clinical protocol for recording the ERG

ERG type	Adaption	Intensity (J)	Flash duration (ms)	Flash rate (Hz)	Sweep duration (ms)	Sensitivity (μV/cm)	Band pass (Hz)
Scotopic	20 min dark	0.1	0.10	0.5	500	10	2–200
Mesopic		1.0	0.10	1	200	20	2–200
Flicker	10 min light	0.5	0.10	33	500	10	2–200
Photopic	10 min light	2.0	0.10	1	200	10	2–200

waves. The waveforms and nomenclature are summarised in *Fig. 3.19*.

Short latency components have latencies of less than 10 ms; long latencies exceed 50 ms; and middle latencies are intermediate between these values.

Long and middle latency components are predominantly generated by PSPs within the auditory cortex. They are greatly affected by the state of the alertness of the subject and by sedation. Both their high variability, which is dependent on the degree of attention and concentration that the subject directs to the stimulus, and uncertainty about the cortical generators strongly affect the clinical utility of these potentials.

By contrast, short latency AEPs are highly consistent across normal subjects, are unaffected by sedation and are only minimally affected by anaesthesia. The earliest potentials (within 2–3 ms of the stimulus) reflect activation of the cochlear hair cells and the auditory nerve, and are described as the electrocochleogram (ECochG). They are best recorded from the external auditory meatus or tympanic membrane but can also be found adjacent to the stimulated ear, e.g. over the mastoid process or on the earlobe. With appropriate ECochG methodology it is possible to record the electrical potentials of the cochlea (cochlear microphonic and summating potentials) and also the VIIIth nerve action potential, which corresponds to wave I of the brainstem auditory evoked potential (BAEP) and previously was used for measuring the cochlear threshold. ECochG methodology using trans-tympanic needle recording (under general anaesthesia in children) is now reserved for very problematic cases.

Otoacoustic emissions (OAEs), which comprise a relatively new physiological technique, are sounds emitted by active mechanical processes in the cochlear outer cell that can be recorded by a small microphone positioned in the external canal (Kemp 1978, 2002). The technique is easy, inexpensive, quick and not invasive, and can give a rough estimation of the presence of a hearing deficit. The central components of the BAEP are generated within the brainstem, and may reflect both action potentials and postsynaptic activity. They are used both for measuring hearing threshold, in the absence of brainstem pathology, and as a test of brainstem function.

Within the latency range of middle latency auditory evoked potentials (MLAEP) and long latency auditory evoked potentials (LLAEP) other potentials can also be recorded that are related to central nervous system (CNS) processing of auditory stimuli: these are the so-called event related potentials (ERPs) (*Fig. 3.19*). In order to obtain these potentials, specific auditory paradigms are necessary. For the methodology and the significance of these potentials the reader is referred to texts such as Binnie et al (2004, Section 3.3.2) and McCallum (1988).

Of all the neurophysiological techniques available for the functional study of the auditory system, only the BAEP is routinely used in paediatric clinical neurophysiology. The technique is described in detail in the following sections.

Normal BAEP waveforms in adults and their generators

BAEPs are composed of six successive peaks, labelled in Roman numerals I–VI (and occasionally VII) (*Fig. 3.20*).

According to Jewett's scheme (Jewett 1970) waves I–VI refer to upward (vertex positive) deflections in the Ai (ipsilateral ear lobe)–Cz montage. Wave I is a localised negative-going change close to the Ai electrode (lead 1), whereas waves II–VI are far-field positivities picked up at

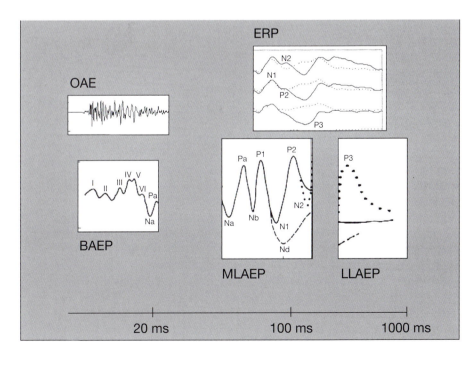

Fig. 3.19 Classification of auditory evoked potentials (AEPs). Note logarithmic scale.

Cz (lead 2); all therefore produce positive deflections (displayed either up or down according to the practice of the laboratory). Due to the complexity of the network of auditory fibres, any attempt to match each of the BAEP waves with a single source is naive and unlikely to be successful. However, specific 'generators' can be attributed to each BAEP wave on the basis of intraoperative recordings and of correlation between abnormal waveforms and lesion sites in patients with discrete cochlear nerve or brainstem lesions (see *Fig. 3.20*) (for a review see Hashimoto 1989).

Wave I originates from the peripheral portion of the cochlear nerve. It peaks with a latency similar to that of the VIIIth nerve action potential of the ECochG recorded directly from the cochlear nerve; it is absent or drastically reduced in recordings from the ear lobe contralateral to stimulation (Ac–Cz). It is the only component to be preserved when conduction is blocked or slowed in the proximal portion of the cochlear nerve, and also in patients with irreversible deterioration of brainstem function and clinical evidence of brain death.

Wave II, although constantly obtained at click intensities of more than 60 dB hearing level (HL), is not commonly used in clinical interpretation of BAEPs. It most probably consists of two separate components overlapping in time and reflecting, respectively, action potentials in the proximal portion of the cochlear nerve and postsynaptic responses of cochlear nucleus cells.

Wave III is, with waves I and V, one of the most constant and robust components of the BAEP complex. It is positive in the Ai–Cz traces, but reverses its polarity or is substantially reduced when the activity is recorded between the contralateral ear lobe and Cz (Ac–Cz). This strongly suggests that wave III is generated by a horizontal dipolar source that, on the basis of intraoperative recordings, has been localised in the pontine portion of the brainstem auditory pathways.

Waves IV and V can be clearly separate or combined to form a single IV/V complex. Wave V reflects the activity of a vertically orientated dipolar source in the midbrain, which is rostral positive and caudal negative. It is most probably generated both by propagation of action potentials in the lateral lemniscus fibres and by postsynaptic responses of midbrain auditory nuclei including the inferior colliculus, in particular that contralateral to the stimulated ear. The origin of wave IV is not firmly established; both a fixed generator in the pons and action potentials ascending in the lateral lemniscus could contribute to this component. Wave V can be detected even at near-threshold levels, and has proved to be the most reliable and useful of BAEP components in clinical practice.

A *wave VI* and even a *wave VII* can be identified in some individual traces. They are present in up to 70% of normal adult BAEPs, but are poorly reproducible. Their origin is unclear and their clinical utility has not yet been established.

BAEPs: methodology

Recording conditions and physiological state affect BAEPs. Subject factors such as age, sex, body temperature and relaxation, as well as technical factors, such as stimulus intensity, repetition rate, polarity, frequency content and mode (monaural/binaural stimuli/white noise masking), all affect the response. However, vigilance, natural sleep, habituation and anaesthesia and most of the sedative and antiepileptic drugs (with the exception of phenytoin) will not significantly affect BAEPs.

BAEP recording environment

Recording of BAEPs in young children may be time consuming, as movements have to be avoided. As the signals are of very small amplitude (a few microvolts) it is important to minimise muscle artefacts. The subject is investigated in a quiet, dimly illuminated room, in a supine or semirecumbent position, to ensure optimal relaxation of the neck and face muscles. In most children aged between 3 months and 3 years, recording is possible only in sleep. This can be obtained by sleep deprivation or after sedation (e.g. by melatonin 3–6 mg, diazepam 0.3 mg/kg or chloral hydrate 50 mg/kg) or under general anaesthesia. Before 3 months of age infants will usually fall asleep naturally (e.g. after feeding). Neonates and children tested in ITUs may be hypothermic; this condition should be avoided during BAEP recordings as it delays neural conduction.

Before testing, any obstruction of the external canal by cerumen (wax) and any middle ear abnormalities should be ruled out by inspection with an otoscope. In case of doubt, the easy and quick technique of tympanometry is available to evaluate middle-ear function. This is of value since middle-ear effusions (glue ear) are very frequent and often asymptomatic, especially in pre-school-age child-

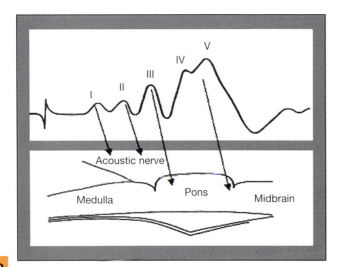

Fig. 3.20   BAEP waveforms and generators.

ren. For tympanometry, a probe inserted into the entrance of the external auditory canal delivers a tone and then measures the ability of the middle ear to transmit sound energy as a function of air pressure in the external canal. The test is completed in a few minutes and does not need special expertise. In cases of middle-ear effusion a variable degree of conductive hearing deficit is expected (never profound). Therefore if the aim of the BAEP investigation is to test brainstem function the test can be performed with an increased stimulus intensity, provided that wave I is elicited. However, if the aim of BAEP recording is a precise estimation of hearing, the test is usually postponed until after the normalisation of middle-ear function.

BAEP stimulators and stimulus parameters

Stimulators The stimulus is usually presented to the infant by a small earphone, which is held gently over the ear whilst the infant lies on its side and has the other ear on the pillow. Alternatively, 'insert' earphones can be used, with the advantage, especially useful in neonatal ITUs, of reducing the effect of ambient noise due to a better coupling of the sound delivery tube to the external auditory canal (Beauchaine et al 1987, Gorga et al 1988). The responses to insert earphones and to more traditional ones seem equivalent provided that a correction for the time taken for the sound to travel through the sound delivery tube is made (~1 ms/ft (1 ms per 30 cm)); hence the need for normative data (see p. 214).

Stimulus types BAEPs are usually evoked by square waves (clicks) of short duration (e.g. 100 μs). Tone bursts of 30–100 ms duration at 1000 Hz are also used. The threshold for clicks is about 30 dB higher than for a continuous tone at 1000 Hz (Stapells et al 1982).

Stimulus polarity A narrow pulse of electrical current passing through the coil of an earphone such that the diaphragm is displaced towards a listener causes a pulse of compressed air ('condensation') to be propagated away from the diaphragm with the velocity of sound (about 340 m/s). It will be heard as a click. If the diaphragm is 'sucked' away from the listener by a current pulse of opposite polarity, the reduced pressure is also propagated, this time as a 'rarefaction'. It will be heard as a similar click. Thus the square waves used can be negative (rarefaction click), positive (condensation click) or alternate rarefaction and condensation clicks giving phase alternation.

From the literature it appears that in adults there are small differences between the averaged brainstem EPs to condensation and rarefaction clicks (Maurer 1985). Rarefaction seems to be preferred, although Stapells et al (1982) showed that in adults there is no difference in threshold for the two types. Many laboratories use alternating clicks in order to reduce the stimulus artefact; however, the nature of the initial phase of the click (rare-

faction or condensation) is an important determinant of BAEP latencies in patients with significant high-frequency hearing loss and is also important in neonates. Stockard and Stockard (1986) obtained results that were more 'useful' clinically in the newborn when evoked by rarefaction than by condensation clicks, since wave I was more easily detectable. They also noted a difference in the interpeak interval (IPI) between waves I and V that was 0.1–0.3 ms shorter with rarefaction clicks. When using alternating condensation and rarefaction clicks, waves I–IV may be attenuated and sometimes undetectable. The use of rarefaction clicks is recommended in the investigation of young infants.

There are cases where it is diagnostically helpful to obtain traces to both rarefaction and condensation clicks (see p. 135).

Intensity The loudness of a sound is determined by the sound pressure (force per unit area, SI units newton/square metre (N/m^2) or pascal (Pa)). The quietest sound at 1000 Hz that can be heard by the average person has a pressure of 20×10^{-6} Pa or 20 μPa, and this pressure level has been adopted as the normal threshold of hearing against which other pressure levels are measured. Unfortunately, there is as yet no standard technique for calibrating the intensity of click stimuli. Both behavioural and acoustical calibrations are used (*Box 3.2*). The sound pressure level (SPL) is defined as the log of the ratio of the measured intensity to the threshold intensity.

There are two methods for standardising recordings using behavioural testing. In the first the threshold of each individual subject is measured by presenting a number of clicks (singly or repetitively at one or two per second) around threshold. This gives the 'sensation level' (SL) in decibels to which the required intensity is added. This is an adequate measure for BAEP testing in neurological (as distinct from audiological) evaluation, but varies to some extent with the ambient noise level and, more importantly, depends on the ability of the

---

Box 3.2 Three standardised scales for defining the intensity of the click in decibels (dB)

- SPL (sound pressure level): the pressure level of the sound, where 0 dB corresponds to a pressure of 20 μPa; this is a physical measure of the sound
- SL (sensory level): the scale zero level is set at the minimal intensity perceived from the subject; this scale can be used in patients able to identify and report their perception of the sound
- nHL (normal hearing level): the scale zero level is set at the mean hearing threshold of a group of normal-hearing subjects (this corresponds to about 30 dB SPL); this scale is used in uncooperative patients where the subjective threshold cannot be obtained

subject to respond to the test procedure. Another method is based on the measurement of the average threshold of ten normally hearing young subjects in the 'standard' environment. This gives a 'normal hearing level' (nHL) against which the hearing of patients can be measured using BAEPs. This standardisation has the disadvantage of being dependent on the ambient noise remaining the same. An acoustic method of measuring click intensity would be very useful, but at present there is no simple technique for calibrating a very brief stimulus.

The choice of the optimal stimulus intensity will depend on the clinical indication. The intensity has a noteworthy effect on BAEP morphology. In a normal subject, at relatively high intensities the major waves I, III and V are always seen, while as the intensity decreases waves I–V disappear. The intensity at which the wave V disappears corresponds to the hearing threshold (*Fig. 3.21*).

Generally, the best approach for neurological applications in infants and children is a two-stimulus-intensity protocol, with one intensity near the threshold value to ensure normal hearing, and one at high intensity at 60–80 dB nHL for evaluation of the central auditory pathways within the brainstem. An intensity of about 70 dB SL is usually adopted in adult patients and it is also suitable for older children.

When BAEP examination is used for audiological indications, more intensities are required. The stimulus intensity is progressively reduced in steps of 10 dB until the response disappears; this allows identification of the hearing threshold and construction of a latency–intensity curve based on wave V.

The ear not being tested can be masked with 30–40 dB white (wide-band) noise to prevent cross-hearing. Although the effect is small, this is important if the tested ear is deaf. The cross-skull attenuation is about 45 dB.

Stimulus rate   The BAEP can be evoked with stimulation rates of 8 Hz up to 200 Hz, but the usually employed frequencies are between 10 and 50 Hz. At faster rates data can be collected quickly but the neural components may be attenuated and delayed. The stimulation rate affects the response more in the neonate than in adults (*Table 3.7*), and therefore normative data have to be collected by age group for the rates used – a tedious but important process.

The faster rates are usually adopted in audiological applications when many trials at different intensities are required. Electrical interference can be reduced by using repetition rates that are not subharmonics of the mains frequencies (e.g. 9 or 11 Hz instead of 10 Hz).

In the author's department the stimulus type is a rarefaction click and the rate is usually 11 Hz. For neurologi-

Table 3.7   Means and standard deviations of increments of wave V latency (in milliseconds at 70 dB HL) calculated by subtracting latency values obtained at 10/s from the latency obtained at 50/s and 90/s.* Note how the latency shift is greater in neonates and infants compared with older children and adults, and the effect of increasing the rate from 5/s to 90/s

Age group	No. of ears	Stimulus rate	
		50–10/s	90–10/s
1 month	23	0.57 ± 0.23	1.01 ± 0.29
3 months	15	0.56 ± 0.12	0.90 ± 0.31
6–12 months	34	0.47 ± 0.16	0.80 ± 0.17
1–3 years	23	0.45 ± 0.19	0.72 ± 0.14
4–6 years	31	0.40 ± 0.19	0.70 ± 0.22
Adults	35	0.41 ± 0.17	0.69 ± 0.17

*From Jiang et al (1991).

SD, standard deviation

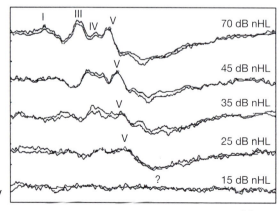

Fig. 3.21   The effect of changing intensity on the BAEP waveform. Note the lengthening of the latency and the progressive disappearance of the waves with reduction of intensity. The last to disappear is wave V at the hearing threshold of the tested subject. Positive is up.

cal assessments the intensity of the clicks is 75 dB nHL (relative to the normal hearing threshold in adults, calibrated in at least ten subjects with a normal audiogram). If information about hearing is not available, responses at 30 dB nHL are obtained to ensure normal cochlear function.

BAEP recording parameters

Analysis time and sampling rates The sweep time should be 10–20 ms, depending on age and clinical indication, with longer sweep being used when testing younger patients for a hearing assessment or when the waveform is clearly abnormal and delayed. A longer sweep time helps to identify the different components of the response.

Filters The frequency spectrum of BAEPs in a normal adult population lies in the range 100–1,200 Hz, and up to five peaks can be identified in the first 6 ms following the stimulus – the band-pass is usually set at 100 or 150 Hz (to reduce mains interference) to 1,500 or 3,000 Hz.

In young children the signal has a lower frequency spectrum, and therefore it is more appropriate to use a wider band-pass set at 30 Hz to 1,500 or 3,000 Hz.

Sensitivity and number of sweeps BAEPs are extremely small (~1 μV), but there is a limit to the amount of gain that can be used without risk of overloading the amplifier with 'noise', due mainly to EMG. One thousand to 2,000 sweeps must be averaged to record BAEPs, but it is sometimes necessary to increase the number of sweeps if there is a poor signal-to-noise ratio. In infants the potentials are of lower amplitude and the ongoing EEG is of higher amplitude than in adults; this means that the time required for a recording session is longer and sometimes unpredictable, with consequences for the organisation of work in the laboratory. If the amplifier is operated at a setting equivalent to 10 μV/cm, a BAEP would have an amplitude of < 1 mm. The amplitude of the display can, however, be 'enlarged' by dividing the sum of the individual trials in the averager not by *N*, their number, but by a submultiple of *N* (Cooper et al 1980).

When doubt arises about the quality of the signals, as can happen in an alert child, it is often useful to run an average without stimulation in order to estimate the recording noise or, better, to average the even- and odd-numbered trials separately and take the difference as a measure of the actual noise during the recording.

Electrode placement and montages

All the brainstem components of BAEPs are far-field potentials picked up on the scalp at a distance from their source. Consequently, they are recorded on the scalp as widespread positivities, of which the majority are maximal over the frontocentral region, due to the median situation and oblique orientation of the brainstem in the cranial cavity (for a review see Starr and Squires 1982).

Only the first peak (wave I) originates at a distance from the brainstem in the peripheral portion of the cochlear nerve, and it has a different orientation. This peak appears at and around the ear ipsilateral to stimulation as a negativity and may also be picked up as a small positive far-field at the vertex. Electrode designations are Cz for the vertex, Ai for the ipsilateral ear lobe, Ac for the contralateral earlobe, Mi for the ipsilateral mastoid and Mc for the contralateral mastoid.

Although each laboratory may have its own preferences, most investigators record BAEP far-fields at Cz and connect this electrode to lead 2 of the amplifier, while Ai or Mi is connected to lead 1. In this type of montage, far-field positive potentials picked up on the scalp are registered as upward deflections (contrary to the convention adopted in most other EP modalities), as the electrode regarded as active, Cz, is connected to lead 2. This method also optimises detection of the earliest component (wave I), which is positive at Cz and negative at the ear lobe or mastoid. The ear lobe is the preferred site in children since it is less contaminated by muscle activity. The scalp distribution of BAEPs in the neonatal period is quite different from that in older subjects, as demonstrated by their representation as vector plots (Picton et al 1986). In this age group additional montages can help the identification of individual peaks. Cz–Ac or Mc and Cz–Oz or neck help in identification of waves IV–V; Ai–Ac helps to identify wave I. Electrode impedance should be maintained under 5 kΩ and balanced.

The following three-channel montage is adopted in the author's laboratory.
- channel 1: Ai (lead 1)–Cz (lead 2)
- channel 2: Ac (lead 1)–Cz (lead 2)
- channel 3: Ai (lead 1)–Ac (lead 2)
- ground electrode at forehead.

In recordings in neonates the vertex electrode is best located anterior to the fontanel, which produces little change in the morphology of the BAEP (McPherson et al 1985).

BAEPs: maturation and normative data

Maturation of the neonatal BAEPs has been investigated in normative studies, which showed that there are several differences between the BAEPs of the newborn and those of older children (*Fig. 3.22*).

The first detectable component in preterm infants is wave I, which has been identified at 30 weeks' CA. Thirty-three weeks after conception, wave V is present. At birth, waves I, III and V are obtained in term infants with markedly prolonged peak latencies and IPIs in comparison with adult values. Wave I is often double-peaked, there is little, if any, wave II, and a prominent negative wave precedes wave III, which is often broad or double-peaked. In normal neonates wave V is smaller than wave I and the I/V ratio is reversed relative to that observed in adults. The I–V IPI is dependent on GA (*Fig. 3.23(a)*); it is also, to a small extent, dependent on stimulus intensity, being decreased by decreasing the stimulus intensity (see *Fig. 3.22*) (Lütschg 1985, Despland 1987).

Amplitudes of all components increase with age, while latencies decrease, especially those of the later components. An adult morphology is attained at around 6 months of age, maturation is largely completed by 10–12 months, with adult values established by the age of 3 years. The evolution of peak latencies with increasing age is not linear (Salamy 1984). The curve of maturational changes of the I–V IPI is shown in *Fig. 3.23*.

Typical waveforms at different ages and the mean IPI between waves I and V for the different age groups (from the normative data of the present author) are shown in *Fig. 3.24*. The reduction in the latencies during the first months and years of life is probably due to myelination and increasing synaptic connections (Yakovlev and Lecours 1967).

Intra- and interindividual variability of the latency of waves I and V recorded under the same conditions is recognised to be very small in all published series of normative values in adults. In children older than 2 years adult normal values can be used as only minimal latency and amplitude changes occur after this age. The upper normal limit for the I–V IPI is about 4.5 ms, while the

I–III IPI is about 2.5 ms and the III–V IPI is 2.4 ms; the right/left asymmetries are < 0.5 ms.

Absolute amplitudes are highly variable even between adult individuals, with standard deviations (SDs) of 30–50% of the mean. The mean amplitude values vary for the different components from 0.2 to 0.5 µV, the highest values being obtained for the peak of wave V or for the IV/V complex. This intersubject variability is even greater in young children as the negative deflections utilised in adults for the amplitude measurements are often poorly defined, in particular those following wave V. To overcome this difficulty the I/V amplitude ratio, expressed as a percentage, can be used. The I/V ratio changes with the click intensity and is less reliable in patients suffering from peripheral hearing loss. In control subjects this ratio is generally less than 100%, and a value over 200% is above the upper limit of normal (mean + 3 SD) for click intensities over 80 dB HL.

In neonates and children under 2 years of age the above criteria have to be slightly modified according to the differences in the waveform. The I–V IPI reference values are given in *Fig. 3.24*. The amplitude ratio V/I values are similar to those of adults by around the age of 1 year; the IFCN 1999 guidelines suggest a lower limit for the V/I ratio of 30% in full term neonates.

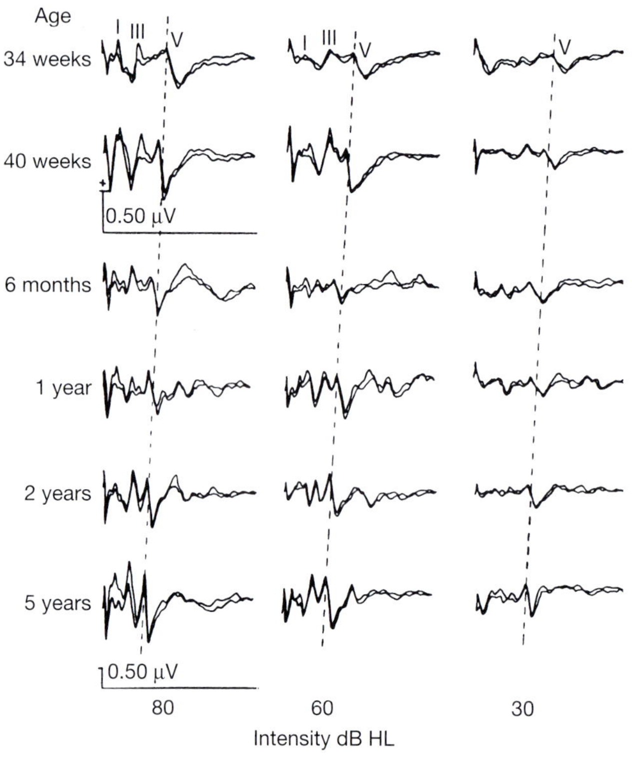

BAEPs: processing

Measurements routinely performed in clinical testing include: peak latencies of waves I, III and V; the interpeak latencies of waves I–III, III–V and I–V; the inter-ear latency differences, using either absolute peak latencies or interpeak latencies; and I/V amplitude ratio.

Up to the age of 6 months, when wave III is double peaked or appears as a broad wave with two or three wavelets superimposed, there are no agreed criteria for identifying the wave III peak. Recording from different montages and knowledge of the normal appearance of this wave at different ages can assist interpretation, but it

Fig. 3.22 Maturation of the BAEP. Recordings with three different stimulus intensities to show the relationship between morphology, wave V latency and age. Each response represents the superimposed responses to 2,048 stimuli (vertex positivity upward). Calibration 0.5 µV, 20 ms. (From Despland (1987), copyright John Wiley & Sons Ltd. Reproduced with permission.)

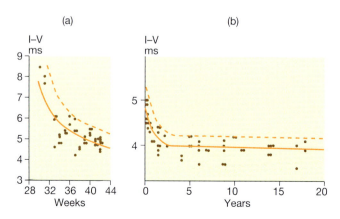

Fig. 3.23 Age dependency of the BAEP I–V IPI of (a) newborns (32–44 weeks' GA) and (b) of children (aged 3 months to 20 years). The continuous and broken lines represent the mean and +2 SDs, respectively.

AGE

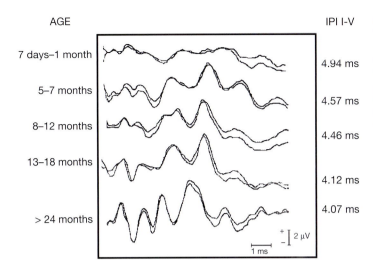

IPI I-V

7 days–1 month   4.94 ms

5–7 months   4.57 ms

8–12 months   4.46 ms

13–18 months   4.12 ms

> 24 months   4.07 ms

2 μV

1 ms

Fig. 3.24 Typical waveforms of BAEPs at different ages and the mean wave I–V IPI for the different age groups from the author's departmental normative data (right column). Note that in infants wave I is sometimes double-peaked and wave II may not be evident. In the first 6 months a wide negativity precedes wave III, which is often broad or double-peaked. Amplitudes of all components increase with age, while latencies decrease, with this effect being greater on wave V. Values similar to those in adults are reached by 2–3 years.

must be noted that I–III or III–V delay in the absence of an I–V IPI abnormality cannot be reported as a pathological result.

Peak-to-peak amplitudes can be measured from the peak of each wave to the trough that immediately follows. However, these measurements have no clinical utility even in inter-ear comparisons. The I/V amplitude ratio, expressed as a percentage, can be used. By comparing the peripheral input in the cochlear nerve (wave I) with the main brainstem component (wave V), this ratio depends both on the synchrony of ascending inputs in the proximal part of the cochlear nerve and the brainstem auditory pathways and on the number of cells activated by the stimulus in the brainstem. However, the amplitude ratio V/I is less reliable under 2 years of age.

The criteria for abnormality of BAEP in audiology (a) and neurology (b) are summarised in *Box 3.3*.

BAEPs: interpretation

The absence of all BAEP waves at high intensities, provided that technical reasons are excluded, is an abnormal finding and implies severe cochlear or auditory nerve dysfunction, at any age. Differentiation between these two entities requires that BAEP testing be supplemented with OAEs or, in very selected cases, with an ECochG.

Sometimes in uncooperative patients where an audiogram cannot be done, when stimulating with rarefaction clicks the presence at high intensities of a positive peak in the latency range of wave I raises the suspicion of a mixed cochlear and brainstem abnormality. This peak needs to be distinguished from the CM (cochlear microphonic), which can be done by showing a polarity reversal of this component with condensation clicks or its abolition with clicks of alternating polarity.

Delay of all BAEP waves without IPI abnormalities implies a middle-ear or mild cochlear hearing deficit. A middle-ear deficit can be easily ruled out by performing tympanometry. Differences in tympanometry have diff-

---

Box 3.3 Criteria for abnormality of BAEPs in audiology and neurology

(a) To evaluate a hearing defect, the following BAEP features are used:
- threshold for the appearance of wave V
- latency and amplitude of wave V
- latency of wave I
- latency–intensity curve

(b) The criteria for BAEP abnormality in neurological evaluation are:
- prolongation of the I–V, I–III and III–V IPIs
- loss of components
- abnormal amplitude ratio V/I

---

erent implications in BAEP waveform and threshold assessment.

It is generally accepted that the latencies of all BAEP waves shift by roughly the same amount, and that inter-peak intervals are affected little or not at all by decreasing click intensities. However, Stockard et al (1979) observed that, despite an increase in latency of all individual peaks at low stimulus intensities, the I–III, III–V and I–V IPIs are shortened by 0.19, 0.16 and 0.34 ms, respectively, at a click intensity of 30 dB SL when compared with IPIs measured in response to 70 dB SL clicks. These small changes are partly attributed to variations in the shape of wave I due to the activation of two different neuronal populations of the cochlear nerve for high and low stimulus intensities (for a review see Chiappa 1983).

The aspects described above have to be considered seriously since they can change the clinical judgement in patients with borderline IPI results. This problem can be overcome by investigating middle-ear function with tympanometry and performing a recording with the stimulus intensity increased to a level that gives a wave I latency in the normal range.

An increased I–V IPI, or absent or dispersed central components are signs of brainstem dysfunction. The localisation of the abnormality within the upper or lower brainstem depending on the III–V or I–III IPI, respectively, has to be considered with caution in young children and neonates for the previously stated reasons (see p. 134). The increased intervals between BAEP waves observed in patients with cochlear nerve or brainstem pathology cannot result from peripheral (conductive or cochlear) hearing loss, which reduces the effective stimulus intensity without affecting conduction in the auditory pathways.

An V/I amplitude ratio of less than 50% indicates excessively small IV–V peaks and arouses suspicion about brainstem integrity, while amplitude ratios greater than 300% indicate an abnormally small wave I and raise the possibility of some peripheral hearing impairment.

The combination of the pattern of BAEP abnormalities and the OAE helps in locating the dysfunction (*Table 3.8*).

BAEPs: clinical role

The application of BAEPs in paediatric neurology, together with other neurophysiological techniques, is included in Chapter 6 in accounts of the integrated neurophysiological investigation of specific groups of disorders. Some audiological and neurological applications are described in Chapter 6 (see p. 324), while clinical applications peculiar to neonatal age are discussed in Chapter 5 (see pp. 209–217). A suggested clinical protocol for BAEP testing is given in *Box 3.4*.

SOMATOSENSORY EVOKED POTENTIALS

Introduction

SEPs allow evaluation of the functional integrity of the somatosensory system from the peripheral nerve to the cerebral cortex. They can be evoked by many types of stimulation (electrical, touch, passive finger movements, CO_2 laser, etc.) applied to different body areas or nerve fibres. In routine clinical work SEPs are elicited by electrical stimulation of the peripheral nerves. In this they differ from other EP modalities in that they are evoked by applying non-physiological stimuli to the skin surface, which depolarise the fibres directly and bypass the peripheral encoding of the physical information (pressure, vibration, joint movement, etc.). SEPs are classified by the nerve being stimulated: upper limb SEP (usually the median nerve at the wrist) and lower limb SEP (usually the posterior tibial nerve at the ankle).

Progress in the methodology and in the understanding of the fibre tracts involved in SEPs and of the intracranial generators of the various SEP components in the last 20 years has led to important advances in the clinical use of the technique in adult neurology and neurosurgery. The application of these methods in paediatrics and neonatology is only just becoming established in many paediatric hospitals, but it appears a promising technique for various reasons. First, evaluation of the tactile and proprioceptive sensory systems is one of the most difficult parts of the neurological examination of a young

---

Table 3.8   Presumed location of auditory dysfunction based on a combination of the OAE and the BAEP pattern

Location	OAE	BAEP wave I	BAEP wave V
Cochlea (inner cells)	Normal	Abnormal	Abnormal (threshold level)
VIIIth nerve	Normal	Abnormal	Abnormal
Brainstem	Normal	Normal	Abnormal

Box 3.4   Suggested clinical protocol for BAEP testing

Except in the neonatal period, BAEP testing should always be preceded by tympanometry. Whenever possible, a clinical examination of hearing should be performed before BAEP testing, possibly including conventional pure-tone audiometry. When documentation of the normal-hearing threshold is not provided, two runs of 2,000 clicks at 35 dB nHL should be performed to ensure an acceptable hearing level.

Monaural rarefaction clicks of 100 μs duration are then delivered at a rate of 11/s, with masking of the non-stimulated ear by continuous white noise. Analysis time 20 ms, filters 20–3,000 Hz.

For neurological applications the click intensity is set at 75 dB nHL (up to 90 dB nHL); increasing or decreasing the click intensity may be used to obtain a better definition of the waveform.

For audiometric testing the click intensity is progressively decreased in steps of 10 dB down to the intensity at which wave V can no longer be identified.

To assess waveform reproducibility, responses to at least two runs of 2,000 clicks are averaged for each ear and for each of the click intensities used. Averaging the same number of sweeps of background activity without stimulation is performed when peaks are of reduced amplitude and are poorly reproducible.

child, and the SEP can make an important contribution to this. Secondly, the SEP has been demonstrated to have a relevant clinical role in many paediatric neurological conditions.

Short, middle and long latency cortical SEP components can all be recorded. As only short latency SEPs have been clinically validated as a test in paediatric and neonatal neurophysiology, the following description of the techniques and applications is limited to these. For a detailed description of the methodology for eliciting the adult SEP see Binnie et al (2004).

Fibre tracts involved in SEPs

Detailed understanding of the fibre tracts involved is important for the clinical use of SEPs. The fibres that subserve the different types of sensation can be characterised by the diameter and thickness of their myelin sheaths, and there is an inverse relationship between their diameter and the conduction velocity, on the one hand, and their threshold to electrical stimulation on the other. Only the most rapidly conducting and heavily myelinated afferent fibres of the somatosensory system are activated by electrical stimuli delivered at low and non-noxious intensities.

In most clinical applications the electrical stimulus is applied to a mixed nerve at intensities of three to four times the sensory threshold. This produces a twitch in the muscles innervated by any stimulated nerve that contains motor fibres. At these stimulus intensities all rapidly conducting large myelinated fibres, including fibres subserving touch, joint sensation and muscle afferents, are activated. The contribution of muscle afferents to the median nerve cortical SEP can be considered weak compared with that of cutaneous afferents.

The situation is quite different after electrical stimulation of lower limb sensory–motor nerves, such as the posterior tibial nerve for which muscle afferents have been shown to have a major contribution to cortical SEP (Burke et al 1981, 1982, Gandevia et al 1984, Macefield et al 1989). After stimulation of the tibial nerve, inputs from muscle afferents, which have the fastest conduction velocity, are able to occlude the response to cutaneous inputs (Burke et al 1982). Because of this gating the cutaneous afferent volley may make little or no contribution to the cerebral potentials evoked by electrical stimulation of lower limb mixed nerves.

Electrical stimulation of pure sensory nerves, such as the digital nerves for the upper limb and the sural nerve for the lower limb, activates exclusively skin and joint peripheral and dorsal column fibres. At the spinal and intracranial levels, the SEP evoked in man by non-painful electrical stimulation of the peripheral nerves has been considered (since the report by Halliday and Wakefield (1963) of normal SEPs in patients with lesions of the spinothalamic tract) to be specifically related to the activity of the dorsal column (DC) system. Moreover, Cusick et al (1978) demonstrated in the monkey that

ablation of the DC abolished SEPs, whereas myelotomy selectively sparing the DC left these responses virtually unaffected. On this basis, potentials recorded over the scalp after a non-painful stimulus are considered to reflect activation of the DC and lemniscal fibres. When high intensity stimuli are used, other ascending tracts in the spinal cord may also be stimulated, especially the anterolateral spinothalamic tracts (Powers et al 1982).

Normal SEP waveforms in adults and their generators

SEPs to upper limb stimulation

The main early SEP components evoked by the stimulation of the median nerve at the wrist that are of proven clinical utility peak before 50 ms. *Figure 3.25* shows a summary of the waveforms at different electrode positions as well as the name and description of the location.

The generators of median nerve SEPs are described below and summarised in *Table 3.9*.

Peripheral components: N9 or brachial plexus volley Stimulation of the median nerve at the wrist elicits a compound action potential corresponding to the peripheral-ascending volley that can be recorded at different levels of the forearm and arm. In most SEP studies, and particularly in those carried out on patients with central lesions, the ascending volley is recorded from the supraclavicular fossa at Erb's point, where the activation of the brachial plexus trunks appears as a triphasic positive–negative–positive waveform, with a negative peak culminating at

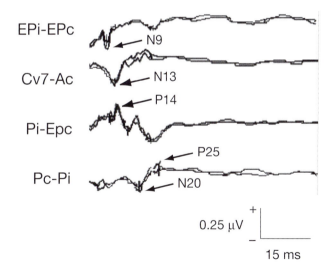

Fig. 3.25 Example of upper limb SEP waveforms recorded at Erb's point (brachial plexus potential), the spinous process of the 7th cervical vertebra (spinal cord response), a centroparietal location ipsilateral to the stimulus (far-field scalp positivities) and a centroparietal location contralateral to stimulation (early cortical components). Recordings from a 6-year-old normal child. At 5–6 years the latencies are much shorter than the nominal adult values (e.g. N20 is at 15 ms). EPi, Erb's point ipsilateral; EPc, Erb's point contralateral; Ac, anterior cervical; Pc, parietal contralateral; Pi, parietal ipsilateral.

Table 3.9 Presumed generators of median nerve SEPs

Component	Origin
N9	Brachial plexus
P11	Cervical cord dorsal column
N13/P13	Compound segmental postsynaptic potential of dorsal horn grey matter
P14	Lower brainstem close to the cervicomedullary junction
N20	Primary somatosensory cortex (SI area)
P25	Sensory cortex

about 9 ms in normal adult subjects (N9 or brachial plexus volley). The N9 is a mixture of motor antidromic and sensory orthodromic responses, and thus is qualitatively different from the sensory SEP components generated in the CNS.

Spinal components: N13 spinal segmental potential Various spinal potentials can be recorded as near-field potentials over the spinous processes of the cervical vertebrae or as far-field scalp positivities within 11–14 ms after stimulation at the wrist. A comprehensive description of the best recording methodology, supposed generators and clinical use of these components is outside the scope of the present book, as the knowledge of their normal characteristics during development, normative data and clinical role have not yet been fully elucidated in young children and infants. In older children the adult data apply. The description here is limited to the spinal segmental N13 potential, which has a clinical relevance in paediatric neurology.

The N13 potential is picked up on the posterior aspect of the neck. Its maximum amplitude is at the Cv5–Cv7 levels, and its amplitude decreases at more rostral or caudal electrode positions. N13 reverses into a P13 when recorded anterior to the cord by an electrode placed on the anterior aspect of the neck (Desmedt and Cheron 1980, Mauguière and Ibañez 1985). The nuchal N13 and its positive P13 counterpart have similar distribution along the craniocaudal axis of the cervical cord. This spatial distribution of the cervical N13/P13 potentials probably arises from a compound postsynaptic dipolar generator in the dorsal horn of the cord, perpendicular to the cord axis, which is triggered by the afferent impulses arriving along fast conducting myelinated fibres. A transverse montage of two electrodes, one located over the Cv7 spinal process and the other anteriorly on the skin above the laryngeal cartilage, is the best arrangement for recording spinal N13/P13 activity. The main advantage of this derivation over the more conventional Cv7–Fz montage is that it records the

spinal potential uncontaminated by SEP components generated above the foramen magnum (Mauguière and Restuccia 1991).

Far-field scalp positivities With an appropriately wide filter bandpass (e.g. 1 Hz to 3 kHz), a fast sampling rate, a large number of averaged trials (at least 1,000) and a non-cephalic reference, four stationary positivities are recorded on the scalp after median nerve stimulation. These are widely distributed, with a midfrontal predominance. In normal adults these potentials peak with mean latencies of 9, 11, 13 and 14 ms, respectively, and are labelled P9, P11, P13 and P14. However, only the P14 far-field potential is consistently recorded in normal subjects (for a review see Restuccia 2000) with some degree of interindividual variability in its shape due to the superimposed P13. The P14, but not the P13, always peaks later than the cervical near-field N13 potential (Mauguière 1987).

The P13–P14 complex is generated after the synaptic relay in the nucleus cuneatus, whereas P9 and P11 are produced by the first-order neurons of the dorsal column pathways.

Early cortical components Many studies have addressed the issue of generators of cortical SEPs, often with conflicting results. Scalp SEPs reflect activities that overlap in time and are generated by radial as well as tangential dipolar sources, simultaneously active in a short latency range. Two sets of early cortical potentials, each made up of two components, recorded on the scalp contralateral to the stimulated arm are usually considered in adult SEP studies. The first is an N20–P27 complex located in the parietal region, and the second is composed of the P22 and N30 potentials recorded in the contralateral central and frontal regions, although the N30 potential often spreads to the frontal region ipsilateral to stimulation.

Only the N20 and P27 components are routinely evaluated in paediatrics and are considered here, but awareness of the presence of the other components is important both for the interpretation of abnormal SEPs and for guiding the choice of the best locations for recording and reference electrodes in clinical practice.

The N20 component is the largest early negative wave, peaking in normal adults at around 20 ms, usually followed by a positive peak at around 24–27 ms. There is now agreement that the parietal N20 is the earliest cortical potential elicited by median nerve stimulation, and this reflects the activity of a dipolar generator in Brodmann's area 3b of the primary somatosensory cortex (area SI) tangential to the scalp surface and situated in the posterior bank of the Rolandic fissure (Broughton 1969, Goff et al 1977, Allison et al 1980).

The peak latency of the parietal P27 potential shows large interindividual variations between 24 and 27 ms. In some subjects two distinct peaks, P24 and P27, can be identified, while only one or other of the two peaks is observed in others. This explains why, according to the polarity–latency nomenclature, the first parietal positive potential following N20 has been given various labels in the literature (P24, P25 or P27). These variations reflect the fact that the activities of several parietal sources overlap in time in this latency range. Generators are supposed to be in the somatosensory cortex in the posterior wall of the central fissure (SI area).

SEPs to lower limb stimulation

Electrical stimulation of the tibial nerve at the ankle is adopted by most workers for testing the sensory pathways of the lower limb. However, electrical stimulation can also be applied to the sural nerve at the ankle, or to the peroneal nerve at the knee, without major changes in the general waveform of the spinal or scalp responses; only the peak latencies will be different because of the different length of the fibres.

The normal tibial nerve SEP waveforms at different electrode positions as well as the name and description of the location are shown in *Fig. 3.26*. The generators of posterior tibial nerve SEPs are described below and summarised in *Table 3.10*.

Peripheral components A compound action potential corresponding to the activation of tibial nerve fibres

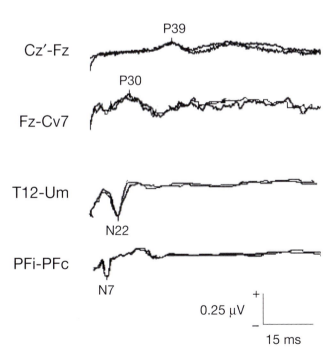

Fig. 3.26 Example of lower limb SEP waveforms recorded at the popliteal fossa (tibial nerve potential), the spinous process of the 12th thoracic vertebra (spinal potential), the frontal scalp location with non-cephalic reference (far-field scalp positivities) and the vertex (early cortical potential). Recordings from an 8-month-old normal child. The latencies are much shorter than the nominal adult values. Um, supra-umbilical region; PFi, popliteal fossa ipsilateral; PFc, popliteal fossa contralateral.

running through the popliteal fossa may be recorded at the posterior aspect of the knee using a bipolar montage. The recording electrode is placed over the tibial nerve about 4 cm above the popliteal crease, midway between the combined tendons of the semimembranous muscle medially and the tendon of the biceps femoris muscle laterally. The reference electrode should be placed on the median surface of the knee, over the median femoral condyle. The negative peak of this near-field compound action potential peaks at a latency of about 7 ms in adults. This 'N7 potential' reflects a mixed response of motor and sensory fibres, which may be clinically useful to assess the integrity of the peripheral segment of the pathway. P17 the afferent volley in cauda equina roots can be recorded using skin electrodes placed at the L5–S1 level, with a distant reference site (e.g. at the knee opposite to the stimulation).

Spinal potentials An electrode situated on the spinal process of T12 or L1 vertebrae and referred to a more distal electrode records a negative potential peaking at between 21 and 24 ms in adult normal subjects (N22); the body height of the subject strongly influences the N22 latency. This lumbar negativity originates in the spinal segment receiving fibres from the S1 root. As with the cervical N13 potential elicited by median nerve stimulation, Desmedt and Cheron (1983) demonstrated a polarity reversal in prevertebral recordings. The field distribution of the N22 potential is consistent with a horizontally orientated dipolar source in the spinal cord, which probably corresponds to the postsynaptic responses of dorsal horn neurons to incoming inputs. Both posterior negativity and anterior positivity show maximal amplitude at the T12–L1 vertebral level, decreasing without any latency shift at more rostral or caudal electrode positions.

Provided that the reference electrode is not situated on the axis of propagation of the peripheral ascending volley, the spinal negativity recorded in the lumbar region is preceded by a small positivity peaking around

Table 3.10 Presumed generators of posterior tibial nerve SEPs

Component	Origin
N7	Tibial nerve action potential
P17	Volley in lumbrosacral trunks
N19	Volley in spinal ascending somatosensory pathway
N22	Lumbar cord grey matter
P30	Lower brainstem close to the cervicomedullary junction
P39	Postcentral somatosensory cortex

17 ms (P17). This P17 positivity is a far-field potential originating in lumbosacral plexus trunks.

Far-field scalp positivities   With appropriate non-cephalic reference montages, two widespread far-field positivities, P27 and P30 (distinct from the P17), can be recorded on the scalp, peaking before the onset of the earliest cortical potential in the latency range 25–32 ms (Yamada et al 1982, Desmedt and Cheron 1983, Kakigi and Shibasaki 1983). The latency of this potential varies according to body height; in this chapter it is called 'P30'.

The P30 potential is generated in the brainstem (Yamada et al 1982, Desmedt and Cheron 1983) and can be viewed as the homologue, for the lower limb, of the far-field P14 recorded on the scalp after median nerve stimulation. The utility of the P30 potential has been validated for clinical applications of tibial SEPs in patients with spinal cord and brainstem lesions (Tinazzi and Mauguière 1995, Tinazzi at al 1996). The P30 potential is widely distributed on the scalp but predominates in the frontal region (Desmedt and Bourguet 1985, Guérit and Opsomer 1991). The most practical reference site for recording the P30 is the spinous process of the 6th or 7th cervical vertebra (Seyal et al 1983, Tinazzi and Mauguière 1995). Due to the time dispersion of afferent impulses of the ascending volley at cervical level, it is exceptional to record a reproducible negativity with a neck electrode at this latency.

Early cortical potentials   The first cortical potential elicited by stimulation of the tibial nerve at the ankle is a positive component labelled P37, P39 or P40, peaking as it does at a mean latency of 37–40 ms in normal adult subjects (for a review see Yamada 2000). In all normal subjects, P39 is present at the vertex, and it can most often be reliably obtained midway between Pz and Cz using a scalp–ear lobe montage. Frequently, the maximum of P39, although close to the midline, is slightly shifted to the side of the scalp ipsilateral to the stimulus (Cruse et al 1982). The reason for this paradoxical lateralisation of P39 is probably the somatotopic representation of the distal lower limb in the somatosensory SI cortex, and in particular the foot area, situated on the medial aspect of the hemisphere.

Despite these topographical variations, the P39 potential is almost invariably found in normals when recording between Pz and the ear lobe ipsilateral to the stimulus (Tinazzi et al 1997). In a few cases the P30 far-field potential is also obtained with this derivation because its amplitude is higher on the scalp than at the ear. An easy way of differentiating P30 and P39 is to use a Pz–Fz derivation, which cancels the widespread P30 far-field, but not the cortical P39.

SEPs: methodology

Since the SEP technique has such varied possibilities, there is no single optimal procedure for all clinical situations. The recording protocol must be designed according to the clinical problem. Once this has been clearly outlined, the second step is to choose the site of the stimulation, then the components to be studied and, finally, choose the most appropriate procedure.

Potentials from the peripheral nerve, spinal cord, and far-field and near-field scalp components can easily be recorded in children from 6 months of age using a method similar to that used in adults. However, a special approach to the child and the parents is needed, since electrical stimulation is perceived, especially by the latter, as potentially noxious. It is first important to explain to the parents that the test will not be particularly disturbing to the child nor will it be painful. It is also important to assess the likely cooperation of the child and the possible need for sedation. Secondly, the clinical question has to be clear before the test is planned, in order to guide the choice between the many recording and stimulating strategies offered by SEP.

In infants under 6 months old and in neonates, lower limb SEPs are only used exceptionally, as they are difficult to elicit and time consuming compared with upper limb SEPs. Reports of recording posterior tibial nerve SEPs in normal neonates shows a success rate of, at best, 93% (White and Cooke 1989), but their clinical role has not yet been fully established. The stimulus and recording parameters are comparable to those for upper limb SEPs.

Upper limb SEPs are less difficult to obtain. It has been demonstrated that the number of stimuli, stimulation rate and filter settings have to be adjusted to take account of habituation and fatigue of the immature neonatal nervous system and the increased low-frequency content of the responses (*Fig. 3.27*) (Bongers-Schokking et al 1989).

De Vries (1993), reviewing the published data on healthy full term infants, showed that the success rate in recording the cortical responses is strongly dependent on the above-mentioned parameters (*Table 3.11*).

Following methodological improvements, a better understanding of normal waveforms and the appreciation of a clear clinical role, particularly in the prognosis of neonatal hypoxic–ischaemic encephalopathy, upper limb SEPs can now be routinely done at the bedside in the neonatal ITU.

Recording environment

Peripheral nerve conduction velocities are affected by changes in limb temperature. The skin temperature of the stimulated limb should be monitored and kept constant throughout the recording session, the limb being warmed if necessary. The core temperature of the body may be low in neonates and in children in ITUs. If recordings have to be done under these circumstances it is important to take into account the temperature effect on the SEP components. During drug-induced hypothermia both the early cortical N20 and the far-field positive potentials have prolonged absolute and interpeak latencies. More-

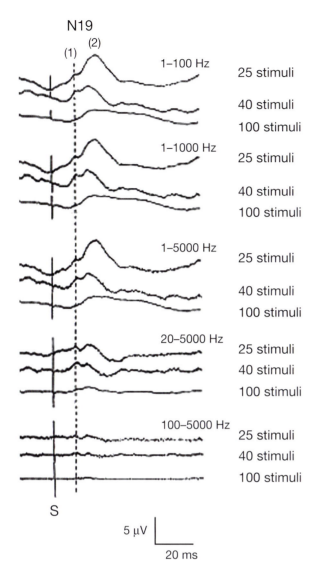

Fig. 3.27 Effect of filters and the number of stimuli on the cortical median nerve SEP in the neonatal period. When the N19 is double peaked the authors labelled them N19(1) and N19(2). Negativity up. S stimulus. (From Bongers-Schokking et al (1989), by permission.)

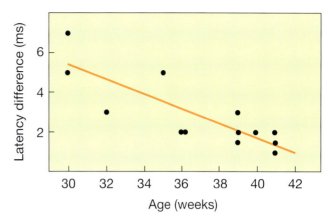

Fig. 3.28 Latency difference of the N20 component of the MN-SEP (the SEP recorded with median nerve stimulation) between quiet and active sleep in preterm infants. The latency of N20 is longer in quiet than active sleep, the difference diminishing with increasing GA.

over the decrease in amplitude with body temperature during hypothermia is not the same for peripheral, spinal brainstem and cortical SEPs (Guérit et al 1990). The N20 component disappears at a temperature between 25°C and 17°C and the P14 between 20°C and 17°C. The peripheral N9 and spinal N13 potentials remain identifiable down to a body temperature of 17°C. SEP changes related to hypothermia deserve special attention in comatose patients, in whom these changes can combine with those induced by CNS depressant drugs.

Sleep and vigilance affect the cortical components of SEPs differently depending on age. The latency of the N20 component is strongly influenced by the behavioural state in neonates and infants (Desmedt et al 1976, Hrbek et al 1969). An increased latency occurs during quiet sleep (QS), the difference being more pronounced in preterm infants, but no differences are apparent during active sleep (AS) and the awake state (*Fig. 3.28*).

In the author's experience, the best state for recording SEPs is during active sleep, when responses are of high amplitude, of more consistent latency and are less affected (especially the spinal response) by muscle artefact, compared with the awake state (*Fig. 3.29*).

The state of awareness should be monitored in patients with a fluctuating state of vigilance, whether related to the disease itself or to sedation. In uncooperative children, peripheral spinal and far-field scalp components are easily obscured by muscle artefact and are best visualised in sleep, while cortical components are modified during natural sleep or sedation and are best recorded whilst awake. The choice of the appropriate vigilance state for each recording is therefore determined by the clinical problem. Any clinical conclusion about N20 abnormality should be avoided if the recording is done in a sleeping young child. When drug sedation is necessary, low anxiolytic doses of oral benzodiazepines or chloral hydrate at 50 mg/kg of body weight, can be used. Ideally, the EEG activity should be monitored during the recording to identify the variations related to vigilance changes.

The technologist should note the use of any drug with CNS effects. Significant differences in central conduction times have been reported between controls and ambulatory epileptic patients treated with phenobarbital and phenytoin (Green et al 1982).

Another source of SEP variation to be taken into account and noted for each patient is body height and arm length. The latencies of SEPs obviously vary according to the distance between the stimulus site and the SEP generators, the effect being more pronounced for lower limb than upper limb SEPs. This variability is less for interpeak than for absolute latencies. In children this

Table 3.11   Success rate in different studies of recording upper limb SEP cortical responses from normal neonates.* Note how the rate is higher in studies where the stimulation rates and number of stimuli were low

Study	N1 (mean ± SD) (ms)	Filter (Hz)	Stimulation rate (Hz)	No. of stimuli	Success rate (%)
Desmedt and Manil (1970)	31.4–33.9	–	0.3	32–256	?
Hrbek et al (1973)	35 ± 8	–	–	–	–
Laget et al (1976)	40 ± ?	–	1	150	100
Pratt et al (1981)	–	30–3,000	?	1,000	10
Willis et al (1984)	24.3 ± 2.1	30–3,000	5	1,000	66
Zhu et al (1987)	24.9 ± 2.7	10–4,000	2	256	100
Klimach and Cooke (1988)	40 ± 6	2–100	0.5	128	100
Laureau et al (1988)	25 ± 2	30–3,000	4	512	85
Laureau and Marlot (1990)	26.7 ± 5.7	20–2,000	2–5	500	87
Bongers-Sch et al (1989)	30.6 ± 3.9	1–100	0.5	25–50	100
Majnermer et al (1990)	24.6 ± 3.1	30–3,000	4	512	85
George and Taylor (1991)	30 ± 6.8	30–3,000 5–1,500	1.1	64	100
Gibson et al (1992)	30 ± 6.8 26.7–44.9	10–3,000	1–5	256–1,024	97.5
Karinski et al (1992)	30.4	1–100	0.9–1.7	300	100

*Adapted from De Vries (1993).

effect on SEP latencies is combined with that of maturation (see p. 145). In addition, there are paediatric disorders associated with short stature, for which special norms are required (Suppiej et al 1991).

SEP stimulators and stimulus parameters

Stimulators   Stimulators can be designed to provide a constant-voltage or a constant-current output. A constant-voltage stimulator will apply the selected voltage to the electrodes irrespective of their contact impedance – they could even be disconnected. A constant-current stimulator passes the selected current through the electrodes whether they are in good contact (low impedance) or poor contact (high impedance). The impedance of the stimulating electrodes should be kept low (as with recording electrodes). For neonates and infants smaller electrodes are required (*Fig. 3.30*).

Stimulus type   In most laboratories, somatosensory potentials are evoked by electrical stimulation. Single monophasic pulses of about 100 µs duration are usually used to evoke the SEP. The cathode should be over the nerve trunk and proximal with respect to the anode.

Stimulus intensity   In most clinical applications electrical stimuli are delivered at intensities equivalent to three to four times the sensory threshold, which also produces a twitch in any muscles innervated by the stimulated nerve.

Stimulus rate   Above 4 months of age stimulus rates of 2–4 Hz, as for adults, can be used. In neonates and young infants it is mandatory to stimulate at rates at or below 1 Hz, usually 0.5 Hz.

Recording parameters

Analysis time and sampling rates   Most SEP components of proven clinical utility peak before 50 and 70 ms, respectively, for upper- and lower limb stimulation. Therefore, analysis times in the region of 100 ms are suitable for children. Longer analysis times are needed only when testing premature infants, since after term the balance between immaturity of conduction and short body and arm length results in latencies similar to those seen in adults (see p. 145). However, a slightly longer analysis time helps in the identification of components with a broader morphology, such as those occurring in the more immature child.

Filters   A broad bandpass, with a high-pass filter set at less than 3 Hz and a low-pass filter set at over 2,000 Hz, is recommended to record without distortion all the fast and slow components of the scalp response. However,

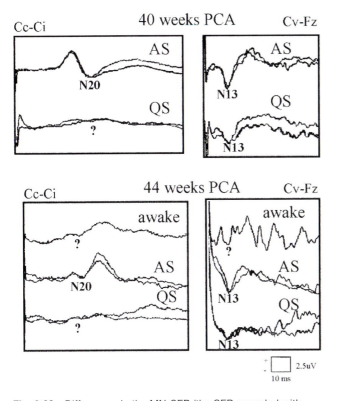

Cc-Ci   40 weeks PCA   Cv-Fz

AS
N20
QS
?

AS
N13
QS
N13

Cc-Ci   44 weeks PCA   Cv-Fz

awake
?
AS
N20
QS
?

awake
?
AS
N13
QS
N13

2.5uV
10 ms

Fig. 3.29 Differences in the MN-SEP (the SEP recorded with median nerve stimulation) waveform between the awake state (AS and QS) in two neonates of 40 and 44 weeks' CA. States were evaluated using clinical scoring. Note: (1) the contamination of both the spinal and the cortical responses by a movement artefact in the awake state; (2) the good responses recorded at the cervical level in both the quiet and AS states, the greater amplitude in active sleep probably being due to cortical activity from the Fz reference; (3) the N20 attenuation in quiet sleep.

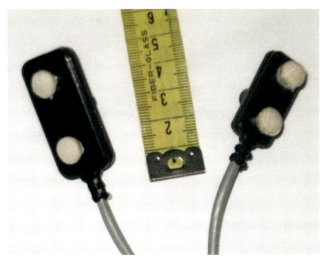

Fig. 3.30 Examples of SEP stimulation electrodes for children and neonates.

are enough to obtain the responses at Erb's point, as well as to extract the main early cortical components, but up to 1,000–2,000 sweeps may need to be averaged for the recording of spinal and scalp far-field SEPs. In children it is easier than in adults to record spinal components, probably due to the thinner bone and subcutaneous structures. Stimulation in neonates should take into account habituation, and therefore the number of sweeps should not be more than 50–100; usually 30–40 are adequate.

Electrode placement and montage   When recording SEPs it is preferable to acquire, if possible all the required components in one montage the design of which depends on the components under study. Activity from peripheral nerve and spinal parts of the sensory pathways can be recorded by cutaneous electrodes placed along the route of the fibre tracts or close to the generators of segmental spinal responses (near-field recording). Moreover, action potential volleys can be picked up at a distance on the scalp surface (far-field recording). On the scalp, the ascending volley in lemniscal and thalamocortical fibres can be recorded only at a distance from their sources as

the following filter settings have both successfully been used in neonates: 5–1,000 Hz and 1–100 Hz (Taylor et al 1996a) (*Fig. 3.31*).

Sensitivity and number of sweeps   Sensitivities of 20–100 µV/cm are adequate. The number of sweeps required varies according to the different SEP components to be recorded. In most instances, 500–800 sweeps

Fig. 3.31 Neonatal upper limb SEP showing the effect of filters on the N20 cortical response. Note: (1) little effect of increasing the HF response from 1 kHz (channel 1) to 5 kHz (channel 2); and (2) reduction of noise in channel 3 with HF response reduced to 100 Hz

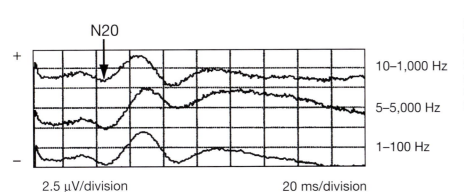

N20
+
10–1,000 Hz
5–5,000 Hz
1–100 Hz
–
2.5 µV/division   20 ms/division

143

far-field scalp positivities, which are widely distributed, while the postsynaptic responses of the cortical receiving areas are picked up as near-field responses located in restricted areas.

The position of the ground electrode is more crucial for SEPs than for the recording of any other type of EP because of the electrical artefact produced by the stimulation. To keep this to a minimum, the ground electrode should be placed on the stimulated limb, between the stimulation site and the recording electrodes. Flexible metal strips covered with saline-soaked cloth wrapped around the limb close and proximal to the stimulus site are recommended.

SEPs to upper limb stimulation   To record all the SEP components reflecting the processing of inputs up to the cortical somatosensory area, a minimum of three active electrodes are needed. They should be placed, respectively, in the supraclavicular fossa (Erb's point), over the spinous process of the sixth or seventh cervical vertebra (Cv6/7) and in the parietal region opposite to the stimulated side (Pc, contralateral parietal). In children, the spinous process of the seventh cervical vertebra (Cv7) is often preferred since it is easier to localise, and Pc is set halfway between parietal and central electrodes instead of the usual adult location of 2 cm behind Cz and 7 cm from the midline; the use of a percentage measurement allows an electrode location independent of head growth during development. It is recommended that the reference electrode be placed in the anterior cervical (Ac) region (above the process of the thyroid cartilage) for spinal recording, and at the ear lobe (A1 or A2) or ipsilateral parietal region for the study of cortical responses.

Alternatively, since interfering muscle and cardiac activities are minimal at the Fz site, this location for the reference electrode has proved to be reliable for a quick exploration in clinical routine as well as for SEP monitoring in hostile environments (Jones 1982, Chiappa 1983).

A non-cephalic electrode, on the dorsum of the hand or on the posterior aspect of the shoulder on the unstimulated side (Cracco and Cracco 1976, Desmedt and Cheron 1980) can also provide a reference site, with the advantage of being neutral for intracerebral generators (which produce virtually no current flow from the head to the body through the neck). However, the use of such an electrode requires control of unwanted interference, arising mainly from muscle and cardiac activity.

In neonates, recordings of the short latency peripheral components of the upper limb SEP are difficult to obtain (George and Taylor 1991), and only the cervical response is usually recorded. However, it has been shown that the brachial plexus potential can best be recorded in newborns at 1 cm above the axilla, i.e. less than the 2 cm used in adults (Cracco and Cracco 1976). The cortical responses are recorded from the central region (C3, C4 of the 10–20 system); the reference electrode is positioned at Fz (or at C-ipsilateral for cortical recordings).

An example of a four-channel montage used by the author for recording upper limb SEPs in children is shown in *Table 3.12*.

SEPs to lower limb stimulation   Usually a four-channel montage is used to record the tibial nerve SEP with four 'active' electrodes placed:

1. over the tibial nerve at the popliteal fossa, 4–6 cm above the popliteal crease, midway between the combined tendons of the semimembranous–semitendonous muscles medially and the tendon of the biceps femoris muscle laterally (popliteal fossa (PF) electrode)
2. over the spinous process of the twelfth thoracic (T12) or first lumbar (L1) vertebra corresponding to the level where the afferent volley enters the cord – the ascending volley in lumbosacral roots and spinal tracts can be recorded as a far-field positivity at the T12/L1 site or by using a supplementary electrode placed over the spinous process of L4 (or L5) vertebra, on the route of the cauda equina roots
3. in the midfrontal region of the scalp at the Fz or Fpz where the cervicomedullary potential 'P30' has its maximal amplitude with minimal contamination by subsequent cortical potentials
4. on the scalp near the vertex, 2 cm behind Cz (Cz') for the recording of the cortical components.

When recording spinal components, account has to be taken of the fact that the conus medullaris is caudally displaced in the preterm period and reaches the adult position by the age of 3 months (Barkovich 1995).

There is no consensus on the optimal site for the reference electrode when recording the lumbosacral response. The iliac crest and the knee contralateral to the stimulated side are the most widely used sites; bipolar recording between L1 and T12, T11 or T10 has also

Table 3.12   Derivations for a four-channel SEP recording: upper limb SEPs with median nerve stimulation (MN-SEP) and lower limb SEPs with posterior tibial nerve stimulation (PTN-SEP). These form a general recommendation but should be used flexibly according to the clinical problem under investigation

Channel	MN-SEP	PTN-SEP
1	Pc–Pi	Cz'–Fz
2	Pi–Epc	Fz–Cv7
3	Cv7–Ac	T12–Um
4	EPi–Epc	PFi–PFc

Ac, anterior cervical; EPc, Erb's point contralateral; EPi, Erb's point ipsilateral; PC, parietal contralateral; PFc, popliteal fossa contralateral; PFi, popliteal fossa ipsilateral; Pi, parietal ipsilateral; Um, supraumbilical region.

been used. As an example, the four-channel montage adopted by the author for recording lower limb SEPs in children is summarised in *Table 3.12*.

SEPs: maturation and normative data

Maturation of neonatal upper limb SEPs has been demonstrated in normative studies (George and Taylor 1991, Taylor et al 1996b) that show the SEP waveform alters with increasing GA (*Fig. 3.32*).

Cortical responses are recorded from 25 weeks' GA and are present in all normal preterm neonates from 27 weeks, provided that strict inclusion criteria are adopted. The first negative component N1, corresponding to the adult N20, has a latency around 120 ms at 27 weeks and 70 ms at 32 weeks (Taylor et al 1996b) and reaches a latency of 24–26 ms in newborns of 40 weeks' GA (*Fig. 3.33*).

Multichannel studies indicate that the negative cortical component corresponding to adult N20 (called N2) is very stable in topography in the neonatal period, while a preceding negative component (N1) is more anterior early in life and becomes more posterior as the infant matures. This is in agreement with anatomical data of posterior movement of the central sulcus as the association areas of the frontal lobe increase in size (Karniski 1992, Karniski et al 1992). There are no differences between term and preterm infants tested at the same GA (Majnermer et al 1990), with the exception of a transitory latency increment in the first week of life due to adaptation to extra-uterine life (Pierrat et al 1990).

The N13 (cervical potential) has a latency of 10 ms in full term newborns (Gibson et al 1992).

Graphs of the postnatal maturation of the components N13 and N20 and of the N13–N20 central conduction

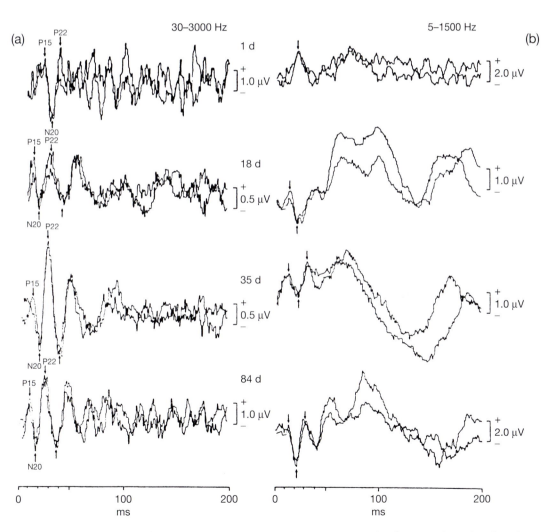

Fig. 3.32 Maturation of the SEP following median nerve stimulation at the wrist. Cross-sectional data from four infants at different ages (1, 18, 35 and 84 days) to show morphological development and the effects of different filter settings. (a) Recording bandpass 30–3,000 Hz. (b) Recording bandpass 5–1,500 Hz. P15, N20 and P22 are present in the neonatal period and further peaks evolve within a few weeks. In (b) the early potentials are less influenced by background noise but are more affected by slower waves. Samples in (b) recorded at half the sensitivity of those in (a). (From George and Taylor (1991), by permission.)

time (CCT) during the first year of life show a progressive decrease in latency that is most evident for the cortical response (*Fig. 3.34*).

Similar processes take place in lower limb SEPs. With maturation the spinal and cortical components decrease in latency compared with the preterm period. The cortical component corresponding to the adult P39 is recorded at a latency of around 80 ms in the youngest preterm infant and reaches about 35 ms by 40 weeks' PMA (Gilmore et al 1987, White and Cooke 1989, Minami et al 1996).

The process of somatosensory development is dominated by two coexisting phenomena with opposite effects on the SEPs. Myelinogenesis, increasing fibre diameter and synaptic modifications cause a progressive increase in conduction velocities and synchronisation of potentials, and thus a latency decrease with age, whereas body growth has the opposite effect, namely latency increase and desynchronisation. The interaction of these two variables explains the complex patterns of SEP development in children (*Figs. 3.35* and *3.36*).

In the first 2–3 years of life the effect of myelination is greatest and this results in an overall latency reduction, which is more evident for the cortical components. After about 3 years of age there is an age-related latency increase, with the dominant effect being due to growth.

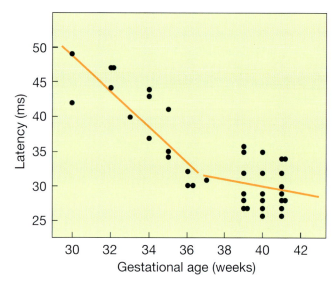

Fig. 3.33 Dependency on GA of the latency of the N20 component of the SEP following median nerve stimulation. Reduction in latency is rapid from the 30th to the 36th week, and thereafter is much slower. The data do not fit a continuous curve; a best-fit procedure was used to produce the angled line.

Differences in sites of stimulation, placement of the electrodes and filter settings make it difficult to compare

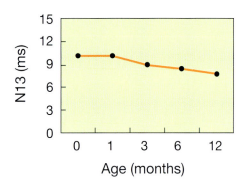

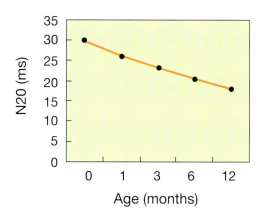

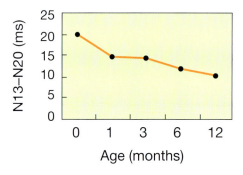

Fig. 3.34 Graphs of maturation during the first year of life of the N13 spinal potential, the N20 cortical potential and of the N13–N20 CCT of upper limb SEPs. Mean values from a group of normal subjects born at term and followed up at 1, 3, 6 and 12 months of age.

the latency and amplitude values reported in the available studies of SEP maturation (Desmedt et al 1976, Hashimoto et al 1983, Cadilhac et al 1985, Tomita et al 1986, Bartel et al 1987, Zhu et al 1987, Laureau et al 1988, Gibson and Levene 1988). Conduction velocities reach adult values before the age of 3 years in the peripheral nervous system (Thomas and Lambert 1960, Desmedt et al 1973), but this acceleration is slower in the central somatosensory pathways (Desmedt et al 1973, 1976, Cracco et al 1979). Spinal cord conduction in fibres from the lower limbs reaches adult values at the age of 5–6 years (Cracco et al 1979). Later, changes in conduction velocity related to fibre maturation begin to interact with those of body growth, and the peak latencies progressively increase to adult values, which are reached at the age of 15–17 years (Allison et al 1983, Tomita et al 1986). To sample normative data in children it is useful to correct absolute latencies by means of an index according to the length of the relevant neural pathways, namely arm length or body height. After this correction the latencies of all central SEPs decrease from birth up to the age of 10 years and stabilise thereafter (Tomita et al 1986). The CCT for the IPI between the N13 and N20 components is less affected by height and arm length than either the N13 or the N20 absolute latencies; it decreases progressively from 1 month to reach adult values at the age of 15 years. This decrease in CCT with age is not steady, but rapid between birth and 6 years of age, and slower thereafter (*Fig. 3.34* and *3.37*).

The rise time between the onsets and the peaks of the cortical N20 and P39 potentials, evoked by median and tibial nerve stimulation, respectively, decreases steadily from birth to the age of 16 years (Zhu et al 1987). There are also some morphological changes in the waveforms, which are age dependent. For instance, in children aged 7–14 years the amplitude of the cervical N13 median nerve SEP recorded with an anterior cervical electrode

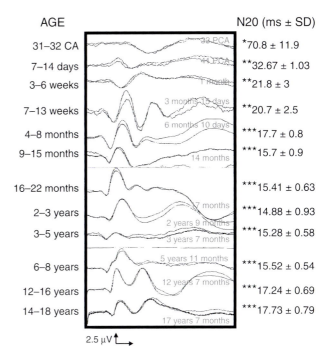

Fig. 3.35 Typical waveforms of the cortical components evoked by median nerve stimulation (MN-SEPs). On the right of the figure group mean values for the N20 component are given from normative data reported by (*) Taylor et al (1996b), (**) George and Taylor (1991) and by (***) Taylor and Fagan (1988) for the age groups in the parallel list on the left of the figure.

tends to be greater than in adults, whereas the P14 component tends to be lower with a greater P9/P14 amplitude ratio.

Normative data have been published for median nerve (Taylor and Fagan 1988, Boor et al 1998a, George and Taylor 1991) and posterior tibial nerve (Boor et al 1998b) SEPs, and can also be found in the IFCN guidelines (Mauguière et al 1999).

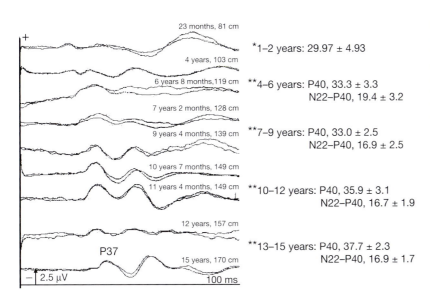

Fig. 3.36 Typical waveforms of the cortical components evoked by posterior tibial nerve stimulation (PTN-SEPs). On the right of the figure group mean values for the P37/P39/P40 component and CCT (N22–P37) are given from normative data reported by (*) Zhu et al (1987) and by (**) Boor et al (1998b) for the age groups and body length indicated on the waveforms.

Typical waveforms of the cortical components, and the mean group values from the normative data reported in the above-cited literature, are shown in *Figs. 3.35* and *3.36* for upper- and lower limb SEPs, respectively.

Normal data of N13 and N20 latency and amplitude and the N13–N20 IPI for upper limb SEPs obtained in the author's laboratory in longitudinal follow-up from birth to 12 months of age in a group of subjects born at term are summarised in *Table 3.13*. The corresponding maturational curves are shown in *Fig. 3.34*. Normal values obtained in the author's laboratory from a group of older children, aged 9–19 years, are summarised in *Table 3.14*.

SEPs: processing
SEPs to upper limb stimulation
Measurements made during clinical testing include the latency and amplitude of the N9, N13 and N20 com-

ponents and the latency of P14. Calculations include: the peripheral conduction velocity, which is obtained by dividing the distance (in millimetres) from the stimulus site at the wrist to the recording electrode at Erb's point by the N9 latency; and the three relevant IPIs, which determine conduction in the proximal dorsal roots up to the dorsal horn of the spinal cord (N9–N13), conduction in the proximal dorsal roots and dorsal columns up to the cervicomedullary junction (N9–P14), and conduction in the intracranial segment (P14–N20).

The following features may be abnormal and the data useful in clinical practice:
1. conduction between wrist and Erb's point N9
2. the N13/P9 amplitude ratio, using peak-to-peak amplitudes (but not in young children)
3. prolongation of the latency of the N13 and N20 components

Table 3.13   Peak latencies, amplitudes and IPIs (mean ± SD) of the N13 and N20 components of upper limb SEPs at various points during the first year of life

Age	N13 peak latency (ms)	N13 amplitude (µV)	N20 peak latency (ms)	N20 amplitude (µV)	N13–N20 IPI (ms)
1 week	10.06 ± 1.22	2.86 ± 0.7	29.87 ± 3.12	1.47 ± 0.8	19.79 ± 3.54
1 month	10.17 ± 1.22	3.11 ± 1.6	25.90 ± 3.06	3.58 ± 3.3	14.84 ± 2.44
3 months	8.88 ± 0.79	4.6 ± 2.3	23.17 ± 5.75	4.21 ± 2.5	14.38 ± 5.64
6 months	8.43 ± 0.92	2.33 ± 1.1	20.36 ± 2.84	3.32 ± 1.5	12.00 ± 2.63
12 months	7.68 ± 0.66	2.76 ± 1.5	17.90 ± 0.66	4.85 ± 1.4	10.23 ± 1.14

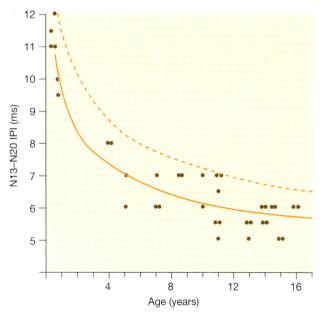

Fig. 3.37 Age dependency of the median nerve SEP N13–N20 IPI over the age range 0–16 years. The continuous and broken lines represent the mean + 2 standard deviations, respectively.

Table 3.14   SEP peak latencies, amplitudes, conduction velocity and IPIs (mean ± SD) in a group of normal children of mean age 14 years (range 9–19 years)

Parameter	Value
N9 peak latency (ms)	9.02 ± 0.74
N9 amplitude (µV)	3.11 ± 1.03
Conduction velocity: wrist to Erb's point (m/s)	63.90 ± 5.06
P14 peak latency (ms)	13.34 ± 1.10
N20 ms peak latency (ms)	18.01 ± 0.96
N20 amplitude (µV)	2.64 ± 1.07
P22 peak latency (ms)	21.39 ± 0.98
N9–N20 IPI (ms)	8.98 ± 0.57
P14–N20 IPI (ms)	4.67 ± 0.69
N20–P22 IPI (ms)	3.37 ± 0.93

4. prolongation of the N9–N13, N9–P14 and P14–N20 IPIs
5. loss of components
6. left–right asymmetry of the components
7. abnormalities of the GA-specific waveforms.

SEPs to lower limb stimulation

Measurements made during clinical testing include the latency and amplitude of the N7, N 22 and P39 components and the latency of the P30 component. Calculations include the peripheral conduction velocity, which is obtained by dividing the distance (in millimetres) from the stimulus site at the ankle to the recording electrode at the popliteal fossa by the N7 latency.

The conduction time from the lumbosacral spinal cord to the cortex is determined from the interval between the peak of the lumbar N22 negativity and the peak of the cortical P39 potential (N22–P39 IPI) (Small and Matthews 1984). The N22–P30 and P30–P39 IPIs provide a reliable evaluation of the intraspinal and intracranial conduction times.

The following features may be abnormal and the data useful in clinical practice:
1. the N7 ankle to popliteal fossa point conduction
2. prolongation of the N22–P39, N22–P30 and P30–P39 IPIs
3. loss of components
4. left–right asymmetry of the components.

SEPs: interpretation

By combining SEP recordings at different levels of the somatosensory pathways and studying the various components of the responses that are consistently obtained in normal subjects, it is possible to assess the transmission of the afferent volley from the periphery up to the cortex. The abnormal SEP waveforms, which reflect interruption, slowed conduction or time dispersion of this volley, are described below according to site of dysfunction, rather than to underlying pathological processes.

SEPs to upper limb stimulation

The peripheral and spinal components reflect the volley of action potentials from receptors up to the cervicomedullary junction, as well as the segmental postsynaptic responses of the dorsal horn neurons in the spinal cord. Components generated inside the skull reflect the ascending volley in the lemniscal and thalamocortical fibres, as well as the postsynaptic responses of the cortical receiving areas. The SEP findings associated with lesions at various sites are summarised in *Table 3.15*.

SEPs to lower limb stimulation

Currently, four components of the SEPs obtained by stimulation of the nerves of the lower limb are commonly used for diagnostic purposes. They reflect mainly the afferent volley in the tibial nerve, the postsynaptic responses from the lumbosacral cord, the afferent volley in the dorsal column at the cervicomedullary junction and the postsynaptic response of dorsal column nuclei, and the postsynaptic response of the primary somatosensory cortex. The SEP findings associated with lesions at various sites are summarised in *Table 3.16*.

SEPs: clinical role

A review of the applications of SEPs in paediatric neurology, together with other neurophysiological techniques, is included with the accounts of the integrated neuro-

Table 3.15 Diagnosis of the presumed site of somatosensory pathway dysfunction based on the pattern of upper limb SEP abnormalities

Dysfunction site	SEP component			
	N9	N13	P14	N20
Peripheral, distal to spinal ganglion	Absent or abnormal	Absent or abnormal	Absent or abnormal	Absent or abnormal
Peripheral, proximal to spinal ganglion	Present	Absent or abnormal	Absent or abnormal	Absent or abnormal
Lower cervical cord	Normal	Absent or abnormal	Absent or abnormal, or normal	Absent or abnormal, or normal
Cervicomedullary junction	Normal	Normal	Absent or abnormal	Absent or abnormal
Upper brainstem, ventroposterolateral (thalamic nucleus) and thalamic–cortical radiations	Normal	Normal	Normal	Absent or abnormal

Table 3.16　Diagnosis of the presumed site of somatosensory pathway dysfunction based on the pattern of lower limb SEP abnormalities

Dysfunction site	SEP component			
	N7	N22	P30	P39
Peripheral, distal to spinal ganglion	Absent or abnormal	Absent or abnormal	Absent or abnormal	Absent or abnormal
Peripheral, proximal to spinal ganglion	Present	Absent or abnormal	Absent or abnormal	Absent or abnormal
Lumbosacral cord	Normal	Absent or abnormal	Absent or abnormal, or normal	Absent or abnormal, or normal
Thoracic and cervical cord	Normal	Normal	Absent or abnormal	Absent or abnormal
Intracranial	Normal	Normal	Normal	Absent or abnormal

physiological investigation of the specific groups of disorders in Chapter 6, which is devoted to the clinical applications of paediatric neurophysiology. Some applications peculiar to SEPs are described on page 326, while clinical applications peculiar to neonatal age are discussed on pages 209–217.

REFERENCES

Allison T, Goff WR, Williamson PD, et al 1980 On the neural origin of early components of the human somatosensory evoked potentials. In: *Clinical Uses of Cerebral, Brainstem and Spinal Somatosensory Evoked Potentials* (ed JE Desmedt). Prog Clin Neurophysiol 7: 51–68.

Allison T, Wood CC, Goff WR 1983 Brainstem auditory, pattern reversal visual and short latency somatosensory evoked potentials: latencies in relation to age, sex, and brain and body size. Electroencephalogr Clin Neurophysiol 55: 619–636.

Apkarian P, Mirmiran M, Tijssen R 1991 Effects of behavioural state on visual processing in infants. Neuropediatrics 22: 85–91.

Arden GB, Carter RM, Hogg CR, et al 1979 A gold foil electrode: extending the horizons for clinical electroretinography. Invest Ophthalmol Vis Sci 18: 421–426.

Arden GB, Vaegan, Hogg, CR 1982 Clinical and experimental evidence that the pattern electroretinogram (PERG) is generated in more proximal retinal layers than the focal electroretinograin (FERG). Ann NY Acad Sci 388: 214–226.

Bach M, Hawlina M, Holder GE, et al 2001 Standard for pattern electroretinography. Doc Ophthalmol 101: 11–18.

Barkovich AJ 1995 Normal development of the neonatal and infant brain, skull, and spine. In: *Pediatric Neuroimaging*, 2nd edn (ed AJ Barkovich). Raven Press, New York, pp 42–47.

Barnet AB, Friedman SL, Weiss IP, et al 1980 VEP development in infancy and early childhood. A longitudinal study. Electroencephalogr Clin Neurophysiol 49: 476–489.

Bartel P, Conradie J, Robinson E, et al 1987 The relationship between somatosensory evoked potential latencies and age and growth parameters in young children. Electroencephalogr Clin Neurophysiol 68: 180–186.

Beauchaine KA, Kaminski JR, Gorga MP 1987 Comparison of Beyer DT 48 and Etymotic Insert Earphones: auditory brain stem response measurements. Ear Hearing 8: 292–297.

Binnie CD, Cooper R, Mauguière F, et al (eds) 2003 *Clinical Neurophysiology*, Vol. 2; *EEG, Paediatric Neurophysiology, Special Techniques and Applications*. Elsevier Science, Amsterdam.

Binnie CD, Cooper R, Mauguière F, et al (eds) 2004 *Clinical Neurophysiology*, Vol. 1; *EMG, Nerve Conduction and Evoked Potentials*. Elsevier Science, Amsterdam.

Blom JL, Barth PG, Visser SL 1980 The visual evoked potential in the first six years of life. Electroencephalogr Clin Neurophysiol 48: 395–405.

Bongers-Schokking CJ, Colon EJ, Hoogland RA, et al 1989 The somatosensory evoked potentials of normal infants: influence of filter bandpass, arousal state and number of stimuli. Brain Dev 11: 33–39.

Boor R, Goebel B, Taylor MJ 1998a Subcortical somatosensory evoked potentials after median nerve stimulation in children. Eur J Paediatr Neurol 2: 137–143.

Boor R, Goebel B, Doepp M, et al 1998b Somatosensory evoked potentials after posterior tibial nerve stimulation – normative data in children. Eur J Paediatr Neurol 2: 145–152.

Bradshaw K, Hansen R, Fulton A 2004 Comparison of ERGs recorded with skin and corneal-contact electrodes in normal children and adults. Doc Ophthalmol 109: 43–55.

Broughton RJ 1969 Methods, results and evaluations. In: *Average Evoked Potentials* (eds E Donchin, DB Lindsley). NASA SP191. US Government Printing Office, Washington, DC, pp 79–84.

Burke D, Skuse NF, Lethlean K 1981 Cutaneous and muscle afferent components of the cerebral potential evoked by electrical stimulation of human peripheral nerves. Electroencephalogr Clin Neurophysiol 51: 579–588.

Burke D, Gandevia SC, McKeon B, et al 1982 Interactions between cutaneous and muscle afferent projections to cerebral cortex in man. Electroencephalogr Clin Neurophysiol 53: 349–360.

Bush RA, Sieving PA 1996 Inner retinal contributions to the primate fast flicker electroretinogram. J Opt Soc Am 13: 557–565.

Cadilhac J, Zhu Y, Georgesco M, et al 1985 La maturation des potentiels évoqués somesthésiques cérébraux. Rev EEG Neurophysiol Clin 15: 1–11.

Celesia GG, Brigell MG 1999 Recommended standards for pattern electroretinograms and visual evoked potentials. In: *Recommendations for the Practice of Clinical Neurophysiology: Guidelines of the International Federation of Clinical Neurophysiology*, 2nd edn (eds G Deuschl, A Eisen). Electroencephalogr Clin Neurophysiol (Suppl 52): 53–68.

Celesia GG, Brigell MG, Peachey N 1999 Recommended standards for electroretinograms. In: *Recommendations for the Practice of Clinical Neurophysiology: Guidelines of the International Federation of Clinical Neurophysiology*, 2nd edn (eds G Deuschl, A Eisen). Electroencephalogr Clin Neurophysiol (Suppl 52): 45–52.

Chiappa KH 1983 *Evoked Potentials in Clinical Medicine*, 1st edn. Raven Press, New York.

Chiozza ML, Suppiej A, Zacchello F 1997 Evoked potentials in paediatrics: economic audit. Childs Nerv Syst 13: 166–170.

Cigánek L 1961 The EEG response (evoked potential) to light stimulus in man. Electroencephalogr Clin Neurophysiol 13: 163–172.

Cooper R, Osselton JW, Shaw JC 1980 *EEG Technology*, 3rd edn. Butterworths, London, pp 200–202.

Cooper R, Binnie CD, Billings R 2005 *Techniques in Clinical Neurophysiology: A Practical Manual*. Elsevier, Oxford.

Cracco RQ, Cracco JB 1976 Somatosensory evoked potentials in man: far field potentials. Electroencephalogr Clin Neurophysiol 41: 460–466.

Cracco JB, Cracco RQ, Stolove R 1979 Spinal evoked potentials in man: a maturational study. Electroencephalogr Clin Neurophysiol 46: 58–64.

Cruse R, Mem G, Lesser R, et al 1982 Paradoxical lateralization of cortical potentials evoked by stimulation of posterior tibial nerve. Arch Neurol 39: 222–225.

Cusick JF, Myklebust J, Larson SJ, et al 1978 Spinal evoked potentials in the primate: neural substrate. J Neurosurg 49: 551–557.

Dawson WW, Trick GL, Litzkow CA 1979 Improved electrode for electroretinography. Invest Ophthalmol Vis Sci 18: 988–991.

Desmedt JE, Bourguet M 1985 Color imaging of parietal and frontal somatosensory potential fields evoked by stimulation of median or posterior tibial nerve in man. Electroencephalogr Clin Neurophysiol 62: 1–17.

Desmedt JE, Cheron G 1980 Central somatosensory conduction in man: neural generators and interpeak latencies of the far-field components recorded from neck and right or left scalp or earlobes. Electroencephalogr Clin Neurophysiol 50: 382–403.

Desmedt JE, Cheron G 1983 Spinal and far-field components of human somatosensory evoked potentials to posterior tibial nerve stimulation analysed with oesophageal derivations and non-cephalic reference recording. Electroencephalogr Clin Neurophysiol 56: 635–651.

Desmedt JE, Manil J 1970 Somatosensory evoked potentials of the normal human neonate in REM sleep, in slow wave sleep and in waking. Electroencephalogr Clin Neurophysiol 29: 113–126.

Desmedt JE, Noel P, Debecker J, et al 1973 Maturation of afferent conduction velocity as studied by sensory nerve potentials and by cerebral evoked potentials. In: *New Developments in Electromyography and Clinical Neurophysiology*, Vol 2 (ed JE Desmedt). Karger, Basel, pp 52–63.

Desmedt JE, Brunko E, Debecker J 1976 Maturation of the somatosensory evoked potentials in normal infants and children with special reference to the early NI component. Electroencephalogr Clin Neurophysiol 40: 43–58.

Despland PA 1987 Evoked reponse audiometry. In: *A Textbook of Clinical Neurophysiology* (eds AM Halliday, SR Butler, R Paul). Wiley, Chichester, pp 399–414.

Deuschl G, Eisen A (eds) 1999 *Recommendations for the Practice of Clinical Neurophysiology: Guidelines of the International Federation of Clinical Neurophysiology*, 2nd revised and enlarged edn. Electroencephalogr Clin Neurophysiol (Suppl 52).

De Vries LS 1993 Somatosensory evoked potentials in term neonates with postasphyxial encephalopathy. Clin Perinatol 20: 463–499.

de Vries LS 1993 Somatosensory-evoked potentials in term neonates with postasphyxial encephalopathy. Clin Perinatol 20: 463–482.

Dubowitz LMS, Mushin J, De Vries L, et al 1986 Visual function in the newborn infant: is it cortically mediated? Lancet 327(4890): 1139–1141.

Dustman RE, Beck EC 1969 The effects of maturation and aging on the waveform of visually evoked potentials. Electroencephalogr Clin Neurophysiol 26: 2–11.

Dustman RE, Schenkenberg T, Lewis EG, et al 1977 In: *Visual Evoked Potentials in Man: New Developments* (ed JE Desmedt). Clarendon Press, Oxford, pp 332–377.

Ellingson RJ 1958 Electroencephalograms of normal, full-term newborns immediately after birth with observations on arousal and visual evoked responses. Electroencephalogr Clin Neurophysiol 10: 31–50.

Ellingson RJ 1960 Cortical electrical responses to visual stimulation in the human infant. Electroencephalogr Clin Neurophysiol 12: 663–677.

Engel R, Butler BV 1963 Appraisal of conceptual age of newborn infants by electroencephalographic methods. J Pediatr 63: 386–393.

Fichsel H 1969 Visual evoked potentials in prematures, newborns, infants and children by stimulation with colored light. Electroencephalogr Clin Neurophysiol 27: 660.

Fishman GA, Birch DG, Holder GE, et al (eds) 2001 *Electrophysiologic Testing in Disorders of the Retina, Optic Nerve, and Visual Pathway*, 2nd edn. Ophthalmology Monograph 2. The Foundation of the American Academy of Ophthalmology, San Francisco, CA.

Flores-Guevara R, Renault F, Ostré C, et al 1996 Maturation of the electroretinogram in children: stability of the amplitude ratio a/b. Electroencephalogr Clin Neurophysiol 100: 422–427.

Fulton AB 1988 The development of scotopic retinal function in human infants. Doc Ophthalmol 69: 101–109.

Fulton AB, Hansen RM, Westall CA 2003 Development of ERG responses: the ISCEV rod, maximal and cone responses in normal subjects. Invest Ophthalmol Vis Sci 107: 235–241.

Gandevia SC, Burke D, McKeon BB 1984 The projection of muscle afferents from hand to cerebral cortex in man. Brain 107: 1–13.

Gastaut H, Régis H 1965 Visually evoked potentials recorded transcranially in man. In: *The Analysis of Central Nervous System and Cardiovascular Data using Computer Methods* (eds LD Proctor, WR Adey). NASA, US Government Printing Office, Washington, DC, pp 7–34.

George SR, Taylor MJ 1991 Somatosensory evoked potentials in neonates and infants: developmental and normative data. Electroencephalogr Clin Neurophysiol 80: 94–102.

Gibson N, Levene MI 1988 Somatosensory evoked potentials. In: *Fetal and Neonatal Neurology and Neurosurgery* (eds M Levene, MJ Bennett, J Punt). Churchill Livingstone, London, pp 213–215.

Gibson NA, Brezinova V, Levene MI 1992 Somatosensory evoked potentials in the term newborn. Electroencephalogr Clin Neurophysiol 84: 26–31.

Gilmore R, Brock J, Hermansen MC, et al 1987 Development of lumbar spinal cord and cortical evoked potentials after tibial nerve stimulation in the pre-term newborns: effect of gestational age and other factors. Electroencephalogr Clin Neurophysiol 68: 28–39.

Goff GD, Matsumiya Y, Allison T, et al 1977 The scalp topography of human somatosensory and auditory evoked potentials. Electroencephalogr Clin Neurophysiol 42: 57–76.

Gorga MP, Kaminski JR, Beauchaine KA 1988 Auditory brain stem responses from graduates of an intensive care nursery using an Insert Earphone. Ear Hearing 9: 144–147.

Green JB, Walcoff MR, Lucke JF 1982 Comparison of phenytoin and phenobarbital effects on far-field auditory and somatosensory evoked potential interpeak latencies. Epilepsia 23: 417–421.

Guérit JM, Opsomer RJ 1991 Bit-mapped imaging of somatosensory evoked potentials after stimulation of the posterior tibial nerves and dorsal nerve of the penis/clitoris. Electroencephalogr Clin Neurophysiol 80: 228–237.

Guérit JM, Soveges L, Baele P, et al 1990 Median nerve evoked potentials in profound hypothermia for ascending aorta repair. Electroencephalogr Clin Neurophysiol 77: 163–173.

Halliday AM, Wakefield GS 1963 Cerebral evoked potentials in patients with dissociated sensory loss. J Neurol Neurosurg Psychiatry 26: 211–219.

Halliday AM, McDonald WI, Mushin J 1973 Visual evoked responses in the diagnosis of multiple sclerosis. BMJ 4: 661–664.

Halliday AM, Barrett G, Carroll WM, et al 1982 Problems in defining the normal limits of the VEP. In: *Clinical Applications of Evoked Potentials in Neurology*. Advances in Neurology, Vol 32 (eds J Courjon, F Mauguière, M Revol). Raven Press, New York, pp 1–9.

Hansen RM, Fulton AB 2005a Recovery of the rod photoresponse in infants. Invest Ophthalmol Vis Sci 46: 764–768.

Hansen RM, Fulton AB 2005b Development of the cone ERG in infants. Invest Ophthalmol Vis Sci 46: 3458–3462.

Harden A 1974 Non-corneal electroretinogram – parameters in normal children. Br J Ophthalmol 58: 811–816.

Harden A 1982 Maturation of the visual evoked potentials. In: *Clinical Applications of Cerebral Evoked Potentials in Pediatric Medicine* (eds GA Chiarenza, DP Papakostopoulos,). International Congress Series 595 – Proceedings of the International Conference on Clinical Application of Cerebral Evoked Potentials in Pediatric Neurology, Milan, Italy. Excerpta Medica, Amsterdam, pp 41–46.

Harden A, Adams GGW, Taylor DSI 1989 The electroretinogram. Arch Dis Childhood 64: 1080–1087.

Harding GFA, Grose J, Wilton A, et al 1989 The pattern reversal VEP in short-gestation infants. Electroencephalogr Clin Neurophysiol 74: 76–80.

Hashimoto I 1989 Critical analysis of short-latency auditory evoked potentials recording techniques. In: *Advanced Evoked Potentials* (ed H Lüders). Kluwer, Boston, MA, pp 105–142.

Hashimoto T, Tayama M, Kiura K, et al 1983 Short latency somatosensory evoked potentials in children. Brain Dev 5: 390–396.

Hawlina M, Konec B 1992 New noncorneal HK-loop electrode for clinical electroretinography. Doc Ophthalmol 8: 253–259.

Heckenlively JR, Arden GB 1991 *Principles and Practice of Clinical Electrophysiology of Vision*. Mosby Year Book, St. Louis, MO.

Hendrickson A, Drucker D 1992 The development of parafo-veal and mid-peripheral human retina. Behav Brain Res 49: 21–31.

Hendrickson AE, Yuodelis C 1984 The morphological development of the human fovea. Ophthalmology 91: 603–612.

Hrbek A, Mares P 1964 Cortical evoked responses to visual stimulation in full-term and premature newborns. Electroencephalogr Clin Neurophysiol 16: 575–581.

Hrbek A, Hrbková M, Lenard H-G 1969 Somatosensory, auditory and visual evoked responses in newborn infants during sleep and wakefulness. Electroencephalogr Clin Neurophysiol 26: 597–603.

Hrbek A, Karlberg P, Olssohn T 1973 Development of visual and somatosensory evoked responses in preterm newborn infants. Electroencephalogr Clin Neurophysiol 34: 225–232.

IFCN 1999 *Recommendations for the Practice of Clinical Neurophysiology: Guidelines of the International Federation of Clinical Neurophysiology*, 2nd edn (eds G Deuschl, A Eisen). Electroencephalogr Clin Neurophysiol (Suppl 52): 45–52.

Jaffe MJ, Hommer DW, Caruso RC, et al 1989 Attenuating effects of Diazepam on the electroretinogram of normal humans. Retina 9: 216–225.

Jewett DL 1970 Volume-conducted potentials in response to auditory stimuli as detected by averaging in the cat. Electroencephalogr Clin Neurophysiol 28: 609–618.

Jiang ZD, Wu YY, Zheng WS, et al 1991 The effect of click rate on latency and interpeak interval of the brain-stem auditory evoked potentials in children from birth to 6 years. Electroencephalogr Clin Neurophysiol 80: 60–64.

Jones SJ 1982 Somatosensory evoked potentials – the abnormal waveform. In: *Evoked Potentials in Clinical Testing* (ed AM Halliday). Churchill Livingstone, London, pp 429–470.

Kakigi R, Shibasaki H 1983 Scalp topography of the short-latency somatosensory evoked potentials following posterior tibial nerve stimulation in man. Electroencephalogr Clin Neurophysiol 56: 430–437.

Kakigi R, Shibasaki H 1987 Generator mechanisms of giant somatosensory evoked potentials in cortical reflex myoclonus. Brain 110: 1359–1373.

Karniski W 1992 The late somatosensory evoked potentials in premature and term infants. I. Principal component topography. Electroencephalogr Clin Neurophysiol 84: 32–43.

Karniski W, Wyble L, Leanse L, et al 1992 The late somatosensory evoked potential in premature and term infants. II. Topography and latency development. Electroencephalogr Clin Neurophysiol 84: 44–54.

Keenan NK, Taylor MJ, Coles JG, et al 1987 The use of VEPs for CNS monitoring during continuous cardiopulmonary bypass and circulatory arrest. Electroencephalogr Clin Neurophysiol 68: 241–246.

Kemp DT 1978 Stimulated acoustic emissions from within the human auditory system. J Acoust Soc Am 64: 1386–1391.

Kemp DT 2002 Otoacoustic emissions, their origin in cochlear function, and use. Br Med Bull 63: 223–241.

Klimach VJ, Cooke RWI 1988 Maturation of the neonatal somatosensory evoked potentials in preterm infants. Dev Med Child Neurol 30: 208–214.

Kriss A 1994 Skin ERGs: their effectiveness in paediatric visual assessment, confounding factors, and comparison with ERGs recorded using various types of corneal electrode. Int J Psychophysiol 16: 137–146.

Kriss A, Jeffrey B, Taylor D 1992 The electroretinogram in infants and young children. J Clin Neurophysiol 9: 373–393.

Laget P, Raimbault J, D'Allest AM, et al 1976 La maturation des potentiels évoqués somesthésiques chez l'homme. Electroencephalogr Clin Neurophysiol 40: 499–515.

Laget P, Flores-Guevara R, D'Allest AM, et al 1977 Maturation of visually evoked potentials in the normal child. Electroencephalogr Clin Neurophysiol 43: 732–744.

Laureau E, Marlot D 1990 Somatosensory evoked potentials after median and tibial nerve stimulation in healthy newborns. Electroencephalogr Clin Neurophysiol 76: 453–458.

Laureau E, Majnermer A, Rosenblatt B, et al 1988 A longitudinal study of short latency somatosensory evoked responses in healthy newborns and infants. Electroencephalogr Clin Neurophysiol 71: 100–108.

Leaf AA, Green CR, Esack A, et al 1995 Maturation of the electroretinogram and visual evoked potentials in preterm infants. Dev Med Child Neurol 37: 814–826.

Lorente de Nò R 1947 Analysis of the distribution of action currents of nerve in volume conductors. Stud Rockefeller Inst Med Res 132: 384–477.

Lupton BA, Wong PK, Bencivenga R, et al 1990 The effect of electrode position on flash visual evoked potentials in the newborn. Doc Ophthalmol 76: 73–80.

Lütschg J 1985 Evozierte Potentiate bei komatösen Kindern. Gustav Fischer, Stuttgart.

Macefield G, Burke D, Gandevia SC 1989 The cortical distribution of muscle and cutaneous afferent projections from the human foot. Electroencephalogr Clin Neurophysiol 72: 518–528.

Mactier H, Dexter JD, Hewett JE, et al 1988 The electroretinogram in preterm infants. J Pediatr 113: 607–612.

Majnermer A, Rosenblatt F, Willis D, et al 1990 The effect of gestational age at birth on somatosensory-evoked potentials performed at term. J Child Neurol 5: 329–335.

Marmor MF, Zrenner E 1999 Standard for clinical electroretinography (1999 update). International Society for Clinical Electrophysiology of Vision. Doc Ophthalmol 97: 143–156.

Mauguière F 1987 Short-latency somatosensory evoked potentials to upper limb stimulation in lesions of brainstem, thalamus and cortex. In: The London Symposia (eds RJ Ellingson, NMF Murray, AM Halliday). Electroencephalogr Clin Neurophysiol (Suppl 39): 302–309.

Mauguière F, Ibañez V 1985 The dissociation of early SEP components in lesions of the cervico-medullary junction: a cue for routine interpretation of abnormal cervical responses to median nerve stimulation. Electroencephalogr Clin Neurophysiol 62: 406–420.

Mauguière F, Restuccia D 1991 Inadequacy of the forehead reference montage for detecting spinal N13 potential abnormalities in patients with cervical cord lesion and preserved dorsal column function. Electroencephalogr Clin Neurophysiol 79: 448–456.

Mauguière F, Allison T, Babiloni C, et al 1999 Somatosensory evoked potentials. In: Recommendations for the Practice of Clinical Neurophysiology: Guidelines of the International Federation of Clinical Neurophysiology, 2nd edn (eds G Deuschl, A Eisen). Electroencephalogr Clin Neurophysiol (Suppl 52): 79–90.

Maurer K 1985 Uncertainties of topodiagnosis of auditory nerve and brain-stem auditory evoked potentials due to rarefaction and condensation stimuli. Electroencephalogr Clin Neurophysiol 62: 135–140.

McCallum WC 1988 Potentials related to expectancy, preparation and motor activity. In: Human Event Related Potentials. EEG Handbook, revised series, Vol 3 (ed TW Picton). Elsevier, Amsterdam.

McPherson DL, Hirasugi Y, Starr A 1985 Auditory brain stem potentials recorded at different scalp locations in neonates and adults. Ann Otol Rhinol Laryngol 94: 236–243.

Mercuri E, Siebenthal KV, Tutuncuoglu S, et al 1995 The effect of behavioural states on visual evoked responses in preterm and full-term newborns. Neuropediatrics 26: 211–213.

Meredith S, Reddy A, Allen L, et al 2003 Full-field ERGs recorded with skin electrodes give valid data for the diagnosis of retinal dystrophy in unsedated children. Proceedings of the 29th Annual Meeting of the European Paediatric Ophthalmological Society, Regensburg, 2–4 October 2003.

Minami T, Gondo K, Nakayama H, et al 1996 Cortical somatosensory evoked potentials to posterior tibial nerve stimulation in newborn infants. Brain Dev 18: 294–298.

Moskowitz A, Sokol S 1983 Developmental changes in the human visual system as reflected by the latency of the pattern reversal VEP. Electroencephalogr Clin Neurophysiol 56: 1–15.

Mushin J 1988 Visual evoked potentials. In: Fetal and Neonatal Neurology and Neurosurgery (eds M Levene, MJ Bennett, J Punt). Churchill Livingstone, London, pp 206–212.

Pampiglione G, Harden A 1977 So-called neuronal ceroid lipofuscinosis. Neurophysiological studies in 60 children. J Neurol Neurosurg Psychiatry 40: 323–330.

Parmelee AH, Wenner WH, Akiyama Y, et al 1967 Sleep states in premature infants. Dev Med Child Neurol 9: 70–77.

Parmelee AH, Schulte FJ, Akiyama Y, et al 1968 Maturation of EEG activity during sleep in premature infants. Electroencephalogr Clin Neurophysiol 24: 319–329.

Peña M, Birch D, Uauy R, et al 1999 The effect of sleep state on electroretinographic (ERG) activity during early human development. Early Hum Dev 55: 51–62.

Picton TW, Taylor MJ, Durieux-Smith A, et al 1986 Brainstem auditory evoked potentials in pediatrics. In: Electrodiagnosis in Clinical Neurology, 2nd edn (ed MT Aminoff). Churchill Livingstone, Philadelphia, pp 505–534.

Pierrat V, De Vries LS, Minami T, et al 1990 Somatosensory evoked potentials and adaptation to extrauterine life: a longitudinal study. Brain Dev 12: 376–379.

Powers SK, Bolger CA, Edwards MSB 1982 Spinal cord pathways mediating somatosensory evoked potentials. J Neurosurg 57: 472–482.

Pratt H, Amlie RN, Starr A 1981 Short latency mechanically evoked peripheral nerve and somatosensory potentials in newborn infants. Pediatr Res 15: 295–298.

Prechtl HFR, Beintema D 1964 The Neurological Examination of the Full Term Newborn Infant. Clinics in Developmental Medicine, No. 12, Spastics Society/Heinemann Medical Books, London.

Pryds O 1992 Stimulus rate-induced VEP attenuation in preterm infants. Electroencephalogr Clin Neurophysiol 84: 188–191.

Pryds O, Greisen G, Trojaborg W 1988 Visual evoked potentials in preterm infants during the first hours of life. Electroencephalogr Clin Neurophysiol 71: 257–265.

Restuccia D 2000 Anatomic origin of P13 and P14 scalp far-field potentials. J Clin Neurophysiol 17: 246–257.

Roy MS, Gosselin J, Hanna N, et al 2004 Influence of the state of alertness on the pattern visual evoked potentials (PVEP) in very young infant. Brain Dev 26: 197–202.

Salamy A 1984 Maturation of the auditory brainstem response from birth through early childhood. J Clin Neurophysiol 3: 293–329.

Scaioli V, Brinciotti M, Di Capua M, et al 2003 Proposta di un database normativo multi-centrico del BAEP in età pediatrica: metodologia per la raccolta ed analisi dei dati. Proceedings of the Annual Meeting of the Italian Society of Clinical Neurophysiology, Florence.

Seyal M, Emerson RG, Pedley TA 1983 Spinal and early scalp-recorded components of the somatosensory evoked potential following stimulation of the posterior tibial nerve. Electroencephalogr Clin Neurophysiol 55: 320–330.

Simons K (ed) 1993 Early Visual Development – Normal and Abnormal. Oxford University Press, New York.

Small M, Matthews WB 1984 A method of calculating spinal cord transit time from potentials evoked by tibial nerve stimulation in normal subjects and patients with spinal cord disease. Electroencephalogr Clin Neurophysiol 59: 156–164.

Sokol S 1986a Visual evoked potentials. In: Electrodiagnosis in Clinical Neurology (ed MJ Aminoff). Raven Press, New York, pp 441–466.

Sokol S 1986b Clinical application of the ERG and VEP in pediatric age group. In: *Evoked Potentials* (eds RQ Cracco, I Bodis-Wollner). Front Clin Neurosci 3: 447–454.

Sokol S, Moskowitz A 1981 Effect of retinal blur on the peak latency of the pattern evoked potential. Vision Res 21: 1279–1286.

Sokol S, Moskowitz A, Paul A 1983 Evoked potential estimates of visual accommodation in infants. Vision Res 23: 851–860.

Stanley OH, Fleming PJ, Morgan MH 1987 Developmental waveform analysis of the neonatal flash evoked potential. Electroencephalogr Clin Neurophysiol 68: 149–152.

Stapells DR, Picton TW, Smith AD 1982 Normal hearing thresholds for clicks. J Accoust Soc Am 72: 74–79.

Starr A, Squires K 1982 Distribution of auditory brainstem potentials over the scalp and nasopharynx in humans. In: *Evoked Potentials*. Ann NY Acad Sci 388: 427–442.

Stockard JE, Stockard J 1986 Clinical applications of brainstem evoked potentials in infants. In: *Evoked Potentials* (eds R Cracco, I Bodis-Wollner). Front Clin Neurosci 3: 455–462.

Stockard JE, Stockard JJ, Westmoreland BF, et al 1979 Brainstem auditory evoked responses: normal variation as a function of stimulus and subject characteristics. Arch Neurol 36: 823–831.

Suppiej A, Casara GL, Boniver C, et al 1991 Somatosensory pathway dysfunction in uremic children. Brain Dev 13: 238–241.

Takashima S, Chan F, Becker LE, et al 1980 Morphology of the developing visual cortex of the human infant. A quantitative and qualitative golgi study. J Neuropathol Exp Neurol 39: 487–501.

Taylor MJ 1992 Visual evoked potentials. In: *The Neurophysiological Examination of the Newborn Infant* (ed JA Eyre). Clinics in Developmental Medicine, No. 120. MacKeith Press, distributed by Cambridge University Press, London, pp 93–111.

Taylor MJ, Fagan ER 1988 SEPs to median nerve stimulation: normative data for paediatrics. Electroencephalogr Clin Neurophysiol 71: 323–330.

Taylor MJ, McCulloch DL 1992 Visual evoked potentials in infants and children. J Clin Neurophysiol 9: 357–372.

Taylor MJ, Menzies R, MacMillan LJ, et al 1987 VEPs in normal full-term and premature neonates: longitudinal versus cross-sectional data. Electroencephalogr Clin Neurophysiol 68: 20–27.

Taylor MJ, Saliba E, Laugier J 1996a Use of evoked potentials in preterm neonates. Arch Dis Child 74: F70–F76.

Taylor MJ, Boor R, Ekert PG 1996b Preterm maturation of the somatosensory evoked potential. Electroencephalogr Clin Neurophysiol 100: 448–452.

Tinazzi M, Mauguière F 1995 Assessment of intraspinal and intracranial conduction by P30 and P39 tibial nerve somatosensory evoked potentials in cervical cord and hemispheric lesions. J Clin Neurophysiol 12: 237–253.

Tinazzi M, Zanette G, Bonato C, et al 1996 Neural generators of tibial nerve P30 somatosensory evoked potential studied in patients with a focal lesion of the cervico-medullary junction. Muscle Nerve 19: 1538–1548.

Tinazzi M, Zanette G, Manganotti B, et al 1997 Amplitude changes of tibial nerve cortical SEPs when using ipsilateral or contralateral ear as reference. J Clin Neurophysiol 14: 217–225.

Thomas JE, Lambert EH 1960 Ulnar nerve conduction velocity and H-reflex in infants and children. J Appl Physiol 15: 1–9.

Thompson DA, Drasdo N 1987 An improved method for using the DTL fibre in electroretinography. Ophthal Physiol Opt 7: 315–319.

Tomita Y, Nishimura S, Tanaka T 1986 Short-latency SEPs in infants and children: developmental changes and maturational index of SEPs. Electroencephalogr Clin Neurophysiol 65: 335–343.

Van Hof-Van Duin J (ed) 1992 Normal and abnormal development in infants and children. Behav Brain Res 49(special issue): 1–148.

Volpe JJ 2000 Neuronal proliferation, migration, organization, and myelination. In: *Neurology of the Newborn*, 4th edn (ed JJ Volpe). WB Saunders, Philadephia, pp 43–92.

Watanabe K, Iwase K, Hara K 1972 Maturation of visual evoked response in premature infants. Dev Med Child Neurol 14: 425–435.

White CP, Cooke RWI 1989 Maturation of the cortical evoked response to posterior nerve stimulation in the preterm neonate. Dev Med Child Neurol 31: 657–664.

Whyte HE, Pearce JM, Taylor MJ 1987 Changes in the VEP in preterm neonates with arousal states, as assessed by EEG monitoring. Electroencephalogr Clin Neurophysiol 68: 223–225.

Willis J, Seales D, Frazier E 1984 Short latency somatosensory evoked potentials in infants. Electroencephalogr Clin Neurophysiol 59: 366–373.

Yakovlev PI, Lecours A-R 1967 The myelogenetic cycles of regional maturation of the brain. In: *Regional Development of the Brain in Early Life* (ed A Minkowski). Blackwell Scientific, Oxford, pp 3–70.

Yamada T 2000 Neuroanatomic substrates of lower extremity somatosensory evoked potentials. J Clin Neurophysiol 17: 269–279.

Yamada T, Machida M, Kimura J 1982 Far-field somatosensory evoked potentials after stimulation of the tibial nerve. Neurology 32: 1151–1158.

Zhu Y, Georgesco M, Cadilhac J 1987 Normal latency values of early cortical somatosensory evoked potentials in children. Electroencephalogr Clin Neurophysiol 68: 471–474.

Chapter 4

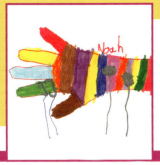

Measurement of Nerve Conduction and the Techniques of Electromyography in Paediatrics

INTRODUCTION

It is assumed that the reader of this chapter will already have experience of nerve conduction and electromyography (EMG) in adults and will have an understanding of the basic physiology and electronics that this implies. This information can be obtained from the first volume of the parent volume in this series of books (Binnie et al 2004) or the first volume of this series of manuals (Cooper et al 2005). The principles that underpin both nerve conduction and EMG are of course identical in adults and paediatrics, but there are many factors peculiar to paediatrics that must be considered when trying successfully to accomplish these examinations. The first and most important challenge is patient cooperation. Using the term EMG to cover both needle EMG and nerve conduction, all children find EMG distressing to an extent, and if the clinician is not able to evolve techniques to overcome this the results will be inaccurate and incomplete. How patient cooperation is obtained is influenced by the mores and traditions of the country in which you practise, modified by other factors such as legal considerations. Thus, in the USA, most patients might insist on sedation of some kind, but the need for qualified anaesthesiologists means that most EMGs in the UK are performed on alert children using various distraction techniques. In Sweden sedation is mandatory. Paediatric EMG is, by necessity, a compromise; make it completely pain free and no voluntary activity is possible, and if the child is too alert you have to accept a more limited sample of investigations than you would want if the patient were an adult. It should be noted that some adults can be less cooperative than children.

Which distraction techniques are used is a very personal decision, but a factor that has been found to be crucial is to not allow any information to be given to the family before they attend. It does not matter in what terms a description of EMG is couched, there are few who do not suffer a degree of concern when electric shocks and needles are discussed. Some people are so anxious that the examination is almost impossible. At the end of examination in these children, it is almost universal to be told that it was not as bad as expected. This strategy is not intended to be deceptive or dishonest to the children and their family, but rather that an explanation and demonstration at the time of the examination is much preferred and most children will pass through the test without distress.

Another point of personal practice is to inform all those concerned that the results will not be made available at the end of the EMG session. This is for two main reasons. The first is that the diagnosis is not always obvious. Secondly, particularly in conditions where multiple tests are being undertaken, one clinician should be responsible for collating the data and then discussing prognosis and, less commonly, treatment. Anticipation that the day of the EMG is when their worst fears may be realised produces significant levels of anxiety in parents, which is always picked up by the children. Knowledge that the results of the examination will not be given until a later time always has a calming effect, which is beneficial to the examination. Anybody who wishes to join the patient should be allowed to watch.

In the sections that follow the technical aspects of the tests are described and points that are of particular relevance in paediatrics are noted and discussed.

CHOICE OF ELECTRODES

Surface recording compared with near-nerve recording

In the UK it has been the norm to use surface recording of both sensory and motor action potentials, the first studies being done by Gilliatt's group at The National Hospital for Nervous Diseases in London (Gilliatt and Sears 1958). It is the same in many countries in the world. Near-nerve recording is painful and only has gained acceptance in the groups linked to the Copenhagen school where Buchthal pioneered its use (Buchthal and Rosenfalck 1966). There it was used particularly to determine sensory nerve function. It is no surprise that this technique is not used in paediatrics.

However, it is premature to consign the near-nerve recording technique to the pool of tests that cannot be used in children without considering what is lost by doing so, and discussing what might be used to replace it. Surface and near-nerve methods both give information on the large diameter fibres, surface recording relying heavily on the amplitude of the sensory nerve action potential to determine the number of nerve fibres. The difference is that near-nerve needle electrodes allow remarkable detail of conduction by smaller fibres (*Fig. 4.1*). Buchthal acknowledged that the positioning of the near-nerve electrode had a very significant influence on amplitude, and found the results so variable that he did not use amplitude to determine nerve pathology; instead he took great care to determine conduction velocity using many thousands of stimuli.

If recordings are made with the near-nerve needle electrodes (Buchthal and Rosenfalck 1966) the responses of very low amplitude (about 0.05 µV) that follow the main early peak can be resolved. With multiple averaging of between 1,000 and 2,000 trials and high intensity stimuli, the very small responses of fibres with conduction velocities as slow as 15 m/s can be recognised in normal nerves, and at even lower velocities in pathologic nerves (see *Fig. 4.1*) (Buchthal et al 1984, Shefner et al 1991).

When the available anatomical data on the distribution of nerve fibres in a typical sensory nerve is reviewed it is easy to understand why near-nerve recording was developed. *Figure 4.2* shows a three-dimensional histogram constructed from the data reported by Jacobs and Love (1985) of the number of nerve fibres, both myelinated and unmyelinated, in the sural nerve. The figure shows a bimodal distribution of the diameters of myelinated fibres, as well as the unmyelinated fibre content of the nerve. This latter group of fibres is obviously the more numerous. The relative proportions of fibre types vary between nerves – studies of the sural nerve show a ratio of myelinated to unmyelinated nerve fibres of the order of 1:4 (Ochoa and Mair 1969).

The small diameter fibres convey sensations of pain and temperature (*Table 4.1*). Thomas (1993) felt that the lack of an easy technique for the detection and classification of disturbances of these fibres was one of the important reasons why there has been less research into small diameter fibre neuropathies than large diameter fibre neuropathies, which are more common. Many of the small fibre neuropathies present in childhood. This deficiency in the electromyographer's tests might even justify the use of near-nerve stimulation under anaesthetic. However, there are indirect means of measuring the psychophysical thresholds of those sensations conveyed by small fibres (e.g. temperature change or vibration), but these can only be used with any confidence in older and mentally aware children. A disadvantage of such quantification of sensation is that it has no localising value in terms of the level of neurological deficit and, even if the pathology is assumed to be in the peripheral

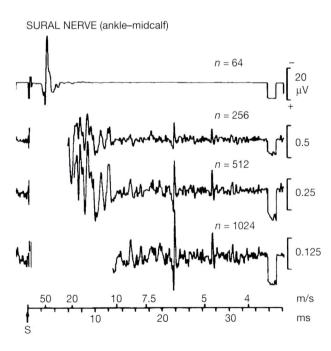

Fig. 4.1 Sural nerve action potential recorded at midcalf with near-nerve needle electrodes, from a patient with Fabry's disease. At low amplification the evoked action potential was normal. At high amplification and electronic averaging of 1,000 and 2,000 responses, several late components of 0.05–0.1 µV had conduction velocities of 2–4 m/s. (From Krarup and Buchthal (1985), by permission.)

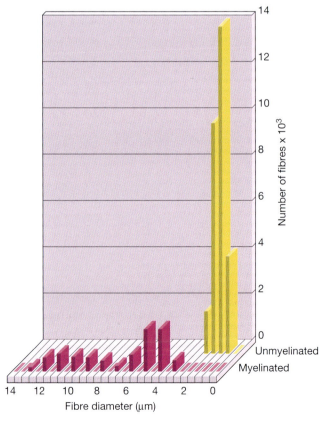

Fig. 4.2 Histogram showing the number of myelinated and unmyelinated fibres in an adult sural nerve. (Data from Jacobs and Love (1985), by permission of Oxford University Press.)

Table 4.1 Sensory modalities conveyed by fibres in primate, distal sensory cutaneous nerves*

	Fibre size		
	Aα/β	Aδ	C
Diameter (μm)	7–15	2–6	0.2–2.6
Conduction velocity (m/s)	35–70	5–30	0.4–2.0
Sensory modality (%):[†]			
mechanoreceptors	100%	30%	–
cold receptors	–	15%	–
warmth receptors	–	–	10%
nociceptors	–	55%	90%

*Courtesy of Dr B Lynn.

[†]The percentage figures indicate approximately the proportion of fibres in each class subserving each sensory modality and were derived from reported studies of human and experimental animal physiology.

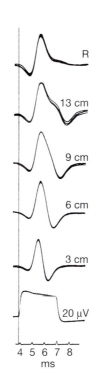

Fig. 4.3 Action potentials from the median nerve (subject RG Willison) to show the effect of electrode separation on the action potential waveform. The stimulus to the nerve at the wrist was unchanged throughout. The lower recording electrode was at the level of the elbow in all cases; the upper electrode was placed above this over the course of the nerve at the distances shown, except in R for which a remote electrode was used. The numerals under the time scale indicate the time after the stimulus. (From Gilliatt et al (1965), with permission of the BMJ Publishing Group.)

nervous system, the methods are necessarily subjective and give only an indirect measure of nerve function. Laser evoked potentials have been used, removing the psychological elements of the test. These primarily test the A delta and C fibres (Purves and Boyd 1993), but the tests are painful and poorly tolerated. Currently, a surface electrode is being developed that can be made to change temperature rapidly and painlessly, and is fast enough to allow averaging of an evoked cortical response. This electrode, by avoiding the need for patient cooperation and by being well tolerated by children, offers great potential in the field of small fibre neuropathy testing.

Recording electrodes

The choice of recording electrodes is very important to a paediatric electromyographer. There are theoretical arguments that encourage the use of electrodes with a fixed inter-electrode separation, as this distance has an important effect on the amplitude and waveform of the response (*Fig. 4.3*) (Gilliatt et al 1965, Andersen 1985). A standard inter-electrode distance of 3–4 cm has been adopted by most adult laboratories, and thus control data can be used from others' experience. In paediatrics, single use surface electrodes are preferred, with separate active and indifferent self-adhesive electrodes. The single use is important to prevent spread of infection, and these electrodes are particularly useful in premature babies as they can be cut down to size. Being self-adhesive is also very useful when attaching the electrodes to small limbs. The relationship between amplitude and inter-electrode distance is still important (3 cm separation is still used whenever possible), but it is further complicated by the nerve conduction velocity changing rapidly in the first 5 years of life. As a result, standardised normal values of the amplitudes of both sensory and motor action

potentials are a rarity, and most practising paediatric electromyographers rely on velocity measurements, for which there is a good deal of reference information, and use their experience to judge whether the amplitudes are significantly reduced (Pitt 2005). Measures on the contralateral limb are useful when unilateral conditions are being investigated, but this is much more uncommon in paediatrics than in adult practice.

Stimulating electrodes

Inter-electrode separation of surface electrode pairs used for stimulation is less critical than for recording. Depolarisation of the nerve takes place under the cathode, not at some point between the cathode and the anode (Gilliatt et al 1965). Studies by Krarup et al (1992) have shown that a delay of about 0.15 ms can reasonably be attributed to 'utilisation' of the depolarising current at the stimulation site. Some manufacturers produce special neonatal stimulating electrodes, which have very small stimulating areas and inter-electrode separation. They are invaluable for smaller patients. Ring electrodes are uncommonly used, as children generally do not like them, but for the youngest children presenting with one of the mucopolysaccharidoses and in whom a carpal tunnel syndrome is being sought, the palm to wrist distance may be too small, in which case the digits must be used.

SENSORY ACTION POTENTIALS

Sensory action potentials (SAPs) will be discussed first. They are always measured first in paediatric practice as they are the least painful to perform and need the greatest cooperation as their amplitude is low. Unlike motor nerves it is possible to stimulate and record directly from over the sensory nerves. If the stimulation and recording match the physiological direction (stimulation distal and recording proximal) the recording is termed orthodromic; in the reverse case it is called antidromic. The potentials obtained with the latter method tend to be bigger. This might be for a variety of reasons. For example, when recording SAPs from the hand it is because the digital nerves are closer to the recording electrodes than are the nerves at the wrist (Buchthal and Rosenfalck 1966). The smaller internal shunting effect of externally recorded depolarising currents in the smaller fascicles of distal nerves may also contribute to this difference. Personal choice influences which is used (*Fig. 4.4*).

The amplitude of SAPs is dependent on the conduction distance over which the recording is made. This phenomenon is much more obvious for sensory than for motor studies, and has been attributed to the efforts of temporal dispersion causing phase cancellation and increased asynchrony (Kimura et al 1986, Horowitz and Krarup 1992). The amplitude of the SAPs is correlated with the number of large diameter sensory nerve fibres. Phase cancellation of the potentials from the smaller diameter fibres prevents any potential being picked up on the surface, as discussed above. Fortunately, most neuropathies affect the largest diameter fibres first.

SAPs that can be recorded in children using surface electrodes are listed in *Box 4.1*. There are important differences when compared to adult practice. In particular, orthodromic recording of the medial plantar nerve can be invaluable in the smallest of subjects because of the distances involved when using other more commonly used nerves such as the sural.

COMPOUND MUSCLE ACTION POTENTIALS

Compound muscle action potentials (CMAPs) are potentials recorded with surface electrodes over muscle. The shape of the CMAP depends on the position of the recording electrodes relative to the motor end-plate region and on the anatomical configuration of the muscle. The most compact and regularly shaped potentials are recorded from small muscles.

The active recording electrode is placed so that it is over the motor entry zone of the nerve into the muscle. The remote electrode is placed distal to the active electrode over some electrically inactive point, such as the muscle tendon. The muscles suitable for such recordings are limited and are listed in *Table 4.2*. Comments made about stimulating electrodes for SAPs (see above) apply equally to motor conduction studies.

Following neuromuscular transmission, depolarisation is conducted from the motor end-plate region by muscle fibres. The shape of a CMAP depends on the placement of the recording electrodes relative to the end-plate, a phenomenon first studied by Henriksen (1956). If the active electrode is directly over the end-plate region, a rapidly rising negative deflection is recorded, followed by a simple biphasic potential. If the active electrode is placed some distance away from the end-plate (*Fig. 4.5*) the onset of the negative deflection is delayed by a preceding initial positive deflection. This can be eliminated by moving the active electrode. It is important to be meticulous in placing the active electrode over the end-point as this is where the amplitude of the CMAP is maximal.

Box 4.1   SAPs recorded using surface electrodes

Upper limb:
- median
- ulnar
- superficial radial

Lower limb:
- sural (proximal and distal)
- plantar (medial)
- superficial peroneal

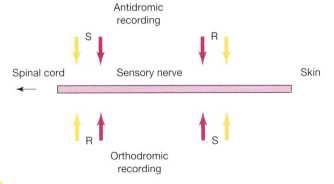

Fig. 4.4 The arrangement of stimulating and recording electrodes used in antidromic and orthodromic recordings.

Table 4.2   Small distal muscles suitable for recording CMAPs with surface electrodes

Muscle	Innervated by
Abductor digiti minimi	Ulnar nerve
First dorsal interosseous	Ulnar nerve
Abductor pollicis brevis	Median nerve
Adductor hallucis	Posterior tibial nerve
Extensor digitorum brevis	Deep peroneal nerve

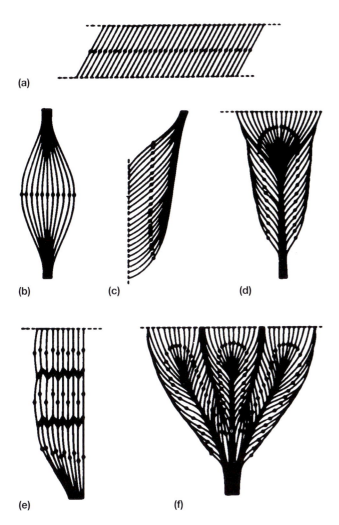

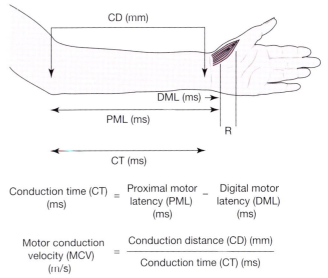

Conduction time (CT) $=$ Proximal motor latency (PML) $-$ Digital motor latency (DML)
(ms)   (ms)   (ms)

Motor conduction velocity (MCV) $= \dfrac{\text{Conduction distance (CD) (mm)}}{\text{Conduction time (CT) (ms)}}$
(m/s)

Fig. 4.6 Measurements and calculations to be made in calculating the motor conduction velocity.

Fig. 4.5 The muscle motor end-plates are indicated by the lines of dots, each situated at the midpoint of a muscle fibre. (a) A muscle (e.g. intercostal) that consists of parallel short fibres. (b) Fusiform muscle (e.g. a head of biceps). (c) Unipennate muscle (e.g. flexor pollicis longus). (d) Arrangement of fibres in a bipennate muscle (e.g. rectus femoris). (e) Rectus abdominis, which is divided by tendinous intersections into several strap-like segments. (f) Multipennate muscle (e.g. deltoid). (From Duchen and Gale (1985), by permission.)

MOTOR CONDUCTION VELOCITY

Stimulation of a peripheral nerve at two points separated by a known distance allows the motor conduction velocity of that nerve to be calculated. The time taken for an impulse to travel from the site of distal nerve stimulation is called the 'distal motor latency' (*Fig. 4.6*). Several different events summate to give the distal motor latency. Some can be estimated but, for safety of the calculation of the motor nerve conduction velocity (MCV), subtraction of the distal motor latency from the proximal latency and then measurement of the separation between the two points of stimulation can be made to allow the MCV to be calculated in the nerve between. In this respect there is little that distinguishes paediatric

from adult practice but errors in the measurement of the distances in children and neonates in particular have a more profound effect on the variability of the MCV than in adults.

F WAVES

The physiology of the F wave

Supramaximal stimulation of a motor nerve produces an impulse that travels both orthodromically and antidromically. The orthodromic impulse stimulates the terminal branches of the motor nerve and activates the motor units, the subsequent depolarisation of which summate to produce the recorded M wave. The antidromic impulse, on reaching the axon hillock of the motor neuron, causes re-excitation in a small proportion of the neurons. This results in an intermittent discharge of a few motor units (*Fig. 4.7*). These late responses were named F waves by Magladery and McDougal (1950), possibly because they were first recorded from the foot. The question was whether F waves were reflex responses or the result of recurrent discharges of antidromically activated motor neurons (Mayer and Feldman 1967). The finding of preserved F waves in baboons after limb deafferentation (McLeod and Wray 1966) and the single-fibre EMG studies of Trontelj (1973) indicated that the latter view was correct.

The use of the F wave in paediatrics

F wave analysis is one of those nerve conduction tests in children, like blink reflexes it might be submitted, where the potential as a diagnostic method is far greater in theory than in practice. This is a great disappointment. This tension between theory and practice begins with the process of recording F waves. Minimum latency is the main measure in children, but multiple stimuli can only be achieved in the most tolerant of subjects. Nobrega et al (1999) suggested that ten stimuli are the minimum needed. Even this reduced number may not be obtainable. Persistence of the F wave may mean that to realise ten stimuli followed by an F response many more stimuli might be needed. It is difficult for these reasons to obtain normative data, and we are fortunate that normative data are available (Pitt 2005). Unfortunately, these data are usually given in relationship to age, but variations in length and height at different ages make these data an estimate rather than definitive, as is wanted. Paradoxically, where large numbers of controls are needed this is the very age group (the neonate, infant and toddler) where ethical permission is most likely to be declined.

Fortunately, this inability to record sufficient F waves for accuracy is not usually a problem for the diagnosis of peripheral neuropathy. The axonal neuropathies, if first suspected by abnormality of sensory conduction, will commonly show EMG changes of denervation. EMG, rather surprisingly it might be thought, is a more effective means of demonstrating motor axonopathy than F waves. There are few cases where it is justifiable not to

perform EMG, even if briefly. In the acquired demyelinating neuropathies, abnormalities of the main motor nerve conduction studies are almost universally present. In these circumstances minor abnormalities of the F waves are supportive rather than diagnostic. However, there are variants of Guillain–Barré syndrome in which the pathological location is very proximal. Prolongation or, more commonly, absence of F waves may be the first and only abnormality. These are often associated with immunoglobulin B antibodies against gangliosides (Kuwabara et al 2000). Absence of a response is one of those situations where it can be most difficult to be certain that this reflects true pathology and not technical difficulty. It is of such importance to the diagnosis that a certain insensitivity on behalf of the examiner can be justified, and repeated stimulation is excusable. It is in this same patient group that very high levels of stimulation must be used. Speed of examination becomes a crucial factor.

H REFLEX

The H reflex is the electrophysiological counterpart of the ankle tendon jerk and results from a monosynaptic reflex. In adults an H reflex can reliably be obtained only by stimulating the tibial nerve and recording from the soleus (*Fig. 4.8*), and explaining its use in the examination of S1 radiculopathy. In contrast, several studies have shown that in paediatrics the presence of the H reflex is more widespread. Thus, Kaeser (1970) recorded an H reflex from almost every muscle in infants. A systematic study in infants and children found H responses in the hypothenar muscles (Thomas and Lambert 1960). One noteworthy use of the H reflex is to calculate the sensory nerve conduction velocity in type 1a fibres (Raimbault 1988). This is calculated by obtaining the H reflex latencies from proximal and distal stimulation, and then using the difference as the divisor of the distance. In his monograph of normative data in children Raimbault (1988) obtained H reflexes for the most common nerves (ulnar, tibial and peroneal) up to the age of 15 years.

REPETITIVE NERVE STIMULATION

Physiology

Repetitive nerve stimulation is the other test that is used in the diagnosis of disorders of the neuromuscular junction. In many centres it is preferred over single-fibre EMG, but it is less sensitive. As the nerve potential propagated down the motor nerve reaches the presynaptic area, it causes vacuoles containing acetylcholine (ACh) to fuse with the presynaptic membrane and to release ACh into the synaptic cleft. These are released in packages called 'quanta'. The end-plate potentials generated are directly correlated with the number of quanta. If the end-

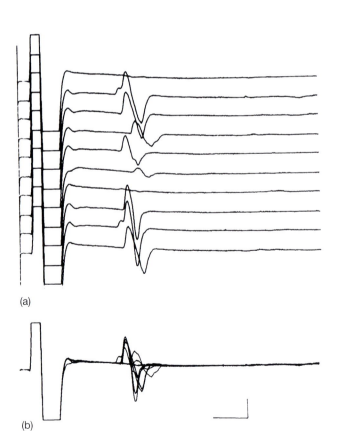

(a)

(b)

Fig. 4.7   Recordings of F waves from abductor pollicis brevis. (The early square complex is the action potential exceeding the limits of the display.) (a) Falling leaf display; (b) superimposed. Vertical bar = 100 μV; horizontal bar = 10 ms. (From Binnie et al (2004), by permission.)

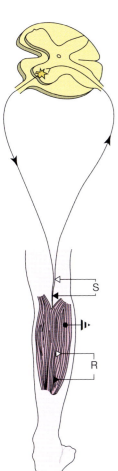

Fig. 4.8 Pathways of the H reflex. This is the electrophysiological counterpart of the ankle reflex and the figure shows the afferent volley entering the dorsal root to excite the anterior horn cells in a monosynaptic pathway.

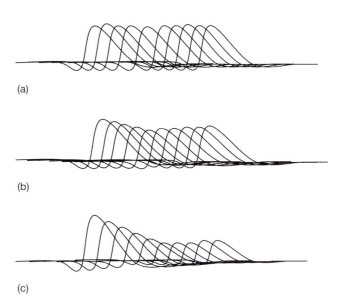

(a)

(b)

(c)

Fig. 4.9 Findings in repetitive stimulation. (a) Normal response. (b) Mild abnormality with a decrement of 20%. (c) Severe abnormality with a decrement of > 50%.

plate potentials are above the threshold for the ACh receptor to depolarise the muscle membrane, an action potential is propagated down the muscle fibre. In a normal subject if the nerve is stimulated at 3 Hz the end-plate potentials are largest at the first stimulation, falling to their nadir around the 4th and 5th stimuli before increasing again to a plateau. The end-plate potentials are normally very much higher than the threshold. This is called having a high safety factor for neuromuscular transmission.

Repetitive nerve stimulation exploits this variation in the end-plate potentials with repeated stimulation. At 3 Hz the maximum depletion of the ACh stores occurs, without the enhancing effects of increased calcium release that will occur if repetition rates are higher. In a normal subject the CMAPs will demonstrate no decrement (*Fig. 4.9(a)*). In an abnormal neuromuscular junction often the transmission will fail around the 4th or 5th stimulus, either because of poor ACh stores in the case of the presynaptic defect, or because of an increased threshold in the postsynaptic abnormality (*Fig. 4.9(b, c)*). Tetanic stimulation is then used to increase the release of quanta by the action of calcium. This will reverse both types of defect, but in the postsynaptic case the increase in the compound motor action potential from the baseline is dramatic, sometimes exceeding 300%, as in

some cases of botulism. In the immediate period afterwards there is a dramatic attenuation or exhaustion of ACh release, which in an abnormal junction will bring out the abnormality that might have been present with the first run at 3 Hz.

Practical aspects

Repetitive stimulation is most likely to be positive in any of the myasthenic syndromes if it is done on a proximal muscle. Of the different choices available the trapezius muscle is one that has been tried clinically, but in children it is difficult to do because of movement. The anconeus muscle is felt to be a good compromise, as it is an extension of the triceps and therefore representative of the proximal muscles. It is difficult in small infants to be sure that one is over this muscle. As a compromise it was decided to use the extensor digitorum communis (EDC) or, perhaps more honestly in the youngest of children, the extensor compartment of the forearm.

The active electrode is placed over the belly of the muscle, with the indifferent electrode placed over the bony part of the wrist. The earth is placed over the medial aspect of the arm. Stimulation of the radial nerve is made above the elbow. In the very smallest babies it is possible to achieve good stability of the recording conditions with only minimal restraint. However, in more agitated babies and older children it is often necessary to hold the arm and the electrode tightly to prevent movement of the arm during stimulation, which is the major cause of a false-positive response. If there is a smooth decrement of the response this is likely to represent a true positive result. When the decrement is less smooth it is much better to disregard that response. As a rule, if you can never obtain a normal response then be highly suspicious that, even allowing for technical difficulties, there is probably a true decrement.

As a screen the protocol aimed for is to do one run of 3/s stimulation, rest, then give approximately 2 s of tetanic stimulation (75 pulses at 30 Hz), immediately do a further train of 3 Hz, and then wait 1 min and repeat the train. This full protocol will be followed only if the suspicion is high. The maturity of the neuromuscular junction is such that even in the neonate and anecdotally in the moderately premature baby there should be no decrement whatsoever using this protocol. In contrast prolonged stimulation at over 10 Hz in normal subjects of this age may show a decrement.

ELECTROMYOGRAPHY

Introduction

EMG is the recording of action potentials generated by contracting muscle fibres. At a molecular level, muscle force is developed by cyclical linkages between the actin and myosin filaments of the sarcomere, which cause the two types of filament to move over one another. This biochemical reaction requires adenosine triphosphate (ATP) and calcium ions, and is brought about by a rapid depolarisation of the muscle membrane. The depolarising potentials, recorded in the extracellular space, form the EMG. The potentials are generated by depolarisation of single muscle fibres, but the innervation of vertebrate muscle is such that under normal physiological conditions a single muscle fibre does not contract on its own, but in concert with other muscle fibres innervated by the same motor neuron. This arrangement forms the basis of the 'motor unit', and a clear understanding of its structure and function is fundamental to EMG.

Characteristics of motor units recorded with a concentric needle electrode

The electrode most commonly used in EMG is the concentric needle electrode (CNE). This electrode was described in a paper by Adrian and Bronk (1929) as being made from 'a central enamel covered wire passed through a hypodermic needle and connected to the amplifier input'. This essential design has endured well and the modern CNE is constructed on very similar lines (*Fig 4.10*).

The recording surface is made by grinding the tip at 15°. All the experimental work has been focused on the standard concentric needle with an external diameter of 0.46 mm (resulting elliptical area 580 × 150 μm) and its use in adults. This picks up activity from fibres that lie within a hemisphere of about 0.5 mm radius. As adult muscle fibres have diameters of between 25 and 100 μm, the needle records activity from about 20 fibres; fibres further away make a minimal contribution to the recorded potentials. The number of motor units recorded depends both on the local arrangement of muscle fibres within the motor unit and on the level of contraction of the muscle. With the needle in a weakly contracted muscle a few motor units are recorded, firing at a rate of around 6–7 Hz.

Most paediatric electromyographers use the smallest concentric needle available, which has an external diameter of 0.3 mm giving a recording area of 0.03 mm².

Needle recording electrodes	Needle tip and recording surface	Pick-up	Needle diameter	Filter settings	Activity recorded
Concentric needle electrode Central insulated platinum wire inside a steel cannula	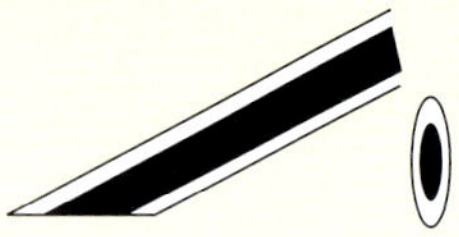	Hemisphere radius 0.5 mm	0.3–0.65 mm	10 Hz to 10 kHZ	Motor units
Single fibre needle electrode Fine platinum wire (25 μm diameter) inside a steel cannula, which records from a steel aperture	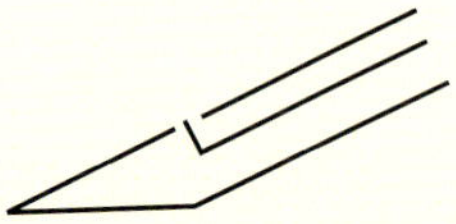	Hemisphere radius 250–300 μm	0.5–0.6 mm	500 Hz to 10 kHZ	Individual muscle fibres of motor units. In health the potentials are either single or pairs; after reinnervation the potentials have multiple components
Monopolar needle electrode Sharpened stainless steel needle insulated down to 25–50 μm from tip Subcutaneous or surface reference	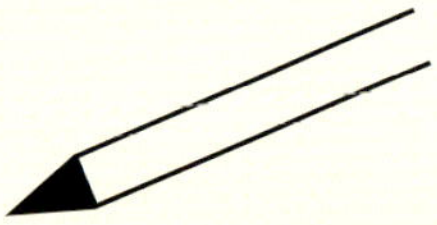	Sphere radius 200 μm	0.3–0.5 mm	2 Hz to 10 kHZ	Motor units of higher amplitude and more complex waveforms than those recorded with a concentric needle, but similar duration

Fig. 4.10   Three different types of recording electrode, their physical characteristics, the filter settings required for use and the nature of the activity that each records. (From Binnie et al (2004), by permission.)

Unfortunately, these needles have not been the subject of similar analyses as the larger ones, and there are bound to be differences.

When the results obtained with CNEs are compared with those obtained using monopolar electrodes, it is found that the motor unit action potential (MUAP) duration and firing rates are similar. However, there are considerable differences in amplitude, rise time and the number of turns (Howard et al 1988). Little can be inferred about how these parameters would differ with two similar needles but with different areas of recording surface. A further complication is that the diameter of muscle fibres increases from the neonate to the adult (Oertel 1988).

The features of the motor unit potential recorded with EMG

Unlike the work discussed in the previous paragraph, much of the experimental and theoretical work on how the motor unit influences the motor unit potential is transferable from adult to paediatric practice. A motor unit potential can be divided broadly into three compo-

nent phases: an initial part, the spike and a late phase (*Fig. 4.11*). The duration of each of these phases is determined by the spatial arrangement of muscle fibres within the unit in relation to the recording electrode, as well as by the conduction velocity in the terminal branches of the motor nerve and the muscle fibres. The initial part is generated by propagation of the muscle action potentials that travel from the motor end-plate region towards the electrode. This part of the potential is therefore not present if the electrode is near the motor end-plate. The spike of the motor unit potential is produced by the sum of the activities of the muscle fibres closest to the electrode, whereas the late phases of the potential result from activity in more spatially distant fibres. The features of a motor unit that are commonly measured are amplitude, duration, number of phases and number of turns (*Fig. 4.12*).

Amplitude

The amplitude of a motor unit potential is largely determined by the activity of those muscle fibres closest to the core of the recording electrode (Thiele and Bohle 1978, Nandedkar et al 1988). The precise relationship between the recording surface and the active fibres is critical in determining amplitude – an adjustment of the needle position from within 0.5 mm of an active fibre to a distance of 1 mm reduces peak-to-peak amplitude by a factor of 10–100 (Ekstedt and Stålberg 1973). Thus, very minor adjustments of the electrode, shifting the recording tip by only a millimetre or so, or rotating it, result in major changes in the detected amplitude of a motor unit. Measurement of the duration of a motor unit potential is much less sensitive to the exact placement of the recording electrode.

Duration

The 'duration' of a motor unit potential is defined as the time interval between the first negative (vertical/upward) deflection and the point at which the waveform finally returns to the baseline (see *Fig. 4.12*). Computer simulation shows that the duration of normal, recorded motor

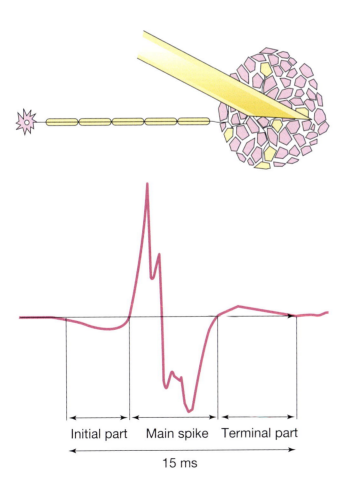

| Initial part | Main spike | Terminal part |

15 ms

Fig. 4.11 Component phases of motor unit potential (lower) recorded with a concentric needle electrode (upper). The spike of the motor unit is produced by the muscle fibres closest to the electrode, whereas the late phases result from activity in more distant fibres.

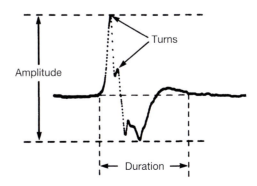

Fig. 4.12 Normal unit potential measurements. (From Binnie et al (2004), by permission.)

unit potentials depends on the number of muscle fibres in the motor unit and is little affected by the proximity of the recording electrode to the nearest fibre (Nandedkar et al 1988). The difficulty with this measurement is in defining the exact point of return to the baseline, as the gradient of potential change after the main spike can be very gradual. Buchthal and his collaborators in the 1950s invested much effort in establishing the control ranges of motor unit parameters for many muscles in subjects of different ages (Buchthal et al 1954, Buchthal and Rosenfalck 1955). In their measurements of the duration of these potentials they paid scrupulous attention to recognising the point at which the late parts of the potential intersected the baseline.

The problem is that this return point may be so ill-defined that the choice of what to measure may sometimes seem somewhat arbitrary. Manual measurement of motor unit potential durations is extremely laborious, which prolongs the discomfort of the examination for the patient and requires considerable additional analysis time at the end of the test. Many modern EMG machines have the facility for automated individual motor unit analysis, which makes formal analysis more reliable. If you have the facility to record the EMG signals onto a playback system, the use of the modern computer analysis programmes to analyse motor units is perfectly possible in children, with the analysis being performed after the event.

Polyphasia

The phases of a motor unit potential are defined by the number of times the potential crosses the baseline, as shown in *Fig. 4.12*. If using an automated method of analysis, the number of phases is derived by counting the number of baseline crossings in both directions and adding one. A unit that has more than four phases is said to be polyphasic. Buchthal et al (1954) found up to 10% of the units from control muscle to be polyphasic.

The interference pattern

This is defined as the electrical activity recorded during maximum voluntary effort (Buchthal and Clemmesen 1941). If healthy muscle is made to exert its full force, the interference pattern should be 'full' and no single unit is discernible. This means that at a sensitivity of 1 mV/division and a time base of 20 ms/division the baseline is completely obscured by motor unit activity. A 'reduced interference pattern' is one in which portions of the baseline remain visible and some individual motor units can still be identified when the full power of the muscle is exerted. With severe motor unit loss, the interference pattern may only contain 2–4 individually recognisable motor units, usually firing at frequencies above 10 Hz, when the pattern may be referred to as 'discrete'. The interpretation of whether the interference pattern is in fact reduced or the force of contraction is not maximal is a problem in adults, but can be overcome with encouragement of the subject. Such methods are less likely to

work in children, if at all, and this is yet another reason for having the whole EMG signal recorded so that it can be played back after the patient has left.

Other normal EMG phenomena
Insertion activity

At complete rest, healthy muscle should be electrically silent or almost so. Any activity that can be recorded is of such low amplitude that a high recording sensitivity of 100 µV/division should be used on needle insertion. On first piercing the muscle with a recording needle (and at subsequent adjustments of the recording tip) there may be transient activity, so-called 'insertion activity'. This consists of bursts of activity, usually less than 500 µV in amplitude, which cease within less than a second of the movement (*Fig. 4.13*). It is caused by mechanical stimulation or injury of muscle fibres.

End-plate noise

Other activity that can be recorded from healthy muscle is 'end-plate' activity. As it is generated at the motor end-plate, this activity is highly localised and disappears with very small adjustments of the recording tip (Wiederholt 1970). Recording it is therefore a chance occurrence, although Buchthal and Rosenfalck (1966) reported this finding in 20% of 200 muscles without further search. The end-plate activity consists of biphasic potentials of moderate amplitude (100–200 µV) and short duration (2–4 ms) that fire in irregular short bursts (*Fig. 4.14*) (Buchthal and Rosenfalck 1966) and can be confused with fibrillation potentials, but this activity is eliminated by minimal repositioning of the needle tip.

Fibrillations

Fibrillations (low amplitude, brief biphasic or triphasic potentials) result from the spontaneous activity of single muscle fibres that have lost their innervation. However, such potentials, often regarded as the hallmark of denervation, may be recorded from healthy muscle. Buchthal carried out the definitive study of this phenomenon. Many hundreds of normal muscles were investigated and single sites found at which biphasic potentials could be recorded. These may have been end-plate spikes, but they

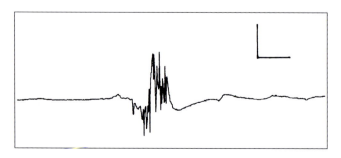

Fig. 4.13   Insertion activity recorded as the needle enters healthy muscle. Vertical scale = 200 µV; horizontal scale = 10 ms. (From Binnie et al (2004), by permission.)

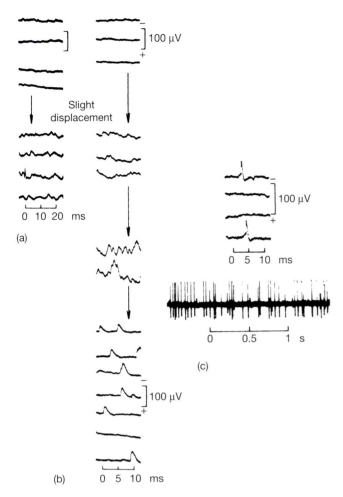

Slight
displacement

0 10 20 ms

(a)

100 μV

100 μV

(c)

0   0.5   1 s

100 μV

(b)   0 5 10 ms

Fig. 4.14 End-plate activity. A concentric needle electrode was inserted in the end-plate zone of a normal biceps and advanced in small steps. (a, b) Recordings of end-plate 'noise'; in the lower part of (b) the discrete negative spikes indicate that the electrode was close to discharging end-plates. It has been suggested that these may be miniature end-plate potentials. (c) The other type of end-plate activity – diphasic potentials. (From Buchthal and Rosenfalck (1966), by permission.)

were indistinguishable from the fibrillations recorded in denervated muscle (Buchthal 1982). The importance of this observation is that before fibrillations can be reported as being a significant abnormality, they must be demonstrated at more than one sampling site in the muscle under study.

Abnormal EMG pattern
Neurogenic changes
To understand the changes that are seen on EMG in neurogenic change it is important to understand the organisation of the motor units in the muscle. Muscle fibres from individual motor units are not discrete, but rather closely intermingled, with motor unit territories overlapping. When one of the anterior horn cells degenerates, the muscle fibres it supplies become denervated. Normally there is only one neuromuscular junction per

muscle fibre. When fibres lose their nerve supply the muscle fibre will develop neuromuscular junctions all over its surface. These are receptive to the ACh which is free in the muscles, causing spontaneous firing. A unit will only be able to innervate other fibres if they overlap. Those fibres that are not in the overlap area will either receive innervation from other motor units or degenerate and atrophy.

The hallmark of acute denervation is recording fibrillation potentials, but this may take some days or weeks to appear. Reinnervation can be (1) by regrowth of sprouts from the proximal stump after complete nerve lesion or (2) by collateral sprouting with the denervated muscle fibres seeking new nerve sprouts from adjacent axons (as seen in partial nerve injury or in anterior horn cell disease). The first is characterised by 'nascent' units (low amplitude, short-duration potentials), which will gradually become increased in complexity and size as reinnervation proceeds. The latter produces alterations in the configuration of the motor unit potentials, with an increase of the duration and number of its phases. The duration is prolonged because there are more fibres to depolarise, and the increase in the number of phases is due to the lack of synchronisation between the original fibres and the newly acquired fibres. The firing rate will increase because there is a reduced number of motor units that can increase the force only by firing at a higher rate.

Myopathic changes
The pathological changes in myopathy produce wide variation in the diameters of the muscle fibres. The small fibres are those affected and some will eventually atrophy completely. Larger than normal fibres may be affected in a different way, but they may also represent normal fibres that have compensated for the other fibres' loss. The motor units are, for these reasons, likely to occupy a smaller volume, and therefore shorter, lower amplitude potentials will be seen. The amplitude of the units is dependent on the fibres nearest to the needle. The chances of a fibre being right next to the needle is higher if the unit is dense, as is the case in reinnervation. The presence of larger fibres raises the odds that a needle will find itself in this situation, and this explains why larger units may be seen in myopathy, particularly if it is chronic. The fullness of the interference pattern to low levels of force is simply explained by the need for motor units to fire faster to sustain the usual levels of force. Interference will therefore be increased.

While the changes in myopathy are well recognised, the classic combination is seen less frequently than might be expected. When any analysis of the correlation between muscle biopsy and EMG is made there is usually a correlation of less than 50%. Higher levels may be found, but this is almost invariably in the studies that have included neurogenic change as well. The mismatch is in both directions, i.e. EMG demonstrating myopathic change when the biopsy does not, and vice versa.

165

Single fibre EMG

An electromyographer competent at concentric or monopolar needle EMG will not find single fibre EMG (SFEMG) difficult, but the practicalities of conducting this examination in children should not be underestimated. The main clinical indication for SFEMG in paediatrics is to diagnose myasthenic syndromes. These have protean clinical presentations, which include ptosis, but also arthrogryposis, apnoeic attacks as well as entering into the differential diagnosis of the 'floppy infant'.

Single fibre needle

The single fibre needle was developed in Uppsala, Sweden, by Ekstedt and Stålberg in the 1960s. It has similar external proportions to a CNE, being made of a steel cannula 0.5–0.6 mm in diameter with a bevelled tip. However, instead of there being a recording surface at the tip, a fine, insulated platinum or silver wire embedded in epoxy resin is exposed through an aperture on the side of the needle, 1–5 mm behind the tip (*Fig. 4.15*). This recording surface is on the side opposite to the bevel to avoid recording from fibres that could have been damaged by insertion. The steel cannula acts as the reference electrode. The platinum wire that forms the recording surface has a diameter of 25 μm and picks up activity from within a hemispherical volume, 300 μm in diameter (Stålberg and Trontelj 1994).

In children, and in particular in the neonate and infant, many electromyographers use a facial concentric needle with a low-frequency filter of 2–3 kHz in preference to a standard SFEMG needle. This is for many reasons. Most SFEMG needles are reusable, but there have been concerns that standard methods of sterilisation may not eliminate the risks of prion diseases. There is a policy of using single use needles in most hospitals, and such an approach should be considered in the light (in Britain) of the Consumer Protection Act 1987, which deals with product liability (Al-Seffar 1990). However, until a disposable SFEMG needle becomes available most paediatric electromyographers will be obliged to use facial needles. Another factor is that the recording area is at the end of the facial needle rather than at a side port a distance from the needle tip. The thinness of an infant's facial muscles occasionally means that the recording area of a conventional SFEMG needle is not embedded in the muscle, which is not a problem with a facial needle. It has been shown that with a high bandpass of 3 kHz facial needles have similar specificity and sensitivity to SFEMG needles (Sarrigiannis et al 2006). Once again in paediatrics, but more so than in any part of EMG, some form of continuous recording of the SFEMG data is essential, whether this be external (e.g. a digital audio tape (DAT) recorder) or internal to the processor. If recorded, the signal can be played back through the SFEMG programme for the calculation of the mean consecutive difference (MCD).

Practical procedure for single fibre EMG

SFEMG using voluntary activation is simply not worth attempting in children of any age. There are very few occasions when it will be successful, and if it is not it will usually not be possible to retrieve the situation and proceed to stimulated SFEMG, which has become the examination of choice. There is no need to do these studies under general anaesthesia. Often it is difficult to obtain time with the anaesthetists and theatres that allow it to be done without undue delay. Anyway, it is best to avoid unnecessary general anaesthesia in children as a matter of principle. If you are unsuccessful you can always fall back on general anaesthesia at a later date. Our experience is that this will be needed in only less than 30% of cases, irrespective of age.

There is a choice of three muscles: the EDC, the frontalis and the orbicularis oculi – presented here in order of increasing sensitivity for the detection of abnormalities (Sanders 1999). It seems most sensible to choose the most sensitive first, and therefore the orbicularis oculi is selected. As with all EMG procedures in children, explanation at the time of the test is to be preferred, but as a small amount of local anaesthetic cream is applied prior to the examination and needs time to be effective, explanation in the broadest of terms, mixed with heavy reassurance, is all that is needed at this stage. After 30 min the child and any accompanying persons are brought in. The test can be done either with the subject on a parent's lap or resting on a couch. The electromyographer approaches the subject from the top of the head, allowing the subject to watch the EMG screen if old enough to appreciate it. This approach means that the needles will not be seen. A surface electrode is placed on the forehead as an earth. The indifferent electrode, also a surface electrode, is positioned halfway on a line joining the tragus and the outer canthus. The stimulating electrode, a monopolar needle, is placed two-thirds of the way along this line, and the final needle, a concentric facial EMG needle, is placed just distal to this. Some operators like to stimulate the orbicularis oculi before placing the final needle, selecting the part of the muscle that is most active. The stimulating current is raised to around 2 mA, and with positioning of the two needles several muscle fibre potentials will appear. If they do not it is important to reposition the

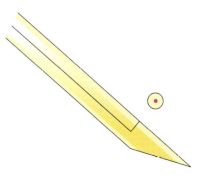

Fig. 4.15 Single fibre recording electrode. The small circle with the central 'dot' shows the recording surface of this type of needle electrode.

stimulating electrode rather than increase the stimulus. In this way it is possible to be confident that there is accurate approximation of the stimulating needle to the nerve fibre and supramaximal stimulation.

Jitter

Jitter, when applied to spontaneous recording of single fibre potentials, is the variation in time between a triggering potential and the firing of a second muscle fibre, when sequential activity of the pair is examined. The time interval between a pair of potentials is called the interpotential interval, and jitter is measured by calculating the mean differences in interpotential interval of consecutive pairs (*Fig. 4.16*), and is alternatively called the mean consecutive difference (MCD). The figure obtained reflects the variation of time in which the second potential falls in relation to the first.

Normative data for the mean MCD or individual MCDs obtained with voluntary contraction exist for children over 10 years old (Stålberg and Thiele 1975). There are no paediatric data for the MCD obtained using axonal stimulation. Theoretically, this MCD value should be lower than the MCD obtained between volun-

tarily activated pairs. The calculated conversion factor is $1/\sqrt{2}$ (0.71), as the jitter is measured between the stimulus and a single muscle fibre action potential (Trontelj et al 1986, 1990). It is recommended for safety of diagnosis that the conversion factor used is 0.8 (Stalberg and Thiele 1975). Used in conjunction with normative data for the orbicularis oculi this gives upper reference limits of 32 µs for the mean MCD and 44 µs for jitter in individual muscle fibre action potentials. A stimulated SFEMG study can be considered to be abnormal if either the MCD of the potentials recorded exceeds 32 µs, or 5% or more of individual units' MCD exceed 44 µs. Extrapolating from the knowledge in routine SFEMG, at least one out of 20 pairs in a normal healthy subject can show increased jitter and even partial blocking (Stålberg and Trontelj 1994).

REFERENCES

Adrian ED, Bronk DW 1929 The discharge of impulses in motor fibres. J Physiol 67: 119–151.

Al-Seffar J 1990 Never mind the quality, what's the cost? An evaluation of EMG needle maintenance. J Electrophysiol Tech 16: 179–191.

Andersen K 1985 Surface recording of orthodromic sensory nerve action potentials in median and ulnar nerves in normal subjects. Muscle Nerve 8: 402–408.

Binnie CD, Cooper R, Mauguière F, et al 2004 *Clinical Neurophysiology*, 2nd edn, Vol 1. Elsevier, Amsterdam.

Buchthal F 1982 Fibrillations: clinical electrophysiology. In: *Abnormal Nerves and Muscles as Impulse Generators* (eds WJ Culp, J Ochoa). Oxford University Press, New York, pp 632–662.

Buchthal F, Clemmesen SV 1941 On the differentiation of muscle atrophy by electromyography. Acta Psychiatr Neurol Scand 16: 143–181.

Buchthal F, Rosenfalck P 1955 Action potential parameters in different human muscles. Acta Psychiatr Neurol Scand 30: 125–131.

Buchthal F, Rosenfalck A 1966 Evoked action potentials and conduction velocity in human sensory nerves. Brain Res 3: 1–122.

Buchthal F, Pinelli P, Rosenfalck P 1954 Action potential parameters in normal human muscle and their physiological determinants. Acta Physiol Scand 32: 219–229.

Buchthal F, Rosenfalck A, Behse F 1984 Sensory potentials of normal and diseased nerve. In: *Peripheral Neuropathy*, 2nd edn (eds PJ Dyck, PK Thomas, EH Lambert, et al). WB Saunders, Philadelphia, pp 981–1029.

Cooper R, Binnie CD, Billings R 2005 *Techniques in Clinical Neurophysiology: A Practical Manual*. Elsevier, Oxford.

Duchen LW, Gale AN 1985 The motor end-plate. In: *Scientific Basis of Clinical Neurology* (eds M Swash, C Kennard). Churchill Livingstone, Edinburgh, pp 400–407.

Ekstedt J, Stålberg E 1973 Single fibre electromyography for the study of the microphysiology of the human muscle. In: *New Developments in Electromyography and Clinical Neurophysiology* (ed JE Desmedt). Karger, Basel, pp 84–112.

Gilliatt RW, Sears TA 1958 Sensory nerve action potentials in patients with peripheral nerve lesions. J Neurol Neurosurg Psychiatry 21: 109–118.

Gilliatt RW, Melville ID, Velate AS, et al 1965 A study of normal nerve action potentials using an averaging technique (barrier grid storage tube). J Neurol Neurosurg Psychiatry 28: 191–200.

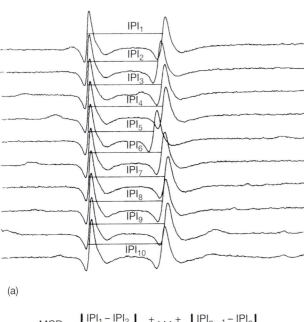

(a)

$$MCD = \frac{|IPI_1 - IPI_2| + \ldots + |IPI_{n-1} - IPI_n|}{n-1}$$

(b)

(c)

Fig. 4.16 Jitter measurements. (a) Interpotential interval (IPI); (b) mean consecutive difference (MCD); (c) superimposed (horizontal bar = 500 µs). (From Binnie et al (2004), by permission.)

Henriksen JD 1956 *Conduction Velocity of Motor Nerves in Normal Subjects and Patients with Neuromuscular Disorders*. MSc Thesis (Physical Medicine), University of Minnesota, MN.

Horowitz SH, Krarup C 1992 Conduction studies of the normal sural nerve. Muscle Nerve 15: 374–383.

Howard JE, McGill KC, Dorfman LJ 1988 Properties of motor unit action potentials recorded with concentric and monopolar needle electrodes: ADEMG analysis. Muscle Nerve 11: 1051–1055.

Jacobs JM, Love S 1985 Qualitative and quantitative morphology of human sural nerve at different ages. Brain 108: 897–924.

Kaeser HE 1970 Nerve conduction velocity measurements. In: *Diseases of Nerves, Part I. Handbook of Clinical Neurology* (eds PJ Vinken, GW Bruyn). North-Holland, Amsterdam, pp 116–196.

Kimura J, Machida M, Ishida T, et al 1986 Relation between size of compound sensory or muscle action potentials, and length of nerve segment. Neurology 36: 647–652.

Krarup C, Buchthal F 1985 Conduction studies in peripheral nerve. Neurobehav Toxicol Teratol 7: 319–323.

Krarup C, Horowitz SH, Dahl K 1992 The influence of the stimulus on normal sural nerve conduction velocity: a study of the latency of activation. Muscle Nerve 15: 813–821.

Kuwabara S, Ogawara K, Mizobuchi K, et al 2000 Isolated absence of F waves and proximal axonal dysfunction in Guillain–Barré syndrome with antiganglioside antibodies. Neurol Neurosurg Psychiatry 68: 191–195.

Magladery JW, McDougal DD 1950 Electrophysiological studies of nerve and reflex activity in normal man. Identification of certain reflexes in the electromyogram and the conduction velocity of peripheral nerve fibres. Bull Johns Hopkins Hosp 86: 265–290.

Mayer RF, Feldman RG 1967 Observations on the nature of the F wave in man. Neurology 17: 147–156.

McLeod JG, Wray SH 1966 An experimental study of the F-wave in the baboon. J Neurol Neurosurg Psychiatry 29: 196–200.

Nandedkar SD, Sanders DB, Stålberg EV, et al 1988 Simulation of concentric needle EMG motor unit action potentials. Muscle Nerve 11: 152–159.

Nobrega JA, Manzano GM, Novo NF, et al 1999 Sample size and the study of F waves. Muscle Nerve 22: 1275–1278.

Ochoa J, Mair WGP 1969 The normal sural nerve in man. Acta Neuropath (Berlin) 13: 197–216.

Oertel G 1988 Morphometric analysis of normal skeletal muscles in infancy, childhood and adolescence: an autopsy study. J Neurol Sci 88: 303–313.

Pitt MC 2005 Maturational changes vis-á-vis neurophysiology markers and the development of peripheral nerves. In: *Clinical Neurophysiology of Infancy, Childhood, and Adolescence* (eds GL Holmes, SL Moshe, HR Jones). Butterworth-Heinemann, Boston, pp 146–167.

Purves AM, Boyd SG 1993 Time-shifted averaging for laser evoked potentials. Electroencephalogr Clin Neurophysiol 88: 118–22.

Raimbault J 1988 *Les Conduction Nerveuses Chez l'Enfant Normal*. Expansion Scientifique Francaise, Paris.

Sanders DB 1999 Electrophysiologic study of disorders of neuromuscular transmission. In: *Electrodiagnosis in Clinical Neurology* (ed MJ Aminoff). Churchill Livingstone, New York, pp 303–321.

Sarrigiannis PG, Kennett RP, Read S, et al 2006 Single-fiber EMG with a concentric needle electrode: validation in myasthenia gravis. Muscle Nerve 33: 61–65.

Shefner JM, Buchthal F, Krarup C 1991 Slowly conducting myelinated fibres in peripheral neuropathy. Muscle Nerve 14: 534–542.

Stålberg E, Thiele B 1975 Motor unit fibre density in the extensor digitorum communis muscle. Single fibre electromyographic study in normal subjects at different ages. J Neurol Neurosurg Psychiatry 38: 874–880.

Stålberg E, Trontelj JV 1994 *Single Fiber EMG: Studies in Healthy and Diseased Muscles*. Raven Press, New York.

Thiele B, Bohle A 1978 [Number of spike-components contributing to the motor unit potential.] Elektroenzephalogr Elektromyogr Verwandte Geb 9: 125–130.

Thomas PK 1993 Hereditary sensory neuropathies. Brain Patholol 3: 157–163.

Thomas JE, Lambert EH 1960 Ulnar nerve conduction velocity and 'H' reflex in infants and children. J Appl Physiol 15: 1–9.

Trontelj JV 1973 A study of the F-responses by single fibre electromyography. In: *New Developments in Electromyography and Clinical Neurophysiology* (ed JE Desmedt). Karger, Basel, pp 318–322.

Trontelj JV, Mihelin M, Fernandez JM, et al 1986 Axonal stimulation for end-plate jitter studies. J Neurol Neurosurg Psychiatry 49: 677–685.

Trontelj JV, Stålberg E, Mihelin M 1990 Jitter in the muscle fibre. J Neurol Neurosurg Psychiatry 53: 49–54.

Wiederholt WC 1970 'End-plate noise' in electromyography. Neurology 20: 214–224.

Chapter 5

Neurophysiology in the Neonatal Period

INTRODUCTION

Major advances in neonatal care over the last few decades have resulted in greatly increased survival for neonates of as few as 24 weeks gestation; present concerns focus on the avoidance of neurological sequelae in survivors. In consequence, the study of the neurology of the neonate has become increasingly important (Eyre 1992, Volpe 2000a, Rennie 2005).

The important contribution of electroencephalography (EEG) has attracted a parallel interest amongst neonatologists. The EEG is an important investigation in the neurologically sick newborn infant, primarily in the diagnosis and treatment of neonatal seizures. The early diagnosis and prognosis of neonatal encephalopathy is emerging as another area in which the EEG can provide useful, non-invasive diagnostic and prognostic information. The contributions of evoked potentials (EPs), nerve conduction studies (NCSs) and electro-myography (EMG) have attracted less interest to date; this is particularly unfortunate as these techniques are of proven value in the assessment of neurological function in the neonate. The role of an integrated approach using the full range of modern neurophysiological techniques is discussed in this chapter, together with an indication of the potential contribution of some newer methods of digital recording and analysis that may well increase the usefulness of clinical neurophysiology in neonates.

The following definitions are used throughout:
- *gestational age* (GA): duration of gestation, expressed in weeks, i.e. the number of complete weeks from the date of the first day of the mother's last menstrual period
- *conceptional age* or *postconceptional age* (CA or PCA): GA at birth plus the number of weeks post-partum
- *full term* (sometimes *term*): neonate born between 37 and 42 weeks GA
- *preterm* or *premature*: neonate born before 37 weeks GA.

The following abbreviations are also used:
- *AS*: active sleep
- *QS*: quiet sleep
- *IBI*: interburst interval.

THE NEONATAL ELECTROENCEPHALOGRAM

Recording techniques
EEG polygraphy
Multichannel EEG–polygraphic recording, according to the latest International Federation for Clinical Neurophysiology (IFCN) recommendations (De Weerd et al 1999), is the gold standard for recording the neonatal EEG. The essentials of the standard techniques are set out in *Table 5.1*, and for a detailed illustration of the methods and findings the reader is recommended to study the classic works by Stockard-Pope et al (1992), Clancy et al (1993), Lamblin et al (1999) and Mizrahi et al (2003).

Before an EEG recording begins, an assessment of the neonate's condition should be obtained from medical and nursing staff. The EEG may be more difficult to obtain if the baby is unstable and ventilated. The EEG assessment of the neonate should always start with a multichannel polygraphic recording in the neonatal intensive care unit (NICU); it should be performed by neurophysiology personnel with experience in neonatal EEG recording. In some cases the neurophysiologist may need to be present during at least part of the recording to guide and interpret the findings in discussion with the neonatal team. Decisions can be made about the need for, and appropriate timing of, follow-up recordings and the use of a suitable form of continuous EEG–polygraphic monitoring to provide immediate warning of any change in cerebral status. In some neonates this might include simultaneous video monitoring to identify subtle seizures that can be easily missed by clinical observation alone.

The recording of other variables with the EEG helps to identify biological artefacts and interactions between cardiac and cerebral function, and is the cornerstone of assessing the relatively short waking and sleep stage cycles of the neonate. This means that a minimum of a 60 min recording is recommended and must include a sleep–waking cycle; long recordings also increase the chances of recording intermittent events such as seizure discharges (*Fig. 5.1*). Sleep cycle patterns provide considerable insight into the maturation of the brain (*Table 5.2*) as well as highlighting events that may be linked to particular phases of these cycles.

The technologist should, so far as possible, observe the neonate continuously and record the sleep–waking state using a structured scale such as that proposed by Prechtl and Beintema (1977). This method is described in *Table 5.3*. The technologist should be aware of other factors that may affect the ongoing EEG and should therefore be noted during the recording. These include: type of respiration (noting the method of assisted ventilation if relevant, e.g. to assist interpretation of unusual artefacts), arterial pH, paO_2, $paCO_2$, arterial blood pressure, intracranial pressure, temperature, oxygen saturation and drug therapy, including muscle relaxants and anticonvulsants.

The NICU environment itself can make it technically demanding to record the small amplitude signals of the neonatal EEG. The technologist must be aware of the wide range of biological and external interference sources (*Table 5.4.*). Modern digital EEG equipment has made dealing with interference somewhat easier, not least because of the lower susceptibility of modern amplifiers

to the electrically noisy environment. Digital equipment also provides the opportunity for convenient review of a prolonged recording and re-montaging to improve interpretation; algorithms are sometimes available for immediate analysis to quantify various features of the ongoing background activities or to identify slow trends.

Certain modifications in recording technique (most commonly a reduction in the number of electrodes) may have to be made when examining very small or very sick neonates. Limitations in physical access for electrode application, the need for minimal handling, and the very considerable nursing and clinical activity involved in the minute-to-minute care of the neonate all add to the difficulties.

Electrodes   The most important factor to consider when applying neonatal EEG electrodes is undoubtedly the preparation of the electrode sites. If possible, the scalp should be cleaned with an alcohol wipe to remove all trace of sweat and birth residues. Each site should then be

---

Table 5.1   Summary of the main polygraphic recording techniques for premature and full term neonates*

Parameter	Transducers	Electrode placement	Gain	Time constant (minimum) (s)	High-pass filter (Hz)	Low-pass filter (Hz)
EEG	Ag/AgCl/Gold EEG[†] cup electrodes secured by tape or a mesh hat. Conductive adhesive paste recommended (not collodion)	Adapted from the International 10–20 System, always including: frontal, central, temporal, occipital and vertex electrodes (minimum 8 leads)	5–10 µV/mm adjustable	0.3	0.5	70
ECG	Disposable ECG electrodes, silver–silver chloride cups	Shoulders, chest or limbs	Adjustable	0.1	1.5	70
Eye movements	Piezo-crystal	Upper eyelid	Adjustable	0.1	1.5	70
	EEG electrodes	Slightly lateral to the outer canthi		0.3	0.5	70
EMG – chin	Surface EEG electrodes	Active: on the chin muscle Reference: on the mandible	Adjustable	0.1	1.5	100 or off
Respiration	Strain gauge	2 cm above umbilicus	Adjustable	1	0.1	15
	Thoracic impedance	Lower thorax				
	Piezo-crystal					
	Thermocouple	Nasobuccal (airflow)	Adjustable	1	0.1	15
Body movements	Piezo–crystals	On the limbs	Adjustable	0.1	1.5	70
Oxygen saturation	Specific transducers	–	DC amplification (parallel recording), manually check the values			

*From Lamblin et al (1999), as adapted from Curzi-Dascalova and Mirmiran (1996), reproduced from the English translation (2000) published by Éditions Scientifiques et Médicales Elsevier SAS, with permission.
[†]Needle electrodes are not recommended by the present authors and are contraindicated in the event of coagulation abnormalities or haematoma.

Table 5.2 Summary of the major and other variables used for sleep-state scoring at various gestational ages*

GA (weeks)	27–34		35–36		37–38		39–41	
State	AS	QS	AS	QS	AS	QS	AS	QS
Major variables								
EEG	Continuous or semi-discontinuous	Discontinuous	Continuous • AS1: slow, high amplitude • AS2: lower amplitude	Discontinuous or semi-discontinuous	Continuous • AS1: slower, higher amplitude • AS2: mixed pattern	Semi-discontinuous or *tracé alternant*	Continuous • AS1: slower, lower amplitude • AS2: mixed pattern	*Tracé alternant* or slow wave
Eye movements	–	–	++	–	+++	–	+++	–
Respiratory rate	Irregular	More regular	Irregular	More regular	Irregular	More regular	Irregular	More regular
Tonic chin	–	▵	–	±	–	±	–	±
Other variables								
EMG – body movements	+++	+	+++	+	+++	+	++	+

*From Lamblin et al (1999), adapted from Curzi-Dascalova and Mirmiran (1996), reproduced from the English translation (2000) published by Éditions Scientifiques et Médicales Elsevier SAS, with permission.

AS, active sleep; AS1, preceding quiet sleep; AS2, following quiet sleep; GA, gestational age; QS, quiet sleep.

–, none; +, some; ++, frequent; +++, abundant; ±, variable.

rubbed gently (cotton buds are ideal) with a conducting gel and the electrode applied using a fixative paste. The preferred electrodes for neonatal recording are silver–silver chloride discs. Collodion is *not* recommended for use in the NICU, especially when the neonate is in an incubator. The fumes from acetone and collodion may be offensive and hazardous in the oxygen-enriched environment of the incubator. It is also possible to attach disc electrodes, in areas where there is no hair, using sticky tape and after careful preliminary cleaning of the skin with an alcohol wipe. Needle electrodes should never be used in the NICU; breaking the skin barrier on the scalp increases the risk of infection in this vulnerable population.

Table 5.4 Common sources of biological and external artefacts on the neonatal EEG

Source	Artefact type
Cardiac	ECG
	Pulse (fontanelle related pulsations)
	Ballistocardiographic (movement of head with heartbeat)
Respiratory	Movement of head and body with respiration
Body	Twitches
	Tremors
Head	Vertical, horizontal and rotatory head movements
Face	Sucking, glossokinaesthetic, eye movements and blinks, and frontotemporal muscle contraction
Other	50 Hz electrical interference
	Infusion pump artefact
	Ventilator artefact
	Overhead cot heater artefact
	Apnoea blanket interference

Table 5.3 System for scoring the state of the newborn infant*

State	Observations
1	Eyes closed, regular respiration, no movements
2	Eyes closed, irregular respiration, no gross movements
3	Eyes open, no gross movements
4	Eyes open, gross movements, no crying
5	Eyes open or closed, crying
6	Other state, e.g. fits, jerks or opisthotonus

*From Prechtl and Beintema (1964, 1977).

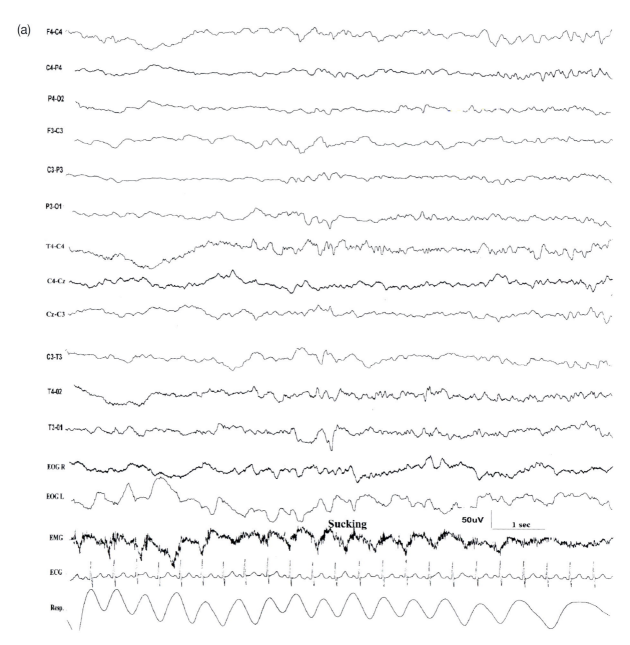

Fig. 5.1 Polygraphic recording in a 1-week-old full term male neonate who had quickly recovered from respiratory difficulties at birth. Recording parameters for all samples are: EEG 50 μV/cm, time constant (TC) 0.3 s; EMG 200 μV/cm, TC 0.03 s; ECG 500 μV/cm, TC 0.3 s; respiration 700 μV/cm, TC 1.0 s. (a) The EEG whilst the child is awake shows continuous diffuse mixed frequencies (*activité moyenne*); note the electrode artefact superimposed on the EMG which is related to sucking. (b–d) See pages 173–175.

Full term neonates should ideally have a full set of electrodes applied to the scalp. However, this may not always be possible or appropriate in a premature, or very sick, unstable neonate. The extra time involved may worsen the neonate's condition (nursing practice usually operates a 'minimal handling' policy). Once electrode application has started there may be frequent interruptions for nursing/medical procedures such as suctioning, blood gas measurement and x-rays. It is often useful to secure all the electrodes on the scalp using a soft net, similar to those used to keep dressings in place after surgery. If a reduced number of electrodes must be used care should be taken to sample widely over all regions of the scalp. The following electrodes should always be included: F4, F3 (or Fp2 and Fp1), C4, Cz, C3, T4, T3, O2 and O1; if possible F8, F7, P4, P3, T6 and T5 (or A2 and A1) should also be used, unless this is precluded by, for example, an infusion line at a designated site (often over C4 or C3).

It is important to appreciate that birth trauma can result in scalp swelling (caput), cranial vault distortions,

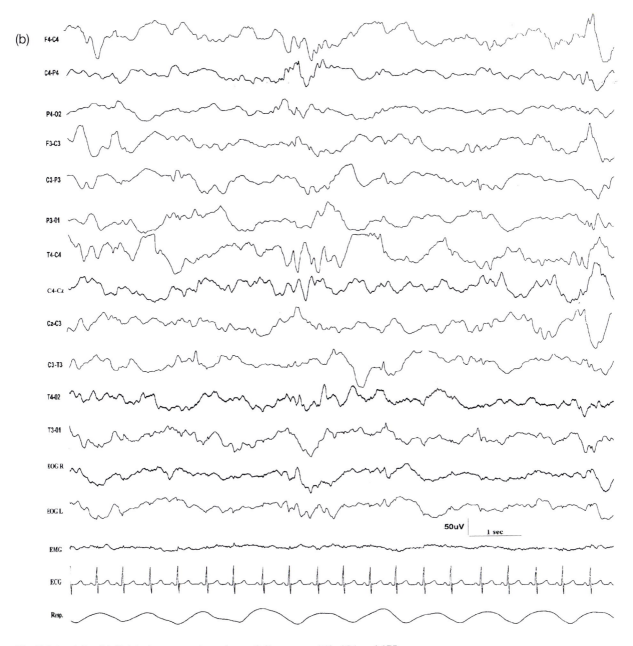

Fig. 5.1 *(contd)* (b) Quiet slow wave sleep. (a, c, d) See pages 172, 174 and 175.

or subdural or epidural fluid collections (cephal haematoma). All of these can alter the interelectrode distance and either amplify or attenuate the recording in localised areas. Cranial fractures can lead to a localised increase in voltage.

Montages Using digital EEG equipment, the choice of montages is not an issue apart from selecting the most appropriate montage to display the ongoing EEG activity. The EEG can be remontaged later if other views are required.

Recording other physiological variables with the EEG

Polygraphy is of even greater importance in the neonate than in older children or adults.

Electrocardiogram (ECG) The ECG is essential for neonatal EEG recording due to the frequent presence of ECG artefact, which may mimic stereotyped neonatal seizure patterns (*Fig. 5.2*). There is also an increased risk of picking up pulse artefact through the fontanelles, which have not yet closed.

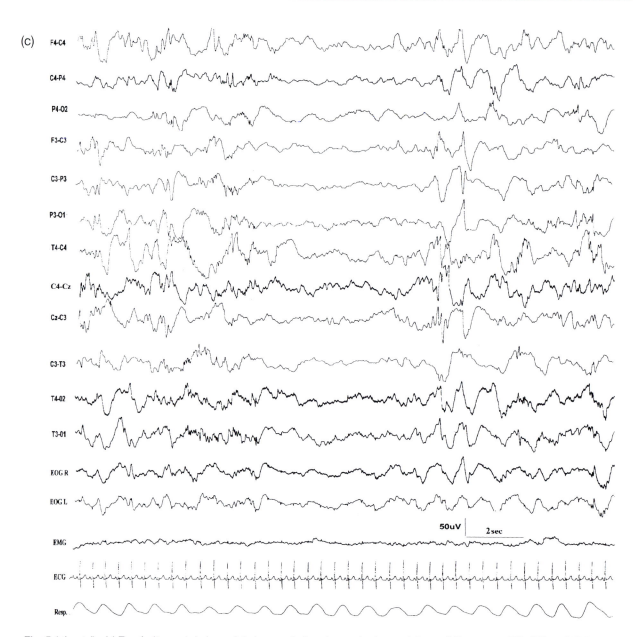

Fig. 5.1 *(contd)*   (c) Tracé alternant during quiet sleep; note the slower chart speed. (a, c, d) See pages 172, 173 and 175.

Lead I of the ECG from the shoulders or arms is commonly used. The apnoeas and bradycardias of prematurity, if prolonged, may cause changes in the ongoing EEG. Standard EEG disc electrodes are suitable for ECG recording and are attached to the skin with surgical tape; alternatively, special neonatal ECG electrodes may be used.

Respiration   Respiration is recorded either by measuring nasal or oral airflow with a thermistor or by sensing changes in chest volume using a strain gauge. Measurement of nasal or oral airflow may be restricted if the neonate is ventilated. Respiratory patterns vary with sleep state and are therefore very useful in its determination. Respiration

monitoring is also extremely useful when trying to eliminate respiratory artefact in the EEG, which may mimic seizure-like activity. In term and preterm neonates with apnoea, respiration monitoring is essential to document the type of apnoea: ictal or non-ictal (*Fig. 5.3*). The time constant (TC) of the amplifier should be set to the maximum available, usually 1 s; the amplifier sensitivity should be adjusted to give a clear (upward) deflection with inspiration.

Electro-oculogram (EOG)   Monitoring the EOG in neonates helps to distinguish different sleep states. In AS rapid eye movement (REM) is seen on the EOG which disappears in QS. An electrode placed 1 cm lateral to and

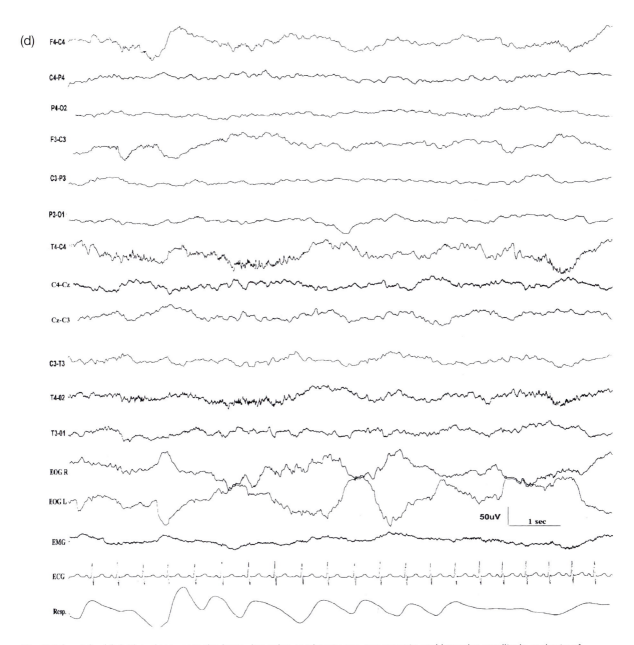

(d)

F4-C4

C4-P4

P4-O2

F3-C3

C3-P3

P3-O1

T4-C4

C4-Cz

Cz-C3

C3-T3

T4-O2

T3-O1

EOG R

EOG L

50uV | 1 sec

EMG

ECG

Resp.

Fig. 5.1 (*contd*) (d) Active sleep; note the large, irregular, conjugate eye movements and irregular amplitude and rate of respiration. (a–c) See pages 172–174.

below the outer canthus of each eye is referred to an electrode slightly lateral to the nasion. If only one channel is available for EOG, two electrodes can be placed such that one is above the outer canthus of one eye and the second below the outer canthus of the other eye.

Electromyogram (EMG) The submental EMG is recorded from a pair of electrodes placed under the chin; activity is present during QS but absent during AS. There may be practical difficulties in attaching submental EMG electrodes if the neonate is ventilated or is dribbling. It may also be necessary to record the EMG from a specific limb, if there are abnormal limb movements that need to be correlated with EEG changes to determine the pres-

ence of seizures. The high-frequency filter settings should be off or as open as possible. A short time constant (TC) of 0.1 or 0.03 s may be used.

Continuous monitoring and related techniques

The condition of the newborn in the NICU can change frequently and events such as seizures need to be recognised immediately and treatment monitored. It is therefore now common to monitor the neonatal EEG over many hours or days.

Cerebral function monitors Monitors of this type are often used in the NICU despite the fact that they were originally designed for adult intensive care use. They are currently used in both term and preterm neonates for seizure detection, prognosis, and to assess the severity of encephalopathy in trials of therapeutic hypothermia (de

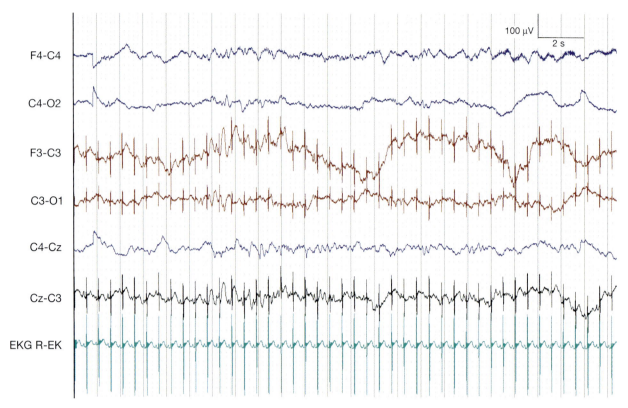

Fig. 5.2   ECG artefact over the left central region. This baby was lying supine with the left hand barely touching the left central electrode (C3).

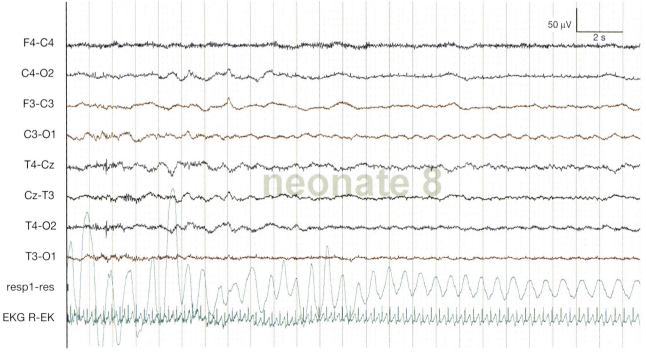

Fig. 5.3   Low voltage rhythmic delta activity in a term neonate with severe HIE. Simultaneous recording of respiration helped to identify this activity as respiration artefact.

Vries and Hellström-Westas 2005, Gluckman et al 2005, Hellström-Westas and Rosen 2005).

The Cerebral Function Monitor (CFM) The CFM (RDM Consultants, Uckfield, Sussex, UK; http://www.cfams.com) (Maynard et al 1969) was first used in the early 1970s and produces a one-channel, amplitude-integrated EEG signal *(Fig. 5.4)*. Despite attempts to develop more sophisticated CFMs, such as the compressed spectral array (CSA) system (Bickford et al 1972, 1973), it is the CFM which remains dominant in the NICU today. There have been criticisms of the CFM because of its limitation to a single EEG channel (plus a simultaneous 'artefact detection' channel), and the lack of detailed information compared with the conventional multichannel EEG, especially when used for detection of neonatal seizure discharges (Eaton et al 1994, Klebermass et al 2001, Toet et al 2002, Rennie et al 2004). Some responsibility for such criticisms about the use of 'inadequate' or 'out-of-date' neonatal monitoring, must be taken by clinical neurophysiologists who find themselves unable to support a full 24 h diagnostic and monitoring service for the neonatal unit. It is essential that clinical neurophysiology teams assist their clinical colleagues in the choice and use of suitable modern monitoring equipment and in reinforcing the need for full diagnostic EEGs as a preliminary and regular adjunct to continuous monitoring.

The Cerebral Function Analysing Monitor (CFAM) The CFAM family of monitors (RDM Consultants, Uckfield, Sussex, UK; http://www.cfams.com) superseded the original CFM system during the late 1970s (Maynard and Jenkinson 1984). Facilities were added for detailed analysis of amplitude and frequency with separate detection of periods of suppression (Prior and Maynard 1986), and in the CFAM2, CFAM3 and CFAM4 series, two- and four-channel recording is also available *(Fig. 5.5)*. These simple, practical devices have been used in neonatal monitoring (Thornberg and Ekström-Jodal 1994, Murdoch-Eaton et al 2001). For a detailed discussion of the advantages and disadvantages of different EEG monitors, see Hellström-Westas et al (2003).

Digital multichannel EEG–polygraphic systems Digital EEG systems can record several additional channels for polygraphy and provide an attractive option for work in the NICU with many advantages over the older 'transportable' or 'portable' analogue systems. Digital systems can display, in real time, multichannel unprocessed EEG and other biological variables, together with various forms of analysis, and have an increasing range of off-line review facilities. Particularly advantageous are their small size, high storage capacity and potential for networking to display or analyse data in the main laboratory system (Thordstein et al 2000). Modern digital EEG systems now have trend displays that can simultaneously display the multichannel EEG and polygraphy channels along with the compressed EEG trend over a 24 h period. An example of an output from one such monitor is shown in *Fig. 5.6*. Digital EEG systems can also provide the possibility of telephone transmission of EEGs carried out in a

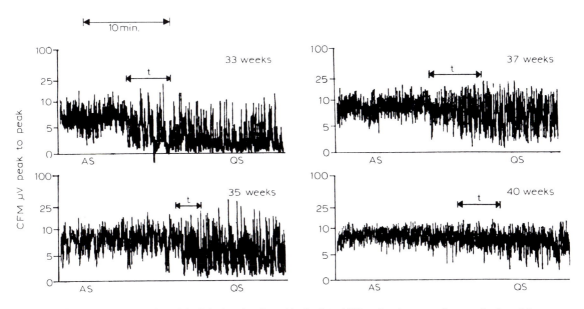

Fig. 5.4 CFM recording on the original device developed in the late 1960s, showing a continuous display of the moment-to-moment amplitude variability of the EEG in a preterm neonate during maturation from 33 weeks GA up to term. The four sequential recordings each show one cycle of active sleep (AS) and quiet sleep (QS) stages separated by a 'transitional' or 'indeterminate' state (t). Note how the periods of QS in the immature neonate show a wide amplitude range (0–10 µV), typical of tracé discontinu, which gradually alters to the narrower range (3–10 µV), typical of the tracé alternant pattern. (From Viniker et al (1984), by permission.)

local general hospital to a regional centre for immediate analysis and interpretation.

Prolonged video–EEG monitoring   This technology is particularly useful for neonates and shows promise for the identification, classification and quantification of neonatal seizures (Mizrahi and Kellaway 1987, Boylan et al 2002). Video–EEG is of particular use in the neonate given the subtle nature of many seizures and in distinguishing epileptic from non-epileptic events (Mizrahi and Kellaway 1998). It is likely that this form of monitoring will become widely used in the NICU in the future, as video–EEG systems have become smaller and very portable.

Automated detection of seizures and sleep state   Automated detection of seizures in the newborn is a natural development of digital techniques and pattern recognition. Algorithms have been designed to assist with detection of seizures in the newborn infant. Liu et al (1992) achieved a sensitivity of 84% and a specificity of 98% using autocorrelation techniques to detect rhythmicity in the EEG. Gotman et al (1997a, b) used a combination of automated methods aimed at increasing detection rates while avoiding false-positive results. Initially, using material from 281 h of recording and 679 seizures, algorithms were developed that extracted features (including rhythmicity, power and stability) from 10 s epochs and compared them with those in earlier epochs. Additional methods were added to identify low-frequency

and arrhythmic seizure discharges. Evaluation showed that 71% of seizures and 78% of seizure clusters were detected with a false detection rate of 1.7/h (Gotman et al 1997a). Subsequent evaluation on new data derived from the same hospitals showed detection rates averaging 66%, with false detection rates of 2.3/h. Variation between the hospitals was considered to have reflected different quality and levels of supervision of the recordings, which ranged from short recordings fully attended by a technologist to overnight recordings, largely unattended (Gotman et al 1997b). Celka and Colditz (2002) used singular spectrum analysis to detect seizures in the neonate by examining the complexity of the EEG signal. In 2005 Faul et al evaluated these three methods of neonatal seizure detection using a new set of neonatal data. Attempts were also made to improve the algorithms. They were unable to achieve the high detection rates reported. Examining the EEG on a complexity basis (Celka et al 2001) proved to have the highest seizure-detection rate. Other techniques that have been tried more recently and with varying degrees of success include synchronisation likelihood (Smit et al 2004), wavelet transform analysis (Kitayama et al 2003) and analysis of movements on video recordings of neonates with seizures (Karayiannis et al 2005). Automated neonatal seizure detection is a rapidly developing field and more and more processing techniques are constantly being applied to the problem of seizure detection, including feature extraction, entropy measures, chaos theory and modelling (Aarabi et al 2005).

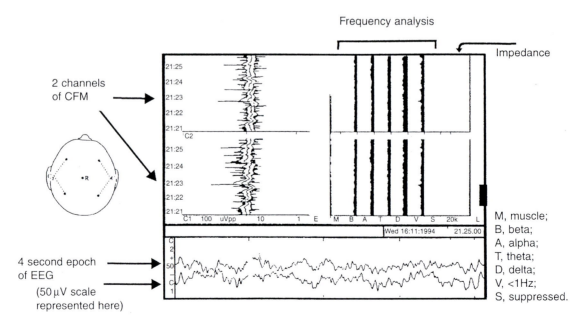

Fig. 5.5   Two-channel CFAM2 recording of 5 min duration from right (F4-P4) and left (F3-P3) frontoparietal regions in a 9-week-old neonate recovering from septicaemia who made a full recovery. Time runs upwards from the lower part of the figure, which shows an initial 4 s sample of conventional ongoing EEG signals from right- and left-sided recording, and then the simultaneous CFAM outputs from right and left hemispheres. Note the comparable EEG activity from each hemisphere. (From Murdoch-Eaton et al (2001), by permission.)

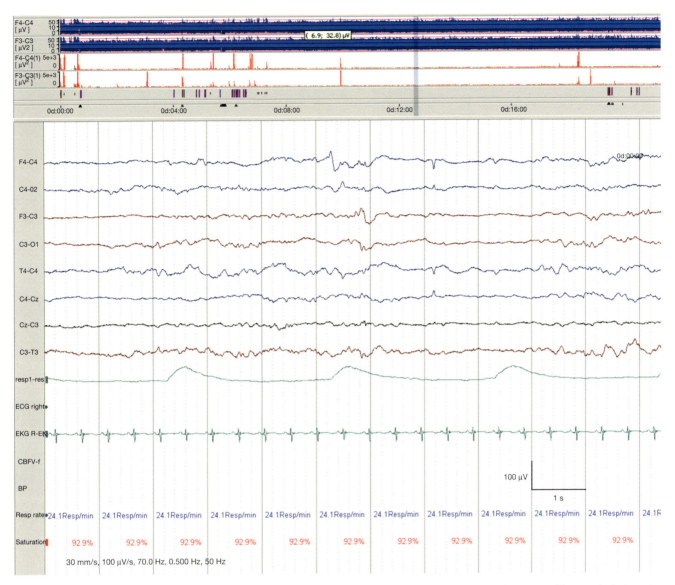

Fig. 5.6 Digital EEG recording showing a continuous low voltage EEG in a full term neonate with a good outcome. The trend window at the top displays the amplitude-integrated EEG from the right and left hemispheres, and also the power from both hemispheres over a 24 h period (F4-C4, F3-C3). Respiration trace, respiration rate ECG and oxygen saturation are also recorded simultaneously.

Signal processing of the neonatal EEG Modern digital EEG systems now provide vast amounts of EEG data that are already digitised, making the signal-processing step much easier. The neonatal EEG signal has been transformed using spectral analysis to provide measures such as spectral edge frequency, spectral entropy (a measure of signal complexity) and the power of the main frequency bands. A range of normal values for spectral edge frequency in preterm neonates has recently been published (Victor et al 2005). Spectral edge frequency was found to be lower in preterm neonates with cerebral white matter injury (Inder et al 2003). In term neonates with hypoxic–ischaemic encephalopathy, low-frequency power was lower in periods of burst suppression when compared with periods of tracé alternant in normal term neonates (Thordstein et al 2004). Neonates with mild encephalopathy may display more pervasive physiological abnormalities, such as poor state regulation and dysmature background EEG patterns. These subtle forms of brain dysfunction may predispose the neonate to later neurodevelopmental problems (Scher et al 2005a). Based on experience of rigorous studies, automated methods for state analysis of the neonatal EEG are increasingly being validated for clinical use (Delapierre et al 1986, Curzi-Dascalova et al 1993, Sahni et al 1995, Scher et al 1996, Turnbull et al 2001, Scher et al 2005b).

Choice of method The choice of method for continuous monitoring is dependent on a dialogue between neurophysiologists and neonatologists as to the objectives, availability of trained staff and cost–benefit considerations. Portable digital EEG–polygraphic equipment operated by experienced neurophysiology technologists will become the universal mainstay of the diagnostic work in the NICU and for regular (often daily) checks on progress. Conventional EEG machines are not generally suitable for unsupervised long term monitoring and consideration has to be given to purpose-built EEG monitoring systems with a minimum of 2–4 channels to ensure that continuous observation of the neonatal EEG is available. Even though most systems now include display of the unprocessed EEG for quality control, these cannot be recommended for use in isolation. A multichannel EEG should always be obtained prior to the commencement of limited channel recording. The distinction between special-purpose monitors and EEG machines is fast becoming very artificial. Machines are now available that can store 32 channels of EEG on the hard drive continuously for 4 days and display trends of a wide range of quantitative measures, seizure detections and other events. As already mentioned, video–EEG recording is almost a necessity for the accurate diagnosis of neonatal seizures, and the most recent generation of machines now record digital images together with the EEG, polygraphy and trends of quantitative measures.

There is a world of difference between a state-of-the-art, 24 h, physician-led, diagnostic EEG service, augmented by suitable multichannel monitors, and the practice of many reputable neonatal units, that have to rely on outdated and often inadequate monitors, unsupported by a proper EEG service. In the UK and Ireland, the areas of which the authors have most personal knowledge, best practice is, regrettably, unusual.

Origins and significance of the neonatal EEG

The EEG of the preterm and term newborn develops in parallel with extensive anatomical, functional and chemical maturational changes in the immature brain. Precise neurophysiological correlates for these processes, which include neuronal and glial multiplication, structural and neurochemical synaptogenesis, and myelination (Sarnat 1984, Holmes 1986, Holmes and Ben-Ari 2001), have not yet been established. Nevertheless, both animal and human studies support the hypothesis that the progressive maturation of the human neonatal EEG between 24 weeks GA and term relates to an increasing cortical modulation of afferent inputs from lower centres, particularly the thalamus and brainstem.

Observations on pathological states in the human newborn suggest a subcortical origin of EEG activity (in the developing brain). EEG activity, albeit abnormal, may still be found in neonates with severe congenital malformations such as hydranencephaly (Ferguson et al 1974, Pampiglione 1974, Engel 1975, Reding et al 1979) and atelencephaly (Danner et al 1985), suggesting an origin in brainstem structures. Studies of neonates with holoprosencephaly have suggested a thalamic origin for high amplitude discharges in the neonatal EEG (DeMyer and White 1964, Watanabe et al 1976).

The investigation of neonates with acquired cerebral lesions has revealed a correlation between excessively discontinuous or paroxysmal EEG activity and extensive cortical destruction or deafferentation at autopsy. This suggests a predominantly subcortical origin for the extremely discontinuous EEG in the early preterm neonate in whom little cortical differentiation has occurred (Dreyfus-Brisac and Larroche 1972). In a study of EEG–neuropathological correlations in newborn infants between 23 and 43 weeks GA, the involvement of thalamic, midbrain and pontine structures in widespread encephalomalacia was associated with electrocerebral inactivity (Aso et al 1989). This study also demonstrated that the overall degree of abnormality of the ongoing background activity of the EEG predicted the severity of brain lesions, ranging from severe multifocal brain damage to more discrete localised lesions.

Factors influencing EEG maturation in the newborn

The EEG has been extensively investigated in term and moderately preterm neonates. There are now studies in neonates of less than 28 weeks gestation who, until 20 years ago, were frequently non-viable and could only rarely be studied in optimal condition (Dreyfus-Brisac 1979). Anderson et al (1985) were the first group to provide normative data on neonates of less than 27 weeks gestation, although the neurological status of these neonates was questionable as all had previously required ventilation and no neuroimaging data were available. It is now possible to study such neonates in good physiological condition, with the additional confirmation of the absence of such major anatomical lesions as provided by serial cranial ultrasound examinations. In these circumstances it has been possible to demonstrate much 'better' EEG activity than previously reported in early prematurity (Biagioni et al 1994, Selton et al 2000, Hayakawa et al 2001, Vecchierini et al 2003). In all these studies it is clear that the EEG has a constant and reproducible pattern in normal very premature neonates. The EEG is always discontinuous with synchronous activity over both hemispheres. *Table 5.5* summarises the main characteristics of the very preterm EEG from a number of studies in which the neurodevelopmental outcome was assessed.

Within these limitations, CA is generally accepted as more influential than birth weight or GA alone in EEG maturation; i.e. EEG evolution normally progresses at the same rate whether the neonate is in utero or ex utero (Parmelee et al 1967, Eisengart et al 1970, Dreyfus-Brisac 1979). Detailed longitudinal studies of EEG maturation from birth and throughout the first year of life in both term and preterm neonates have not demonstrated any significant differences, except for minor variations in

Table 5.5 Features of the preterm EEG from 21–28 weeks*

Study	No. of subjects	PCA	Normal outcome	Mean maximum IBI (s)	Mean burst duration (s)	Periods of continuity	Periods of discontinuity
Hahn et al (1989)	5	26–27	3 years	32	–	Yes	Yes
	15	28–29	3 years	21	–	Yes	Yes
Biagioni et al (1994)	7	27–28	1 year	30.7	1.5	Excluded from analysis	Yes
Selton et al (2000)	4	26	2 years	46.4	1–131	Yes	Yes
	9	27	2 years	36	1–179	Yes	Yes
	4	28	2 years	26.6	1.4–159	Yes	Yes
Hayakama et al (2001)	3	21–22	Most normal	126	4.3	Brief	Yes
	7	23–24		86	5	Yes	Yes
	5	25–26		44	5.8	Yes (up to 48%)	Yes
Vecchierini (2003)	10	24–26	9 at 3 years	NA	NA	Yes (up to 55%)	Yes

*Only studies from neonates considered 'normal' at the time of the EEG and at follow-up (generally a normal cranial ultrasound examination and neurological examination) are included.

IBI, interburst interval; NA, not available.

sleep-cycle maturation (Ellingson and Peters 1980a, b, Watanabe et al 1980).

Clear cyclical patterns of activity can be detected in the EEG from 25 weeks onwards, but correlation with specific behavioural states is not complete until between 32 and 35 weeks gestation (Scher et al 2005c). During this later stage of prematurity there is a relative decline in active or REM sleep and an increase in quiet or non-REM sleep (Dreyfus-Brisac et al 1957, Parmelee et al 1967, Prechtl 1974). Interpretation of the EEG, unless it is seriously abnormal, especially close to and at term, may be difficult without reference to a polygraphic recording of variables reflecting the behavioural state, such as respiration, ECG, EOG and EMG (Schulte 1970). Indeed, the official guidelines of IFCN (De Weerd et al 1999) continue to emphasise polygraphic studies, suggesting a total of 21 channels comprising EEG (17 channels) and other polygraphic variables (four channels) as necessary to maintain minimum standards of paediatric and neonatal recording. If only fewer channels are available, it is advisable to sacrifice some EEG capacity to permit registration of at least the ECG and respiration.

Specific features of the maturing neonatal EEG

The latest IFCN definitions (Noachtar et al 1999) are quoted here verbatim, followed by any terms and definitions in common usage in neonatal practice (Stockard-Pope et al 1992, Lamblin et al 1999, De Weerd et al 1999):

- *Anterior slow dysrhythmia*. Short bursts of mono- or polyphasic delta waves at 1–3 Hz with a voltage of 50–100 μV over the frontal region appear at CA 36–37 weeks in AS.
- *Burst suppression*. Pattern characterised by bursts of theta and/or delta waves, at times intermixed with faster waves, and intervening periods of low amplitude (< 20 μV). *Comment*: an EEG pattern that indicates either severe brain dysfunction or is typical for some anaesthetic drugs at certain levels of anaesthesia (Noachtar et al 1999). This activity shows no lability or reactivity, and suggests a poor outcome. In pure phenomenological terms the burst-suppression pattern is somewhat similar to normal age-related patterns in neonates, and terminology is often used interchangeably in the published literature. However, because of the association with very poor prognosis, this term must be distinguished from those used to describe normal patterns, such as tracé discontinu and tracé alternant, and abnormal discontinuous patterns, such as the 'dysmature' EEG pattern where an EEG shows 'normal' maturation features for a younger CA (Hahn and Tharp 1990). Thus, in neonatal EEG the term burst-suppression should only be used when the bursts lack all normal activity for CA, are invariant and non-reactive (Tharp et al 1981, Stockard-Pope et al 1992).
- *Delta brushes*. These consist of high amplitude slow waves (0.3–1.5 Hz, 50–300 μV) with superimposed bursts of fast activity (> 8 Hz). They are first

observed around CA 28 weeks, have a peak incidence at around 32 weeks, with progressive diminution between 34 and 37 weeks. At first delta brushes are diffuse, then predominant over the temporal and occipital region and are always occipital in the term neonate (Engel 1975, Dreyfus-Brisac 1979, Lombroso 1985, Torres and Anderson 1985).

- *Frontal sharp transients (encoche frontale)*. Diphasic sharp waves in the frontal area with a voltage of 50–200 µV and a duration of 0.5–0.75 s and appear at about CA 35 weeks. They can be unilateral or bilateral and may occur at random and in variable frequency (Monod et al 1972, Torres and Anderson 1985).
- *Mixed frequencies (activité moyenne)*. Continuous, irregular, diffuse 1–10 Hz activity (mainly 4–7 Hz) at 25–50 µV during wakefulness and during AS that occurs at the onset of sleep (Werner et al 1977). It appears at CA 36 weeks or later.
- *Premature temporal theta (thêta temporal)*. Brief rhythmic bursts of 4–7 Hz, 100–250 µV, sharp theta activity in the temporal regions, sometimes referred to as 'temporal saw-tooth', which are maximal at CA 29 weeks but can be seen less frequently up to CA 32 weeks (Werner et al 1977, Torres and Anderson 1985, Hughes et al 1987).
- *Tracé discontinu*. EEG pattern of preterm neonates below 36 weeks CA characterised by mixed frequency high voltage bursts separated by periods of very low voltage background (Noachtar et al 1999).
- *Tracé alternant*. An alternating pattern of non-REM (quiet) sleep seen in term neonates of 36 weeks CA or older which can persist up to 3–4 weeks after birth in full term neonates. The pattern is characterised by bursts of predominantly slow waves (1–3 Hz, 50–100 µV) appearing approximately every 4–5 s, and intervening periods of low voltage activity (< 50 µV, 4–7 Hz) (Noachtar et al 1999).
- *Tracé continu*. Continuous activity, replacing a previously marked intermittent record during evolution of EEG in preterm neonates (Noachtar et al 1999).
- *Reactivity*. Changes seen in the neonatal EEG associated with stimulation. These changes have been described associated with pain, photic and auditory stimulation in premature neonates of CA 27–31 weeks. The most dramatic change in response to stimulation is seen in term newborns when a transient flattening or augmentation of activity can occur.

Current knowledge of specific developmental features of the EEG in preterm and term neonates has been documented in detail by several authors (Dreyfus-Brisac 1979, Clancy et al 1993, Holmes and Lombroso 1993; Lamblin et al 1999). The main EEG features are discussed in detail separately, although in practice they are often interrelated. The possibility has to be borne in mind that the characteristics of the ongoing EEG may be modified in the sick neonate, particularly the neonate with seizures, who may have been treated with sedative drugs.

The interpretation of the neonatal EEG is, as in older children and adults, based on analysis of the ongoing background activity and also on the age dependent EEG features. The patterns of brain activity seen in the neonatal period mirror the rapid maturational changes taking place in the brain. Waveforms appear that are not present at any other time of life. Sleep states are varied, change rapidly and are very different from those seen in older children and adults. Abnormalities of the EEG may manifest as altered or unexpected features. For example, a true discontinuous pattern is normal at 30 weeks CA, may be acceptable during sleep at 34 weeks, but presents as an abnormal feature at term. Thus, it is impossible to interpret accurately the neonatal EEG without knowledge of the physiological factors influencing the background activity, such as maturation and the sleep–wake state. The background activity of the neonatal EEG can be described in terms of:

- continuity, i.e. the presence of a continuous or discontinuous pattern
- amplitude and frequency
- synchrony and symmetry, i.e. the degree with which activity appears simultaneously over both hemispheres and the symmetry between the two sides
- state, i.e. the presence of sleep–wake cycles
- reactivity, i.e. whether the stimulus evokes a change in background activity
- specific maturational features, e.g. delta brush activity, temporal saw-tooth activity or frontal sharp transients
- other transients.

Normal maturation of the EEG

Brain growth and differentiation in preterm neonates are reflected by a rapid change in the EEG findings. The progressive development of continuity of activity provides perhaps the most immediate visually striking aspect of EEG maturation. There are no absolute criteria available to determine the appropriate amount of continuity and discontinuity in the EEG of the premature neonate. Difficulties relate to the definition of the interburst interval (IBI), the effects of sedative and anticonvulsant medications, and the 'normality' of the subjects (e.g. whether neonates were neurologically normal for CA at the time of the investigation and at follow-up). The very preterm EEG shows long periods of quiescence (< 15 µV) interrupted by short bilateral bursts of high voltage, mixed frequency waves. This normal background pattern is referred to as 'tracé discontinu' or 'discontinuous pattern'. The lengths of the IBIs and the duration of burst activity are directly proportional to the degree of prematurity. The definition of the normal preterm neonate has to be considered. Quantitative analysis of the relative proportions of continuous and discontinuous activity during the total recording period can also be of maturational significance. A study of 461 neonates demonstrated a persistent rise in periods of activity and a commensurate decline in periods of quiescence as a percentage of total recording time during maturation (Hughes et al 1983). Similar findings were reported in an earlier but more limited study of 23 neonates between 28 and 41 weeks gestation (Graziani et al 1974), and these have been confirmed with very strict criteria in a study of 104 EEGs from 36 healthy premature neonates (Hahn et al 1989). In early studies IBIs in premature neonates of up to 8 min have been reported, but no reference was made to the neurological status at the time of recordings or the eventual outcome of the neonates. Neurological sequelae often only become evident after the first or second year of life, and therefore these values cannot be considered 'normative', leading to great variation in standards for both IBI and burst duration. Recent studies in which subjects had a normal neurodevelopmental outcome (Hahn et al 1989, Selton et al 2000, Vecchierini et al 2003) reported much shorter IBIs, and even at 24 weeks IBIs of more than 60 s are probably abnormal.

Definitions of tracé discontinu and tracé alternant are varied and do not specify a clear amplitude cut-off for the IBIs between the maximum voltage of tracé discontinu and the lower limit of tracé alternant. This appears to lie in the range 10–15 μV for tracé discontinu and 20–25 μV for tracé alternant. For example, Lamblin et al (1999) define the periods of inactivity in tracé discontinu as < 10 μV and the continuous activity in tracé alternant as > 25 μV. Noachter et al (1999) defined the low voltage activity of tracé alternant as < 50 μV, but did not give a lower limit or defined limits for tracé discontinu. Thus there is a grey area in which the rest of the background activity has to be taken into account in order to differentiate between a normal or abnormal pattern for GA in an individual patient. Agreed definitions are needed to guide neonatologists regarding treatment and prognosis. With maturation there is a gradual reduction in peak burst amplitude from more than 400 μV at 24 weeks to around 200 μV at term, accompanied by an increase in the dominant frequency from extremely slow delta activity of 0.5–1.0 Hz to mixed frequencies (Dreyfus-Brisac 1979, Stockard-Pope et al 1992, Vecchierini et al 2003).

Other aspects are the presence of interhemispheric synchrony, the development of wake–sleep stages, and age-specific waveforms and pattern (transients). After an initial period of hypersynchrony at 24–27 weeks, the EEG becomes relatively asynchronous up to around 32 weeks. Later, the activity over the occipital areas becomes more synchronous again, and at term there are only minor asynchronies present. The characteristic features of normal EEG maturation in preterm neonates are described below according to the CA and are summarised in *Table 5.6*. The characteristic transients of the neonatal EEG at different GAs are given in *Table 5.7*.

22 and 23 weeks CA There are very few reports on reference criteria for neonates less than 24 weeks old. However, Hayakawa et al (2001) have reported the EEG findings from three neonates aged 21–22 weeks. The recordings were reported as discontinuous, but with some short continuous periods totalling a mean of 5.5%. The maximum IBI was 126 s, with a mean of 25.8 s, and the mean burst duration was 4.3 s. They also reported the EEG findings of seven neonates at 23–24 weeks. In this group the recording was discontinuous, but periods of continuity had increased to 23.8%. The maximum IBI had reduced to 86 s, with a mean of 18.4 s, and the mean burst duration had increased to 5 s.

24–26 weeks CA
Continuity. The background activity is characterised by tracé discontinu with IBIs between 10 and 60 s (Biagioni et al 1994, Hayakawa et al 2001). Brief periods of more continuous activity may occur.

Amplitude and frequency. Smooth, diphasic delta waves of high amplitude (300–400 μV), low frequency delta of 0.5–1.0 Hz and theta waves of variable amplitude are predominant.

Interhemispheric synchrony. High amplitude delta activity is usually synchronous over both hemispheres.

Characteristic transients. Unilateral or bilateral sharp activity at 5–6 Hz maximal over the occipital regions (sharp theta on the occipitals of prematures (STOPs)) may be seen.

Behavioural state. No differentiation of sleep states is possible at this age, although cycling on the EEG has been described even at 25 weeks and is often thought to represent rudimentary sleep cycles (Scher et al 2005c).

Reactivity. Probably none.

27–29 weeks CA
Continuity. The background activity is characterised by tracé discontinu. IBIs at 28 weeks are generally 10–40 s (Hahn et al 1989, Selton et al 2000). Continuous runs of delta activity of > 60 s duration may occur, but will occupy less than 10% of a 24 h period.

Amplitude and frequency. Occipital preponderance of delta activity (0.3–1 Hz). Bursts consist of a polymorphic mixture of activity in the 0.3–14 Hz range.

Interhemispheric synchrony. Burst activity is often synchronous over the two hemispheres (so-called 'hypersynchrony').

Characteristic transients. Premature temporal theta (PTθ) at 4–6 Hz, sharp activity (50–100 μV) with a saw-toothed waveform that is generally bilateral and maximal in temporal areas starts to appear (*Fig. 5.7*) (Torres and Anderson 1985, Hughes et al 1987). The bursts last up to 2 s and may be variably asymmetrical. They persist until 33 weeks gestation. From 28 weeks onwards diffuse delta brush activity (*Fig. 5.8*) may be seen. These are the combination of very slow delta waves (0.3–1.5 Hz, 50–300 μV) with superimposed fast activity at 10–20 Hz.

Table 5.6 Maturation of the neonatal EEG

CA (weeks)	Background activity	Characteristic transients	Behavioural state
24–26	Discontinuous EEG, IBI 10–60 s; bursts 'hypersynchronous'	STOPS (sharp theta on the occipitals of prematures)	None
27–29	Discontinuous EEG (IBI at 28 weeks 10–40 s); bursts consist of polymorphic mixture of activity in the 0.3–14 Hz range, 'hypersynchronous'	Occipital preponderance of delta activity. PTθ (4–6 Hz sharp activity bilateral in temporal areas); diffuse delta brush activity (delta with superimposed fast activity of 10–20 Hz)	None
30–32	Discontinuous EEG with occipital delta and a mix of delta, theta and alpha activity in centrotemporal regions in QS	Abundance of occipital delta activity and delta brush activity, more prominent in Rolandic and occipital areas	Poorly differentiated AS and QS
	Tracé continu starts to appear in AS	Preponderance of PTθ (disappears after 32 weeks)	
33–34	Discontinuous activity in QS; long periods of quiescence are no longer characteristic; tracé continu in AS	Delta brush activity over Rolandic and occipital areas, multifocal sharp transients common; anterior frontal delta activity and frontal sharp transients (encoche frontale); response to sensory stimulation	More definite periods of AS and QS; AS predominates
35–37	Awake: continuous, poorly sustained, low voltage, mixed frequency activity	Frontal sharp transients and anterior delta activity most prominent	QS, AS and wakefulness
	AS: continuous 0.1–0.5 Hz activity, primarily in occipital regions	Anterior slow dysrhythmia (delta activity of 1–3 Hz of 50–100 μV) appear in active sleep	
	QS: discontinuous, but tracé alternant replaces tracé discontinu	Delta brush activity much less frequent	
38–42	Mixed activity (activité moyenne): mixed frequency during wakefulness and AS that occurs at sleep onset	Scattered sharp transients are common in Rolandic and temporal areas; frontal sharp transients and anterior slow dysrhythmia persist	Fully developed sleep cycles
	Low voltage intermittent activity: irregular activity seen in AS that follows QS, AS periods reduced	Short bursts of alpha and theta activity in QS; delta brushes are rarely seen	
	Mixed: activité moyenne with superimposed delta (2–4 Hz, < 100 μV) during AS that occurs at the onset of sleep	The record reaches almost 100% synchrony	
	High voltage, slow wave sleep: diffuse delta activity (25–100 μV) during portions of QS		
	Tracé alternant		
43–44	High voltage, slow wave sleep: during QS; by 46 weeks GA tracé alternant has disappeared	Frontal sharp waves persist; rudimentary sleep spindles may appear at 42 weeks GA	

AS, active sleep; CA, conceptional age; GA, gestational age; IBI, interburst interval; PTθ, premature temporal theta; QS, quiet sleep.

Behavioural state. Clear cyclical activity showing phases of tracé discontinu and more continuous activity that some authors believe represent AS and QS.

Reactivity. Inconsistent and variable.

30–32 weeks CA

Continuity. Tracé discontinu is seen with IBIs usually lasting 3–15 s, sometimes up to 20 s in QS; bursts have a duration of 3 s or more. Continuous or semicontinuous activity (tracé continu) may occur during wakefulness and AS. This activity consists of theta and delta activity and lasts > 1 min.

Amplitude and frequency. Delta activity predominates over the occipital and temporal region and a mix of delta, theta and alpha activity over the centrotemporal regions in QS. Amplitude is 250 μV or more.

Interhemispheric synchrony. Partially synchronous (> 60%).

Characteristic transients. PTθ is now maximal and starts to disappear after 32 weeks (Hughes et al 1987). Delta brushes are more frequent, most prominent during continuous periods and are predominant in the Rolandic and occipital areas (*Fig. 5.9*).

Behavioural state. Body movements are the first param-

Table 5.7 Normal transient features of the maturing neonatal EEG

Name	GA when feature is maximal	Type of activity
Delta brush activity	28-32 weeks	
STOPs (sharp theta on the occipitals of prematures)	23-28 weeks	
PTθ (premature temporal theta)	27-32 weeks	
(a) Encoches frontale	34-40 weeks	
(b) Anterior slow dysrhythmia (bifrontal delta)	36-40 weeks	

eter to become reliable indicators of state. REMs are seen during AS from 32 weeks (Parmelee and Stern 1972). More recent studies have shown that, using EOG, AS and QS can be clearly identified as early as 27 weeks (Curzi-Dascalova et al 1993).

Reactivity. Transient appearance of continuous activity.

33–34 weeks CA

Continuity. The background activity remains discontinuous in QS (tracé discontinu). Long periods of quies-

cence are no longer characteristic and should be under 10 s at 34 weeks CA. The activity is now continuous in the AS and wake state (tracé continu).

Amplitude and frequency. High voltage anterior frontal delta activity (up to 250 μV) is still predominant, but some faster activity may be intermixed.

Interhemispheric synchrony. More synchronous, particularly over the occipital areas.

Characteristic transients. Delta brush activity is still abundant, particularly over the Rolandic and occipital

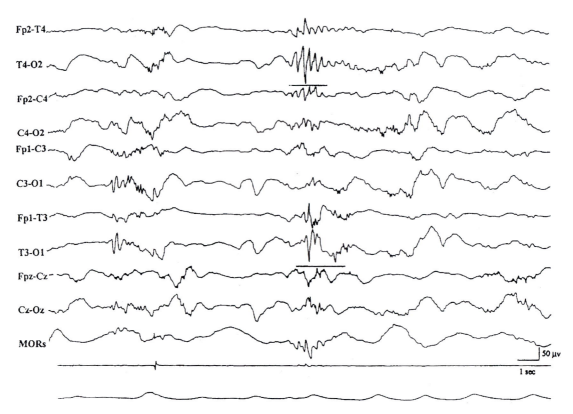

Fig. 5.7   Posterior temporal theta bursts in a 33-weeks GA neonate. On the left are asynchronous bursts, then synchronous (bihemispheric) burst diffusions in REMs. (From Lamblin et al (1999) EEG in premature and full term neonates: developmental features and glossary. *Neurophysiologie Clinique,* **29**; English translation (2000) published by Éditions Scientifiques et Médicales Elsevier SAS, by permission.)

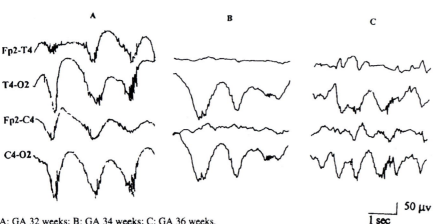

A: GA 32 weeks; B: GA 34 weeks; C: GA 36 weeks.

Fig. 5.8   Delta brush pattern progression with GA. (From Lamblin et al (1999) EEG in premature and full term neonates: developmental features and glossary. *Neurophysiologie Clinique,* **29**; English translation (2000) published by Éditions Scientifiques et Médicales Elsevier SAS, by permission.)

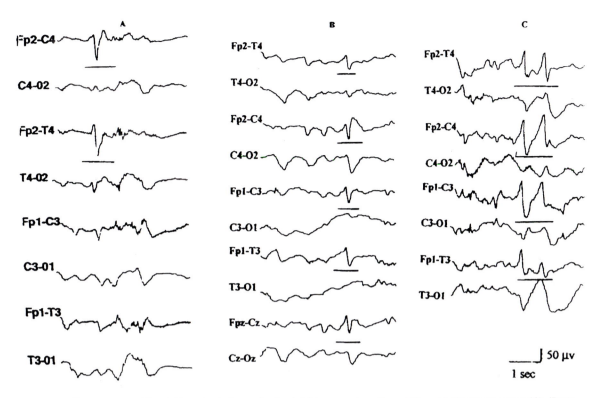

Fig. 5.9 Different types of frontal transients (encoche frontale) in premature (A and B) and full term neonates (C). (From Lamblin et al (1999) EEG in premature and full term neonates: developmental features and glossary. *Neurophysiologie Clinique*, **29**; English translation (2000) published by Éditions Scientifiques et Médicales Elsevier SAS, by permission.)

areas, and is more prominent during periods of discontinuity. Multifocal sharp transients are common. (Pre)frontal sharp transients, often asymmetrical (encoche frontale), may start to appear at 34 weeks (see *Fig. 5.9*). At this age the transients are still incomplete and broad. They are more prominent during the transition from AS or QS. Temporal theta has disappeared by 34 weeks.

Behavioural state. More definite periods of AS and QS. AS predominates, occupying about 60% of sleep time.

Reactivity. The EEG now shows a strong response to sensory stimulation (a diffuse and transient decrease in activity during AS, and a transient appearance of continuous slow wave activity during QS), but this cannot always be obtained.

35–37 weeks CA

Continuity. The record is continuous during wakefulness and AS (tracé continu). In QS the record remains discontinuous or semi-discontinuous. The interburst activity is less than 10 s and often of low amplitude and not flat, i.e. tracé alternant starts to replace tracé discontinu. By 37 weeks nearly all neonates will show a continuous pattern in QS (tracé alternant or slow continuous tracing).

Amplitude and frequency. (1) When awake there is continuous, poorly sustained, low voltage, mixed frequency activity (activité moyenne). (2) In AS the EEG shows continuous 0.1–0.5 Hz activity, primarily in occipital regions. (3) In QS diffuse delta waves are predominant.

Interhemispheric synchrony. May be asynchronous during the onset of QS; during AS the activity is now synchronous.

Characteristic transients. Diphasic frontal sharp transients (encoche frontale) are more prominent and may persist until term. They are uni- or bilateral, sometimes sharp, with a voltage of 50–200 μV. These transients may occur either singly or in series of two or three; they may be distinguished from eye movements by their lack of synchrony. Anterior delta activity of 1–3 Hz of 50–100 μV (slow anterior dysrhythmia) appears in AS. Delta brush activity is much less frequent and is always occipital.

Behavioural state. The EEG can now be used as a reliable predictor of state, but the EMG (chin) is not a reliable predictor of state until 40 weeks (Parmelee and Stern 1972). There is clear definition of three sleep–wake states: QS, AS and wakefulness.

Reactivity. The EEG shows a strong response to sensory stimulation. If high voltage, slow wave activity is present stimulation will produce generalised attenuation of all activities for up to 5 s (usually in AS). If a low voltage background is present, a train of slow waves may be produced (usually in QS).

Normal pattern in full term neonates

Continuity   The EEG activity is now continuous during wakefulness and all sleep stages.

Amplitude and frequency   At term, variations in amplitude and frequency are particularly related to behavioural states. Amplitude is lower in wakefulness and AS and increases during QS.

Mixed (activité moyenne). Continuous 1–10 Hz activity at 40–100 µV during wakefulness and during the AS that occurs at the onset of sleep (Werner et al 1977).

Low voltage intermittent (LVI). Low voltage (20–50 µV) irregular activity seen in AS following QS.

High voltage slow wave sleep. Diffuse, often high voltage delta activity (0.5–2 Hz, 25–150 µV) during parts of QS.

Tracé alternant. Continuous activity (4–7 Hz, 25–50 µV) alternating with bursts of bilateral delta waves (1–3 Hz, 50–100 µV), lasting for 5–6 s and occurring at 3–5 s intervals, is present during QS.

Interhemispheric symmetry and synchrony   The record reaches almost 100% synchrony. At the onset of QS asymmetry and asynchrony, followed by a more discontinuous tracing, may be observed (O'Brien et al 1987).

Characteristic transients   Scattered sharp transients are common in Rolandic and temporal areas. Frontal sharp transients (encoche frontale) and anterior slow dysrhythmia persist. Delta brushes are rarely seen. Short bursts of alpha and theta activity (< 5 s) over the Rolandic region may be observed during AS and QS. Spikes and sharp waves may occur in the traces of apparently normal term neonates without adverse long term outcome, and should not automatically be regarded as an abnormal feature, especially when infrequent, multifocal, non-repetitive and occurring during QS (Harris and Tizard 1960, Rose and Lombroso 1970, Torres and Anderson 1985). A particularly detailed longitudinal EEG and developmental study showed sporadic focal and bilaterally synchronous sharp waves in two-thirds of a cohort of term neonates, all of whom had a normal outcome (Hagne 1972). To be accepted as normal these features should be random and without consistent focus, occurring anywhere on the scalp, but mainly in temporal and occipital regions.

Behavioural state   By term, four distinct EEG sleep–wake states are recognised (see *Fig. 5.1*). Scoring is based on both behavioural observations and polygraphic data. AS and QS can be considered as antecedents of REM sleep and non-REM sleep; in addition, a special type of sleep referred to as 'indeterminate' (sometimes called 'transitional sleep') occurs. Indeterminate sleep is scored when the tracing does not completely meet the criteria for either the AS or QS patterns described above.

Wakefulness is associated with irregular respiration, open eyes and spontaneous movements. The EEG shows a continuous irregular background of activity of 1–10 Hz and 25–50 µV (activité moyenne) (see *Fig. 5.1(a)*).

Quiet sleep (non-REM) is accompanied by regular respiration, submental EMG activity and an absence of eye movements. The corresponding EEG may show one of two patterns: (a) high voltage continuous slow activity, mainly delta activity at 0.5–2 Hz and 25–100 µV (see *Fig. 5.1(b)*); or (b) the tracé alternant pattern of irregular bursts of high voltage delta activity (1–3 Hz, 50–100 µV) alternating with periods of low voltage theta activity (4–7 Hz, 25–50 µV) occurring in 3–5 s intervals (see *Fig. 5.1(c)*). This pattern is present up to 6 weeks post-term. Tracé alternant should not be confused with the discontinuous pattern of prematurity, in which the low amplitude periods are not more than 15 µV.

Active sleep (REM) occupies around 50% of sleep time at term, and is associated with irregular respiration, the absence of submental EMG potentials and the occurrence of rapid eye movements and facial grimaces. The two typical EEG patterns of AS are: (a) continuous mixed irregular theta activity of intermediate amplitude with some delta frequencies present, often referred to as a mixed pattern of AS (2–4 Hz, < 100 µV) and seen at the onset of sleep and then referred to as 'AS onset'; or (b) LVI theta and alpha activities with some delta activity, which is the type of pattern seen after a period of QS (see *Fig. 5.1(d)*).

Sleep states vary in a cyclical manner, the mean cycle duration in the full term neonate being 60 min with a range of 30–70 min. Cyclical activity is, however, present from as young as 25 weeks (Scher et al 2005c). In a group of 33 neonates studied between 25 and 30 weeks CA a mean cycle duration of 68 ± 19 min with a range of 37–100 min was recorded. It is therefore necessary to perform EEG recordings for a minimum duration of 1 h so that periods of state cycling can be adequately described and an accurate assessment of normality or otherwise made (Prechtl 1974). A simple classification of QS and AS can be obtained with monitors such as the CFM and CFAM (Viniker et al 1980, 1984). The methodology for the automatic analysis of polygraphic recordings of sleep cycle patterns in the preterm and full term newborn has received some attention (Goto et al 1992, Scher et al 2005c).

Reactivity   It is very important to study reactivity in all neonates > 35 weeks CA. Auditory and tactile stimulation should be used and must also be tested during QS to elicit a response. The EEG may show either a transient generalised flattening of background activity, a transient sharp and slow wave, or diffuse bursts of theta or delta activity. If stimulation is performed during AS or wakefulness, startles, grimaces or stretches are usually seen and movement artefact is recorded on the EEG.

Abnormal patterns

Unfortunately, there is no standardised method of categorising the abnormal EEG of the neonate, and most authors adopt their own method. Interpretation of the EEG in the sick neonate is based on the same principles as in the normal neonate. Basic concepts such as CA and state are as important as in normal neonates, and abnormal background activity, disturbances in state, abnormal transients and seizure pattern (epileptiform discharges) may be associated with a variety of underlying diseases. Thus, a thorough knowledge of all aspects of the EEG in healthy neonates is indispensable for the assessment of the abnormal EEG (De Weerd et al 1999). Several factors need to be taken into consideration when assessing the neonatal EEG, and these are described in the following sections.

Disturbance of continuity

Discontinuity is abnormal if it occurs for an excessive time for the GA, e.g. neonates with metabolic acidosis or ischaemic lesions may have a discontinuous EEG for more than 90% of the total recording time. Prolonged IBIs are also an indication of cerebral dysfunction. This may occur as a transient finding, when due to respiratory acidosis that is later corrected, or as a more prolonged finding, as may be seen in neonates with ischaemic or haemorrhagic lesions.

Any discontinuous tracing in wakefulness or AS at term is abnormal. In QS tracé alternant is seen, but this is not considered a discontinuous tracing because the periods of lower voltage are over 25 μV. Occasional short, suppressed periods of 2–6 s (< 20 μV) may occasionally be seen in QS of normal full term neonates (*Fig. 5.10*).

The prognostic value of the continuity of the background activity EEG in full term neonates has been well documented (Monod et al 1972, Watanabe et al 1980, Pezzani et al 1986, Selton and Andre 1997). Inactive background recordings, electrocerebral inactivity or burst-suppression pattern are generally associated with major sequelae or death (see p. 190). Discontinuous activity with IBIs of more than 30 s is more common in neonates with poor outcome (Menache et al 2002). In full term neonates, a predominant IBI duration of more than 30 s has been found to correlate with the occurrence of both unfavourable neurological outcome and subsequent epilepsy (Menache et al 2002). There is still uncertainty about the significance of moderately abnormal EEGs. There is the possibility that the effects of certain drugs (e.g. phenobarbital) may also influence the continuity (Bell et al 1993).

More cautious interpretation may be necessary in earlier prematurity. In neonates of 27–34 weeks, Tharp et al (1981) found markedly discontinuous activity to be highly correlated with adverse outcome and with diffuse brain lesions at autopsy. Ellison et al (1985) have found an IBI greater than 1 min to be predictive of severe sequelae in neonates of less than 30 weeks gestation. However, Engel (1975) found an IBI of this length to have a less serious prognosis in early prematurity than at term, a finding which has been confirmed by others (Hughes et al 1983, Lombroso 1985, Van Sweden et al 1991).

Disturbance of amplitude and frequency

The association of low amplitude traces during the first week of life with later neurological abnormality was recognised more than 50 years ago (Hughes et al 1948). Subsequent studies have uniformly confirmed the adverse prognostic significance of persistent, generalised, low amplitude, inactive or isoelectric traces (Monod et al 1972, Engel 1975, Tharp et al 1981, Dehkharghani 1984, Lombroso 1985, Hellström-Westas et al 1995).

Electrocerebral silence is present when there is absence of EEG activity above 2 μV. Some authors have described the 'inactive EEG' as unreactive EEG activity less than 5–10 μV despite long interelectrode distances

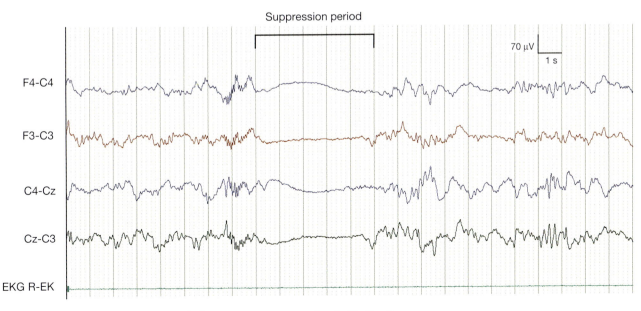

Fig. 5.10 Normal full term neonate in QS. Spontaneous suppressed period lasting 5 s.

and maximum amplification (Stockard-Pope et al 1992). The prognosis is generally very poor when either electro-cerebral silence or an inactive trace is present. However, Pezzani et al (1986) and Pressler et al (2001) have reported favourable outcomes in neonates with initial inactive EEGs that recovered within 24 h. Reversible depression of up to 10 min duration may also be found in response to the onset and correction of acute hypoxia (Roberton 1969). Low voltage plus theta tracing consists of theta frequencies of 5–30 μV, which are either continuous or present in discontinuous bursts against an inactive or very low voltage background of < 15 μV, with no particular location or lability (Monod et al 1972). This pattern is associated with poor prognosis.

Electrocerebral inactivity

Electrocerebral inactivity (previously termed 'electro-cerebral silence' or an 'isoelectric EEG') is characterised by the absence of EEG activity over 2 μV under maximal amplification and long interelectrode distance. To be indicative of cortical death the most rigorous technical standards are required. Electrocerebral inactivity has to be recorded on two consecutive examinations at an interval of at least 12 h (Engel 1975). However, consider-able caution should be exercised to ensure the correction of factors such as hypothermia, electrolyte imbalance, hypoglycaemia or drug overdose. Certainly, the neonatal EEG is not generally regarded as a reliable guide to brain death, a concept which is not in any case usually applied to the newborn (Moshé and Alvarez 1986, Vernon and Holzman 1986, Task Force on Brain Death in Children 1987, British Paediatric Association 1991, Working Group of the Royal College of Physicians 1995). Con-versely, Pasternak and Volpe (1979) reported a case of full recovery with normal outcome at 1 year in a 35-week-old neonate with prolonged clinical signs of brainstem failure but a virtually normal EEG.

Burst-suppression pattern

This pattern is widely regarded as being of adverse prog-nostic significance (Monod et al 1972, Dehkharghani 1984, Stockard-Pope et al 1992). Bursts of high-fre-quency activity or sharp and slow wave activity are interrupted by periods of silence (< 20 μV). This type of pattern is usually unreactive to stimuli of any sort and does not show any evidence of sleep cycling (*Fig. 5.11*). This unreactive type of burst-suppression pattern is commonly seen in neonatal encephalopathies and inborn errors of metabolism, which have a poor progno-sis. A reactive burst-suppression pattern (i.e. a burst-

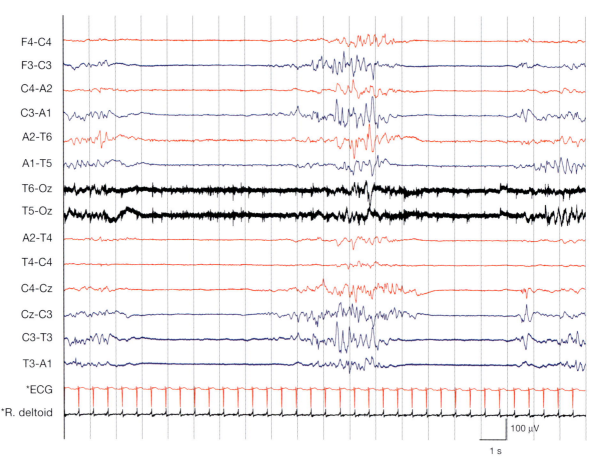

Fig. 5.11   Non-reactive burst-suppression pattern in a full term neonate on the NICU.

suppression pattern that can be interrupted by stimulation) and a modified burst-suppression pattern have been described and are associated with more favourable outcomes (Sinclair et al 1999, Douglass et al 2002). Burst suppression is more difficult to define in the very preterm neonate because the normal tracé discontinu pattern may appear not unlike burst suppression to the inexperienced eye. True burst suppression in the neonate lacks variability in pattern, is unreactive, cyclical patterns are absent and bursts may contain high amplitude multifocal epileptiform discharges. Thus, the term burst suppression should be used with caution when describing the neonatal EEG, particularly when applied to that of preterm neonates. Specifically, it should never be used to describe the normal discontinuous pattern in preterm neonates or to describe an immature or discontinuous pattern in term neonates. Tracé discontinu typically shows variability of IBIs and bursts, which are also of variable morphology; reactivity to stimuli can be elicited and, as has been recently established, cycling of EEG states can be seen from as early as 25 weeks (Curzi-Dascalova et al 1993, Scher et al 2005c).

Interhemispheric asymmetry and asynchrony

Asynchrony is excessive if more than half the delta bursts occur asynchronously (*Fig. 5.12*). This obviously applies only for full term neonates, because in preterm neonates (after a period of hypersynchrony up to around 28 weeks CA) only 50–60% of bursts are synchronous at 31–32 weeks gestation (Lombroso 1979). Nevertheless, persistent asynchrony is generally regarded as abnormal, although Monod et al (1972) found it to be a major adverse factor in term neonates only. In preterm neonates asynchrony of ongoing rhythms (i.e. bursts) for more than 75% of the time was usually associated with major sequelae or death (Tharp et al 1981). An abnormal asymmetry is taken as one with an amplitude ratio greater than 2:1 between the hemispheres. Some degree of asymmetry is regarded as normal in both term and preterm neonates. In term neonates an overall interhemispheric amplitude ratio is reported as abnormal if greater than 2:1, as equivocal if between 4:3 and 2:1, and as normal if less than 4:3 (Varner et al 1977). Similar criteria were found to apply in premature neonates between 30 and 33 weeks gestation (Peters et al 1981). Even at 27 weeks a mean of 85% of bursts were of symmetrical amplitude, rising to 91% at 31 weeks (Anderson et al 1985).

Consistent asymmetry is associated with adverse neurological outcome or death in a cohort of preterm neonates in whom it could be correlated with various diffuse lateralising brain pathologies on subsequent neuropathology (Tharp et al 1981). Lombroso (1985) has also confirmed the adverse prognostic significance of EEG amplitude asymmetry and its association with lateralised structural hemisphere lesions on the side of lower voltage. Caution must be used for those neonates who have had an instrumental delivery (vacuum and/or forceps), as

the scalp may have localised oedema which may lead to persistent asymmetries in the EEG.

Abnormal transients

Spikes and sharp waves may be frequent in the EEG of the newborn and their interpretation requires careful attention. Isolated sharp waves may be abnormal if they seem clearly more frequent than usual or persistently focal. Multifocal sharp waves are often seen in neonates who have experienced a previous central nervous system (CNS) insult (e.g. hypoxia). In preterm neonates the significance of spikes and sharp waves is less certain.

Positive Rolandic sharp waves

Positive Rolandic sharp waves (PRSWs) (*Fig. 5.13*) are characteristic electropositive sharp waves over the Rolandic (central) regions (C3–C4) of the brain that can occur as isolated sharp waves, or with greater frequency in trains (Cukier et al 1972, Blume and Dreyfus-Brisac 1982, Marret et al 1992, Vecchierini-Blineau et al 1996a, Hayakawa et al 1999, Okumura et al 1999, Biagioni et al 2000, Vermeulen et al 2003). The designation of a PRSW requires a single, but clear and definite, paroxysmal discharge, positive in polarity and with a steep initial slope and sharp peak, suddenly arising out of the background rhythm with a duration of 60–200 ms but usually 60–100 ms (Hughes et al 1991). PRSWs may be unilateral or bilateral over the central regions, with their maximum amplitude over the vertex.

Blume and Dreyfus-Brisac (1982) described two types of PRSW: type A, consisting of an isolated positive sharp wave, the amplitude and morphology of which clearly differ from the background; and type B, which are of smaller amplitude than type A and occur in short runs of 3–5 s.

Type A PRSWs are associated with white matter damage, such as periventricular leucomalacia (see p. 198), and poor prognosis. A poor outcome has been found in preterm neonates with more than two PRSWs per minute (Marrett et al 1992, Vermeulen et al 2003).

Periodic lateralised epileptiform discharges

Periodic lateralised epileptiform discharges (PLEDs) consist of discharges of variable morphology, which are repeated in an (almost) identical form at regular intervals (see p. 58). This pattern is rare in full term neonates but has been described in cases of neonatal stroke (Rando et al 2000). A periodic pattern has also been described in neonatal herpes simplex encephalitis (Mizrahi and Tharp 1982).

Disturbances of sleep state

Sleep–wake disturbances have been investigated in both the term and preterm neonate. Rudimentary sleep states have been reported in neonates from 24 weeks (Vecchierini et al 2003). Absence of any form of cyclicity in the neonatal EEG at any age is a poor prognostic sign (Scher et al 2003, Scher et al 2005c). In term newborns sleep states are well established and easily identifiable. Their absence is abnormal and sleep-state disturbances have been used to classify encephalopathy (Scher 1994).

Apparently normal cycles of continuous and discontinuous activity may still be pathological if inappropriate to behavioural state. Immature, regressive or poorly differentiated states may also in themselves be abnormal

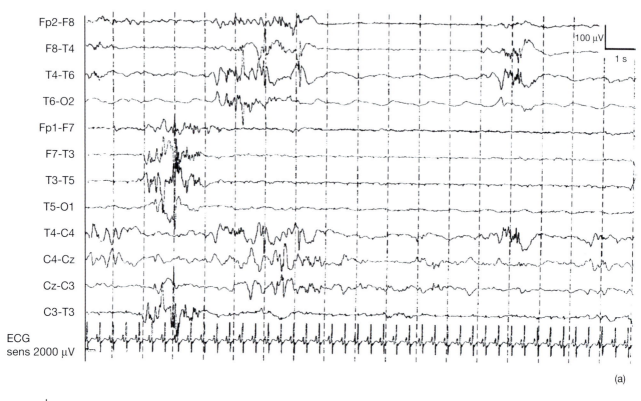

(a)

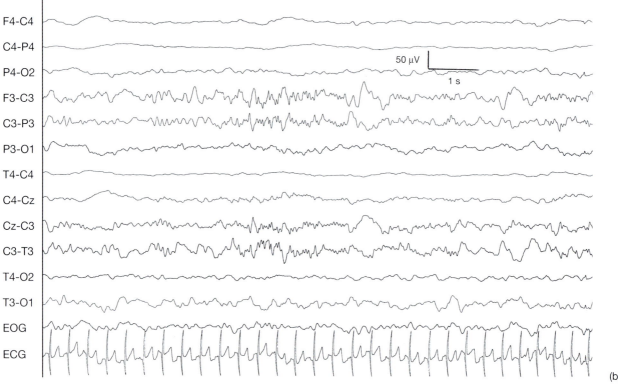

(b)

Fig. 5.12 (a) Gross interhemispheric asynchrony 2 weeks after severe birth depression. (b) Gross asymmetry between the right and left hemispheres in a 1-month-old baby with cortical dysplasia.

(Haas and Prechtl 1977, Shirataki and Prechtl 1977). The only large study to find burst suppression to be prognostically insignificant at term was that by Torres and Blaw (1968), but this work has been criticised because of its failure to take adequate account of sleep states (Watanabe et al 1980).

Electrographic seizure discharges

Definition of the electrographic neonatal seizure EEG seizure phenomena are often the most dramatic of all abnormalities and are identified by a repetitive discharge with gradual build up and termination. There is controversy over what constitutes a true electrographic neonatal seizure. Some workers have chosen an arbitrary duration of 10 s, while others, notably Shewmon (1990), have chosen 5 s. The minimum duration, according to the latest IFCN recommendations (De Weerd et al 1999), is 5 s if the background activity is normal and 10 s if it is abnormal.

The electrographic neonatal seizure has been described as an event in which abnormal EEG patterns and waveforms evolve in time and location (Clancy 1996). These abnormal waveforms are sudden, repetitive, evolving, stereotyped waveforms with a definite beginning, middle and end. Their length is variable, from 5–20 s to many minutes. The most frequent EEG features during seizures are:

- rhythmic sharp waves or spikes of variable frequency
- monomorphous delta or theta waves
- sequence of alpha or beta frequencies (less common)
- transient flattening (rare)
- PLEDs (rare).

Each electrographic seizure can be described in terms of location, duration, frequency, morphology and amplitude.

Location of the electrographic neonatal seizure The neonatal seizure is known to arise focally and is by nature a partial seizure. Generalised onset spike-and-wave seizure discharge is extremely rare (Clancy 1996). Multifocal seizures are common in severe encephalopathies and are more often associated with neurological sequelae than are unifocal seizure discharges. The exact site of origin of the neonatal seizure may vary within an individual, but the most common site of origin is the midtemporal region (T4 and T3) (Clancy 1996), hence the need to use these electrodes in any reduced array for neonates. Neonates can also display simultaneous independent focal electrographic seizures. Illustrations of this type of neo-

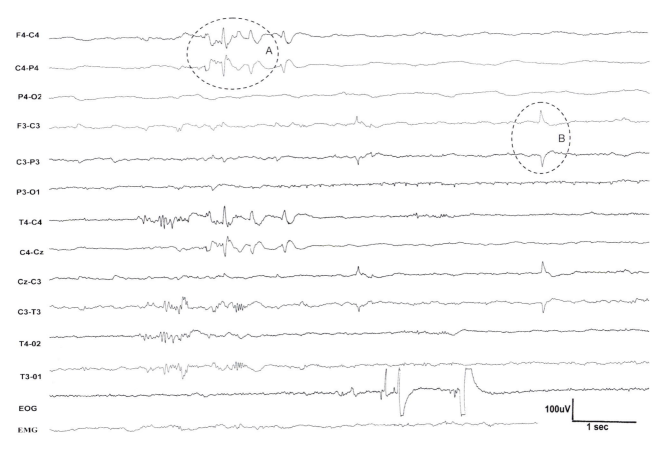

Fig 5.13 PRSWs in a 37-week-old monozygotic twin. He developed multicystic leucoencephalomalacia and is severely handicapped. Note: the background activity is discontinuous. (A) Surface positive sharp waves localised at C4. (B) Positive broad-based sharp wave with phase reversal in C3. PRSWs are recorded during an IBI.

natal seizure are given in *Fig. 5.14*. Neonatal seizures can also migrate, i.e. they leave their site of origin and move to a remote location, which may even be the contralateral hemisphere.

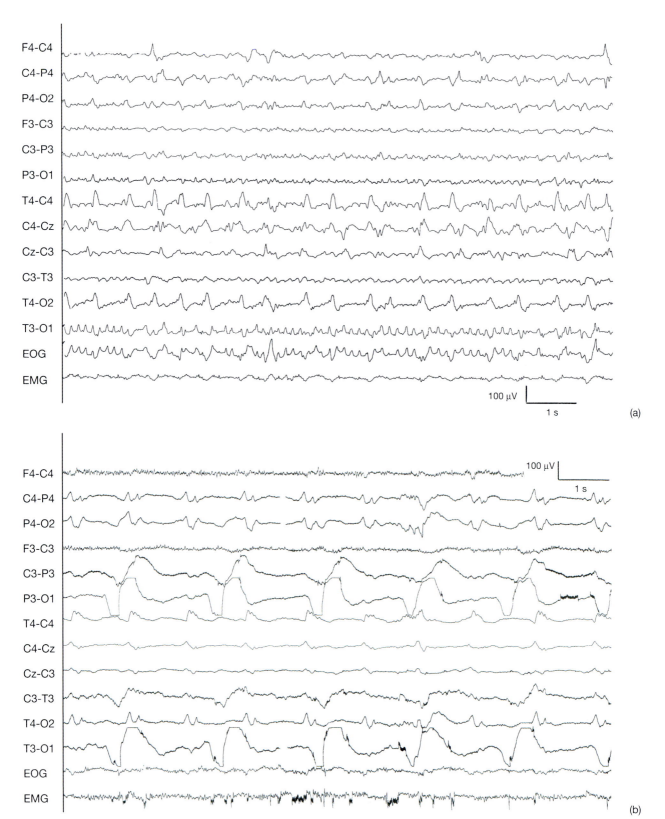

Fig. 5.14 Independent but simultaneous seizures in a neonate. (a) Full term baby with hypoxic–ischaemic encephalopathy. There is a fast 'alpha'-like seizure over the left posterior region and a slower right-hemisphere seizure of higher amplitude. (b) Preterm infant (30 weeks GA) with bilateral intraventricular haemorrhages. Seizure discharges over the right and left hemispheres are at different frequencies and have different morphologies.

Duration of the electrographic neonatal seizure The typical duration of the electrographic neonatal seizure is 2–3 min (Clancy and Legido 1987). Scher et al (1993a) found that the mean ictal duration was 14.2 min for full term neonates as compared with 3.1 min for preterm neonates. A more recent study by Patrizi et al (2003) reported a mean seizure duration of 1 min 37 s in preterm neonates and 2 min in full term neonates. The duration of neonatal seizures can vary and most likely depends on the severity of the underlying aetiology. Discharges shorter than 10 s have been studied, and are described as brief intermittent rhythmic discharges (BIRDs) rather than short duration seizures (*Fig. 5.15*). However, their significance is still unclear. One study has demonstrated an association between BIRDs and a clinical history of hypoxic–ischaemic encephalopathy and poor outcome (Oliveira et al 2000).

There is no generally accepted definition of status epilepticus in the neonate. In adults, status epilepticus is defined as a period in which there is an uninterrupted seizure lasting 30 min or more, or a series of recurrent seizures between which there is no mental recovery. It is difficult to assess the mental status of the neonate between seizures. A definition accepted by many is that used in the study by Scher et al (1993b) in which status epilepticus was considered present if 50% or more of the recording contained seizures. Electrographic status epilepticus may be more common than previously thought.

Morphology of the electrographic neonatal seizure The morphology of the electrographic neonatal seizure varies tremendously, as can be seen from *Figs 5.14* and *5.16*, which also demonstrate that morphology can vary within a single seizure discharge.

Amplitude of the electrographic neonatal seizure A characteristic of the neonatal seizure is that it generally evolves in amplitude. The electrographic neonatal seizure gradually builds up in amplitude, but it may suddenly end when it has reached a maximum or it may continue and gradually wane. As the amplitude of the electrographic seizure increases, the frequency decreases (see *Fig. 5.16*).

Evolution of EEG abnormalities An abnormal EEG may show a combination of one or more of the features described above, or patterns inappropriate for the CA of the neonate (e.g. the 'dysmature' EEG pattern reported by Hahn and Tharp (1990) in neonates with broncho-

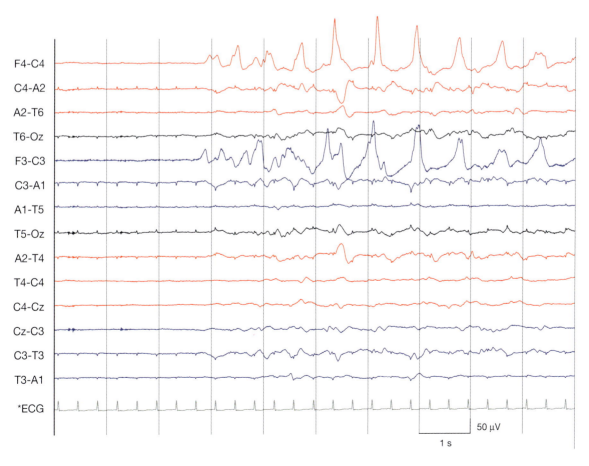

Fig. 5.15 BIRDs in a term infant with neonatal pneumonia (same child as in *Fig. 2.39*, but now 20 days old) on extracorporeal membrane oxygenation. The EEG shows runs of bifrontal rhythmic sharp waves lasting up to 7 s. No clinical correlate was seen. Note the discontinuous background activity.

pulmonary dysplasia). The timing of the EEG is important, as abnormality may be transient and an initial recording taken after normalisation may be misleading as to future outcome. The first recording should ideally be obtained within the first 24 h following birth asphyxia and, at most, within the first 72 h in premature neonates at risk of intraventricular haemorrhage. If the initial EEG is abnormal, a follow-up recording within the first week can be helpful to predict outcome. Periventricular leucomalacia may occur over a longer time period, and recommendations for the timing of EEG recordings have been proposed (Vermeulen et al 2003). This is a situation where continuous monitoring techniques have a clear advantage, since they may be employed over prolonged periods until the record has normalised (*Fig. 5.17*), or improvement has ceased. If only intermittent recordings are possible, it would be reasonable to perform these at least weekly, but this has to be decided on an individual basis with the neonatologist.

Effect of drugs on the neonatal EEG

There is little information available on the effects of drugs on the background EEG activity in the neonate. Phenobarbital is still the first-line treatment for neonatal seizures, despite many studies demonstrating that it suppresses electrographic seizures in not more than half of treated neonates (Connell et al 1989a, Painter et al 1999, Boylan et al 2004). The appearance or increase of discontinuity and, sometimes, a complete suppression of

the background EEG has been described (Staudt et al 1982, Radvanyi-Bouvet et al 1985). Others found a reduction in the amount of AS but no abnormal EEG patterns (Gabriel and Albani 1977). Ashwal and Schneider (1989) reported that phenobarbital levels > 25 mg/ml in sick preterm and term neonates suppressed EEG activity. However, this correlation was not confirmed in preterm neonates (Benda et al 1989). Bell et al (1993) compared the effects of phenobarbital and morphine on the EEG activity of preterm neonates in a retrospective study. The mean maximum IBI was significantly higher in neonates treated with phenobarbital or morphine in comparison to a control group. Lignocaine can produce background EEG suppression in some neonates (Hellström-Westas et al 1988). Midazolam (Bye et al 1997, Ter Horst et al 2004), morphine (Holmes and Lombroso 1993, Young and da Silva 2000) and pethidine (Eaton et al 1992) may all increase discontinuity. Pancuronium, used to paralyse ventilated neonates, has been found to produce no significant change (Staudt et al 1981).

Patterns of uncertain significance

Positive temporal sharp waves may occur isolated or in short bursts in preterm neonates of 31–33 weeks, and may still be seen at term. They are usually maximal at T3 and/or T4 (Chung and Clancy 1991, Scher et al 1994, Vecchierini-Blineau et al 1996b).

Type B PRSWs (see p. 191) are positive sharp waves similar to type A PRSWs, but are of smaller amplitude

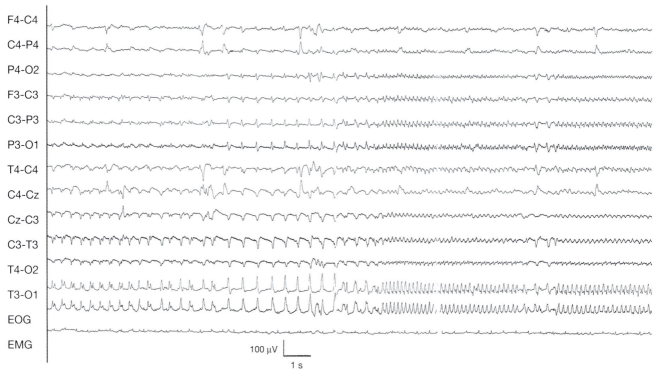

Fig. 5.16 Seizure discharge in a preterm neonate with bilateral intraventricular haemorrhages. Note the change in morphology, amplitude and frequency during the seizure discharge.

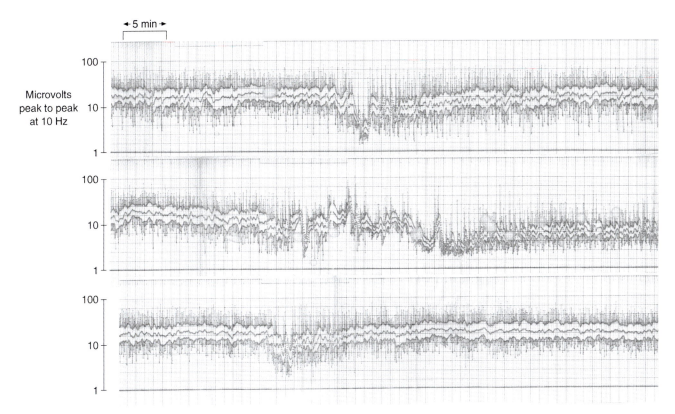

Fig. 5.17 Three examples of transient abnormalities, each lasting a few minutes, during a 24 h period of monitoring with a CFAM1 in an neonate born 4 months earlier after 28 weeks gestation (44 weeks CA at time of recording). Each sample is of 80 min duration; frequency analysis not shown. This very sick neonate, who died 2 days later, had several episodes of 'collapse', with respiratory arrest and bradycardia and possible seizures (none were observed during the period of monitoring). In the first episode, reintubation and ventilation were required to restore adequate oxygen saturation. The second and third episodes, while the neonate was still being ventilated, consisted of bradycardia, the third occurring at the time of handling the neonate's head in the course of changing its position during nursing care. Note the considerable variation in patterns of short term variability of the amplitude recording and, after each episode, the different rates and degrees of recovery to the previous ongoing pattern.

and occur in short runs of 3–5 s. This pattern has been described after 34 weeks CA (Blume and Dreyfus-Brisac 1982) and has been associated with a normal outcome (Vermeulen et al 2003).

Tracing with long alpha and/or theta bursts consists of long bursts of alpha or theta waves, lasting 3–10 s. This activity may be difficult to differentiate from seizure activity (Hrachovy and O'Donnell 1999).

Clinical applications of the neonatal EEG
Intraventricular haemorrhage

Germinal matrix/intraventricular haemorrhage (IVH) is the most frequent form of intracranial haemorrhage at less than 32 weeks gestation. At this GA bleeding arises from the fragile capillaries of the subependymal germinal matrix. Potential predisposing factors include disturbances of cerebral perfusion, which may be related to mechanical or chemical complications of the respiratory distress syndrome (Pape and Wigglesworth 1979, Levene and De Vries 1984). The outcome depends on the extent of the haemorrhage, and hydrocephalus, epilepsy and learning difficulties may develop.

The contributions of EEG to diagnosis and prognosis in IVH have both been specifically investigated. There do not seem to be any specific EEG features associated with IVH in preterm neonates. In 1972, Cukier et al reported PRSWs (see p. 191) in the EEGs of preterm neonates with IVH or periventricular haemorrhage. However, these were later found to be more specific for white matter damage in preterm neonates. EEG depression has been reported by several authors in neonates developing IVH, and the degree of background abnormality correlates well with the grade of haemorrhage (Clancy et al 1984, Greisen et al 1987, Benda et al 1989, Hellström-Westas et al 1991, Aso et al 1993). Other findings that have been described in neonates with IVH are electrical seizures and positive Rolandic spikes or PRSWs (Anderson et al 1985, Lombroso 1985, Benda et al 1989, Marrett et al 1997).

Germinal matrix/intraventricular haemorrhage is the most frequent cause of seizures in premature neonates (Kohelet et al 2004). Scher et al (1993b) found that 45% of preterm neonates with seizures had IVH. Seizures in preterm infants are generally associated with a poor outcome (d'Allest et al 1997).

There are reports of the prognostic value of the EEG background activity relative to imaging, mainly with reference to computerised tomography (CT) scanning. In a comparative study of 19 neonates Watanabe et al (1983) found that the grade of haemorrhage diagnosed by CT scan or at autopsy generally correlated well with EEG assessment of severity. Where discrepancies were found, the EEG proved the more reliable predictor. Neonates with a severely abnormal EEG had the worst outcome, even with milder grades of haemorrhage, whereas those with larger lesions had a better outcome if their EEG was well preserved. Similar conclusions were reached by Clancy et al (1984) in relation to CT scan diagnosis in 44 neonates. Neither of these studies found any particular EEG feature to be diagnostic, but important adverse prognostic features were severe background depression, discontinuity and seizures. In contrast, Lacey et al (1986), comparing standard EEG with haemorrhage diagnosed on CT scan, were unable to find any significant relationship, although there was a trend for neonates with severe ongoing abnormalities to have a poor outcome. Hellström-Westas et al (2001) used EEG to study neonates with grade III–IV IVH and found the maximum burst rate per hour during the first 24–48 postnatal hours useful for the prediction of outcome. In addition, the presence of sleep–wake cycling during the first weeks of life was also a good prognostic sign.

Systematic comparison of serial EEGs and ultrasound monitoring, in addition to confirming the prognostic value of the EEG, has permitted timing of the relative development of ultrasound and electrical changes. Both Greisen et al (1987) using the CFM, and Connell et al (1988), using a cassette monitoring system, have reported the emergence of EEG abnormality prior to IVH in some neonates, suggesting that the EEG is sensitive to the fundamental insult that precipitates the haemorrhage rather than solely to the effects of the bleeding itself.

Periventricular leucomalacia

Periventricular leucomalacia (PVL) is the term used to describe necrotic and/or gliotic lesions of perinatal origin occurring in the periventricular ring of the telencephalic white matter. PVL is found in 4–10% of neonates born before 33 weeks gestation. Cranial ultrasound imaging can detect cysts in the periventricular region during the first weeks of life. The cysts resolve, but are followed by a permanent reduction of myelin in the affected regions, which correspond to areas of PVL at autopsy or signal abnormalities on magnetic resonance imaging (MRI). The outcome is poor; most children develop cerebral palsy, typically spastic diplegia, epilepsy and poor vision (Rennie 1997).

There have been many studies of the EEG aspects of leucomalacia. PRSWs are early markers of white matter damage. Type A PRSWs (see p. 191) have been associated with white matter changes and subsequent poor outcome (Blume and Dreyfus-Brisac 1982). It has also

been shown that in preterm neonates a PRSW frequency of more than 1/min is associated with the development of PVL and cysts; most of these cases had poor motor development (Marret et al 1986, Baud et al 1998). Although some contradicting findings have been published (Clancy and Tharp 1984, Lombroso 1985) there is clearly enough evidence that PRSWs are a pathological finding in preterm neonates, particularly if they occur as isolated spikes and with a frequency of more than 1–2/min (Novotny et al 1987). The detection of PRSWs has been found to improve the sensitivity of ultrasound scanning for the diagnosis of white matter damage (Marret et al 1997). Thus, the EEG has value in the assessment of functional severity and prediction of neurological outcome. Marret et al (1986) found PRSWs to provide an early indication of PVL in four neonates studied within the first week of life. A similar suggestion has been made by Tharp (1987) regarding the significance of PRSWs in PVL. Continuous monitoring with a cassette recorder has confirmed the adverse prognostic significance of major ongoing abnormalities, with or without seizures, in PVL. It has also demonstrated the particular functional severity of subcortical leucomalacia, which has always, in the acute phase, been associated with maximal background amplitude depression. As with IVH, EEG abnormalities of prognostic value may precede those on ultrasonography (Connell et al 1987), suggesting that the EEG is more sensitive to the acute cerebral insult producing the lesion than to the structural effects thereof. Kubota et al (2002) found that the EEG was more sensitive than cranial ultrasound imaging for the detection of non-cystic PVL in neonates.

More recently, Okumura et al (2003) found the presence of frontal and/or occipital sharp transients in the EEG of preterm neonates to be a more specific marker than PRSWs for PVL (*Fig. 5.18*). Inder et al (2003) have also described a lower spectral edge frequency in the EEGs of preterms with white matter damage.

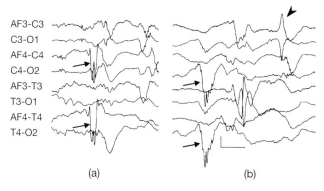

Fig. 5.18 Samples of frontal and occipital sharp waves in the EEG at day 6 (31 weeks CA) of infants with severe PVL. Calibration: 100 μV, 1 s. (a) Frontal sharp waves (arrows). (b) Occipital sharp waves (arrows). A PRSW is also observed (arrowhead). (From Okumura et al (2003), by permission.)

Neonatal cerebral stroke

Clonic seizures presenting in term neonates with normal APGAR scores are often due to focal lesions, most commonly middle cerebral artery infarction. Seizures are usually focal; the neonate remains alert between seizures and the background EEG may be normal or show only mild abnormalities (*Fig. 5.19*). EEG is also useful for prognosis and should be performed in the first 24 h, as some neonates with stroke will have seizures early in the postnatal period, usually between 12 and 72 h (Estan and Hope 1997).

Hypoxic–ischaemic encephalopathy

Hypoxic–ischaemic encephalopathy (HIE) is an encephalopathy seen in neonates usually as the culmination of a series of intrapartum processes. The primary disturbance to neural tissue is a deficit in oxygen supply. It is now clear from experimental studies that the time of occurrence of most cerebral damage is the period of reperfusion following the hypoxic–ischaemic event. The incidence of HIE has been estimated at between 2–4/1000 full term births (Volpe 2000b). The risk of death or severe neurological impairment following hypoxic–ischaemic injury is estimated as 0.5–1.0/1000 live births (Levene and Evans 2005). HIE is the commonest cause of early-onset seizures in full term neonates. The prognosis following HIE is dependent on the severity of the encephalopathy. Neonates with grade I HIE all have a normal outcome. The majority of neonates with HIE, however, have a grade II encephalopathy and outcome is more difficult to predict. The prognosis following grade III HIE is very poor, and cerebral palsy, severe learning disability and epilepsy are common in the event of survival.

The prognostic value of the EEG in HIE has been studied since 1972 (Monod et al 1972, Sarnat and Sarnat 1976, Watanabe et al 1980, Holmes et al 1982). A normal EEG was strongly associated with a normal outcome, while severely abnormal background activity (very discontinuous with IBIs of more than 10–20 s; burst suppression; very low voltage or inactive) was highly predictive of severe neurological sequelae (*Figs 5.20* and *5.21*). The few studies that have systematically correlated the prognostic value of the EEG with other methods of prediction have shown it to be at least as effective, or even more so, than clinical examination. The study by Sarnat and Sarnat (1976) provided a prognostically significant three-tier staging system for severity of HIE. Stage 1 was essentially one of clinical hyperactivity during

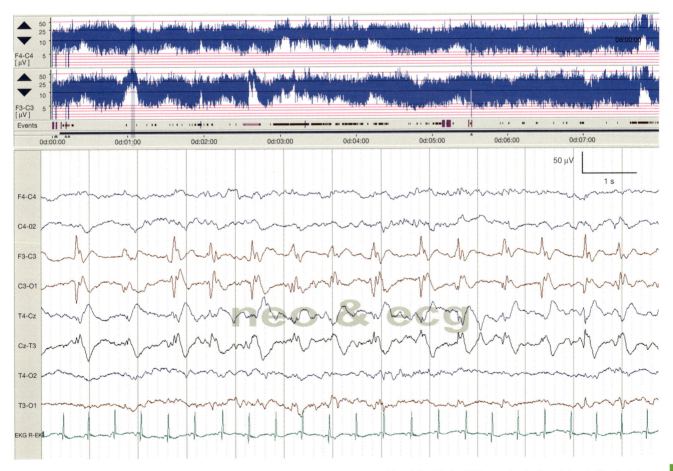

Fig. 5.19   Left-sided rhythmic spike-and-wave activity in a full term neonate with a left-sided middle cerebral artery infarct. Clonic jerks of the right hand were seen synchronous with this activity. Each vertical line represents 1 s.

which the EEG was normal. In stage 2 depressed consciousness and hypotonia were accompanied by seizures and a discontinuous EEG pattern. Stage 3 was characterised by severe clinical neurological depression, with an isoelectric, unreactive or extremely discontinuous EEG. Neonates who remained in stage 1 had a good prognosis and those in stage 3 a poor prognosis. For neonates in stage 2 the most important factor was the duration of

EEG abnormalities, the outcome being good if it persisted for less than 5 days and poor thereafter. The authors recommended EEG examinations on the second and sixth days, with further studies every 3–4 days up to 2 weeks in those neonates whose sixth day EEG was still abnormal. Abnormalities of the background activity are more predictive of outcome than the presence of seizures (Sarnat and Sarnat 1976, Watanabe et al 1980). Similar

(a)
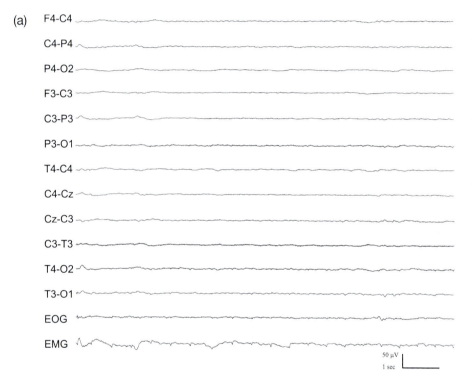

(b)
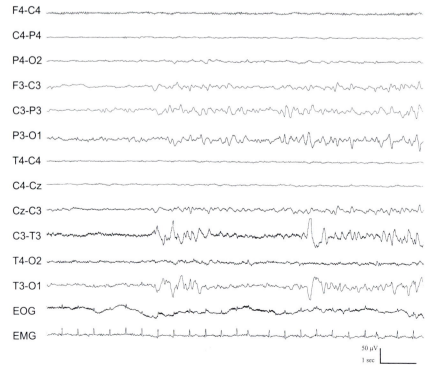

Fig. 5.20 Term neonate with severe HIE, born by caesarean section after 24 h of reduced fetal movements; umbilical cord pH 6.8. The outcome was poor, with spastic quadriplegic cerebral palsy and severe developmental delay. (a) The initial EEG recorded at 6 h of life shows background activity of < 5 µV (inactive). (b) The EEG on the second day of life shows major abnormalities; background activity is asymmetric and discontinuous, with IBIs up to 40 s. Positive sharp waves, as seen over T3, are of uncertain significance.

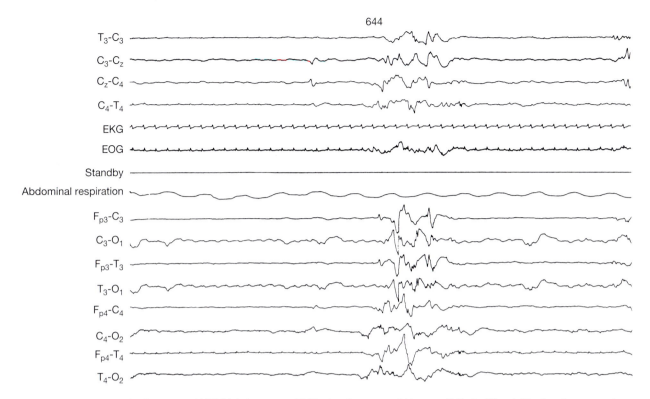

644

Fig. 5.21 Burst suppression in neonatal HIE (high frequency 70 Hz, low frequency 1 Hz, sensitivity 5 μV/mm). The burst-suppression pattern, defined as 'a background with bursts of high amplitude delta and theta activity with intermixed or superimposed sharp waves and spikes lasting 1–10 s and alternating with periods of marked background attenuation (voltage consistently below 5 μV)', was associated with a worse long term outcome than the 'modified burst-suppression pattern', which the authors defined as 'one in which the pattern is not continuous throughout the recording and/or the periods of background attenuation are at times incomplete (i.e. voltage is higher than 5 μV)', in this study of 23 neonates followed up for 9–44 months (mean 26 months). Neonates with 'toxic' levels of antiepileptic drugss had been excluded from the study. (From Sinclair et al (1999), by permission.)

conclusions have been reached using EEG monitors (Hellström-Westas et al 1995, Toet et al 1999).

More recent studies have concentrated on the early EEG (within 3 days of birth), and a strong correlation between early EEG and outcome has been demonstrated even for minor sequelae (Van Lieshout et al 1995; Pressler et al 2001). The characteristic time of seizure onset is within 24 h of birth; seizures often begin in the first 12 h, but are rare in the first 6 h unless there has been an antenatal insult (Filan et al 2005). Poor correlation between clinical and electrographic seizures in the neonate means that clinical seizure onset time cannot be used to 'time' the insult. The electrographic seizure onset time may be more useful, but it may also vary according to the severity and nature of the insult (Filan et al 2005) (*Fig. 5.22*). In severe insults the EEG can evolve over a period of time. Very suppressed activity is seen for a number of hours after birth (*Fig. 5.23(a)*). This is often then followed by a period of seizure activity that, depending on the severity of the insult, can be very difficult to control with anticonvulsants (*Fig. 5.23(b)*). When seizures resolve, a burst-suppression pattern is often seen for a number of days (*Fig. 5.23(c)*). This is very similar to the sequential pattern of activity reported by Gunn et al (1992) in the lamb model.

The relationship between clinical and EEG seizure manifestations may also be predictive of outcome: electroclinical dissociation has been associated with severe sequelae (Monod et al 1969). Clinical seizures with an inconstant relationship to EEG features were correlated with a higher incidence of diffuse encephalopathy and a worse short term outcome on discharge from the neonatal unit (Mizrahi and Kellaway 1987).

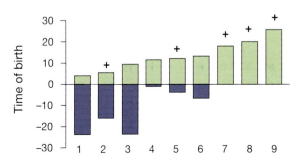

Fig. 5.22 Electrographic seizure onset time (in hours) in nine babies with HIE who had continuous EEG monitoring from soon after birth (Filan et al 2005).

201

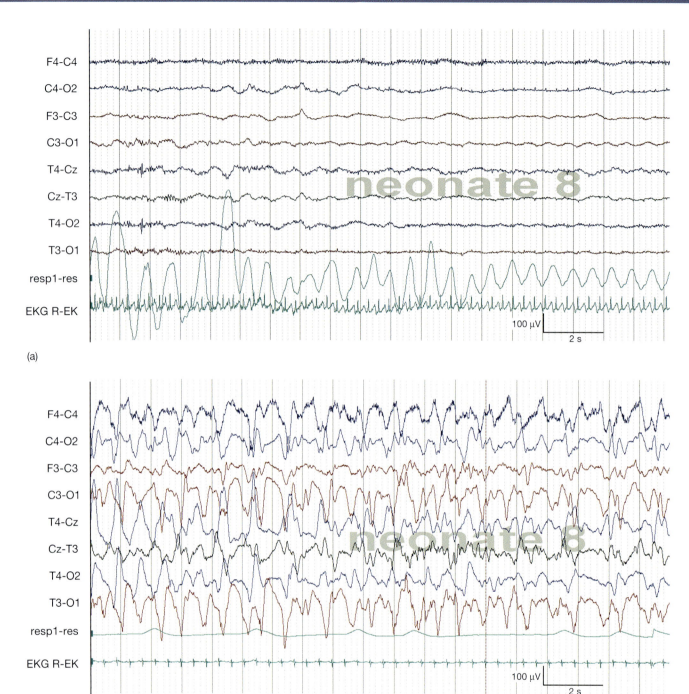

(a)

(b)

Fig. 5.23 Sequence of EEGs in a full term neonate born at home. Born flat, APGAR score 0 at 1 min, 0 at 5 min and 3 at 10 min of age. The child was resuscitated by the midwife, and given free flow oxygen and cardiac compressions; a heart rate of 50 bpm was detected at 10 min. The child arrived in the NICU at 2 h after birth: arterial pH 7.1, pink and breathing. He had lots of tremors, but his EEG was very suppressed at 4 h. (a) Seizures developed at 18 h which (b) progressed to 8 h of status epilepticus that was difficult to control with antiepileptic medication. A burst-suppression pattern developed when the seizures were controlled (c). This infant had no suck, required tube feeding and was in a severely abnormal neurological state. He died at the age of 8 months.

Another important factor used to assess prognosis from the EEG in HIE is the duration of abnormalities. Persistence of moderate abnormality for 2 weeks and of mild abnormality for 3 weeks was indicative of poor outcome (Watanabe et al 1980). However, EEG background activity that was initially abnormal can improve progressively with recovery of acute brain damage. The cause of this recovery is unknown and is seen even when permanent neuronal damage has occurred (Takeuchi and Watanabe 1989). If the initial EEG is performed as close as possible to the peak of symptomatic neurological impairment, an accurate prognostic statement can usually

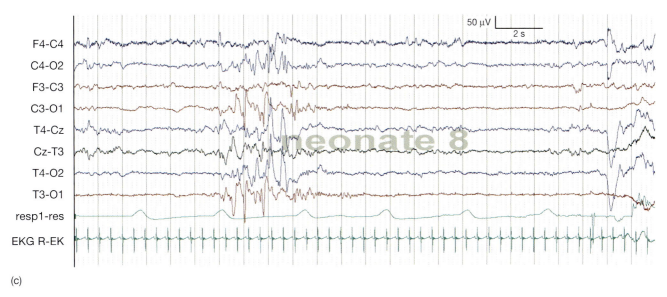

(c)

Fig. 5.23 *Continued.*

be made (Tharp 1987). A normal EEG recorded 2–3 weeks after a severe hypoxia could lead to a falsely optimistic prognosis.

Herpes simplex encephalitis

Neonatal herpes simplex encephalitis is a rare but devastating neurological condition that requires a high degree of clinical suspicion and rapid initiation of antiviral therapy (acyclovir). It commonly presents with seizures, lethargy and dysthermia (Toth et al 2003). Skin lesions may (unlike later cases) or may not be present. A periodic focal or multifocal background EEG pattern with periodic or quasi-periodic complexes has been described by several investigators (*Fig. 5.24*) (Mizrahi and Tharp 1982, Sainio et al 1983, Mikati et al 1990). The diagnosis should always

be considered in neonates that have had an uneventful delivery but who develop seizures and lethargy later in the neonatal period. This is one of the few cases in which a specific EEG pattern can be helpful in the diagnosis of disease in the neonatal period. However, a significant proportion of neonates with herpes simplex encephalitis will show non-specific abnormalities and focal or multifocal seizures, not necessarily located in the temporal regions.

Neonatal seizures

The brain of premature infants may have a more limited capacity to generate and propagate seizure activity than the more mature brain. The clinical features of neonatal seizures differ considerably from those seen in older children and adults, and seizures in preterm infants differ

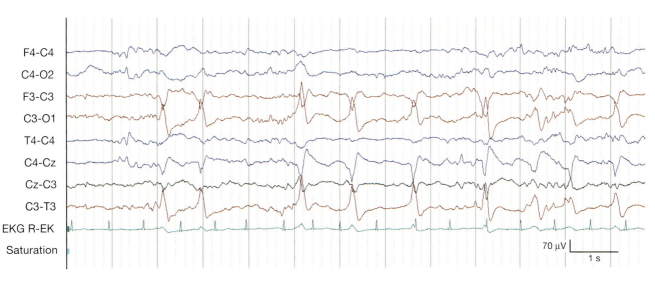

Fig. 5.24 Herpes simplex encephalitis. A 7-day-old neonate presented with lethargy, rash on his trunk and seizures. The EEG shows periodic sharp waves over the left hemisphere.

from those in term neonates. In the immature brain there is a relative excess of excitatory neurotransmitters and receptors, in part explained by the fact that γ-amino-butyric acid (GABA) is excitatory in neonates (for a review see Holmes and Ben-Ari 2001). The arborisation of axons and dendritic processes, as well as myelinisation, are incomplete in the neonatal brain, resulting in weakly propagated, fragmentary seizures the electrical activity of which may not spread to surface EEG electrodes. The development within the limbic system of connections to the midbrain and brainstem is more advanced than the cerebral cortical organisation, leading to a higher frequency of mouthing, eye deviation and apnoea in neonates than in seizures of adults. The importance of subcortical centres has been suggested by reports of both EEG and clinical seizures in neonates with complete absence of the cerebral hemispheres. These reports suggest that lower centres may be involved not only in propagation but also in the actual genesis of at least some types of seizure activity in the newborn (Ferguson et al 1974, Pampiglione 1974, Reding et al 1979, Danner et al 1985).

There have been very few studies of seizure treatment for neonates under EEG control. Most studies have continued treatment until clinical seizures were abolished. We now know that clinical estimation of seizure burden in neonates is inaccurate. It is still unclear if treatment of neonatal seizures to electrographic quiescence improves outcome. Painter et al (1999) were the first group to use EEG to establish seizure control in neonates given phenobarbitone and phenytoin. The response was poor and seizures were not controlled in a third of all neonates studied. Boylan et al (2004) used video–EEG to measure the response to commonly used second-line antiepileptic drugs in neonates. Again the response to second-line treatment was poor, and in some cases the seizures seemed to simply 'burn out' over a number of days.

Recognition of seizures is important in order to institute appropriate investigations, counsel parents about prognosis and monitor treatment. It is often far from certain whether some episodic phenomena of the neonatal period are epileptic or not. Extensor spasms as part of a seizure may be difficult to distinguish from the decerebrate posturing seen during acute neurological illness. Abnormal movements, whether seizures or not, may be recurrent but infrequent. Video–EEG monitoring is of considerable value in determining whether electrographic seizures have any clinical concomitants, and vice versa. In neonates paralysed to aid ventilation there will be no clinical evidence of seizures. In neonates treated with multiple anticonvulsants, seizures may continue without any clinical signs (*Fig. 5.25*), and continuous EEG monitoring is therefore essential to document the effects of treatment. It is important to consider a differential diagnosis in neonates.

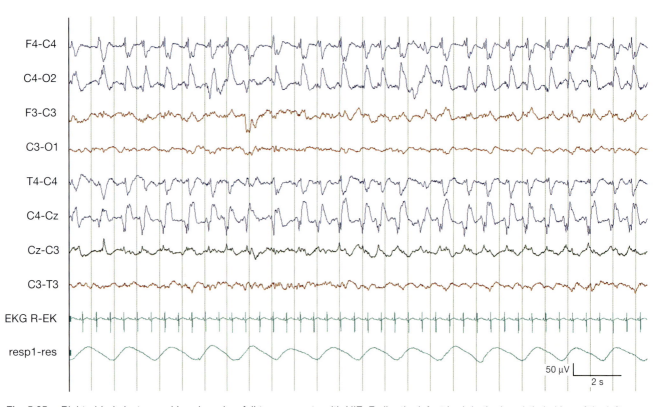

Fig. 5.25 Right-sided electrographic seizure in a full term neonate with HIE. Earlier the infant had rhythmic subtle jerking of the left arm and was treated with phenobarbitone. This recording shows electrographic seizure activity after administration of the anticonvulsant. No clinical signs were visible on simultaneous video recording.

Classification and semiology Seizures in the newborn period pose serious diagnostic and therapeutic problems. The overall incidence of neonatal seizures is relatively low at 1.5–3.0/1000 live births, although it has been reported as being as high as 15% in neonates of less than 2,500 g birthweight (Volpe 2000c). These figures are probably an underestimate, as they only include clinical and electroclinical seizures. The exact incidence of electrographic, clinically silent seizures is not known. The vast majority of neonatal seizures start on the first day, and 70% of all cases have been diagnosed by the fourth day. Aetiological factors have changed with patterns of neonatal care, particularly in the management of premature neonates. Only a few conditions account for most seizures (*Table 5.8*). Very few seizures are idiopathic. At term, HIE remains the most common underlying factor. In preterm neonates, intracranial haemorrhage is the most common cause. Meningitis, focal cerebral infarction, metabolic disorders and congenital abnormalities of the brain can cause seizures at any point in gestation (for a review see Tharp 2002).

Volpe's (2000c) classification of neonatal seizures has become widely adopted and has the merit of simplicity. It combines clinical signs and whether or not the seizure is associated with EEG seizure activity. For the benefit of the present text the classification has been adapted to give more neurophysiological information (*Table 5.9*). There is some overlap with the more complex classification of Mizrahi and Kellaway (1998). Neonatal seizures can be

Table 5.8 Causes of neonatal seizures*

Cause	Frequency[†] (%)
Hypoxic–ischaemic encephalopathy	30–53
Intracranial haemorrhage	7–17
Cerebral infarction	6–17
Cerebral malformations	3–17
Meningitis/septicaemia	2–14
Metabolic:	
hypoglycaemia	0.1–5
hypocalcaemia, hypomagnesaemia	4–22
hypo-/hypernatraemia	–
inborn errors of metabolism	3–4
pyridoxine dependency	–
kernicterus	1
Maternal drug withdrawal	4
Idiopathic	2
Benign idiopathic neonatal seizures	1
Neonatal epileptic syndromes	–
Congenital infections	–
Unintentional injection of local anaesthetic during labour	–

*Data from Bergman et al (1982), Levene and Trounce (1986), Andre et al (1988), Estan and Hope (1997) and Mizrahi and Kellaway (1998).
[†]Frequencies not listed have not been reported but are rare.

Table 5.9 Classification of seizures in the newborn*

Type	Clinical signs	EEG
Subtle	Ocular, oral, autonomic, limb posturing and movements	Common; background activity often abnormal; usually accompanied by ictal EEG change
Clonic	Repetitive rhythmic jerking, distinct from jittering, can still be felt in the restrained limb	Usually accompanied by ictal EEG changes
Focal	Well-localised, involvement of face, limb or axial structures	Usually has time-synchronised ictal EEG changes
Multifocal	Jerking of several body parts simultaneously or in sequence	Usually has time-synchronised ictal EEG changes
Tonic	Sustained posturing, stiffening	Variable
Focal	Sustained posturing of a limb or asymmetric posturing of trunk and/or neck. Rare	Ictal changes more common than in generalised tonic seizure, background activity often abnormal
Generalised	Stiffening, 'decerebrate' or 'decorticated' posturing	Usually no consistent EEG correlate, background activity often severely abnormal
Myoclonic	Rapid isolated (single or multiple) jerks, distinguish from benign sleep myoclonus	Variable, background activity may be abnormal
Focal, multifocal	Well-localised, migrating jerks of limbs	Usually not accompanied by EEG changes
Generalised	Bilateral synchronous jerks of limbs (upper > lower)	Suppression burst or hypsarrhythmia may be seen

*Adapted from Volpe (2000c).

subtle (50%), clonic (25–30%), myoclonic (15–20%) or tonic (5%) (Volpe 2000c).

Subtle seizures This is the most common seizure type in both preterm and term neonates. It is more common in preterm than in term neonates. Subtle seizures can be difficult to recognise clinically. Manifestations include:

- ocular phenomena (staring, blinking, eye deviation, sustained eye opening)
- oral–buccal–lingual phenomena (lip smacking, mouthing, chewing, sucking, smiling)
- autonomic phenomena (change in blood pressure and/or heart rate, pallor, increased salivation or secretions)
- central apnoea
- fragmentary body movements (limb posturing, boxing, cycling, stepping, swimming, pedalling).

Video–EEG has shown that the majority of neonates with subtle seizures will, at least during some seizures, exhibit rhythmic epileptiform activity. EEG changes are more commonly associated with ocular manifestations. The absence of ictal EEG discharges, however, does not necessarily rule out seizures, if no typical episodes have been recorded.

Most apnoeas in preterm neonates are non-epileptic in origin. However, in term neonates apnoea as a seizure manifestation may be more common than previously thought (Watanabe et al 1982). Usually other subtle phenomena will occur at some point during the apnoeic seizure. However, apnoeic seizures have been specifically associated with paroxysmal alpha bursts on the EEG, and have been found to be more likely to occur during AS than QS (Watanabe et al 1982).

Clonic seizures Clonic movements are rhythmic, rather slow (1–3 Hz) and can be focal, multifocal migrating from limb to limb or, rarely, hemiconvulsive; Jacksonian march is exceptional in neonates. This seizure type is most consistently associated with synchronised EEG changes (*Fig. 5.26*). It is the most common type of seizure seen in neonatal stroke. It should be possible to differentiate from non-epileptic movements such as jitteriness, tremors, shudders or spontaneous clonus. These non-epileptic movements can be suppressed by gentle restraint or passive movement of the limb, may be enhanced by sensory stimuli and are obviously not accompanied by EEG seizure activity. Clonic seizures, by contrast, are usually asymmetrical and are frequently accompanied by synchronous eye movements.

Myoclonic seizures Erratic, fragmentary or more generalised myoclonic jerks are more rapid than the movements of clonic seizures and show a predilection for flexor muscle groups. They may be associated with tonic spasms, with multifocal clonic or tonic patterns, or with mixed seizure types. Ictal EEG changes are usually not seen in myoclonic seizures as a group, but are more common in generalised myoclonias. This seizure type can easily be distinguished from benign neonatal sleep myoclonus.

Benign neonatal sleep myoclonus is still often mistaken for true seizures in the neonate and may even

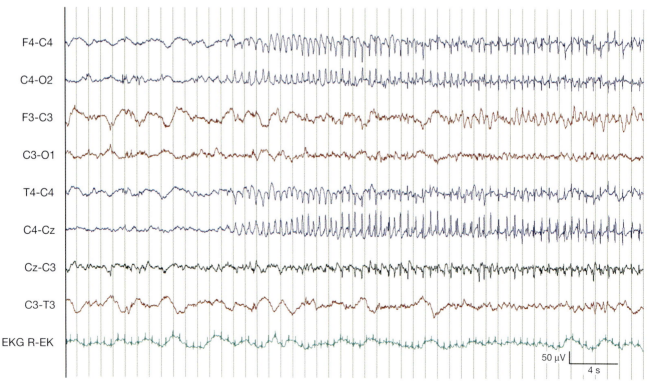

Fig. 5.26 Sudden onset of seizure discharge over the right hemisphere in a full term neonate with HIE. This was accompanied by subtle clonic right arm and leg jerks.

mimic status epilepticus (Daoust-Roy and Seshia 1992, Ramelli et al 2005). Benign neonatal sleep myoclonus is characterised by the occurrence of repetitive myoclonic movements (which may or may not be regular) of the limbs exclusively during non-REM sleep in the early postnatal period. The onset is within the first 15 days of life and resolves within 6 months. Long term video–EEG monitoring has shown that the EEG is always normal during these events (Di Capua et al 1993).

Tonic seizures *Focal tonic seizures* manifest as stereotyped, abrupt or slower tonic posturing of limb and/or trunk or eyes, often accompanied by apnoea, flushing or mild cyanosis. The EEG background is often abnormal and ictal discharges are relatively common compared with generalised tonic seizures.

Generalised tonic seizures manifest as symmetric tonic postures that mimic decerebrate or decorticate posturing. Those that can be triggered by stimulation and show no ictal EEG correlates may not represent epileptic seizures, but a brainstem release phenomenon (Mizrahi and Kellaway 1998). EEG background activity is almost always severely abnormal. This seizure type is usually associated with a poor prognosis. If seizure activity is seen in the EEG, clinical phenomena are usually associated with autonomic symptoms.

The EEG in neonatal seizures Ictal EEG patterns are described on pages 193–196. Harris and Tizard (1960) demonstrated the association of clinical seizures with particular EEG discharges such as repeated stereotyped sharp wave or spike complexes. The diagnostic correlation of such EEG and clinical abnormalities has been confirmed in several other large-scale studies (Monod et al 1969, Rose and Lombroso 1970).

Continuous monitoring using standard EEG equipment can provide a much higher rate of seizure detection than is possible by intermittent recording, and is particularly applicable to paralysed neonates (Coen et al 1982, Goldberg et al 1982). However, a discrepancy between the clinical and EEG manifestations of seizures had already been noted by intermittent standard recording (Harris and Tizard 1960, Monod et al 1969, Rowe et al 1985): not all electrographic seizures are accompanied by clinically obvious seizure manifestations.

Conversely, some clinical seizures may not consistently be accompanied by paroxysmal EEG changes. This characteristic feature of neonatal seizures is the phenomenon of *electroclinical dissociation* (*Fig. 5.27*); seizures can be electroclinical, electrographic (sometimes called subclinical) or clinical only (Mizrahi and Kellaway 1987). There is an ongoing controversy as to whether 'clinical-only' seizures (i.e. those that have clinical expression but for which no EEG concomitant is found) do or do not have an epileptogenic pathophysiology (Mizrahi and Kellaway 1998) and whether or not electrical-only seizures have an impact on long term outcome and thus do or do not require treatment. Mizrahi and Kellaway (1987) described a group of neonates with clinical-only

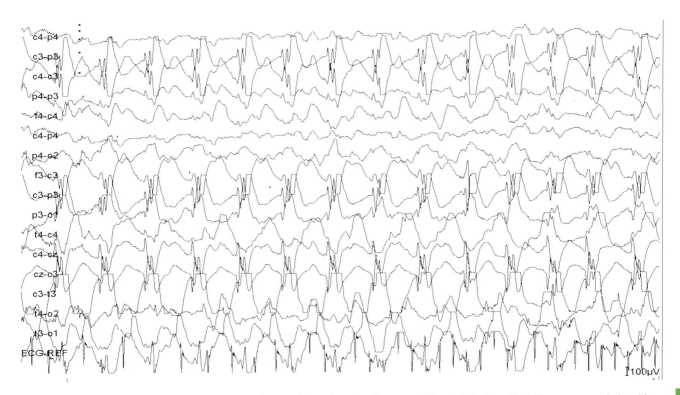

Fig. 5.27   Electroclinical dissociation in a neonate born at 38 weeks GA with severe HIE. No clinical manifestation was seen during this generalised seizure lasting 55 s.

seizures and severely abnormal background. In these neo-nates, the tonic posturing and motor automatisms could be induced by sensory stimulation, and were inhibited by gently restraining the limb. These movements may indeed not be epileptic in nature. Continuous simultan-eous video and EEG recording has been particularly important in allowing further detailed comparison of clinical and electrical phenomena. Shewmon (1983) has reported the frequent dissociation of discharges in the EEG and clinical ictal tonic movements and suggested, as did Sarnat (1975), that they may actually be of brainstem origin. There is evidence that electrographic seizures have the same impact on the cerebral perfusion as electroclini-cal seizures (Boylan et al 1999a) and are associated with poor neurodevelopmental outcome (McBride et al 2000).

If events are epileptic in neonates with severe back-ground abnormality, prolonged EEG monitoring with appropriate derivations will eventually reveal electroclin-ical seizures in most cases. There may be a further group of neonates with clonic seizures in whom the background EEG remains normal, who have normal neuroimaging and who appear entirely well between attacks. Their prognosis appears to be excellent, even without anticon-vulsant treatment (Boylan et al 1999b).

The interictal EEG   The interictal EEG can vary in a neonate with seizures from normal to extremely abnormal. The interictal background reflects the severity of the underlying pathology. A study by Laroia et al (1998) showed that an early EEG could predict which high-risk babies would go on to develop seizures in the newborn period. A normal or immature EEG back-ground strongly predicted the absence of electrographic seizures in the subsequent 18–24 h. A suppressed EEG background was strongly associated with concomitant seizures or the development of seizures within 18–24 h. This finding, although sensitive for seizures, was not specific, as neonates can have a suppressed background EEG without seizures. Many studies have found that the interictal EEG rather than the seizure characteristics themselves are more predictive of outcome. Holmes et al (1998) have used the neonatal rat brain to show that repetitive short-duration seizures are associated with long term neurological sequelae.

Response to anticonvulsants   Anticonvulsant response has usually been assessed on the basis of cessation of clinical seizures rather than on EEG data. However, several studies have reported a poorer response than was apparent on clinical grounds alone, with persistence of EEG seizure activity after clinical manifestations had ceased. This was reported following barbiturates and diazepam by Monod et al (1969), who felt that con-tinuing neonatal seizures were unlikely to respond to treatment once reversible underlying causes had been rectified. Continuous monitoring has been particularly useful in demonstrating poor anticonvulsant response,

whether with cassette recording (Connell et al 1989a), CFM monitoring (Hellström-Westas et al 1988, 2003), conventional EEG recording (Painter et al 1999) or simultaneous multichannel video and EEG recording (Mizrahi and Kellaway 1987, Bye and Flanagan 1995, Boylan et al 2002). Objective documentation of seizures is inherently more reliable than clinical observation in the neonate, as long as the attacks are sufficiently frequent for monitoring to be practical. However, the clinical sig-nificance in terms of determining clinical management and outcome remains uncertain.

Drug withdrawal in neonates   Abuse of narcotics in mothers may lead to withdrawal symptoms in the new-born. Clinical manifestations are seen in about 50–75% of neonates born to heroin- or methadone-addicted mothers, and include jitteriness, irritability, hyperactivity, high-pitched cry and myoclonic or clonic seizures. Seizures are more common in methadone addiction (10–20%). Cocaine and phencyclidine abuse does not usually produce drug-withdrawal symptoms in newborn infants.

Seizures usually present in the first week of life but may take several weeks to develop because of the prolonged excretion of the drugs in neonates. While the neurologi-cal examination of neonates with withdrawal symptoms is usually normal, altered sleep cycles with a decrease in AS and an increase in QS may be seen (Manfredi et al 1983, Pinto et al 1988). During seizures the EEG shows ictal discharges as described on pages 193–196.

Epileptic syndromes in the neonatal period

In contrast to seizures in later infancy and childhood, most neonatal seizures are not part of an epileptic syn-drome, but an non-specific reaction of the brain to an acute insult such as HIE, haemorrhage or metabolic disturbances. However, several circumscribed syndromes have to be considered when assessing neonates with seizures, even if they account for only a small percentage of patients (see *Table 6.1*). The most recent classification of epileptic syndromes of the International League Against Epilepsy (ILAE) (Commission on Classification and Terminology 1989; not as yet updated) describes the following neonatal syndromes.

Benign idiopathic neonatal seizures   Also known as 'fifth day' fits, these seizures of clonic (never tonic) and apnoeic type occur in full term neonates in good condit-ion and with a normal birth history. The seizures are usually partial but of changing distribution (*Fig. 5.28*). They may be so frequent as to represent status epi-lepticus, but respond readily to medication. The inter-ictal EEG may be normal, show generalised or multifocal discharges, or show the characteristic bursts of central sharp waves (thêta pointu alternant) in 60% (Dehan et al 1977). Ictal discharges consist of rhythmic spikes or slow waves without postictal suppression. Treatment may not be necessary, but the diagnosis is by exclusion. Seizures resolve often within days. The outcome is good, but an increased risk of minor neurological impairment has been reported (Dehan et al 1977). Apart from an initial series of reports, mainly from Australia, this type of neonatal seizure is now rarely seen (Pryor et al 1981).

Benign familial neonatal seizures This is a rare disorder with autosomal dominant inheritance (mutations in the voltage-gated potassium channel genes, in most cases *20q13.3*, but in a few families *8q24*) (see the review by Honovar and Meldrum 1997). This disorder presents on the second or third day in otherwise healthy neonates, with clonic or apnoeic seizures, and tends to persist longer than benign idiopathic neonatal seizures. The EEG may be normal or abnormal; no specific pattern is described. No causative factors are found on history or investigation. Psychomotor development is normal, but some may develop secondary epilepsy at some point in childhood or early adulthood, which is generally mild.

Early myoclonic encephalopathy This presents with myoclonic seizures in the neonatal period or early infancy (Aicardi and Goutières 1978). It is often associated with inborn errors of metabolism, but cerebral malformations have also been reported. Ictal manifestations include:
- partial or fragmented myoclonus
- massive myoclonias
- partial motor seizures
- tonic spasms.

Myoclonias may shift from one part of the body to another and usually persist in sleep. Normal background activity is absent. The interictal EEG shows a burst-suppression pattern with complex bursts of spikes and irregular, arrhythmic sharp waves and slow waves lasting for 1–5 s alternating with flat periods of 3–10 s in both wake and sleep. The bursts can be bilateral or independent in the two hemispheres. The fragmented myoclonias usually have no EEG correlate, whereas massive myoclonias may be synchronous with the bursts. The EEG eventually evolves towards atypical hypsarrhythmia. All neonates are severely neurologically abnormal and half of them die before the age of 1 year.

Early infantile epileptic encephalopathy with burst-suppression pattern (Ohtahara syndrome) Onset is in the neonatal period (in some cases up to 3 months), with frequent tonic spasms (100–300/day), often in clusters. Partial motor seizures may also occur. This syndrome is usually associated with cerebral malformations (e.g. Aicardi syndrome, hemimegalencephaly or other structural abnormality of the brain). The EEG is characterised by a true burst-suppression pattern (Ohtahara 1978), both in sleep and waking. Bursts consist of irregular, high amplitude (150–350 μV) slow waves mixed with spikes

and of 1–3 s duration. Suppression, which usually lasts for 3–4 s, may be asymmetric, asynchronous or even unilateral (*Fig. 5.29*). During seizures, desynchronisation is seen. The prognosis is serious, with death (in about 50%) or severe developmental delay. Evolution into infantile spasms or other severe epilepsies is seen. This age dependent epileptic syndrome should be differentiated from cases of severe HIE in which a similar EEG pattern may be seen. These neonates are usually comatose and have myoclonic, rather than tonic, seizures.

Inborn errors of metabolism
Inborn errors of metabolism are described in detail in Chapter 6 (see pp. 300 ff.) and include EEG findings in the neonatal period.

The EEG and clinical prognosis
A normal background EEG correlates well with normal outcome (Monod et al 1972, Holmes and Lombroso 1993). Severe abnormalities, such as inactive, low amplitude or extremely discontinuous traces, are strongly associated with major adverse sequelae. Several other studies have supported the general consensus as to the prognostic value of the ongoing EEG in both term and preterm neonates (Rowe et al 1985, Connell et al 1989b, Ortibus et al 1996).

Concluding remarks
In conclusion, the EEG is recognised as a safe, non-invasive guide to the assessment of cerebral function in the newborn. It acts as an index of cerebral functional maturation, provides valuable information regarding the timing, severity and prognostic significance of cerebral lesions, and is of particular value in the diagnosis of seizures and monitoring their response to therapy. Recent advances in recording techniques now allow monitoring for prolonged periods, with minimal interference, even in very sick neonates, and provide an exciting opportunity for involvement of the clinical neurophysiologist in a rapidly expanding field of medical practice.

NEONATAL EVOKED POTENTIALS

Sensory evoked potentials (EPs) have applications in newborn infants that differ considerably from those of children and adults. They are commonly used in screening for, or specific assessment of, possible sensory defects. Evaluation of maturity in the newborn infant is another useful application. EPs are also becoming increasingly important in the investigation of various pathologies of the brain and spinal cord encountered in the neonatal period.

Techniques of and normative data for electroretinograms (ERGs), visual evoked potentials (VEPs), brainstem auditory evoked potentials (BAEPs) and somatosensory evoked potentials (SEPs) have already been

(a)

(b)

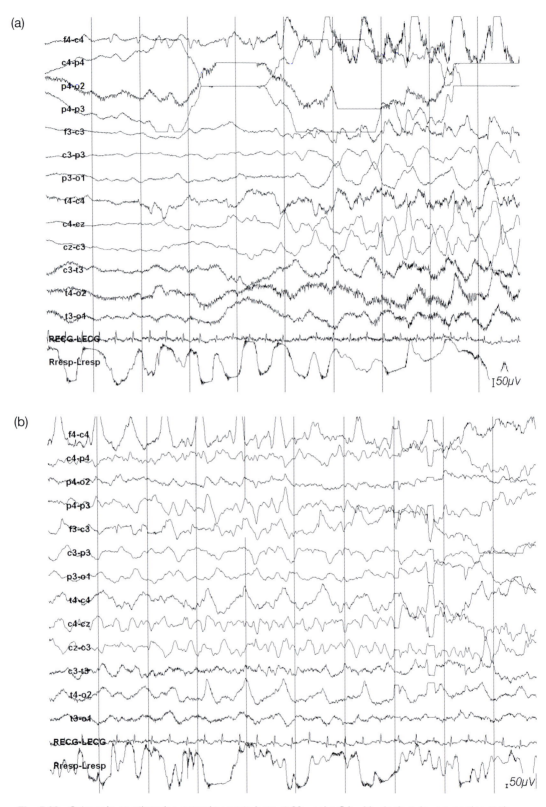

Fig. 5.28   Seizure in an otherwise normal neonate born at 39 weeks GA with clonic seizures starting at the fifth day of life. Rhythmic activity starts over the right frontal region and then spreads to the right centroparietal and left central regions. There are changes in morphology, frequency and amplitude as the seizure progresses. Note the change in sensitivity during the evolution of the seizure.

described in Chapter 3. The maturational features in preterm and term newborn infants have been published

elsewhere and several reports include illustrative normative data (Hrbek and Mares 1964, Starr et al 1977,

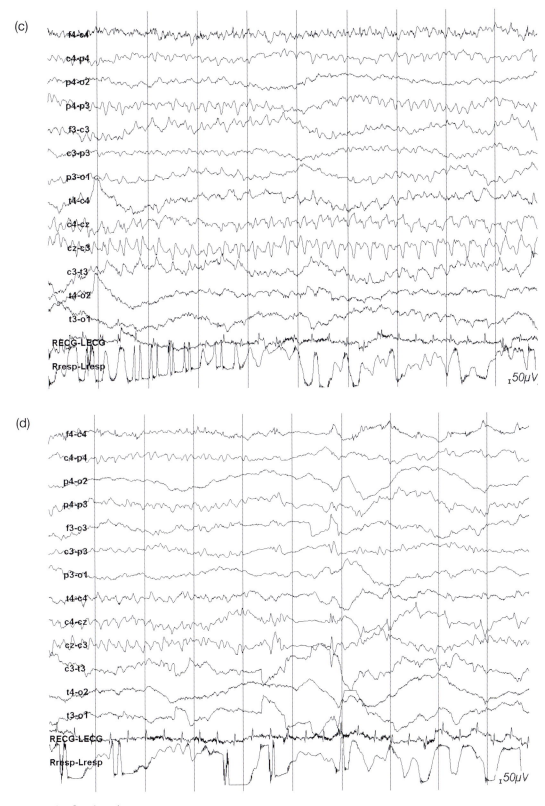

Fig. 5.28 *Continued.*

Klimach and Cooke 1988, *Journal of Clinical Neurophysiology* (special issue on EPs in neonates and children) 1992). In this section we review the clinical applications of EP recording as currently used in neonatal medicine.

Some of the difficulties of electrode application in small newborn infants have already been addressed in Chapter 3 and in the section on the neonatal EEG (see pp. 170–173). Here it is important to consider the practical problems of delivery of the stimulus, if necessary in the incubator, and the special need to monitor the state of the neonate. This is done firstly to ensure that stimulation is not unduly disturbing, and secondly because changes in behavioural state occur rapidly in the small neonate. The awake and different sleeping states may not appear markedly different to the inexperienced, yet may influence cortical EPs as well as the amplitude of the neonatal ERG which is larger in AS compared with QS, with a clearer effect at 36–40 weeks CA than at 56 weeks (Peña et al 1999). The technologist needs expertise in distinguishing clinically between the awake, AS and QS states. Whenever possible the best behavioural state for the specific EPs to be recorded should be chosen: awake for flash and pattern VEPs; AS for SEPs; asleep for BAEPs when the threshold is to be tested. The neonate should be observed continuously, preferably using the guidelines for scoring sleep and wakefulness in newborn infants detailed in *Table 5.3*.

Stimulus and recording parameters have to take into account the receptor organ and neurological immaturity.

The stimulus rate generally has to be slower than in older children, particularly when recording cortical responses. Signal transmission is slower and permits lower high-pass filters (reducing the noise) and a longer analysis time. It is important to consider that fatigue and habituation may affect responses, particularly at the cortical level. It is therefore advisable to do many averages of a few trials; this has the further advantage of reducing the risk of acquisition during different behavioural states. If an unfavourable signal-to-noise ratio is suspected, it is helpful to do an average without stimulation in order to estimate the background noise from EEG and EMG.

The need for individual laboratories to establish their own normal ranges for EP measurements because of the effects of different stimuli, equipment and recording conditions is just as necessary in neonates as in adults. This important task can prove time-consuming, as normative data are needed throughout the period of rapid development, from the youngest preterm to the mature neonate. All normative data should be collected in the appropriate state.

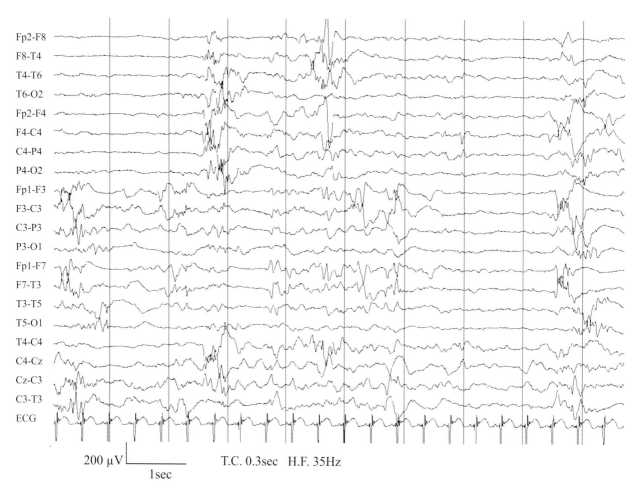

Fig. 5.29   Ohtahara syndrome in a term neonate. The child presented with hypertonia and frequent tonic and apnoeic seizures 2 weeks after birth. The EEG shows an asynchronous burst-suppression pattern. Periods of suppression last 1–3 s and bursts consisting of high amplitude slow waves mixed with spikes and sharp waves last 2–4 s.

The ERG and VEPs in neonatal neurology

The ERG and VEPs can be achieved with the neonate in the incubator. It is not essential to disturb a sick preterm neonate by taking him or her to the laboratory, even though it is easier there to ensure high-quality standardised recordings. Stimulation through closed eyelids may not be perfect and may require increased stimulus intensity (Hobley and Harding 1988, Kriss 1994), but no significant differences in latency and amplitude of the b-wave between the eyes-open and eyes-closed states were found by Harding (1992) in 30 neonates studied longitudinally between 32 and 52 weeks CA, with the stimulus intensity being increased for periods of eye closure (*Fig. 5.30*).

The ERG has been used mainly as a tool to assess the development of retinal function and for the exclusion of ocular disease as a cause of impaired vision. There is an increased incidence of retinopathy of prematurity (previously designated 'retrolental fibroplasia') in very low birthweight neonates (Ng Yin et al 1988). Reports suggest that the ERG in the neonatal period is not generally abnormal in this condition (Mactier et al 1988). However, a rod photoreceptor dysfunction has been described (Fulton et al 2001).

There have been attempts to assess the role of nutritional factors in visual development in the preterm and term neonate by means of electrophysiological techniques; this has been an area of active research in several centres (e.g. Mactier et al 1988, Neuringer et al 1988, Uauy et al 1990, 1992). Nutritional factors, including deficiencies of long-chain fatty acids or their precursors and vitamins A and E, may be relevant in visual development. As with the ERG there is evidence that the VEP may be abnormal with nutritional deficiency in the fetus. Stanley et al (1989) have shown abnormal development of visual function following intrauterine growth retardation. A study of the ERG findings in fetal alcohol syndrome by Hug et al (2000) has shown reduced b-wave amplitude and increased a- and b-wave implicit time. The scotopic ERG was more severely affected than the photopic (Hug et al 2000).

VEPs are classified as abnormal if:
- an abnormal waveform (including one of low amplitude) is obtained
- the latency is prolonged
- an asymmetry between the two hemispheres is detectable by conventional two- or three-channel recording or by mapping techniques
- an abnormal interocular latency difference between responses is recorded in Oz after monocular stimulation, provided recording is done in the same behavioural state.

Flash VEPs rarely add much to clinical observation. However, in several studies with flash VEPs an association between the EP results and visual acuity and long term visual outcome has been shown (Kurtzberg 1982, Placzek et al 1985, Norcia et al 1987, McCulloch et al

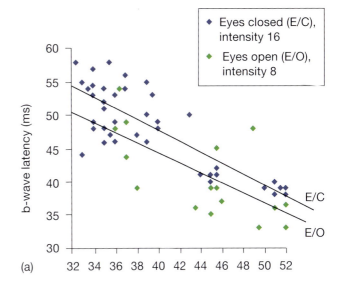

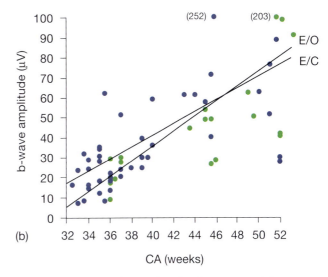

Fig. 5.30 The effect of eyes-closed and eyes-open states on (a) the latency and (b) the amplitude of the b-wave of the flash ERG in neonates; group data from 32–52 weeks CA. The stimulus intensity was increased to compensate for stimulation through closed eyelids. (From Harding (1992), by permission.)

1991, Pike et al 1994). In general, an absent VEP in the neonatal period indicates a poor prognosis for vision. Neonates with developmental visual agnosia provide an exception to this rule: an initially absent VEP may become normal in the second 6 months of life in association with normal visual behaviour. However, for technical and other reasons, the VEP cannot determine visual function accurately in neonates with grossly abnormal EEGs or severe electrolyte abnormalities, or in neonates under anaesthesia or heavy sedation. These difficulties in the main relate to the problem of separating out an EP waveform from the ongoing activity of a severely abnormal EEG, or when the peripheral component is also affected by the pathological process. Both ERG and VEP recordings may be affected in the latter case. Clearly, the level of sophistication of the techniques used and that of the associated database of experience from both normal and sick neonates in various diagnostic categories will govern success. For example, it has been suggested that VEP mapping in a group of children suffering from visual impairment of cortical origin gives more information than VEPs derived by the conventional technique (Whiting et al 1985).

The assessment of visual acuity by means of VEPs to patterned stimuli can be very accurate (Taylor 1992). Visual thresholds can be determined directly from the parameters of the stimuli that elicit just detectable VEPs. A common technique is to determine the smallest pattern size that elicits an identifiable response. In healthy newborn infants of 38–40 weeks GA a reproducible VEP is obtained by checkerboard stimulation with check sizes down to 1° of arc.

Abnormal VEPs are seen more frequently in lesions of the midbrain and the subcortical visual pathway than in cortical lesions (Placzek et al 1985). In very low birth-weight neonates, persistent absence or abnormality of the flash VEP has been shown to predict an abnormal neurological state at the age of 1 year (Kurtzberg 1982). Asymmetries tend to follow the lateralisation of structural abnormalities.

Several authors have also studied VEPs in children with hydrocephalus. The aim of these investigators was to find early involvement of the visual pathways by the enlarged ventricles. The VEP results in these cases were inconsistent. Guthkelch et al (1982) found increased latencies in cases with hydrocephalus and increased head circumference, but George and Taylor (1987) could not confirm this association. However, both groups found a normalisation of previously abnormal VEPs soon after shunting. Thus it seems that VEPs can be informative, not in the clinical diagnosis, but in monitoring the efficiency of the shunting procedure.

VEPs have been correlated with bilirubin-induced neurotoxicity (Chen and Kang 1995) and with neurological signs or brain abnormalities of neonates exposed to cocaine in utero (Salamy and Eldredge 1994). They have also been used in the assessment of the extent and severity of brain lesions in newborns suffering from HIE (see below).

BAEPs in audiological assessment and neonatal neurology

Recording techniques in the neonate are generally similar to those in older children or adults, but for audiological screening particular attention must be paid to appropriate and strictly standardised methodology (see p. 133). The stimulus is usually presented to the neonate by a small earphone which is held gently over the ear whilst the neonate lies on his or her side and has the other ear on the pillow. Care must be taken not to occlude or collapse the external auditory canal because this may result in appearances simulating a conductive lesion.

For the evaluation of a hearing defect, the following BAEP features are used:
- threshold for the appearance of wave V
- latency and amplitude of wave V
- latency of wave I
- latency/intensity curve.

The criteria for BAEP abnormality in neurological evaluation are:
- I–V, I–III and III–V interpeak interval prolongation
- loss of components
- amplitude ratio V/I (for full term neonates the lower limit is 30% (Pratt et al 1999)).

BAEPs in audiological assessment

Early identification of significant hearing loss is the basis for early intervention with amplification and thus is very important for speech, language and cognitive development. If deficits are undetected the serious consequences will last throughout life.

Since their introduction, BAEPs have been the diagnostic tool of choice in the evaluation of neonates at risk of sensorineural hearing loss. More recently, otoacoustic emissions (OAEs) have also been used (see p. 128). This is a cheaper and quicker method, but with many false-positive results (Smyth et al 1999).

The most commonly used criterion for threshold estimation, based upon BAEPs, is the minimal intensity required to detect wave V (Picton and Durieux-Smith 1988). Hearing threshold has been found to be at 40 dB (hearing level (HL)) in preterm neonates of 28–34 weeks GA, at 30 dB in neonates of 35–38 weeks and below 20 dB in term neonates (Lary et al 1985). Deficits can be classified as mild, moderate or severe. In term neonates, a mild hearing defect will range from 20 to 40 dB, a moderate one from 40 to 60 dB and a severe defect will exceed 70 dB.

Diagnostic information in identifying the type of hearing impairment comes from determination of the curve of the latency–intensity function (LIF), which represents the relationship between the latency of wave V and its changes with intensity. In conductive hearing disorders, the effective stimuli reaching the cochlea are reduced. Thus the LIF slope is normal, but the latency is abnormally prolonged. The pathological curve is parallel to, but above, the normal curve. In sensorineural hearing impairment the LIF shows a rapid decrease in wave V latency from an elevated threshold at low stimulus intensities to near-normal latencies at the higher stimulus intensities (Galambos and Hecox 1977, 1978).

The Joint Committee of Infant Hearing (2000) and the American Academy of Paediatrics Task Force on Newborn and Infant Hearing (1999) have recommended that early identification of hearing loss based on clinically assessed risk factors be abandoned. This is because approximately 50% of infants with hearing loss will be missed, with consequences throughout life in terms of speech, language and cognitive development. As a result, neonatal and paediatric units have started to implement protocols for neonatal audiological screening based on the combination of OAE recordings, done as a first step, and BAEPs, done in those who fail the first step. The automated auditory brainstem response (A-ABR) test is preferred to conventional BAEP because of its low cost and ease of use; this is a modification of the conventional technique based on single-stimulus intensity and automated detection of the response, wherein the waveform is tested statistically thus eliminating the need for specialist interpretation. However, the risk of a false-negative outcome (i.e. the occurrence of permanent hearing loss) in those who pass the test has been recently documented with the A-ABR technique (Johnson et al 2005).

In all neonates with abnormal BAEPs in the neonatal period, repeat examinations are mandatory as normalisation may occur in the first few months of life (Stein et al 1983). This is particularly observed in neonates with hypoxic–ischaemic encephalopathy, in whom a reversible elevation of threshold has been described, which is possibly explained by a 'critical period' of particular sensitivity to the effect of hypoxia during the development of the human auditory system, some time prenatally and shortly after birth, as demonstrated by Jiang (1995, 1998).

Therefore, in planning hearing tests for this group of patients, it has to be considered that if testing is done before discharge from the NICU the failure rate is about 10–20%, while after the first 3 months it is 2–5% (for a review see Suppiej 2001.) Conversely, it should be remembered that, in congenital cytomegalovirus infection and with some genetic causes of sensorineural hearing loss, the hearing is normal initially but then deteriorates, usually during the first year.

BAEPs in neonatal neurology

Several authorities have confirmed a relationship between BAEP abnormalities and neonatal risk factors such as low birthweight, asphyxia, hyperbilirubinaemia, intracranial haemorrhage, apnoea and ototoxic drug administration.

In neonates with facial or cranial malformations, abnormal BAEPs are frequently found and indicate the coexistence of abnormalities in the CNS.

In newborns with apnoea, single BAEPs have no value in screening for sudden infant death syndrome (SIDS). They may reveal a brainstem lesion as the cause of recurrent apnoea, but generally the central conduction time falls within the normal range.

BAEPs have been studied in neonates with hyperbilirubinaemia with the goal of distinguishing the neonates who will develop neurological sequelae and auditory impairment, but results are still inconsistent (for a review see Taylor et al 1996).

In preterm neonates with a GA of 34 weeks or less, statistically significant risk factors for abnormal BAEPs at follow-up examination (2–24 months) were hypoxia, hypercapnia and acidosis persisting for at least 1 week and abnormal BAEPs on discharge from hospital (Lary 1988).

In neonates with HIE the BAEP does not appear to be helpful in outcome prediction (see below).

SEPs in neonatal neurology

In birth palsies abnormal SEP findings may be observed and, depending on the location of the trauma at nerve, plexus or nerve root level, different patterns of abnormalities are found (for EMG findings see p. 219).

An extensive spinal cord injury may be presumed in neonates after a difficult delivery in which abnormal N13 SEP components are found. The SEP may be of limited value for determining the level of the injury. However, an absence of reproducible potentials above midthoracic level on posterior tibial stimulation, in contrast to normal median nerve potentials, may help to locate a lesion below the seventh cervical segment (Gilmore 1988, 1989).

In spite of the fact that SEPs cannot give direct evidence about the state of the motor pathways, they have proved to be a valuable early indicator of severe motor impairment if they remain abnormal in follow-up examinations during the first 4 weeks of life. In neonates SEPs probably only reflect neurological damage that includes motor defects, such as in cerebral palsy. They may, however, make a contribution to the monitoring of treatment, for example of congenital hypothyroidism (Laureau et al 1986).

SEPs have become a promising tool in the prognostic evaluation of neonatal HIE (see page 199).

Multimodal EPs in neonatal hypoxic–ischaemic encephalopathy and intracranial haemorrhage

Cortical and white matter lesions may be found in children with severe neonatal HIE. Neonatal HIE continues to be a major cause of neonatal mortality and is associated in some cases with an increased risk of chronic neurological disability in childhood (Taylor 1992). The main purpose of clinical and electrophysiological studies is to determine the neurological prognosis of newborn infants (for the role of the EEG in HIE see p. 199).

Experimental and anatomical studies show that visual, auditory and somatosensory pathways pass through brain areas selectively vulnerable to asphyxia (Volpe 2000b). Therefore, since the 1970s, EPs have been extensively studied in neonates affected by HIE and intracranial haemorrhage. Most of the early studies show contradictory results and/or make a poor contribution to prognostication, due to methodological factors and the selection of patients. With the introduction of stricter methodology appropriate to CNS immaturity, with more rigorous selection criteria for asphyxia and with distinction between the EP data recorded in preterm and full term neonates, more recent reports have demonstrated the important role of multimodal EPs in HIE (for a review see Suppiej 2001).

The results are very helpful in *term neonates*. In a study of 93 neonates, 39 with abnormal VEPs (abnormal waveform) throughout the first week of life, or with absent VEPs, either died or were left with neurological sequelae (Whyte et al 1986, Muttit et al 1991, Whyte 1993). Eken et al (1995) found the SEP useful for prognostication of motor outcome in HIE, with a positive predictive value of 81.8% and a negative predictive value of 91.7%. In a study by Mandel et al (2002), the bilateral absence of the N20 wave on short latency sensory EPs had a positive predictive value of unfavourable outcome of 100% (sensitivity 63%). In a study of 34 newborns with asphyxia, the bilateral absence of cortical SEPs predicted neurological abnormalities in 100% of cases, but in those with only an increased latency predicted only 73% of neurological abnormalities (De Vries et al 1991). In contrast, when both the SEP and VEP were normal after the first week of life, all newborns studied by Scalais et al (1998) and 97% of those reported by Taylor et al (1992) were neurodevelopmentally normal on follow-up. Likewise, in neonates who were only examined with SEPs, those with normal responses within the second week of life studied by De Vries et al (1991), De Vries (1993) and by Gibson et al (1992) all had a normal neurological outcome. The authors emphasise the importance of these data as their prognostic reliability proved highest in the moderately asphyxiated group in which prognostication is more difficult clinically. The predictive value of SEPs seems superior to that of EEG and transfontanellar ultrasound examination (Harbord and Weston 1995).

In *preterm neonates* the prognostic role of EPs has not yet been fully established. It is more difficult in this group of patients to separate the role of hypoxia from other factors of neurological risk. Furthermore, in distinction from term newborns, the brain areas more susceptible to insult, in the periventricular white matter, are easily visualised by ultrasound. The best results in terms of prognostic accuracy are obtained when median or tibial nerve SEPs and ultrasound both demonstrate an abnormality; if the SEPs are normal and the ultrasound is normal or normalises, neuromotor development is normal in nearly all cases. The SEP is of prognostic value in the early stages when ultrasound abnormalities are still evolving (Klimach and Cooke 1988, De Vries et al 1992, White and Cooke 1994, Ekert et al 1997a, Pierrat et al 1997).

The role of VEPs is controversial in preterm neonates: Ekert et al (1997b) reported that VEPs recorded within 3 weeks of life in neonates of < 32 weeks GA were not predictive of abnormal development, while Shepherd et al (1999) in high-risk neonates of < 35 weeks GA evaluated at around the third day of life, reported a sensitivity of 60% and a specificity of 92% for cerebral palsy as an outcome. Thus it seems that in early recordings the division by risk factors and a comparison with appropriate reference values increase the prognostic value of VEPs in the preterm population.

When VEPs and ultrasound are considered together, agreement is found between absent VEP and persistent subcortical cysts (De Vries et al 1987).

The VEPs are abnormal or absent in about 85% of neonates with severe HIE. Persistence of these abnormalities at the age of 2 months is likely to be associated with later clinical evidence of brain damage. In neonates with cystic leucomalacia, the presence, nature and severity of VEP abnormalities are also dependent on the site of the lesion, being most pronounced if the lesions are around the occipital horn, especially in the region of the trigone (Mushin 1988). The majority of studies report EP findings in the perinatal period and compare results with later neurodevelopmental outcome, but little is known about the evolution of EP findings during follow-up. It is a common observation that in most patients with cerebral palsy the VEP and SEP are normal. Preliminary results of a neurophysiological follow-up study of term neonates with perinatal HIE (A Suppiej, unpublished data, 2003) indicate that VEP and SEP abnormalities recover after perinatal brain damage, approaching normal values after the sixth month (*Fig. 5.31*).

Several studies have assessed the role of BAEPs in high-risk neonates; few have evaluated specifically BAEPs in neonates with HIE, more emphasis being placed on abnormalities at a cochlear level than at brainstem level. The prognostic role of BAEP abnormalities is controversial. Most of the studies report that the majority of abnormalities in the amplitude and latency of the central components of the BAEP are potentially reversible within 3–4 weeks (Stockard et al 1983) or 6–12 months (Jiang and Tierney 1996), and are associated with normal neurological outcome. However, when a pattern of craniocaudal loss of components is found, neonates either do not survive the neonatal period or are left with major neurological sequelae. In six asphyxiated neonates dying in the neonatal period, the loss of BAEP components was principally due to severe neuronal necrosis in several nuclei of the auditory pathway. Only in asphyxiated neonates without intracranial haemorrhage was loss of components of diagnostic and prognostic significance. A loss of components was seen in about 50% of neonates with intracranial haemorrhage, but many survived without neurological deficit (Lütschg 1986). The SEP has a poor prognostic value also in conditions such as isolated subarachnoid haemorrhage, in which either a latency prolongation or loss of the N20 component has been associated with a complete recovery (Willis et al 1987). However, prolongation of the N20 component predicted neurological deficits, especially a motor handicap in patients with extensive germinal matrix haemorrhage (Willis et al 1984, Gorke 1986).

VEPs may be useful in assessing the brain damage of neonates suffering from an intraventricular or periventricular (germinal matrix) haemorrhage. The VEPs in cases with a grade I or II haemorrhage are mostly normal, but are abnormal in those with grade III or IV haemorrhage

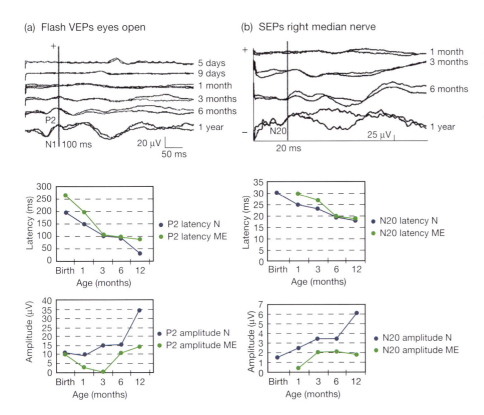

(a) Flash VEPs eyes open

(b) SEPs right median nerve

Fig. 5.31 Maturation of (a) VEP and (b) SEP cortical responses in a group of normal term neonates (N) compared with one who suffered from asphyxia metabolic acidosis and clinical signs of moderate encephalopathy (ME), as shown in graphs in the lower part of the figure. Above, this child's VEPs and SEPs during follow-up are shown.

(Watanabe et al 1981, Pryds et al 1989). Neonates suffering from germinal matrix or intraventricular haemorrhage have also been found to have immature VEPs. The delay of VEP maturation tends to be most marked in cases where the parenchymal haemorrhage extends down towards the thalamus. The abnormality is not correlated with ventricular dilatation (Placzek et al 1985).

Intellectual deficit and mild neuromotor abnormalities at school age have been observed in NICU survivors who had exhibited transient upper limb SEP and BAEP abnormalities in the neonatal period (Majnemer and Rosenblatt 2000). The utility of the BAEP in neurological prognostication appears to increase when special techniques, including high click frequencies, are adopted (Jiang et al 2000).

ELECTROMYOGRAPHY IN NEWBORNS

Neuromuscular disorder in the neonatal period presents most often as a 'floppy baby'. Peripheral nerve, anterior horn cell, neuromuscular junction and muscle diseases can all present in the neonatal period with hypotonia; so also do many metabolic, systemic and CNS disorders (Dubowitz 1980). EMG plays a very useful part in the examination of hypotonic neonates, but must always be considered as a complement to the clinical history and physical examination.

Techniques

Technical aspects and findings have been discussed by Alexander and Turk (1980), Payan (1991), Brown et al (2002), Jones et al (2003) and Holmes et al (2006).

With modern, reusable, self-adhesive electrodes, which can be cut down to whatever size is needed, there is really now no real impediment to the electromyographer doing as full an examination as is wished. This is not to diminish the technical difficulties of this policy, but with practice it is possible to achieve this goal. The biggest impediment to study is the use of the limbs for other clinical management, particularly in the NICU where many of the infants will come from. All of us, at some stage, will have been asked to do an EMG in the NICU setting, where the child has only one limb free of either monitoring equipment or intravenous or intra-arterial lines. Recording sensory nerve potentials and the EMG may be difficult because of electrical artefact in this environment. This is not usually a problem with motor nerve studies. Even if the wiring in your NICU has been shielded, and it often has in modern hospitals, it is important to try turning off non-vital equipment in rotation. The heated beds can be particularly troublesome. Disconnection of the plug from the wall socket, not simply turning them off, may be needed and can be done just as the tests are being performed to prevent any danger to the neonate. Meticulous cleaning of the skin, which may have some kind of lotion on it, and mild abrasion with one of the proprietary products used to prepare the skin for EEG, may also yield pleasing results.

Sensory nerve testing

This is first attempted in the legs, and the use of the medial plantar response can be particularly productive. The sural and superficial peroneal nerves are also possibilities even in the neonate. Whatever technique is used it is common to not be able to record any sensory responses in the very youngest normal neonates. There is no clear explanation for this. It is possible that the immaturity of the nerve and the poor formation of the myelin sheath mean that the largest diameter fibres do not summate their action potentials. Instead their responses are mixed in with those from the other nerve fibres and phase cancellation is more likely to occur. In the hands the median and, less commonly, the ulnar sensory responses can be measured. Ring electrodes around the digits can be usefully employed here. In other age groups they are poorly tolerated.

Motor nerve testing

As with the sensory testing the legs are chosen first. Both the lateral and medial popliteal nerves can be used. The latter has the advantage of being normally associated with large compound muscle action potentials (CMAPs) being elicited from abductor hallucis and, unlike in adults, stimulation proximally in the popliteal fossa is not as painful or difficult. In the arms a personal preference is to use the ulnar rather than the median nerve. It is easier to place recording electrodes over either the first dorsal interossesous or the adductor digiti minimi than on the abductor pollicis, and also the proximal stimulation of the ulnar nerve above the elbow presents less technical difficulties. F waves can be studied as well.

Normal control values for motor and sensory conduction velocities for most nerves, ranging from the preterm to adolescence, have been reported over the years (Thomas and Lambert 1960, Gamstorp 1963, Dunn et al 1964, Baer and Johnson 1965, Smit et al 1999, Garcia et al 2000). Control values for maturation of motor and sensory nerve conduction in preterm neonates are given by Bouglé et al (1990). A monograph by Raimbault (1988) presents one of the most comprehensive studies. The conduction velocities of sensory nerves are faster at all ages than those of motor nerves. There is a rapid increase in velocity over the first 5 years of life, by which time near-adult values will have been attained. In the neonatal period, very slow velocities may be encountered. In the neonatal period there are similar velocities in lower and upper limbs, with the increase in the median and ulnar velocities not being seen until middle childhood (Gamstorp and Shelbourne 1965). The range is wide due in part to differences in intrauterine nutrition. Hopes that values could be correlated with corrected GA (Dubowitz et al 1968) were not realised.

Over the last decade the recognition of the many different presentations of the congenital myasthenic syndromes (CMSs) (Engel et al 2003) has meant that these conditions, many of which may be treatable, should be sought if the routine screening, as described above, is normal or the EMG is myopathic. Repetitive nerve stimulation (RNS) should be performed in such cases and, if clinical suspicion remains high but RNS is negative, stimulated single-fibre EMG should be done. Even then we have the experience of a case showing normal studies, including single-fibre EMG, but later proven to have CMS. It is important for there to be a very close collaboration between the neurophysiologist and the clinical teams, and the tests may need to be repeated to prevent these important, but very rare, conditions being overlooked. In practice it is rare not to do RNS in our practice in every infant under the age of 1 year.

It is important to try to choose a proximal muscle to do RNS because of the recognised propensity of nearly all neuromuscular disorders for proximal, as opposed to distal, musculature (Costa et al 2004). Carried to its logical conclusion this policy would mean that the alae nasalis or trapezius muscles should be preferred, but the real difficulties of getting satisfactory, and therefore reliable, results from these sites means that the extensor compartment of the forearm is chosen. The protocol used is for one run at 3 Hz to be followed by a burst of 2 s at 30 Hz, to emulate tetanic stimulation, and then a further run at 3 Hz immediately and then 1 min later. A decrement of > 10% is abnormal. In most cases of CMS the decrement may be much more, except for CMS with episodic apnoea. This presynaptic condition, caused by an abnormality of choline acetyl transferase (Engel and Lambert 1987), may only show a decrement with prolonged repetitive stimulation at 10 Hz for several minutes. This is painful and often cannot be done on an unanaesthetised baby. The method used for stimulated single-fibre EMG includes stimulation at 10 Hz and may avoid the need for prolonged RNS.

Needle EMG

This is the final part of the examination, and there are few occasions when it is not done in the weak neonate. Needle EMG may produce very revealing changes, which may not have been possible to predict from the NCSs themselves. In these tiny patients a facial concentric needle (0.3 mm external diameter, 0.03 mm^2 recording area) must be used. The tibialis anterior is selected first. The plantar withdrawal reflex is particularly useful here, as this muscle can be made to contract by stroking the sole of the foot. If a proximal muscle is needed in the leg, the iliopsoas is much easier to activate using this same reflex than is the vastus medialis. In the arms the extensor digitorum communis may be used. This muscle will be activated when the baby grasps a little finger placed in its palm. Unfortunately, even in normals the interference pattern can be a little unusual, so overinterpretation has to be guarded against. The biceps is also useful but, because of the discomfort the use of this muscle appears to provoke, it can be very difficult to activate satisfactorily.

Changes suggestive of neurogenic change are easy to identify, those of myopathy less so. The latter is because the fibre size in a normal neonate is smaller than than that in adults, which means that the interference pattern is dense and the fibre responses within it are of short duration and low amplitude. The rate of increase of the amplitude and duration of the motor unit potential during infancy and childhood is not exactly the same in proximal and distal limb muscles (Sacco et al 1962) because the increase in diameter of the muscle fibres and their expansion in the innervation zone do not occur at the same rates. The proper way to examine a baby with a suspected myopathy is by DNA analysis and muscle biopsy, and the role of the EMG is to show other causes, such as anterior horn cell disease.

There are occasions when it is essential to sample a bulbar muscle, particularly to exclude spinal muscular atrophy. The genioglossus, approached from the sub-mental route, and the masseter provide no significant technical difficulties. These muscles have even smaller diameter fibres than the limb muscles. The interference pattern in the genioglossus should not exceed 1.5 mV (Renault and Raimbault 1992).

EMG investigations in neonatal neurology
The 'floppy infant' syndrome

Neuromuscular disease can produce hypotonia, and the associated neurophysiological abnormalities are often pronounced (Jones 1990). The most common condition causing hypotonia in the neonate is SMA1 (spinal muscular atrophy 1 or Werdnig–Hoffman disease; see p. 328), a form of anterior horn cell disease characterised by a reduced number of large motor unit potentials, which is usually obvious on EMG sampling. It is almost invariably associated with a deletion of the survival motor neuron (SMN) gene, making referrals for EMG less common.

Myopathies are the second most common cause of floppy infant syndrome. Rapidly recruited short-duration and low amplitude motor unit potentials, usually referred to as the 'myopathic pattern', are very suggestive of this condition, but it is often difficult to distinguish these changes from normal motor unit potentials in the infant, as has been discussed above. Abnormalities such as positive sharp waves, complex repetitive discharges or myotonic bursts are sometimes present at rest in some congenital myopathies and usually in congenital myotonic dystrophy. These two conditions represent the most frequent myopathic syndromes encountered in the neonatal period.

Congenital myasthenic syndromes are interesting and potentially treatable conditions. They have been discussed as a group in the section on techniques of nerve conduction and EMG (see p. 165). The only CMS that might not be seen in the newborn are slow channel syndrome or familial limb-girdle myasthenia. EMG, to include stimulated single-fibre EMG and RNS, while it can demonstrate an abnormality of the neuromuscular junction, is often not able to advance the diagnosis further, this being done by a combination of clinical and genetic studies. One potential EMG marker, the repetitive CMAP, which is seen on motor nerve testing and found in end-plate acetylcholinesterase deficiency and slow channel disease (Engel 1990), is unfortunately not invariably seen.

Infant botulism should be considered in a neonate or infant of a few months of age who becomes weak after a period of normal development (Pickett et al 1976). Additional features are external ophthalmoplegia and constipation. Infant botulism is thought to be due to ingestion of spores that germinate into bacteria, which produce botulinum toxin. Additional risk factors for developing the illness appear to be breast-feeding and possibly the consumption of honey. Small CMAPs are recorded with incremental responses at stimulation rates of 20 or 50 Hz and mild incremental responses at stimulation rates of 2 or 5 Hz (Cornblath et al 1983). The EMG shows spontaneous fibrillations and myopathic-like, short-duration, low amplitude motor units.

In the floppy infant, benign congenital hypotonia or disorders of the CNS may be suggested by normal power and normal EMG and nerve conduction velocity. Depending on the EMG findings, muscle or nerve and muscle biopsy may be indicated. This should be performed in the intercostal region or in a homologous muscle contralateral to that most affected, as judged by EMG sampling.

Birth palsies

Palsies arising from trauma to nerves, nerve roots and plexuses complicate a small percentage of difficult deliveries. Hopes that obstetric brachial plexus palsy should be less commonly encountered with improvements in obstetric procedures (Sunderland 1978, Brett 1997) have not been realised. In most developed countries the incidence has not dropped (Adler and Patterson 1967, Greenwald et al 1984, Sjoberg et al 1988, Bager 1997, Ecker et al 1997, Donnelly et al 2002, Evans-Jones et al 2003). This is almost certainly due to increases in the birthweight of babies, particularly in the developed world, and the associated increase in shoulder dystocia. Forces during traction on the shoulders or arms, or extreme lateral flexion, rotation or traction of the neck, can damage the brachial plexus or cause root avulsion. The incidence of different clinical syndromes is given by Tudehope and Vacca (1988) as 80% affecting the upper part of the plexus (C5 and C6; Erb–Duchenne palsy), 7.5% the whole plexus (C5–T1; Erb–Duchenne– Klumpke paralysis) and 2.5% the lower part (C8 and T1; Klumpke's paralysis). Other forms of obstetric palsy, for example due to damage to the sciatic or peroneal nerves, are, with the exception of facial palsy, extremely uncommon.

EMG investigation is complicated by the discovery of luxury innervation of the muscles supplied by the brachial plexus (Vredeveld et al 2000), which means that at birth many muscles may have dual innervation. For example, the biceps will receive input from C7 as well as C5. In normal babies the adult system of innervation will evolve, but in obstetric brachial plexus palsy this may not happen. This makes it more difficult for EMG results to determine the roots that are involved, although they can still give invaluable information on the type of nerve damage. Smith (1996) has evolved a robust method of classifying the EMG findings that correlates well with the findings at surgery. The timing of the investigation is linked closely to the timing of surgery. There is a great deal of experience (Gilbert et al 1988, Birch et al 2005) that pushes surgeons to operate around the age of 3 months. Arguing from a theoretical basis, Pitt and Vredeveld (2005) encourage a much more aggressive use of EMG, for example twice within the first week of birth. They also argue that surgery to correct those palsies where avulsion and neurotmesis have occurred should be contemplated as early as possible, and that results in these cases will be improved in the long term. This belief is justified from experience with the correction of other nerve injury in adults, and the emerging evidence of central remodelling that takes place after birth; if surgery is delayed, even if there is a good peripheral repair, there still may be apraxia of the limb (Brown et al 2000).

Recent evidence (Anand and Birch 2002) indicates that microsurgical repairs, including nerve grafting and transfer of intercostal nerves into the brachial plexus in cases of root avulsion, may prove effective in reducing disability and restoring accurate sensory function.

Facial nerve palsy

Congenital facial nerve palsies may be another source of referral in this age group. Some palsies may have an obvious relationship to birth trauma, particularly if related to forceps delivery, but many may not (Falco and Eriksson 1990). A proportion of these may be prenatal in origin, caused by pressure on the facial nerve when the fetus was lying awkwardly in relation to the pelvis, a situation more likely to occur when the liquor is reduced. A congenital aplasia of the facial nucleus may be the underlying cause in Möbius syndrome.

Various techniques may be employed, of which needle EMG of the facial muscles is the most informative and robust. Direct stimulation of the facial nerve and comparison of the amplitudes of the CMAPs between the good and affected side is often disappointing. Blink reflexes may also fail to fulfil to their potential. In this age group, the later waves R2, both contralateral and ipsilateral, may be profoundly affected by the state of the infant. In the traumatic palsies the presence of an R1 response indicates a good prognosis.

Möbius syndrome is thought to be due to a congenital aplasia of the facial nucleus. If this is the case it might be expected that little motor activity might be seen on EMG, with the muscles failing to develop. However, studies have shown a pattern indicative of axonal loss (Renault 2000), thus providing evidence to suggest there has been denervation and partial reinnervation. This would suggest that the cases of Möbius syndrome, while perhaps associated with a loss of or decrease in the facial nuclei size, may have originated by a process such as ischaemia (Govaert et al 1989), and in this respect might be described as being an acquired rather than a congenital process. The key to the diagnosis of Möbius syndrome is to demonstrate that other cranial nuclei in the brainstem have been affected, either clinically (Meyerson and Foushee 1978) or by demonstrating EMG changes in other bulbar muscles.

The role of EMG in the management of facial palsy is quite limited in the UK as plastic surgeons here are reluctant to operate early, preferring to wait many years to allow recovery to occur to whatever extent is possible before going on to make the definitive repair when the residual defect has become established. A point of observation is that it is not uncommon to find a fairly full and normal interference pattern established some months after a birth-related injury, but associated with poor voluntary function. This may be another example, as discussed in the context of brachial plexus palsy of the newborn, where the failure of central modelling that occurs when the normal connections are not restored early can produce a degree of dyspraxia, or even apraxia in the worst cases.

Anomalies of the pelvic floor

Malformations of the lower bowel (high anorectal) can be treated surgically, but satisfactory outcome in terms of bowel control depends on preoperative assessment of the anal sphincter. 'Sphincter mapping' techniques have been used by Boyd and Kiely (1986) and Boyd et al (1987) to document the presence and functional significance of the external anal sphincter. This allows planning of the corrective surgical procedures so that the anal orifice is created distally to the sphincter. A combination of transcutaneous electrical stimulation of the sacral roots and a sphincter EMG has also been found helpful in the assessment of neurological problems in infants with sacral anomalies (Boyd 1989).

REFERENCES

Aarabi A, Wallois F, Grebe R 2005 Automated neonatal seizure detection: a multistage classification system through feature selection based on relevance and redundancy analysis. Clin Neurophysiol 117(2): 328–340.

Adler JB, Patterson RL 1967 Erb's palsy: long-term results in treatment in eighty-eight cases. J Bone Joint Surg 49: 1052.

Aicardi J, Goutiéres F 1978 Encéphalopathie myoclonique néonatale. Rev Electroencéphalogr Neurophysiol Clin 8: 99–101.

Alexander MA, Turk M 1980 Pediatric considerations. In: *Practical Electromyography* (ed EW Johnson). Williams and Wilkins, Baltimore, OH, pp 260–275.

American Academy of Pediatrics 1999 Newborn and infant hearing loss: detection and intervention. Task Force on Newborn and Infant Hearing. Pediatrics 103: 527–530.

Anand P, Birch R 2002 Restoration of sensory function and lack of long term chronic pain syndromes after brachial plexus injury in human neonates. Brain 125: 113–122.

Anderson CM, Torres F, Faoro A 1985 The EEG of the early premature. Electroencephalogr Clin Neurophysiol 60: 95–105.

Andre M, Matisse N, Vert P, et al 1988 Neonatal seizures – recent aspects. Neuropaediatrics 19: 201–207.

Ashwal S, Schneider S 1989 Brain death in the newborn. Pediatrics 84: 429–437.

Aso K, Scher MS, Abdab-Barmada M 1989 Neonatal electro-encephalography and neuropathology. J Clin Neurophysiol 6: 103–123.

Aso K, Abdab-Barmada M, Scher MS 1993 EEG and the neuropathology in premature neonates with intraventricular hemorrhage. J Clin Neurophysiol 10: 304–313.

Baer RD, Johnson EW 1965 Motor nerve conduction velocities in normal children. Arch Phys Med Rehab 46: 698.

Bager B 1997 Perinatally acquired brachial plexus palsy – a persisting challenge. Acta Paediatr 86: 1214–1219.

Baud O, d'Allest AM, Lacaze-Masmonteil T, et al 1998 The early diagnosis of periventricular leukomalacia in premature infants with positive Rolandic sharp waves on serial electroencephalography. J Pediatr 132: 813–817.

Bell AH, Greisen G, Pryds O 1993 Comparison of the effects of phenobarbitone and morphine administration on EEG activity in preterm babies. Acta Paediatr Scand 82: 35–39.

Benda GI, Engel RC, Zhang YP 1989 Prolonged inactive phases during the discontinuous pattern of prematurity in the electroencephalogram of very-low-birthweight infants. Electroencephalogr Clin Neurophysiol 72: 189–197.

Bergman I, Painter MJ, Crumrine PK 1982 Neonatal seizures. Semin Perinatol 6: 54–67.

Biagioni E, Bartalena L, Boldrini A, et al 1994 Background EEG activity in preterm infants: correlation of outcome with selected maturational features. Electroencephalogr Clin Neurophysiol 91(3): 154–162.

Biagioni E, Bartalena L, Boldrini A, et al 2000 Electroencephalography in infants with periventricular leukomalacia: prognostic features at preterm and term age. J Child Neurol 15: 1–6.

Bickford RG, Billinger TW, Fleming NI, et al 1972 The compressed spectral array (CSA) – a pictorial EEG. Proc San Diego Biomed Symp 11: 365–370.

Bickford RG, Brimm J, Berger L, et al 1973 Compressed spectral array in clinical EEG. In: *Automation of Clinical Electroencephalography* (eds P Kellaway, I Petersén). Raven Press, New York, pp 55–64.

Birch R, Ahad N, Kono H, et al 2005 Repair of obstetric brachial plexus palsy: results in 100 children. J Bone Joint Surg Br 87: 1089–1095.

Blume WT, Dreyfus-Brisac C 1982 Positive Rolandic sharp waves in neonatal EEG; types and significance. Electroencephalogr Clin Neurophysiol 53: 277–282.

Bouglé D, Denise P, Yaseen H, et al 1990 Maturation of peripheral nerves in preterm infants. Motor and proprioceptive nerve conduction. Electroencephalogr Clin Neurophysiol 75: 118–121.

Boyd SG 1989 Clinical neurophysiological assessment of children with sacral anomalies. Neuropediatrics 20: 207–210.

Boyd SG, Kiely EM 1986 Pre-operative neurophysiological studies of external anal sphincter and puborectalis in children with ano-rectal malformations. Pediatr Surg Intl 1: 184–185.

Boyd SG, Kiely EM, Swash M 1987 Electrophysiological studies of puborectalis and external anal sphincter in incontinent children with corrected high ano-rectal anomalies. Pediatr Surg Intl 2: 110–112.

Boylan GB, Panerai RB, Rennie JM, et al 1999a Cerebral blood flow velocity during neonatal seizures. Arch Dis Child 80: F105–F110.

Boylan GB, Pressler RM, Rennie JM, et al 1999b Outcome of electroclinical, electrographic, and clinical seizures in the newborn infant. Dev Med Child Neurol 41: 819–825.

Boylan GB, Rennie JM, Pressler RM, et al 2002 Phenobarbitone, neonatal seizures, and video-EEG. Arch Dis Childhood, Fetal Neonatal Ed 86: F165–F170.

Boylan GB, Rennie JM, Chorley G, et al 2004 Second-line anticonvulsant treatment of neonatal seizures: a video-EEG monitoring study. Neurology 62: 486–488.

Brett E 1997 Cerebral palsy, perinatal injury to the spinal cord and brachial plexus birth injury. In: *Paediatric Neurology*, 3rd edn (ed E Brett). Churchill Livingstone, Edinburgh, pp 291–332.

British Paediatric Association 1991 *Diagnosis of Brain Stem Death in Infants and Children*. A Working Party Report. British Paediatric Association, London, pp 1–6.

Brown T, Cupido C, Scarfone H, et al 2000 Developmental apraxia arising from neonatal brachial plexus palsy. Neurology 55: 24–30.

Brown WF, Bolton CF, Aminoff MJ 2002 *Neuromuscular Function and Disease: Basic, Clinical and Electrodiagnostic Aspects*. WB Saunders, Philadelphia, PA.

Bye AM, Flanagan D 1995 Spatial and temporal characteristics of neonatal seizures. Epilepsia 36: 1009–1016.

Bye AM, Lee D, Naidoo D, et al 1997 The effects of morphine and midazolam on EEGs in neonates. J Clin Neurosci 4: 173–175.

Celka P, Colditz P 2002 A computer-aided detection of EEG seizures in infants: a singular-spectrum approach and performance comparison. IEEE Trans Bio-Med Eng 49: 455–462.

Celka P, Boashash B, Colditz P 2001 Preprocessing and time–frequency analysis of newborn EEG seizures. IEEE Eng Med Biol Mag 20(5): 30–39.

Chen YJ, Kang WM 1995 Effects of bilirubin on visual evoked potentials in term infants. Eur J Pediatr 154: 662–666.

Chung HJ, Clancy RR 1991 Significance of positive temporal sharp waves in the neonatal electroencephalogram. Electroencephalogr Clin Neurophysiol 79: 256–263.

Clancy RR 1996 The contribution of EEG to the understanding of neonatal seizures. Epilepsia 37(Suppl 1): S52–S59.

Clancy RR, Legido A 1987 The exact ictal and interictal duration of electroencephalographic neonatal seizures. Epilepsia 28: 537–541.

Clancy RR, Tharp BR 1984 Positive Rolandic sharp waves in the electroencephalograms of premature neonates with intraventricular haemorrhage. Electroencephalogr Clin Neurophysiol 57: 395–404.

Clancy RR, Tharp BR, Enzman D 1984 EEG in premature infants with intraventricular hemorrhage. Neurology 34: 583–590.

Clancy RR, Chung HJ, Temple JP 1993 *Neonatal Encephalography. Atlas of Electroencephalography*, Vol 1 (eds MR Sperling, RR Clancy). Elsevier Science, Amsterdam.

Coen RW, McCutcheon CB, Werner D, et al 1982 Continuous monitoring of the electroencephalogram following perinatal asphyxia. J Paediatr 100: 628–630.

Commission on Classification and Terminology of the International League Against Epilepsy 1989 Proposal for revised classification of epilepsies and epileptic syndromes. Epilepsia 30: 389–399.

Connell J, Oozeer R, Regev R, et al 1987 Continuous four-channel EEG monitoring in the evaluation of echodense ultrasound lesions and cystic leucomalacia. Arch Dis Child 62: 1019–1024.

Connell J, De Vries L, Oozeer R, et al 1988 Predictive value of early continuous electroencephalographic monitoring in ventilated preterm infants with intraventricular hemorrhage. Pediatrics 82: 337–343.

Connell J, Oozeer R, De Vries L, et al 1989a Continuous EEG monitoring of neonatal seizures: diagnostic and prognostic considerations. Arch Dis Child 64: 452–458.

Connell J, Oozeer R, De Vries L, et al 1989b Clinical and EEG response to anticonvulsants in neonatal seizures. Arch Dis Child 64: 459–464.

Cornblath DR, Sladky JT, Sumner AJ 1983 Clinical electrophysiology of infantile botulism. MuscleNerve 6: 448–452.

Costa J, Evangelista T, Conceicao I, et al 2004 Repetitive nerve stimulation in myasthenia gravis – relative sensitivity of different muscles. Clin Neurophysiol 115: 2776–2782.

Cukier F, André M, Monod N, et al 1972 Apport de l'EEG au diagnostic des hemorragies intra-ventriculaires du prématuré. Rev Electroencéphalogr Neurophysiol Clin 2: 318–322.

Curzi-Dascalova L, Mirmiran M 1996 *Manuel des Techniques d'Enregistrement et d'Analyse des Stades de Sommeil et de Veille Chez le Prématuré et le Nouveau-né à Terme*. Éditions Inserm, Paris, p 180.

Curzi-Dascalova L, Figueroa JM, Eiselt M, et al 1993 Sleep state organization in premature infants of less than 35 weeks gestational age. Pediatr Res 34: 624–628.

d'Allest AM, Navelet Y, Nedelcoux H, et al 1997 [Intraventricular hemorrhage and parenchymatous ischemia in the newborn at term. Report of five cases.] Neurophysiol Clin 27: 129–138.

Danner R, Shewmon A, Sherman MP 1985 Seizures in an atelencephalic infant. Is the cortex necessary for neonatal seizures? Arch Neurol 42: 1014–1016.

Daoust-Roy J, Seshia SS 1992 Benign neonatal sleep myoclonus: a differential diagnosis of neonatal seizures. Am J Dis Child 146: 1236–1241.

Dehan M, Quillerou D, Navelet Y, et al 1977 Les convulsions du cinquième jour de vie: un nouveau syndrome? Arch Franc Pédiatr 34: 730–742.

Dehkharghani F 1984 Application of electroencephalographical and evoked potential studies in the neonatal period. In: *Topics in Neonatal Neurology*. Grune and Stratton, New York, pp 257–287.

Delapierre G, Dreano E, Samson-Dollfus D, et al 1986 Mise au point chez le nourrisson d'un critère de dètection automatique des fuseaux de sommeil. Rev Electroencèphalogr Neurophysiol Clin 16: 311–316.

DeMyer W, White PT 1964 EEG in holoprosencephaly (arhinencephaly). Arch Neurol 11: 507–520.

De Vries LS 1993 Somatosensory evoked potentials in term neonates with postasphyxial encephalopathy. Clin Perinatol 20: 463–499.

de Vries LS, Hellström-Westas L 2005 Role of cerebral function monitoring in the newborn. Arch Dis Child Fetal Neonatal Ed 90(3): F201–F207.

De Vries LS, Connell JA, Dubowitz LM, et al 1987 Neurological, electrophysiological and MRI abnormalities in infants with extensive cystic leukomalacia. Neuropediatrics 18: 61–66.

De Vries LS, Pierrat V, Eken P, et al 1991 Prognostic value of early somatosensory evoked potentials for adverse outcome in full-term infants with birth asphyxia. Brain Dev 13: 320–325.

De Vries LS, Eken P, Pierrat W, et al 1992 Prediction of neurodevelopmental outcome in the preterm infant: short latency cortical somatosensory evoked potentials compared with cranial ultrasounds. Arch Dis Child 67: 1177–1181.

De Weerd AW, Despland PA, Plouin P 1999 Neonatal EEG. In: *Recommendations for the Practice of Clinical Neurophysiology* (eds G Deuschl, A Eisen). Clin Neurophysiol (Suppl 52): 149–157.

Di Capua M, Fusco L, Ricci S, et al 1993 Benign neonatal sleep myoclonus: clinical features and video-polygraphic recordings. Move Disord 8: 191–194.

Donnelly V, Foran A, Murphy J, et al 2002 Neonatal brachial plexus palsy: an unpredictable injury. Am J Obstet Gynecol 187: 1209–12.

Douglass LM, Wu JY, Rosman NP, et al 2002 Burst suppression electroencephalogram pattern in the newborn: predicting the outcome. J Child Neurol 17: 403–408.

Dreyfus-Brisac C 1979 Neonatal electroencephalography. Rev Perinatal Med 3: 397–471.

Dreyfus-Brisac C, Larroche J-C 1972 Discontinuous EEGs in premature and full-term neonates. Electroencephalogr Clin Neurophysiol 32: 575.

Dreyfus-Brisac C, Samson-Dollfus D, Saint-Anne Dargassies S 1957 Veille, sommeil et réactivité sensorielle chez le prématuré et le nouveau-né et le nourrisson. In: *Conditionement et Réactivité en Electroencéphalographie*. Electroencephalogr Clin Neurophysiol (Suppl 6): 418–424.

Dubowitz V 1980 *The Floppy Infant*. Spastics International Medical Publications/Lavenham Press, Lavenham.

Dubowitz V, Whittaker GF, Brown BH, et al 1968 Nerve conduction velocity – an index of neurological maturity of the newborn infant. Dev Med Child Neurol 10: 741–749.

Dunn HG, Buckler WSJ, Morrison GCE, et al 1964 Conduction velocity of motor nerves in infants and children. Pediatrics 34: 708.

Eaton DG, Wertheim D, Oozeer R, et al 1992 The effect of pethidine on the neonatal EEG. Dev Med Child Neurol 34: 155–163.

Eaton DM, Toet M, Livingston J, et al 1994 Evaluation of the Cerebro Trac 2500 for monitoring of cerebral function in neonatal intensive care. Neuropediatrics 25: 122–128.

Ecker JL, Greenberg JA, Norwitz ER, et al 1997 Birth weight as a predictor of brachial plexus injury. Obstet Gynecol 85: 643–647.

Eisengart M, Gluck L, Glaser GH 1970 Maturation of the electroencephalogram of infants of short gestation. Dev Med Child Neurol 12: 49–55.

Eken P, Toet MC, Groenendaal F, et al 1995 Predictive value of early neuroimaging, pulsed Doppler and neurophysiology in full term infants with hypoxic–ischaemic encephalopathy. Arch Dis Child Fetal Neonatal Ed 73: F75–F80.

Ekert PG, Taylor MJ, Keenan NK, et al 1997a Early somatosensory evoked potentials in preterm infants: their prognostic utility. Biol Neonate 71: 83–91.

Ekert PG, Keenan NK, Whyte HE, et al 1997b Visual evoked potentials for prediction of neurodevelopmental outcome in preterm infants. Biol Neonate 71: 148–155.

Ellingson RJ, Peters JF 1980a Development of EEG and daytime sleep patterns in normal fullterm infants during the first three months of life: longitudinal observations. Electroencephalogr Clin Neurophysiol 49: 112–124.

Ellingson RJ, Peters JF 1980b Development of EEG and daytime sleep patterns in low risk premature infants during the first year of life: longitudinal observations. Electroencephalogr Clin Neurophysiol 50: 165–171.

Ellison PH, Franklin S, Brown P, et al 1985 A simplified method for interpretation of electroencephalograms in the preterm neonate. Ann Neurol 18: 381–382.

Engel R 1975 *Abnormal Electroencephalograms in the Neonatal Period*. CC Thomas, Springfield, IL.

Engel AG 1990 Congenital disorders of neuromuscular transmission. Semin Neurol 10: 12–26.

Engel AG, Lambert EH 1987 Congenital myasthenic syndromes. Electroencephalogr Clin Neurophysiol 39(Suppl): 91–102.

Engel A, Ohno K, Sine SM 2003 Congenital myasthenic syndromes: progress in the past decade. Muscle Nerve 27: 4–25.

Estan J, Hope PL 1997 Unilateral neonatal cerebral infarction in full term infants. Arch Dis Child 76: F88–F93.

Evans-Jones G, Kay SP, Weindling AM, et al 2003 Congenital brachial palsy: incidence, causes, and outcome in the United Kingdom and Republic of Ireland. Arch Dis Child Fetal Neonatal Ed 88: F185–F189.

Eyre JA (ed) 1992 *The Neurophysiological Examination of the Newborn Infant*. Clinics in Developmental Medicine, No. 120. MacKeith Press/Cambridge University Press, London.

Falco NA, Eriksson E 1990 Facial nerve palsy in the newborn: incidence and outcome. Plast Reconstr Surg 85: 1–4.

Faul S, Boylan G, Connolly S, et al 2005 An evaluation of automated neonatal seizure detection methods. Clin Neurophysiol 116: 1533–1541.

Ferguson JH, Levinson MW, Derakshan I 1974 Brainstem seizures in hydranencephaly. Neurology 24: 1152–1157.

Filan P, Boylan GB, Chorley G, et al 2005 The relationship between the onset of electrographic seizure activity after birth and the time of cerebral injury in utero. BJOG 112: 504–507.

Fulton AB, Hansen RM, Petersen RA, et al 2001 The rod photoreceptors in retinopathy of prematurity: an electroretinographic study. Arch Ophthalmol 119: 449–505.

Gabriel M, Albani M 1977 Rapid eye movement sleep, apnea and cardiac slowing influenced by phenobarbitol administration in the neonate. Pediatrics 60: 426–430.

Galambos R, Hecox K 1977 Clinical applications of the brainstem evoked potentials. In: *Auditory Evoked Potentials in Man* (ed JE Desmedt). Karger, Basel, pp 1–19.

Galambos R, Hecox KE 1978 Clinical application of the auditory brainstem response. Otolaryngol Clin North Am 11: 709–722.

Gamstorp I 1963 Normal conduction velocity of ulnar, median, and peroneal nerves in infancy, childhood and adolescence. Acta Paediatr Scand (Suppl 146): 68.

Gamstorp I, Shelburne SA Jr 1965 Peripheral sensory conduction in ulnar and median nerves of normal infants, children and adolescents. Acta Pediatr Scand 54: 309–313.

Garcia A, Calleja J, Antolin FM, et al 2000 Peripheral motor and sensory nerve conduction studies in normal infants and children. Clin Neurophysiol 111: 513–520.

George SR, Taylor MJ 1987 VEPs and SEPs in hydrocephalic infants before and after shunting. Clin Neurol Neurosurg 89(Suppl I): 96.

Gibson NA, Brezinova V, Levene MI 1992 Somatosensory evoked potentials in the term newborn. Electroencephalogr Clin Neurophysiol 84: 26–31.

Gilbert A, Razaboni R, Amar-Khodja S 1988 Indications and results of brachial plexus surgery in obstetrical palsy. Orthop Clin North Am 19(1): 91–105.

Gilmore R 1988 Use of somatosensory evoked potentials in infants and children. Neurol Clin 6: 839–859.

Gilmore R 1989 The use of somatosensory evoked potentials in infants and children. J Child Neurol 4: 3–19.

Gluckman PD, Wyatt JS, Azzopardi D, et al 2005 Selective head cooling with mild systemic hypothermia after neonatal encephalopathy: multicentre randomised trial. Lancet 365(9460): 663–670.

Goldberg RN, Goldman SL, Ramsay RE, et al 1982 Detection of seizure activity in the paralysed neonate using continuous monitoring. Pediatrics 69: 583–586.

Gorke W 1986 Somatosensory evoked cortical potentials indicating impaired motor development in infancy. Dev Med Child Neurol 28: 633–641.

Gotman J, Flanagan D, Zhang J, et al 1997a Automatic seizure detection in the newborn: methods and initial evaluation. Electroencephalogr Clin Neurophysiol 103: 356–362.

Gotman J, Flanagan D, Rosenblatt B, et al 1997b Evaluation of an automatic seizure detection method for the newborn EEG. Electroencephalogr Clin Neurophysiol 103: 363–369.

Goto K, Wakayama K, Sonoda H, et al 1992 Sequential changes in electroencephalogram continuity in very premature infants. Electroencephalogr Clin Neurophysiol 82: 197–202.

Govaert P, Vanhaesebrouck P, De Praeter C, et al 1989 Moebius sequence and prenatal brainstem ischemia. Pediatrics 84: 570–573.

Graziani LJ, Katz L, Cracco RQ, et al 1974 The maturation of inter-relationships of EEG patterns and auditory evoked responses in premature infants. Electroencephalogr Clin Neurophysiol 36: 367–375.

Greenwald AG, Schute PC, Shiveley JL 1984 Brachial plexus birth palsy: a 10-year report on the incidence and prognosis. J Pediatr Orthop 4: 689–692.

Greisen G, Hellström-Westas L, Lou H, et al 1987 EEG depression and germinal layer haemorrhage in the newborn. Acta Paediatr Scand 76: 273–275.

Gunn AJ, Parer JT, Mallard EC, et al 1992 Cerebral histologic and electrocorticographic changes after asphyxia in fetal sheep. Pediatr Res 31: 486–491.

Guthkelch AN, Schabassi RJ, Vries JK 1982 Changes in the visual evoked potentials of hydrocephalic children. Neurosurgery ii: 599–602.

Haas GH, Prechtl HFR 1977 Normal and abnormal EEG maturation in newborn infants. Early Hum Dev 1: 69–90.

Hagne I 1972 Development of the EEG in normal infants during the first year of life. A longitudinal study. Acta Paediatr Scand 232(Suppl): 1–53.

Hahn JS, Tharp BR 1990 The dysmature EEG pattern in infants with bronchopulmonary dysplasia and its prognostic implications. Electroencephalogr Clin Neurophysiol 76: 106–113.

Hahn JS, Monyer H, Tharp BR 1989 Interburst interval measurements in the EEGs of premature infants with normal neurological outcome. Electroencephalogr Clin Neurophysiol 73: 410–418.

Harbord MG, Weston PF 1995 Somatosensory evoked potentials predict neurologic outcome in full-term neonates with asphyxia. J Paediatr Child Health 31: 148–151.

Harding GFA 1992 Visual electrophysiology. In: *The Neurophysiological Examination of the Newborn Infant*. Clinics in Developmental Medicine, No. 120 (ed JA Eyre). MacKeith Press/Cambridge University Press, London, pp 112–123.

Harris R, Tizard JP 1960 The electroencephalogram in neonatal convulsions. J Pediatr 57: 501–520.

Hayakawa F, Okumura A, Kato T, et al 1999 Determination of timing of brain injury in preterm infants with periventricular leukomalacia with serial neonatal electroencephalography. Pediatrics 104(Pt 1): 1077–1081.

Hayakawa M, Okumura A, Hayakawa F, et al 2001 Background electroencephalographic (EEG) activities of very preterm infants born at less than 27 weeks gestation: a study on the degree of continuity. Arch Dis Child Fetal Neonatal Ed 84: F163–F167.

Hellström-Westas L, Rosen I 2005 Electroencephalography and brain damage in preterm infants. Early Hum Dev 81(3): 255–261.

Hellström-Westas L, Rosen I,Svenningsen NW 1991 Cerebral function monitoring during the first week of life in extremely small low birthweight (ESLBW) infants. Neuropediatrics 22(1): 27–32.

Hellström-Westas L, Rosén I, Svenningsen NW 1995 Predictive value of early continuous amplitude integrated EEG recordings on outcome after severe birth asphyxia in full term infants. Arch Dis Child 72: F34–F38.

Hellström-Westas L, Westgren I, Rosén I, et al 1988 Lidocaine for treatment of severe seizures in newborn infants. I. Clinical effects and cerebral electrical activity monitoring. Acta Paediatr Scand 77: 79–84.

Hellström-Westas L, Klette H, Thorngren-Jerneck K, et al 2001 Early prediction of outcome with a EEG in preterm infants with large intraventricular hemorrhages. Neuropediatrics 32(6): 319–324.

Hellström-Westas L, De Vries LS, Rosén I 2003 An Atlas of Amplitude-Integrated EEGs in the Newborn. Parthenon Press, Boca Raton, FL.

Hobley AJ, Harding GFA 1988 The effect of eye closure on the flash visually evoked response. Clin Vis Sci 3: 273–278.

Holmes GL 1986 Morphological and physiological maturation of the brain in the neonate and young child. J Clin Neurophysiol 3: 209–238.

Holmes GL, Ben-Ari Y 2001 The neurobiology and consequences of epilepsy in the developing brain. Paediatr Res 49: 320–325.

Holmes GL, Lombroso CT 1993 Prognostic value of background patterns in the neonatal EEG. J Clin Neurophysiol 10: 323–352.

Holmes GL, Rowe J, Hafford J, et al 1982 Prognostic value of the electroencephalogram in neonatal asphyxia. Electroencephalogr Clin Neurophysiol 53: 60–72.

Holmes GL, Gairsa JL, Chevassus-Au-Louis N, et al 1998 Consequences of neonatal seizures in the rat: morphological and behavioral effects. Ann Neurol 44: 845–857.

Holmes GL, Moshe SL, Jones HR (eds) 2006 Clinical Neurophysiology of Infancy, Childhood, and Adolescence. Butterworth-Heinemann, Boston, MA.

Honovar M, Meldrum BS 1997 Epilepsy. In: Greenfield's Neuropathology, 6th edn (eds DI Graham, PL Lantos). Arnold, London, pp 931–971.

Hrachovy RA, O'Donnell DM 1999 The significance of excessive rhythmic alpha and/or theta frequency activity in the EEG of the neonate. Clin Neurophysiol 110: 438–444.

Hrbek A, Mares P 1964 Cortical evoked responses to visual stimulation in full-term and premature newborns. Electroencephalogr Clin Neurophysiol 16: 575–581.

Hug TE, Fitzgerald KM, Cibis GW 2000 Clinical and electroretinographic findings in fetal alcohol syndrome. J AAPOS 4: 200–204.

Hughes JR, Ehemann B, Brown UA 1948 Electroencephalography of the newborn. IV. Abnormal electroencephalograms of the neonate. Am J Dis Child 76: 643–647.

Hughes JR, Fino JJ, Gagnon L 1983 Periods of activity and quiescence in the premature EEG. Neuropediatrics 14: 66–72.

Hughes JR, Fino JJ, Hart LA 1987 Premature temporal theta. Electroencephalogr Clin Neurophysiol 67: 7–15.

Hughes JR, Kuhlman DT, Hughes CA 1991 Electroclinical correlations of positive and negative sharp waves on the temporal and central areas in premature infants. Clin Electroencephalogr 22: 30–39.

Inder TE, Buckland L, Williams CE, et al 2003 Lowered electroencephalographic spectral edge frequency predicts the presence of cerebral white matter injury in premature infants. Pediatrics 111: 27–33.

Jiang ZD 1995 Long-term effect of perinatal and postnatal asphyxia on developing human auditory brainstem responses: peripheral hearing loss. Int J Pediatr Otorhinolaryngol 33: 225–238.

Jiang ZD 1998 Maturation of peripheral and brainstem auditory function in the first year following perinatal asphyxia: a longitudinal study. J Speech Lang Hear Res 41: 83–93.

Jiang ZD, Tierney TS 1996 Long-term effect of perinatal and postnatal asphyxia on developing human auditory brainstem responses: brainstem impairment. Int J Pediatr Otorhinolaryngol 34: 111–127.

Jiang ZD, Brosi DM, Shao XM, et al 2000 Maximum length sequence brainstem auditory evoked responses in term neonates who have perinatal hypoxia-ischemia. Pediatr Res 48: 639–645.

Johnson JL, White KR, Widen JE, et al 2005 A multisite study to examine the efficacy of the otoacoustic emission/automated auditory brainstem response newborn hearing screening protocol: introduction and overview of the study. Am J Audiol 14(2): S178–S185.

Joint Committee of Infant Hearing 2000 Year 2000 Position statement: principles and guidelines for early hearing detection and intervention programs. Am J Audiol 9: 9–29.

Jones HR 1990 EMG evaluation of floppy infant: differential diagnosis and technical aspects. Muscle Nerve 13: 338–437.

Jones HR, De Vivo DC, Darras BT 2003 Neuromuscular disorders of infancy, childhood and adolescence. A clinician's approach. Butterworth Heinemann, Philadelphia.

Journal of Clinical Neurophysiology 1992 Special issue. Evoked Potentials in Infants and Children. J Clin Neurophysiol 9: 323–414.

Karayiannis NB, Tao G, Xiong Y, et al 2005 Computerized motion analysis of videotaped neonatal seizures of epileptic origin. Epilepsia 46: 901–917.

Kitayama M, Otsubo H, Parvez S, et al 2003 Wavelet analysis for neonatal electroencephalographic seizures. Pediatr Neurol 29: 326–333.

Klebermass K, Kuhle S, Kohlhauser-Vollmuth C, et al 2001 Evaluation of the Cerebral Function Monitor as a tool for neurophysiological surveillance in neonatal intensive care patients. Child Nerv Syst 17: 544–550.

Klimach VJ, Cooke RWI 1988 Maturation of the neonatal somatosensory evoked potentials in preterm infants. Dev Med Child Neurol 30: 208–214.

Kohelet D, Shochat R, Lusky A, et al 2004 Risk factors for neonatal seizures in very low birthweight infants: population-based survey. J Child Neurol 19: 1–9.

Kriss A 1994 Skin ERGs: their effectiveness in paediatric visual assessment, confounding factors, and comparison with ERGs recorded using various types of corneal electrode. Int J Psychophysiol 16: 137–146.

Kubota T, Okumura A, Hayakawa F, et al 2002 Combination of neonatal electroencephalography and ultrasonography: sensitive means of early diagnosis of periventricular leukomalacia. Brain Dev 24: 698–702.

Kurtzberg D 1982 Event-related potentials in the evaluation of high-risk infants. In: Evoked Potentials (ed I Bodis-Wollner). Ann NY Acad Sci 388: 557–571.

Lacey DJ, Topper WH, Buckwald S, et al 1986 Preterm very-low-birth-weight neonates: relationship of EEG to intracranial hemorrhage, perinatal complications, and developmental outcome. Neurology 36: 1084–1087.

Lamblin MD, d'Allest AM, André M, et al 1999 EEG in premature and full-term neonates: developmental features and glossary. Neurophysiol Clin 29: 123–219. English translation (2000), Éditions Scientifiques et Médicales Elsevier.

Laroia N, Guillet R, Burchfield J, et al 1998 EEG background as predictor of electrographic seizures in high-risk neonates. Epilepsia 39: 545–551.

Lary S 1988 Brainstem evoked potentials. In: Fetal and Neonatal Neurology and Neurosurgery (eds M Levene, MJ Bennet, J Punt). Churchill Livingstone, London, pp 197–205.

Lary S, Briassoulis G, De Vries L, et al 1985 Hearing threshold in preterm and term infants by auditory brainstem response. J Pediatr 107: 593–599.

Laureau E, Vanasse M, Hebert R, et al 1986 Somatosensory evoked potentials and auditory brainstem responses in congenital hypothyroidism. I: A longitudinal study before and after treatment in six infants detected in the neonatal period. Electroencephalogr Clin Neurophysiol 64: 501–510.

Levene M, Evans DJ 2005 Hypoxic ischaemic brain injury. In: *Roberton's Neurology of the Newborn* (ed J Rennie). Churchill Livingstone, Philadelphia, PA, pp 1128–1148.

Levene MI, De Vries LS 1984 Extension of neonatal intraventricular haemorrhage. Arch Dis Child 59: 631–636.

Levene MI, Trounce JQ 1986 Causes of neonatal convulsions. Arch Dis Child 61: 78–79.

Liu A, Hahn JS, Heldt GP, et al 1992 Detection of neonatal seizures through computerized EEG analysis. Electroencephalogr Clin Neurophysiol 82: 30–37.

Lombroso CT 1979 Quantified electrographic scales on 10 preterm healthy newborns followed up to 40–43 weeks of conceptional age by serial polygraphic recording. Electroencephalogr Clin Neurophysiol 46: 460–474.

Lombroso CT 1985 Neonatal polygraphy in full-term and premature infants: review of normal and abnormal findings. J Clin Neurophysiol 2: 105–155.

Lütschg J 1986 Evozierte Potentiale bei neonataler Asphyxie. In: *Entwicklungsstörungen des ZNS* (ed Neuhäuser). Kohlhammer, Stuttgart, pp 48–149.

Mactier H, Dexter JD, Hewett JE, et al 1988 The electroretinogram in preterm infants. J Pediatr 113: 607–612.

Majnermer A, Rosenblatt B 2000 Prediction of outcome at school age in neonatal intensive care unit graduates using neonatal neurologic tools. J Child Neurol 15: 645–651.

Mandel R, Martinot A, Delepoulle F, et al 2002 Prediction of outcome after hypoxic–ischemic encephalopathy: a prospective clinical and electrophysiologic study. J Pediatr 141: 45–50.

Manfredi LG, Rocchi R, Panerai AE, et al 1983 EEG sleep patterns and endogenous opioids in infants of narcotic addicted mothers. Rev Electroencéphalogr Neurophysiol Clin 13: 199–206.

Marret S, Parain D, Samson-Dollfus D 1986 Positive Rolandic sharp waves and periventricular leukomalacia in the newborn. Neuropediatrics 17: 199–202.

Marret S, Parain D, Jeannot E, et al 1992 Positive Rolandic sharp waves in the EEG of the premature newborn: a five year prospective study. Arch Dis Child 67: 948–951.

Marret S, Parain D, Ménard J-F, et al 1997 Prognostic value of neonatal electroencephalography in premature newborns less than 33 weeks of gestational age. Electroencephalogr Clin Neurophysiol 102: 178–185.

Maynard DE, Jenkinson JL 1984 The cerebral function analysing monitor. Initial clinical experience, application and further development. Anaesthesia 39: 678–690.

Maynard D, Prior PF, Scott DF 1969 Device for continuous monitoring of cerebral activity in resuscitated patients. BMJ 4: 545–546.

McBride MC, Laroia N, Guillet R 2000 Electrographic seizures in neonates correlate with poor neurodevelopmental outcome. Neurology 55: 506–513.

McCulloch DL, Taylor MJ, Whyte HE 1991 Visual evoked potentials and long term visual prognosis following perinatal asphyxia. Arch Ophthalmol 105: 229–233.

Menache CC, Bourgeois BF, Volpe JJ 2002 Prognostic value of neonatal discontinuous EEG. Pediatr Neurol 27: 93–101.

Meyerson MD, Foushee DR 1978 Speech, language and hearing in Moebius syndrome: a study of 22 patients. Dev Med Child Neurol 20: 357–365.

Mikati MA, Feraru E, Krishnamoorthy K, et al 1990 Neonatal herpes simplex meningoencephalitis: EEG investigations and clinical correlates. Neurology 40: 1433–1437.

Mizrahi EM, Kellaway P 1987 Characterisation and classification of neonatal seizures. Neurology 37: 1837–1844.

Mizrahi EM, Kellaway P 1998 *Diagnosis and Management of Neonatal Seizures*. Lippincott-Raven, Philadelphia, PA.

Mizrahi EM, Tharp BR 1982 A characteristic EEG pattern in neonatal herpes simplex encephalitis. Neurology 32: 1215–1220.

Mizrahi EM, Hrachovy RA, Kellaway P 2003 *Atlas of Neonatal Electroencephalography*. Lippincott Williams and Wilkins, Philadelphia, PA.

Monod N, Dreyfus-Brisac C, Sfaello Z 1969 Dé pistage et pronostic de l'état de mal néonatal (d'aprés l'étude électroclinique de 150 cas). Arch Franc Pédiatr 26: 1085–1102.

Monod N, Pajot N, Guidasci S 1972 The neonatal EEG: statistical studies and prognostic value in full-term and pre-term babies. Electroencephalogr Clin Neurophysiol 32: 529–544.

Moshé SL, Alvarez LA 1986 Diagnosis of brain death in children. J Clin Neurophysiol 3: 239–249.

Murdoch-Eaton D, Darowski M, Livingston J 2001 Cerebral function monitoring in paediatric intensive care: useful features for predicting outcome. Dev Med Child Neurol 43: 91–96.

Mushin J 1988 Visual evoked potentials. In: *Fetal and Neonatal Neurology and Neurosurgery* (eds M Levene, MJ Bennett, J Punt). Churchill Livingstone, London, pp 206–212.

Muttitt SC, Taylor MJ, Kobayashi JS, et al 1991 Serial visual evoked potentials and outcome in term birth asphyxia. Pediatr Neurol 7(2): 86–90.

Neuringer M, Anderson GJ, Connor WE 1988 The essentiality of n-3 fatty acids for the development and function of the retina and brain. Ann Rev Nutr 8: 517–541.

Ng YK, Fielder AR, Shaw DE, et al 1988 Epidemiology of retinopathy of prematurity. Lancet ii: 1235–1238.

Noachtar S, Binnie CD, Ebersole J, et al 1999 A glossary of terms most commonly used by clinical electroencephalographers and proposal for the report form for the EEG findings. In: *Recommendations for the Practice of Clinical Neurophysiology* (eds G Deuschl, A Eisen). Clin Neurophysiol (Suppl 52): 21–40.

Norcia AM, Tyler CW, Piecuch R, et al 1987 Visual acuity development in normal and abnormal preterm human infants. J Pediatr Ophthalmol Strabis 24: 70–74.

Novotny EJ Jr, Tharp BR, Coen RW, et al 1987 Positive Rolandic sharp waves in the EEG of the premature infant. Neurology 37: 1481–1486.

O'Brien MJ, Lems YL, Prechtl HRF 1987 Transient flattenings in the EEG of newborns – a benign variation. Electroencephalogr Clin Neurophysiol 67: 16–26.

Ohtahara S 1978 Clinico-electrical delineation of epileptic encephalopathies in childhood. Asian Med J 21: 7–17.

Okumura A, Hayakawa F, Kato T, et al 1999 Positive Rolandic sharp waves in preterm infants with periventricular leukomalacia: their relation to background electroencephalographic abnormalities. Neuropediatrics 30: 278–282.

Okumura A, Hayakawa F, Kato T, et al 2003 Abnormal sharp transients on electroencephalograms in preterm infants with periventricular leukomalacia. J Pediatr 143: 26–30.

Oliveira AJ, Nunes ML, Haertel LM, et al 2000 Duration of rhythmic EEG patterns in neonates: new evidence for clinical and prognostic significance of brief rhythmic discharge. Clin Neurophysiol 111: 1646–1653.

Ortibus EL, Sum JM, Hahn JS 1996 Predictive value of EEG for outcome and epilepsy following neonatal seizures. Electroencephalogr Clin Neurophysiol 98: 175–185.

Painter MJ, Scher MS, Stein AD, et al 1999 Phenobarbital compared with phenytoin for the treatment of neonatal seizures. N Engl J Med 341: 485–489.

Pampiglione G 1974 What may we learn from cerebral malformations? Proc R Soc Med 67: 337–342.

Pape KE, Wigglesworth JS 1979 *Haemorrhage, Ischaemia and the Perinatal Brain*. Clinics in Developmental Medicine, No. 69/70. Spastics International Medical Publications/Heinemann Medical Books, London.

Parmelee AH, Stern E 1972 Development of states in infants. In: *Sleep and the Maturing Nervous System* (eds CD Clemente, DP Purpura, FE Mayer). Academic Press, New York, pp 199–228.

Parmelee AH, Wenner WH, Akiyama Y, et al 1967 Sleep states in premature infants. Dev Med Child Neurol 9: 70–77.

Pasternak JF, Volpe JJ 1979 Full recovery from prolonged brainstem failure following intraventricular hemorrhage. J Pediatr 95: 1046–1049.

Patrizi S, Holmes GL, Orzalesi M, et al 2003 Neonatal seizures: characteristics of EEG ictal activity in preterm and fullterm infants. Brain Dev 25: 427–437.

Payan J 1991 Clinical electromyography in infancy and childhood. In: *Paediatric Neurology*, 2nd edn (ed EM Brett). Churchill Livingstone, Edinburgh, pp 797–829.

Peña M, Birch D, Uauy R, et al 1999 The effect of sleep state on electroretinographic (ERG) activity during early human development. Early Hum Dev 55: 51–62.

Peters JF, Varner JL, Ellingson RJ 1981 Interhemispheric amplitude symmetry in the EEGs of normal full term, low risk premature, and trisomy-21 infants. Electroencephalogr Clin Neurophysiol 51: 165–169.

Pezzani C, Radvanyi-Bouvet MF, Relier JP, et al 1986 Neonatal electroencephalography during the first twenty four hours of life in full-term newborn infants. Neuropediatrics 17: 11–18.

Pickett J, Berg B, Chaplin E, et al 1976 Syndrome of botulism in infancy: clinical and electrophysiological study. N Engl J Med 295: 770–772.

Picton TW, Durieux-Smith A 1988 Auditory evoked potentials in the assessment of hearing. Neurol Clin 6: 791–808.

Pierrat V, Eken P, De Vries LS 1997 The predictive value of cranial ultrasound and of somatosensory evoked potentials after nerve stimulation for adverse neurological outcome in preterm infants. Dev Med Child Neurol 39: 398–403.

Pike MG, Holmstrom G, de Vries LS, et al 1994 Patterns of visual impairment associated with lesions of the preterm infant brain. Dev Med Child Neurol 36: 849–862.

Pinto F, Torrioli MG, Casella G, et al 1988 Sleep in babies born to chronically heroin addicted mothers. A follow up study. Drug Alcohol Depend 21: 43–47.

Pitt M, Vredeveld JW 2005 The role of electromyography in the management of the brachial plexus palsy of the newborn. Clin Neurophysiol 116: 1756–1761.

Placzek M, Mushin J, Dubowitz LMS 1985 Maturation of the visual evoked response and its correlation with visual acuity in preterm infants. Dev Med Child Neurol 27: 448–454.

Pratt H, Aminoff M, Nuwer MR, et al 1999 Short latency auditory evoked potentials. In: *Recommendations for the Practice of Clinical Neurophysiology: Guidelines of the International Federation of Clinical Neurophysiology*, 2nd edn (eds G Deuschl, A Eisen). Electroencephalogr Clin Neurophysiol (Suppl 52): 69–77.

Prechtl HFR 1974 The behavioural states of the newborn infant (a review). Brain Res 76: 185–212.

Prechtl HFR, Beintema DJ 1964 *The Neurological Examination of the Full Term Newborn Infant*. Clinics in Developmental Medicine, No. 12. Spastics Society/Heinemann Medical, London.

Prechtl HFR, Beintema D 1977 *The Neurological Examination of the Full-Term Newborn Infant*. Clinics in Developmental Medicine (2nd edn), No. 63. Spastics International Medical Publications/Heinemann Medical, London.

Pressler RM, Boylan GB, Morton M, et al 2001 Early serial EEG in hypoxic ischaemic encephalopathy. Clin Neurophysiol 112: 31–37.

Prior PF, Maynard DE 1986 *Monitoring Cerebral Function*. Elsevier, Amsterdam.

Pryds O, Trojaborg W, Carlson V, et al 1989 Determinants of visual evoked potentials in preterm infants. Early Hum Dev 19: 117–125.

Pryor DS, Don N, Macourt DC 1981 Fifth day fits: a syndrome of neonatal convulsions. Arch Dis Child 56: 753–758.

Radvanyi-Bouvet MF, Vallecalle MH, Morel-Kahn F, et al 1985 Seizures and electrical discharges in premature infants. Neuropediatrics 16: 143–148.

Raimbault J 1988 *Les Conductions Nerveuses chez l'Enfant Normal*. Expansion Scientifique Française, Paris.

Ramelli GP, Sozzo AB, Vella S, et al 2005 Benign neonatal sleep myoclonus: an under-recognized, non-epileptic condition. Acta Paediatr 94: 962–963.

Rando T, Ricci D, Mercuri E, et al 2000 Periodic lateralized epileptiform discharges (PLEDs) as early indicator of stroke in full-term newborns. Neuropediatrics 31: 202–205.

Reding MJ, Kader FJ, Pellegrino RJ, et al 1979 Seizures in hydranencephaly: a report of 2 cases. Electroencephalogr Clin Neurophysiol 47: 27.

Renault F 2000 Cranial nerve studies in 22 children with Möbius syndrome. Muscle Nerve 23: 1631.

Renault F, Raimbault J 1992 Electromyographie faciale, linguale et pharyngée chez l'enfant: une methode d'etude des troubles des succion-deglutition et de leur physiopathologie. Neurophysiol Clin 22: 240–280.

Rennie JM 1997 Cranial ultrasound imaging and prognosis. In: *Neonatal Cerebral Ultrasound* (ed JM Rennie). Cambridge University Press, Cambridge, pp 210–234.

Rennie JM 2005 *Neurological Problems in the Newborn*, 4th edn (ed JM Rennie). Elsevier, Amsterdam, pp 1093–1203.

Rennie JM, Chorley G, Boylan GB, et al 2004 Non-expert use of the cerebral function monitor for neonatal seizure detection. Arch Dis Child Fetal Neonatal Ed 89(1): F37–F40.

Roberton NRC 1969 Effect of acute hypoxia on blood pressure and electroencephalogram of newborn babies. Arch Dis Child 44: 719–725.

Rose AL, Lombroso CT 1970 Neonatal seizure states. A study of clinical, pathological and electroencephalographic features in 137 full-term babies with a long-term follow-up. Pediatrics 45: 404–425.

Rowe JC, Holmes GL, Hafford J, et al 1985 Prognostic value of the electroencephalogram in term and preterm infants following neonatal seizures. Electroencephalogr Clin Neurophysiol 60: 183–196.

Sacco G, Buchthal F, Rosenfalck P 1962 Motor unit potentials at different ages. Arch Neurol 6: 366–373.

Sahni R, Schulze KF, Stefanski M, et al 1995 Methodological issues in coding sleep states in immature infants. Dev Psychobiol 28: 85–101.

Sainio K, Granstrom ML, Pettay O, et al 1983 EEG in neonatal herpes simplex encephalitis. Electroencephalogr Clin Neurophysiol 56: 556–561.

Salamy A, Eldredge L 1994 Risk for ABR abnormalities in the nursery. Electroencephalogr Clin Neurophysiol 92: 392–395.

Sarnat HB 1975 Pathogenesis of decerebrate seizures in the premature infant with intraventricular haemorrhage. J Paediatr 87: 154–155.

Sarnat HB 1984 Anatomic and physiologic correlates of neurologic development in prematurity. In: *Topics in Neonatal Neurology*. Grune and Stratton, New York, pp 1–25.

Sarnat HB, Sarnat MS 1976 Neonatal encephalopathy following fetal distress. Arch Neurol 33: 696–705.

Scalais E, Adant AF, Nuttin C, et al 1998 Multimodality evoked potentials as a prognostic tool in term asphyxiated newborns. Electroencephalogr Clin Neurophysiol 108: 199–207.

Scher MS 1994 Neonatal encephalopathies as classified by EEG-sleep criteria: severity and timing based on clinical/pathologic correlations. Pediatr Neurol 11: 189–200.

Scher MS, Hamid MY, Steppe DA, et al 1993a Ictal and interictal electrographic seizure durations in preterm and term neonates. Epilepsia 34: 284–288.

Scher MS, Aso K, Beggarly ME, et al 1993b Electrographic seizures in preterm and full-term neonates: clinical correlates, associated brain lesions, and risk for neurologic sequelae. Pediatrics 91: 128–134.

Scher MS, Bova JM, Dokianakis SG, et al 1994 Positive temporal sharp waves on EEG recordings of healthy neonates: a benign pattern of dysmaturity in pre-term infants at post-conceptional term ages. Electroencephalogr Clin Neurophysiol 90: 173–178.

Scher MS, Dokianakis SG, Sun M, et al 1996 Computer classification of sleep in preterm and full-term neonates at similar postconceptional ages. Sleep 19: 18–25.

Scher MS, Jones BL, Steppe DA, et al 2003 Functional brain maturation in neonates as measured by EEG-sleep analyses. Clin Neurophysiol 114: 875–882.

Scher MS, Waisanen H, Loparo K, et al 2005a Prediction of neonatal state and maturational change using dimensional analysis. J Clin Neurophysiol 22: 159–165.

Scher MS, Turnbull J, Loparo K, et al 2005b Automated state analyses: proposed applications to neonatal neurointensive care. J Clin Neurophysiol 22: 256–270.

Scher MS, Johnson MW, Holditch-Davis D 2005c Cyclicity of neonatal sleep behaviors at 25 to 30 weeks' postconceptional age. Pediatr Res 57: 879–882.

Schulte FJ 1970 Neonatal EEG and brain maturation: facts and fallacies. Dev Med Child Neurol 12: 396–399.

Selton D, Andre M 1997 Prognosis of hypoxic–ischaemic encephalopathy in full-term newborns; value of neonatal electroencephalography. Neuropediatrics 28: 276–280.

Selton D, Andre M, Hascoet JM 2000 Normal EEG in very premature infants: reference criteria. Clin Neurophysiol 111: 2116–2124.

Shepherd AJ, Saunders KJ, McCulloch DL, et al 1999 Prognostic value of flash visual evoked potentials in preterm infants. Dev Med Child Neurol 41: 9–15.

Shewmon DA 1983 Dissociation between cortical discharges and ictal movements in neonatal seizures. Ann Neurol 14: 368.

Shewmon DA 1990 What is a neonatal seizure? Problems in definition and quantification for investigative and clinical purposes. J Clin Neurophysiol 7: 315–368.

Shirataki S, Prechtl HFR 1977 Sleep state transitions in newborn infants: preliminary study. Dev Med Child Neurol 19: 316–325.

Sinclair DB, Campbell M, Byrne P, et al 1999 EEG and long-term outcome of term infants with neonatal hypoxic–ischemic encephalopathy. Clin Neurophysiol 110: 655–659.

Sjoberg I, Erichs K, Bjerre I 1988 Cause and effect of obstetric (neonatal) brachial plexus palsy. Acta Paediatr Scand 77: 357–364.

Smit BJ, Kok JH, De Vries LS, et al 1999 Motor nerve conduction velocity in very preterm infants. Muscle Nerve 22: 372–377.

Smit LS, Vermeulen RJ, Fetter WP, et al 2004 Neonatal seizure monitoring using non-linear EEG analysis. Neuropediatrics 35: 329–335.

Smith SJM 1996 The role of neurophysiological investigations in traumatic brachial plexus injuries in adults and children. J Hand Surg (Br) 21: 145–148.

Smyth V, McPherson B, Kei J, et al 1999 Otoacoustic emission criteria for neonatal hearing screening. Int J Pediatric Otorhinolaryngol 48: 9–15.

Stanley OH, Fleming PJ, Morgan MH 1989 Abnormal development of visual function following intrauterine growth retardation. Early Hum Dev 19: 87–101.

Starr A, Amlie RM, Martin WH, et al 1977 Development of auditory function in newborn infants revealed by auditory brainstem evoked potentials. Pediatrics 60: 831–834.

Staudt F, Roth JG, Engel RC 1981 The usefulness of electroencephalography in curarised newborns. Electroencephalogr Clin Neurophysiol 51: 205–208.

Staudt F, Scholl ML, Coen RW, et al 1982 Phenobarbital therapy in neonatal seizures and the prognostic value of the EEG. Neuropediatrics 13: 24–33.

Stein L, Ozdamar O, Kraus N, et al 1983 Follow-up of infants screened by auditory brainstem response in the neonatal intensive care unit. J Pediatr 103: 447–453.

Stockard JE, Stockard JJ, Kleinberg F, et al 1983 Prognostic value of brainstem auditory evoked potentials in neonates. Arch Neurol 40: 360–365.

Stockard-Pope JE, Werner SS, Bickford RG, et al (eds) 1992 *Atlas of Neonatal Electroencephalography*. Raven Press, New York.

Sunderland S 1978 *Nerves and Nerve Injuries*, 2nd edn. Churchill Livingstone, Edinburgh, pp 878–880.

Suppiej A 2001 Ruolo dei potenziali evocati nell'encefalopatia ipossico-ischemia neonatale: revisione della letteratura. Ann Ist Superiore Sanità 37: 515–525.

Takeuchi T, Watanabe K 1989 The EEG evolution and neurological prognosis of neonates with perinatal hypoxia. Brain Dev 11: 115–120.

Task Force on Brain Death in Children 1987 Guidelines for the determination of brain death in children. Pediatrics 80: 298–300.

Taylor MJ 1992 Visual evoked potentials. In: *The Neurophysiological Examination of the Newborn Infant*. Clinics in Developmental Medicine, No. 120 (ed JA Eyre). MacKeith Press/Cambridge University Press, London, pp 93–111.

Taylor MJ, Saliba E, Laugier J 1996 Use of evoked potentials in preterm neonates. Arch Dis Child 74: F70–F76.

Taylor MJ, Murphy WJ, Whyte HE 1992 Prognostic reliability of somatosensory and visual evoked potentials of asphyxiated term infants. Dev Med Child Neurol 34: 507–515.

Ter Horst HJ, Brouwer OF, Bod AF 2004 Burst suppression on amplitude-integrated electroencephalogram may be induced by midazolam: a report on three cases. Acta Paediatr 93: 559–563.

Tharp BR 1987 The electroencephalographic aspects of ischaemic hypoxic encephalopathy and intraventricular haemorrhage. In: *Neonatal Brain and Behavior* (eds H Yabuchi, K Watanabe, S Okada). University of Nagoya Press, Japan, pp 71–85.

Tharp BR 2002 Neonatal seizures and syndromes. Epilepsia 43(Suppl 3): 2–10.

Tharp BR, Cukier F, Monod N 1981 The prognostic value of the electroencephalogram in premature infants. Electroencephalogr Clin Neurophysiol 51: 219–236.

Thomas JE, Lambert EH 1960 Ulnar nerve conduction velocity and H-reflex in infants and children. J Appl Physiol 15: 1–9.

Thordstein M, Bågenholm R, Andreasson S, et al 2000 Long-term EEG monitoring in neonatal and pediatric intensive care. In: *Clinical Neurophysiology at the Beginning of the 21st Century* (eds Z Ambler, N Nevšmalová, PM Rossini). Clin Neurophysiol 53(Suppl): 76–83.

Thordstein M, Flisberg A, Lofgren N, et al 2004 Spectral analysis of burst periods in EEG from healthy and post-asphyctic full-term neonates. Clin Neurophysiol 115: 2461–2466.

Thornberg E, Ekström-Jodal B 1994 Cerebral function monitoring: a method of predicting outcome in term infants after severe perinatal asphyxia. Acta Paediatr 83: 596–601.

Toet M, Hellström-Westas L, Groenendaal F, et al 1999 Amplitude integrated EEG 3 and 6 h after birth in full term neonates with hypoxic–ischaemic encephalopathy. Arch Dis Child Fetal Neonatal Ed 81: 19–23.

Toet MC, Van der Meij W, De Vries LS, et al 2002 Comparison between simultaneously recorded amplitude integrated electroencephalogram (cerebral function monitor) and standard electroencephalogram in neonates. Pediatrics 109: 772–779.

Torres F, Anderson CM 1985 The normal EEG of the human newborn. J Clin Neurophysiol 2: 89–103.

Torres F, Blaw ME 1968 Longitudinal EEG-clinical correlations in children from birth to 4 years of age. Pediatrics 41: 945–954.

Toth C, Harder S, Yager J 2003 Neonatal herpes encephalitis: a case series and review of clinical presentation. Can J Neurol Sci 30: 36–40.

Tudehope DI, Vacca A 1988 Traumatic injury to the nervous system. In: *Fetal and Neonatal Neurology and Neurosurgery* (eds M Levene, MJ Bennett, J Punt). Churchill Livingstone, Edinburgh, pp 393–404.

Turnbull JP, Loparo KA, Johnson MW, et al 2001 Automated detection of trace alternant during sleep in healthy full-term neonates using discrete wavelet transform. Clin Neurophysiol 112: 1893–1900.

Uauy RD, Birch DG, Birch EE, et al 1990 Effect of dietary omega-3 fatty acids on retinal function of very-low-birth-weight neonates. Pediatr Res 28: 485–492.

Uauy R, Birch E, Birch D, et al 1992 Visual and brain function measurements in studies of n-3 fatty acid requirements of infants. J Pediatrics 120(Symp 4, Pt 2): S168–S180.

Van Lieshout HB, Jacobs JW, Rotteveel JJ, et al 1995 The prognostic value of the EEG in asphyxiated newborns. Acta Neurol Scand 91: 203–207.

Van Sweden B, Koenderink M, Windau G, et al 1991 Long-term EEG monitoring in the early premature: developmental and chronobiological aspects. Electroencephalogr Clin Neurophysiol 79: 94–100.

Varner JL, Ellingson RJ, Danahy T, et al 1977 Interhemispheric amplitude symmetry in the EEGs of full-term newborns. Electroencephalogr Clin Neurophysiol 43: 846–852.

Vecchierini MF, D'Allest AM, Verpillat P 2003 EEG patterns in 10 extreme premature neonates with normal neurological outcome: qualitative and quantitative data. Brain Dev 25: 330–337.

Vecchierini-Blineau MF, Nguyen The Tich S, Debillon T, et al 1996a [Severe periventricular leukomalacia: characteristic electroencephalographic features.] Neurophysiol Clin 26: 102–108.

Vecchierini-Blineau MF, Nogues B, Louvet S, et al 1996b Positive temporal sharp waves in electroencephalograms of the premature newborn. Neurophysiol Clin 26: 350–362.

Vermeulen RJ, Sie LT, Jonkman EJ, et al 2003 Predictive value of EEG in neonates with periventricular leukomalacia. Dev Med Child Neurol 45: 586–590.

Vernon DD, Holzman BH 1986 Brain death: considerations for pediatrics. J Clin Neurophysiol 3: 251–265.

Victor S, Appleton RE, Beirne M, et al 2005 Spectral analysis of electroencephalography in premature newborn infants: normal ranges. Pediatr Res 57: 336–341.

Viniker DA, Bromley IE, Maynard DE 1980 Monitoring the neonate with the cerebral function monitor. In: *Fetal and Neonatal Physiological Measurements* (ed P Rolfe). Pitman Medical, London, pp 297–306.

Viniker DA, Maynard DE, Scott DF 1984 Cerebral function monitor studies in the neonate. Clin Electroencephalogr 15: 185–192.

Volpe JJ 2000a *Neurology of the Newborn*, 4th edn (ed JJ Volpe). WB Saunders, Philadelphia, PA.

Volpe JJ 2000b Hypoxic–ischemic encephalopathy: neuropathology and pathogenesis. In: *Neurology of the Newborn*, 4th edn (ed JJ Volpe). WB Saunders, Philadelphia, PA, pp 279–313.

Volpe JJ 2000c Neonatal seizures. In: *Neurology of the Newborn*, 4th edn (ed JJ Volpe). WB Saunders, Philadelphia, PA, pp 129–159.

Vredeveld JW, Blaauw G, Slooff BA, et al 2000 The findings in paediatric obstetric brachial palsy differ from those in older patients: a suggested explanation. Dev Med Child Neurol 42: 158–161.

Watanabe K, Iwase K, Hara K 1972 Maturation of visual evoked response in premature infants. Dev Med Child Neurol 14: 425–435.

Watanabe K, Hara K, Iwase K 1976 The evolution of neurophysiological features in holoprosencephaly. Neuropediatrics 7: 19–41.

Watanabe K, Miyazaki S, Hara K, et al 1980 Behavioral state cycles, background EEGs and prognosis of newborns with perinatal hypoxia. Electroencephalogr Clin Neurophysiol 49: 618–625.

Watanabe K, Hara K, Miyazaki S, et al 1981 The value of EEG and cerebral evoked potentials in the assessment of neonatal intracranial hemorrhage. Eur J Pediatr 137: 177–184.

Wanatabe K, Hara K, Miyazaki S, et al 1982 Apneic seizures in the new born. Am J Dis Child 136: 980–984.

Watanabe K, Hakamada S, Kuroyanagi M, et al 1983 Electroencephalographic study of intraventricular haemorrhage in the preterm newborn. Neuropediatrics 14: 225–230.

Werner SS, Stockard JE, Bickford RG 1977 *Atlas of Neonatal Electroencephalography*. Raven Press, New York.

White CP, Cooke RWI 1994 Somatosensory evoked potentials following posterior tibial nerve stimulation predict later motor outcome. Dev Med Child Neurol 36: 34–40.

Whiting S, Jan JE, Wong P, et al 1985 Permanent cortical visual impairment in children. Dev Med Child Neurol 27: 730–739.

Whyte HE 1993 Visual evoked potentials in neonates following asphyxia. Clin Perinatol 20: 451–461.

Whyte HE, Taylor MJ, Menzies R, et al 1986 Prognostic utility of visual evoked potentials in full term asphyxiated neonates. Pediatr Neurol 2: 220–223.

Willis J, Seales D, Frazier E 1984 Short latency somatosensory evoked potentials in infants. Electroencephalogr Clin Neurophysiol 59: 366–373.

Willis J, Duncan C, Bell R 1987 Short latency somatosensory evoked potentials in perinatal asphyxia. Pediatr Neurol 3: 203–207.

Working Group of the Royal College of Physicians 1995 Criteria for the diagnosis of brain stem death. J R Coll Phys London 29: 381–382.

Young GB, da Silva OP 2000 Effects of morphine on the electroencephalograms of neonates: a prospective, observational study. Clin Neurophysiol 111: 1955–1960.

Chapter 6

Neurophysiology in Paediatrics

INTRODUCTION

This chapter addresses various general and specific aspects of clinical neurophysiology that are peculiar to children. The reader is also referred to the chapters describing the normal maturation of the electroencephalogram (EEG) (Chapter 2, pp. 31–34), evoked potentials (EPs) (Chapter 3) and the electromyogram (EMG) (Chapter 5, pp. 180–188). Apart from these considerations, an important reason for regarding paediatric clinical neurophysiology as a distinct subspecialty is that there are many diseases, often with characteristic neurophysiological features, that occur only in childhood. Examples include infantile spasms, which is correlated with a hypsarrhythmic EEG, and neuronal ceroid lipofuscinosis, which is associated with a specific EEG response to low rates of stroboscopic stimulation and distinctive EPs.

METHODOLOGICAL PROBLEMS AND NORMATIVE DATA

Electroencephalography

With mutual confidence established (see pp. 1 ff.), the technologist should be able to record an EEG while the child is awake, with open and closed eyes, activated by photic stimulation and hyperventilation and, whenever possible, during spontaneous drowsiness and (at least) light sleep. Spontaneous sleep is preferred, but this is time consuming and can be difficult to obtain even in cooperative children. Sometimes it will be necessary to administer melatonin or sedative drugs to obtain a readable EEG at all or if a sleep recording is necessary, for example if Landau–Kleffner syndrome is suspected (see p. 258). The risks of activation procedures (see pp. 61–76) and the need for informed consent (see pp. 64–66) must be considered as in adults. In general the adverse consequences of precipitating a seizure in a child in hospital are less significant than in an adult.

A prolonged recording may be needed, particularly for the investigation of seizures. All-night recording may be required to capture nocturnal attacks, and for long term recording by day or over periods of more than 24 h an ambulatory cassette recorder or telemetry system may be used (Binnie et al 2003). Techniques that allow correlation between clinical and EEG data, such as combined video and EEG recording, are of particular value in the investigation of seizure disorders in children.

Interpretation of an EEG from a child depends very much on reports of the child's behaviour, level of vigilance and observations of movement, especially chewing, swallowing, sucking, etc., and, in the infant, even breathing (see pp. 22–31). Thus it is extremely important before trying to interpret an EEG to make sure that the technologist is satisfied with the technical quality of the recording and to discuss in detail specific aspects of the procedure that might lead to ambiguity or confusion. For example, misplacement of electrodes or resistances high enough to produce apparent asymmetries may be unavoidable – but *must* be documented. Nothing can replace first-hand observation, and a team approach is recommended, whereby the clinical neurophysiologist joins the technologist to observe particularly problematic EEGs whilst they are being obtained. If a child refuses to lie down on the couch, a technologist skilled in paediatric work will consider recording while the child is lying in its mother's arms or sitting on a nurse's lap. It is often of great help to perform simultaneous video recording of the child and the EEG in order to correlate changes in the EEG (including possible artefacts) with behaviour.

Evaluation of the composition of the ongoing rhythms has to take into account both the age of the child and its level of vigilance. This is especially important because, in general terms, the maturation of the EEG with age involves replacement of slower (delta–theta) components by an emergent alpha rhythm. In contrast, generalised pathological processes follow the opposite sequence, leading to the slowest components of the ongoing activity being excessive for the child's age. Thus, it may become extremely difficult to differentiate minor changes in the ongoing EEG from the range of normality for a given age group or from spontaneous variation within the individual patient (see pp. 31–34). Slow activity, which is abnormal and reflects pathology, at one stage of development may be normal in an earlier phase of maturation. Given the considerable variance, both of the normal EEG and of the rate of maturation, it may be impossible to detect minor but significant changes in the frequency composition of the individual child's EEG unless serial recordings are made.

The organisation of sleep phenomenology and architecture also changes with age (see pp. 34–43 and

180–188), and various normal transients or episodic features become more prominent. Their appearance may cause confusion when identifying certain genetically determined EEG features, which can be expected with a changing probability at different ages. This means that, to exclude the influence of genetic traits, there must be sufficient information about the child's EEG in the awake and sleeping states, if necessary with repeated recordings as the child matures. To correlate clinical findings with EEG phenomena legitimately, sufficient data must always be available to allow differentiation of pure coincidences from causal relationships.

Evoked potentials

EP testing in children requires an understanding of the methodological differences with adults as well as within the various age groups. Up to the age of 6 months the methods already described for neonates (see p. 169–180 and Chapter 3) are used. In older children the general approach to the child and to the often anxious parents has first to be considered. The EP recording room should be equipped with toys for different ages, small books, children's music and a television for showing cartoons (especially useful during somatosensory evoked potential (SEP) recording). It should be possible to darken the room completely as well as to have bright illumination. Oxygen, bronchial aspiration and cardiorespiratory monitors should be available for dealing safely with sedated children. Small electrodes for recording and SEP stimulation may be needed. The test should be planned with a knowledge of age, mental status and degree of cooperation of the child. Possible sedation should be prepared. Knowledge of the clinical problem guides the strategy of the neurophysiological examination, first gathering the essential and robust data, and then, if the cooperation and patience of the child will allow, other complementary data.

The techniques applied need to take into account the degree of maturation at receptor, peripheral and central nervous system (CNS) levels, in relation to the age of the child. The appropriate environment, stimulus and recording parameters for recording EPs in the visual, auditory and somatosensory modalities, the maturation of the waveforms and normative data are described in Chapter 3. EP testing in paediatric practice is less difficult than might be thought and can be very rewarding, as it is often possible to obtain, very quickly and easily, important clinical information that is unobtainable by other means.

Electromyography

The fundamentals of the technique and normative data applicable to both the neonatal and paediatric patient are discussed in Chapters 4 and 5 (see p. 217) and by Payan (1991).

SEIZURE DISORDERS

It is often easier to classify epilepsy by seizures in adults (International Classification of Epileptic Seizures) and by syndromes in childhood. The classification of epilepsies and epileptic syndromes according to that of the International League Against Epilepsy (ILAE) (Commission on Classification and Terminology of the ILAE 1981, 1985, 1989) is based on the seizure type, EEG and clinical picture. It is not necessarily equivalent to neurobiological criteria, such as aetiology or genetic factors.

Recently the Task Force on Classification and Terminology of the ILAE has published a new proposal for a diagnostic scheme (Engel 2001). This scheme includes other syndromes that are currently not included in the approved ILAE classification (1989), such as benign infantile seizures, benign familial infantile seizures, autosomal dominant nocturnal frontal lobe epilepsy, familial temporal lobe epilepsy, hemiconvulsion–hemiplegia syndrome and startle epilepsy. Several genetic syndromes, such as generalised epilepsies with febrile seizures plus, have now been identified and are included. In addition, the scheme mentions a few syndromes that are still being defined, such as familial focal epilepsy with variable foci, migrating partial seizures of early infancy, and myoclonic status in non-progressive encephalopathies. These pathologies are not considered further here.

The more recent opinion regarding those aspects pertaining to childhood epilepsy have been reviewed in: *Epilepsy through the Life Cycle* (*Epilepsia* 2002), and *Catastrophic Epilepsies* (*Epilepsia* 2004), *Myoclonic Epilepsies in Childhood* (*Advances in Neurology* 2005) and *Epileptic Encephalopathies in Childhood* (*Journal of Clinical Neurophysiology* 2003).

Genetic factors are of predominant importance in those epilepsies that are not associated with neurological abnormalities, but may also play a role in epilepsies associated with demonstrable brain damage. The inheritance of most idiopathic epilepsies is thought to be multifactorial and result from the interaction of complex genetic influences and environmental factors. However, several single gene mutations in certain inherited epilepsy syndromes have been discovered over the last 5 years. Often the phenotype is similar to those of common idiopathic epilepsies (for reviews see George 2004, Gourfinkel-An et al 2004). The mode of inheritance is variable with the type of epilepsy. A further 1–2% of cases can be attributed to neurological disorders presenting with seizures with a single-locus trait such as tuberous sclerosis, Angelman's syndrome or metabolic errors.

Because brain growth and neurophysiological maturation are incomplete in children, epileptic seizures and epilepsy pose specific problems, as symptomatology is often dependent on age. For example, infantile spasms present only between 3 and 18 months of life, which is a period of maximal cerebral development, while absence seizures usually occur after the age of 3 years. As age of onset is also one of the most prominent clinical features, this chapter is structured according to the age of onset of the syndromes.

The effective use of the EEG in childhood epilepsy requires a detailed knowledge of the electroclinical patterns underlying the various syndromes, which may not be familiar to the general clinical neurophysiologist. Indeed, for the enthusiastic electroencephalographer, with an interest in EEG phenomenology, electroclinical correlations in the childhood epilepsies provide a particularly rewarding area of study. In the account that follows, therefore, each syndrome is introduced by a brief summary of the main clinical features. Some essential details of epilepsies and epileptic syndromes in infants and children are also given in *Table 6.1*.

Epileptic syndromes in neonates

The syndromes presenting in the neonatal period are:
- benign neonatal convulsions (fifth-day fits)
- benign familial neonatal convulsions
- early myoclonic encephalopathy
- early infantile epileptic encephalopathy with burst suppression (Ohtahara's syndrome).

They are considered in the section on neonatology (see pp. 193–196).

Epileptic syndromes of infancy and early childhood

The following syndromes are discussed here:
- infantile spasms (West syndrome, Blitz–Nick–Salaam Krämpfe)
- febrile seizures
- benign myoclonic epilepsy in infancy
- severe myoclonic epilepsy in infancy (Dravet's syndrome)
- Lennox–Gastaut syndrome
- epilepsy with myoclonic–astatic seizures.

Infantile spasms (West syndrome, Blitz–Nick–Salaam Krämpfe)

Infantile spasms are brief axial contractions of the head, trunk, and limbs (flexor–extensor, flexor, or extensor), presenting at the age of 3–12 months; only 3% present after the first year of life (Vigevano et al 2001). The incidence is between 24 and 42/100,000 births with a male preponderance (2:1). The condition may be cryptogenic (in 30–40%) or symptomatic of structural cerebral disease, notably tuberose sclerosis. Delay in psychomotor development or regression accompanies onset of spasms. The condition often evolves into other seizure disorders, including the Lennox–Gastaut syndrome. Learning disability results in 60% of cryptogenic and in 90% of symptomatic cases. Classical attacks (salaam) occur in clusters of 10–20 in a row and consist of rapidly bowing the head, flexion of the trunk and hips, and extension of the arms, often followed by a cry. Infantile spasms occur

as a response of the immature brain to many different types of insults, including hypoxic–ischaemic, infective, haemorrhagic and traumatic, metabolic and toxic (Aicardi 1994). Around 10% of children with infantile spasms have tuberous sclerosis (see p. 266). West syndrome refers to the combination of infantile spasms, hypsarrhythmia and learning disability.

Hypsarrhythmia ('hypselos' Greek for 'high'; and 'arrhythmia') is the typical electrophysiological picture in infantile spasms. Hypsarrhythmia as an interictal EEG pattern is defined as a diffuse, high voltage (exceeding 200 μV), irregular, largely chaotic mixture of slow waves (1–7 Hz) with sharp waves and spikes. Superimposed low voltage faster components may be present (*Fig. 6.1(a)*). The amount of spike and sharp wave activity varies, as does the amplitude, morphology, topographic distribution (multifocal spikes) and organisation. The localisation of sharp waves and spikes is not constant but shifts from one moment to another, although some children may have in addition a more consistent focus. In contrast to Lennox–Gastaut syndrome, the discharges are usually of highest amplitude over the posterior areas. Variations in the typical hypsarrhythmia are common (modified hypsarrhythmia) and include:
- asymmetric or even unilateral hypsarrhythmia (*Fig. 6.2*)
- association of a constant focus of abnormal discharges
- appearance similar to burst-suppression pattern with alternation of high amplitude discharges and episodes of voltage attenuation (*Fig. 6.3*)
- hypsarrhythmia with increased bilateral synchrony
- preservation of normal ongoing activity.

The younger the child, the more disorganised is the EEG and the fewer the spikes seen. Modified hypsarrhythmia occurs in later infancy and is indicative of cerebral pathology.

Usually hypsarrhythmia is seen both when the infant is awake and during non-rapid-eye-movement (non-REM) sleep, although periods of relative suppression often become more evident in sleep (*Fig. 6.4*). During REM sleep there is a marked attenuation or transient disappearance of hypsarrhythmia and the trace can be nearly normal. While hypsarrhythmia evolves or vanishes during therapy, multifocal spikes and sharp waves, together with moderate slowing of the background activity, may become the prominent features. Disappearance of the abnormal pattern occurs first during waking periods and only subsequently during sleep. The development of the hypsarrhythmic EEG usually coincides with the onset of spasms, but patterns of abnormality relating to underlying pathology or multifocal abnormalities may precede hypsarrhythmia (Watanabe et al 1973).

The ictal pattern varies; according to Vigevano et al (2001) the most characteristic patterns consist of:
- a positive wave over the vertex–central region
- a medium amplitude, fast activity at 14–16 Hz, called 'spindle-like'

Table 6.1   Chronological overview of epilepsy syndromes in infancy and childhood

Syndrome	Onset	Aetiology	Seizures	EEG	Prognosis
Epileptic syndromes in neonates					
Benign idiopathic neonatal convulsions (fifth day fits)	5 days (2–7 days)	Unknown	Clonic, mostly partial clonic, often partial and/or apnoeic	Theta pointu alternant in 60%, otherwise normal or discontinuous, ictally rhythmic spikes or slow waves	Good, normal development
Benign familial neonatal convulsions	2–3 days	Channelopathy (K$^+$) (KCNQ2 on 20q13 and KCNQ3 on 8q24)	Clonic and/or apnoeic	Normal background, no specific pattern	Development normal, may develop epilepsy later in life
Early myoclonic encephalopathy	0–2 months	Associated with inborn errors of metabolism	Fragmented myoclonus, massive myoclonus, partial seizures, tonic spasms persist in sleep	Burst suppression with bursts of discharges (1–5 s) alternating with flat periods (3–10 s), persist in sleep, may develop atypical hypsarrhythmia	Severe neurological abnormality in all, death in 50% within 1 year, may develop infantile spasms
Early-infantile epileptic encephalopathy with burst-suppression pattern (Ohtahara syndrome)	0–3 months	Associated with cerebral malformations	Frequent tonic spasms, partial motor seizures	Burst suppression unilateral/asynchronous. Ictal: desynchronisation, later atypical hypsarrhythmia	Poor prognosis, death in 50%, severe learning difficulties and subsequent infantile spasms common
Epileptic syndromes of infancy and early childhood					
Infantile spasms	3–12 months (3% after 12 months)	One-third cryptogenic, two-thirds symptomatic	Brief axial contractions of head/trunk/limbs	Interictal: hypsarrhythmia. Ictal: attenuation, fast activity	Learning difficulties, arrest of psychomotor development
Febrile seizures	6 months to 5 years	Fever from extracerebral infection and genetic factors	Bilateral tonic–clonic, unilateral in 4%	Postictal: diffuse or asymmetrical slow. Interictal: normal, excess theta, or epileptiform discharge	Risk of more febrile convulsions 33%, risk of later epilepsy 5%
Severe myoclonic epilepsy of infancy	2–12 months	Channelopathy (Na$^+$) (SCN2A on 2q24)	Initially febrile seizures. Subsequently myoclonias, complex partial seizures, atypical absences	Initially normal. Fast generalised spike/multiple-spike waves, focal abnormalities, early photosensitivity	Poor prognosis, early death in 20%, severe learning difficulties, ataxia
Benign myoclonic epilepsy in infancy	4–24 months	Genetic predisposition	Single generalised myoclonic seizures, brief bursts of myoclonus	Spike or multiple spike-and-slow-wave complexes at about 3 c/s, increase in sleep	Good prognosis
Lennox–Gastaut syndrome	2–10 years	25% cryptogenic, birth asphyxia, CNS disease	Atypical absences, myotonic, tonic, astatic, GTCS, partial, non-convulsive status epilepticus	Slow background activity, interictal 2–2.5 c/s sharp-and-slow-wave focal and multifocal	Often not or only partially responsive to treatment, learning difficulties common, poor

	Age	Aetiology	Seizure type	EEG	Prognosis
Myoclonic astatic epilepsy of early childhood	0–5 years	Genetic predisposition	GTCS, myoclonic and astatic, myoclonic–astatic seizures	sharp waves or spikes, in sleep rhythmic bilaterally synchronous 10–20 c/s discharges	May have unfavourable course, learning difficulties in 50%
Epileptic syndromes of childhood					
Childhood absence epilepsy	3–12 years (peak prevalence 6–7 years)	Idiopathic with genetic predisposition, female predominance	Frequent simple or complex absences, photosensitivity in 20%, GTCS in 40%	Initially normal, prominent abnormal theta at 4–7 c/s. Irregular fast spike-and-wave or multiple spike-and-wave	Generally good, with 80% becoming seizure-free
Epilepsy with myoclonic absences	5–10 years	Male predominance	Absences with marked myoclonus	As for childhood absences	Less favourable, other types of seizures often develop
Benign childhood epilepsy with centrotemporal spikes (BECTS, Rolandic epilepsy)	3–12 years	Idiopathic with possible genetic predisposition	Focal (somatosensory, motor, orofacial, speech arrest), GTCS during sleep	Normal background, unilateral or bilateral centrotemporal spikes, may cluster, increase in sleep, up to 20% have generalised spike-wave	Spontaneous remission during puberty
Childhood epilepsy with occipital paroxysms	1–17 years (early variant 2–5 years, late variant 7–9 years)	Idiopathic with genetic predisposition	*Early variant*: nocturnal adversive seizures with eye deviation, vomiting, hemi- or generalised seizures *Late variant*: visual seizures with simple/complex visual hallucinations, amaurosis, migrainous headache, complex partial seizures	Normal background, high-amplitude spikes, sharp wave over occipital or posterior temporal lobe, usually blocked by eye opening, fixation-off sensitivity	Spontaneous remission during puberty, but in late variant seizures may persist into late adolescence or adulthood
Atypical benign partial epilepsy	2–8 years	Idiopathic with genetic component	Simple partial motor, atypical absences, myoclonias, atonic, GTCS (often nocturnal)	Rolandic sharp waves, multifocal spikes, activation and generalisation during sleep	Cognitive and behavioural problems common, spontaneous remission of seizures during puberty. Outcome re. cognition? Variant of BECTS?
Other forms of benign epilepsy of childhood	–	Idiopathic with genetic predisposition	Partial seizures, other seizures	Frontal, midtemporal, centrotemporoparietal or occipital spikes, high voltage EPs in some	Variant of BECTS?
Epilepsy with continuous spike and wave during slow-wave sleep (CSWS), aka electrical status epilepticus during slow wave sleep (ESES)	1–12 years	Unclear	Partial/generalised motor seizures, atypical absences, atonic	Bilateral and diffuse slow spike-and-wave discharges in > 85% slow wave sleep	Progressive neuropsychological deterioration, spontaneous remission, but variable degrees of learning difficulties may persist

Continued

Table 6.1   Continued

Epileptic syndromes of childhood – *continued*

Syndrome	Onset	Aetiology	Seizures	EEG	Prognosis
Landau–Kleffner syndrome (epileptic aphasia)	3–7 years	Unclear	GTCS, tonic seizures, absences, one-third have no seizures	Focal/generalised sharp waves or spikes, maximum over posterior temporal and centro-temporal lobes, ESES in slow wave sleep	Language disorder with auditory verbal agnosia and aphasia, behavioural disorders common. Spontaneous remission during adolescence, but some language deficits often persist
Epilepsia partialis continua of childhood	1–10 years	Rasmussen's encephalitis	(Semi)continuous focal motor or somato-motor seizures or myoclonias	Focal or multifocal discharges, lateralised background abnormalities	Depends on lesions, progressive hemiparesis

Epileptic syndromes of later childhood and adolescence

Syndrome	Onset	Aetiology	Seizures	EEG	Prognosis
Juvenile absence epilepsy	7–16 years (peak at 10–12 years)	Idiopathic with a strong genetic predisposition	Absences, also myoclonic and GTCS	Normal background generalised 3–4 c/s (multiple) spike-and-wave discharges	Seizures may persist into adulthood, recurrence of seizures after AED withdrawal high
Juvenile myoclonic epilepsy	8–26 years (peak 12–18 years)	Idiopathic with a strong genetic predisposition	Irregular bilateral myoclonic seizures, GTCS and absences in 50%	Normal background, irregular, fast multiple spike-and-wave, 50% photosensitivity	Seizures usually persist into adulthood, recurrence of seizures after AED withdrawal very high
Epilepsy with GTCS on awakening	9–25 years	Idiopathic with a strong genetic predisposition	GTCS, especially within 2 h of waking	Normal background, generalised regular or irregular 2.5–4 c/s spike-and-wave activity	Recurrence of seizures after AED withdrawal high
Autosomal dominant nocturnal frontal lobe epilepsy	1–18 years	Autosomal dominant	Clusters of brief nocturnal partial seizures	Normal background, focal or bilateral slow waves, spikes or sharp waves	Usually respond well to AEDs but often persist throughout adulthood
Progressive myoclonic epilepsies	–	Heterogeneous group of unrelated disorders	Myoclonic jerks, other seizures	Generalised slowing; bursts, spikes, spike-and-wave, multiple spike-and-wave	Progressive mental deterioration in most, variable neurology
Ring chromosome 20	1–6 years	Mosaic r(20), sporadic	Complex partial seizures, GTCS, non-convulsive status epilepticus	Interictal: high amplitude slow waves. Ictal: prolonged periods of high voltage slow waves with occasional spikes, predominantly frontal	Seizures always drug resistant, various degrees of learning difficulties and behavioural problems

AED, antiepileptic drug; GTCS, generalised tonic–clonic seizures.

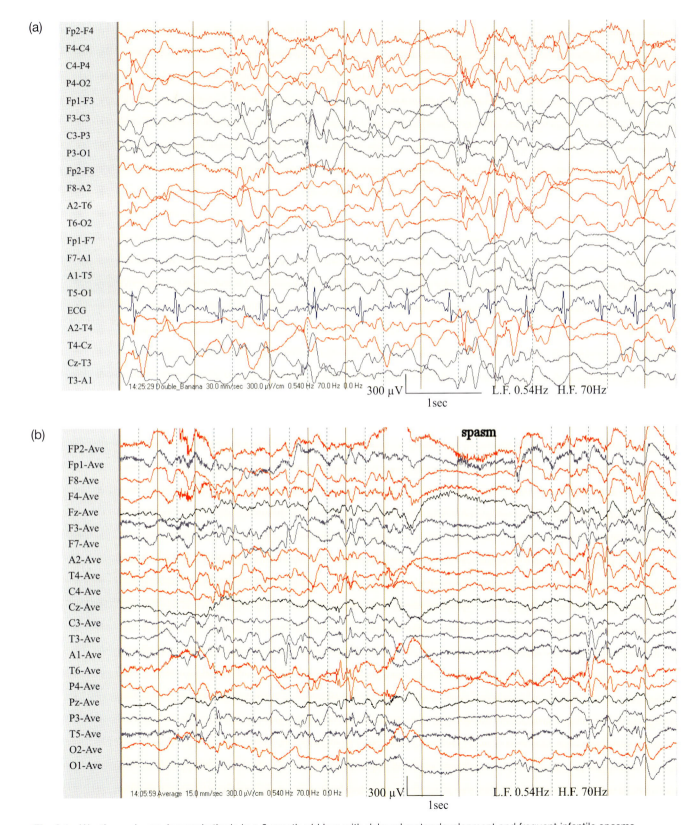

Fig. 6.1 West's syndrome: hypsarrhythmia in a 6-month-old boy with delayed motor development and frequent infantile spasms. (a) The EEG, during sleep, shows high voltage (note sensitivity: 300 µV/cm), chaotic slow wave activity with multifocal sharp waves and superimposed low voltage, fast activity over the frontal areas. (b) During a spasm, with the infant awake, an electrodecremental event with superimposed fast activity is seen. (From Binnie et al (2003), by permission.)

235

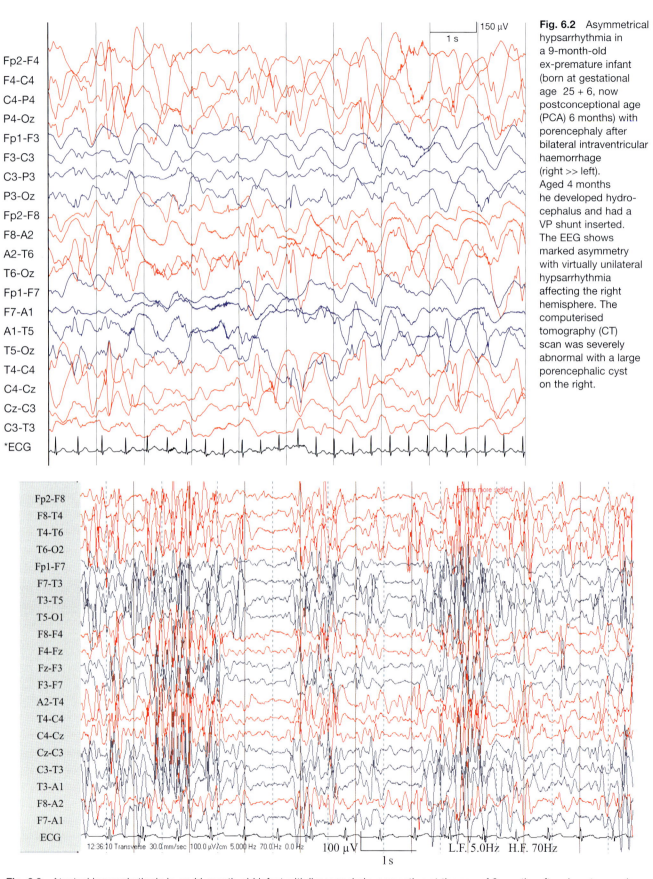

Fig. 6.2 Asymmetrical hypsarrhythmia in a 9-month-old ex-premature infant (born at gestational age 25 + 6, now postconceptional age (PCA) 6 months) with porencephaly after bilateral intraventricular haemorrhage (right >> left). Aged 4 months he developed hydrocephalus and had a VP shunt inserted. The EEG shows marked asymmetry with virtually unilateral hypsarrhythmia affecting the right hemisphere. The computerised tomography (CT) scan was severely abnormal with a large porencephalic cyst on the right.

Fig. 6.3 Atypical hypsarrhythmia in an 11-month-old infant with lissencephaly, presenting at the age of 6 months after almost normal development with infantile spasms and regression of psychomotor abilities. The infant had become drowsy during recording and slept later. The EEG shows high amplitude, irregular slow bursts of about 2 s duration and suppressions of about 1.5 s; multifocal sharp waves occur. During suppressions rhythmic activity at 10–14/s is visible. (From Binnie et al (2003), by permission.)

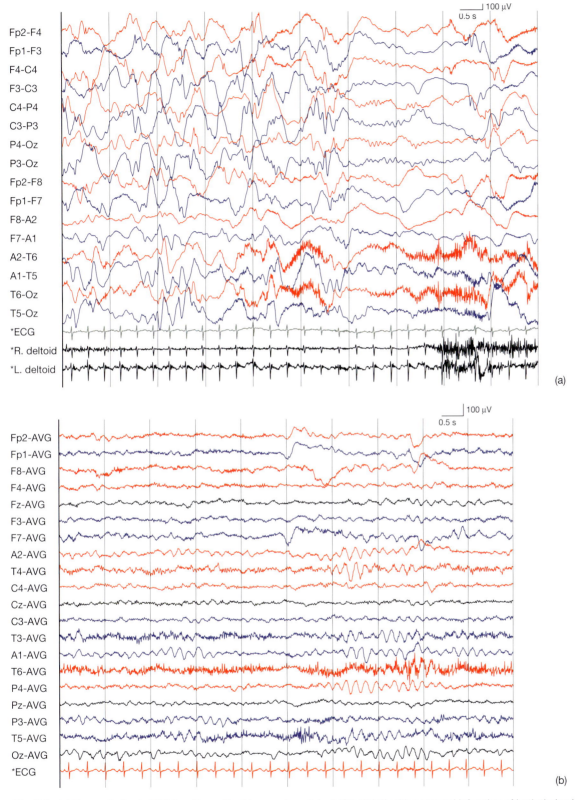

100 μV

0.5 s

Fp2-F4

Fp1-F3

F4-C4

F3-C3

C4-P4

C3-P3

P4-Oz

P3-Oz

Fp2-F8

Fp1-F7

F8-A2

F7-A1

A2-T6

A1-T5

T6-Oz

T5-Oz

*ECG

*R. deltoid

*L. deltoid

(a)

100 μV

0.5 s

Fp2-AVG

Fp1-AVG

F8-AVG

F4-AVG

Fz-AVG

F3-AVG

F7-AVG

A2-AVG

T4-AVG

C4-AVG

Cz-AVG

C3-AVG

T3-AVG

A1-AVG

T6-AVG

P4-AVG

Pz-AVG

P3-AVG

T5-AVG

Oz-AVG

*ECG

(b)

Fig. 6.4 (a) Hypsarrhythmic EEG in a 7-month-old ex-preterm infant (GA 32/40) with postnatal infarction of both thalami. She had neonatal seizures treated successfully with phenobarbital. She presented aged 6 months with infantile spasms occurring in clusters, 10–20 times a day. The EEG is dominated by high voltage (note calibration), chaotic, slow wave activity, with sharp waves and spikes intermixed. Note the electrodecremental event with superimposed fast activity associated with spasms of the upper limbs (EMG channels). (b) The same patient aged 9 months: spasms have ceased and the EEG normalised considerably with sodium valproate and prednisolone treatment. The EEG now shows mild excess of slow activity over the right posterior temporal and parietal areas, but no epileptiform discharges.

237

- a diffuse flattening, called an 'electrodecremental event'.

Spasms may be accompanied by one or a combination of these patterns (see *Figs 6.1(b)* and *6.4(a)*). In some children, high amplitude slow waves and spikes or multiple spikes and slow waves accompany the spasm.

The interictal EEG pattern does not enable one to differentiate between the aetiologies (symptomatic or idiopathic) of infantile spasms, and is of doubtful value in prognosis. Some authors, however, consider that the prognosis may be more favourable in the absence of gross abnormality of the ongoing activity, without focal slow waves and with preserved or restored sleep organisation, including symmetrical sleep spindles. Both infantile spasm and hypsarrhythmia gradually resolve or, more often, evolve into a different seizure pattern, usually by the age of 2–3 years.

In those who respond to steroid therapy, spikes, sharp waves and paroxysmal bursts become less frequent or vanish, the composition of the ongoing activity becoming faster, and the amplitude lower (*Fig. 6.5*). Sustained low amplitude recordings, however, do not necessarily indicate that the response to treatment will be permanent.

It should especially be stressed that hypsarrhythmia may occur occasionally without infantile spasms and even without recognisable epileptic seizures; it reflects an age dependent reaction of the brain to some form of functional or morphological impairment. The EEG is always abnormal in infantile spasms but may show abnormalities other than hypsarrhythmia in 10–15% of cases, more so in older children or if the EEG is done later in the course of the disorder (Hrachovy and Frost 1984). Occasionally, and for not more than 2 weeks after the onset of spasms, a normal EEG may be found in the awake state. Two consecutive normal EEGs, including one sleep recording, are considered to exclude the diagnosis of infantile spasms.

Unilateral 'hemi-hypsarrhythmia' indicates a gross morphological lesion. If the lesion is moderately severe, but more so on one side, hemi-hypsarrhythmia will be ipsilateral, but if one hemisphere is almost totally destroyed or is replaced by a large cystic lesion, it may be unable to display the characteristic pattern, thus producing a contralateral hemi-hypsarrhythmia.

The syndromes described below are associated with infantile spasms and/or hypsarrhythmia and may have a characteristic electroclinical pattern.

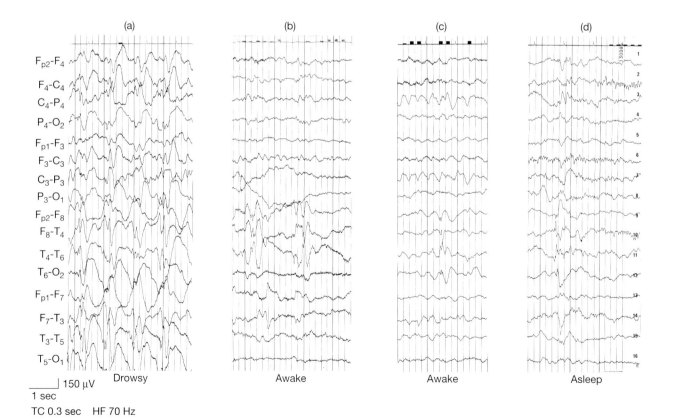

Fig. 6.5 Serial EEGs in a child with infantile spasms. (a) Hypsarrhythmic EEG at 5 months of age during drowsiness; the EEG activity shows little change with sleep, even at the time of small abnormal movements. (b) At 6 months, awake: no spasms since 36 h after starting adrenocorticotrophin hormone (ACTH) followed by clonazepam; the EEG pattern is markedly improved, but some right frontotemporal complex discharges remain. (c) At almost 7 months, awake: postcentral rhythms have increased but focal right-sided discharges are still evident, although less complex. (d) The latter become more widespread during sleep; note diminution of fast activity in channel 2 versus channel 6. (From Binnie et al (2003), by permission.)

Aicardi's syndrome This syndrome (Chevrie and Aicardi 1986, Harding and Copp 1997, Aicardi 2005) is caused by a dominant mutation linked to the X chromosome (Xp22) and consists of total or partial agenesis of the corpus callosum, chorioretinal lacunae, frequently other brain malformations and infantile spasms. The EEG is characterised by a burst-suppression pattern occurring independently over each hemisphere, rather than hypsarrhythmia, and is attributable to agenesis of the corpus callosum and multiple epileptogenic cortical areas (*Fig. 6.6*). In a series of 95 children with such defects, Chevrie and Aicardi (1986) found that only 18% exhibited hypsarrhythmia. Boyd and Harden (1991) found that abnormalities of the electroretinogram (ERG) and visual evoked potential (VEP) were variable according to the extent of retinal and other abnormalities affecting the visual pathway.

Miller–Dieker syndrome While the EEG may show atypical hypsarrhythmia, diffuse, high amplitude rhythmic alpha or beta activity is more characteristic (*Fig. 6.7*) (see pp. 283–287).

Hemimegalencephaly This is a brain malformation characterised by hypertrophia and architectural distortion of an entire hemisphere (see pp. 283–287). It can occur in isolation or be associated with various neurocutaneous syndromes (see pp. 286–287).

Febrile seizures

Febrile seizures or febrile convulsions are usually tonic–clonic seizures, which occur with fever due to an extracerebral infectious disease in early childhood, from age 6 months up to 5 years. Sometimes the seizures are prolonged, lateralised or followed by a Todd's paresis. The incidence is about 3–10%, but the risk is increased four-fold if siblings or parents have had febrile seizures. About one-third have subsequent febrile seizures, and up to 5% develop epilepsy later in life. Risk factors for subsequent epilepsy are prolonged seizures (> 20 min), focal features in or after the convulsion, recurrence within 24 h, prior abnormal development or neurological abnormalities, and a family history of non-febrile seizures. There is no evidence that febrile seizures are a risk for

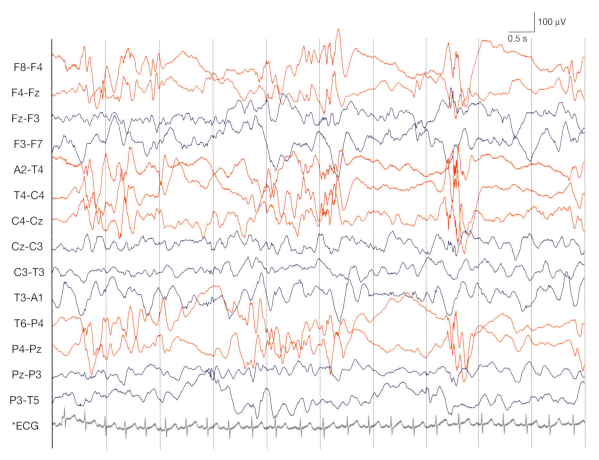

Fig. 6.6 Girl aged 8 months with early-onset seizures, developmental delay and absent corpus callosum on magnetic resonance imaging (MRI) (Aicardi syndrome). The EEG is extremely asymmetric, with a burst-suppression pattern seen over the right, and more continuous independent activity over the left, which shows an excess of slow activity and epileptiform discharges, particularly over the frontal area.

subsequent intellectual and behavioural development (Verity et al 1998).

On the rare occasions that an ictal EEG can be obtained it shows the typical pattern of a generalised tonic–clonic seizure. In those children, whose seizures show focal symptomatology, asymmetrical EEG features or even a focal onset can be seen.

The longer the seizure lasts, the more severe is the generalised slowing of the background activity post-ictally, and the longer it takes before the usual ongoing activity is restored. Febrile seizures are largely genetically determined; some patients also have a family history of idiopathic generalised epilepsy or benign epilepsy of childhood. These genetic associations appear to be reflected in the interictal EEG which, a week or more after the seizure, often shows abnormal theta rhythms, which are considered to be a sign of a genetic liability to epilepsy. In some children spike-and-slow-wave discharges may be seen during the awake state and/or during photic stimulation. Focal Rolandic sharp waves are found in some patients.

Strictly, febrile seizures are not epilepsy as they do not represent a liability to recurring *spontaneous* seizures;

unfortunately, this may be established only by hindsight. However, given the incidence of abnormal EEG findings (theta rhythms and spike wave discharges) on serial recording, it is also argued that febrile seizures are a generalised form of early childhood epilepsy, usually with a good prognosis because they are self-limiting.

Although an EEG is not a necessary investigation in most children with febrile seizures it is sometimes requested to address diagnostic or prognostic issues. In the acute phase it needs to be established whether the seizures and fever are both symptoms of a cerebral infection. Here a relatively normal EEG is reassuring, but note that a prolonged convulsion may lead to considerable postictal slowing, possibly of several days' duration. Similarly, high fever can cause generalised synchronous or asynchronous slow waves and slowing of alpha rhythm in children. Subsequently, it may be asked whether febrile seizures are likely to recur and whether epilepsy will develop. The overall risks are about 33% and 3–5%, respectively; clinical predictive criteria are long established, but do not include EEG findings (Frantzen et al 1968, Maher and McLachlan 1995). It has been claimed that interictal epileptiform activity often appears only

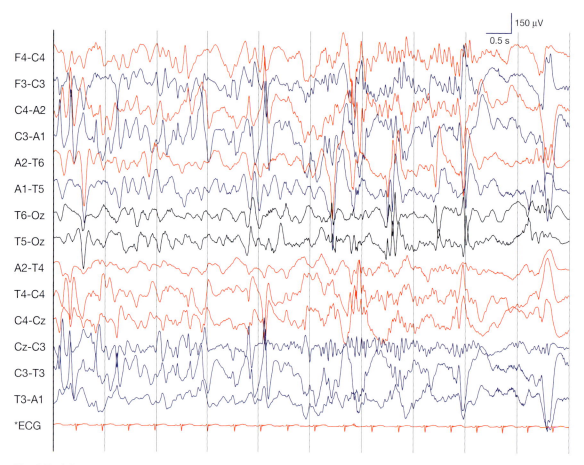

Fig. 6.7 A 3-year-old boy with lissencephaly type I with the LIS1 mutation, presenting with infantile spasms and severe developmental delay. The EEG shows widespread multifocal, high amplitude spikes and multiple spike-and-wave discharges superimposed on a rather slow background with short periods of relative suppression.

after epilepsy has developed. However, the more serial follow-up recordings (including light sleep) that are done, the greater the chance of finding warning spike-and-slow-wave activity.

Recently, genetic syndromes have been described in which children with febrile seizures subsequently have afebrile seizures later in life. Generalised epilepsy with febrile seizures plus (GEFS+) may be caused by mutations of SCN1A, SCN1B, SCN2A (subunits of the voltage-gated sodium channel) or GABRG2 (a subunit of the γ-aminobutyric acid A (GABA$_A$) receptor) (Baulac et al 2004). Severe myoclonic epilepsy in infancy (see below) shares similar genetic origins and febrile seizures, and GEFS+ is more common than expected in families of children with severe myoclonic epilepsy in infancy, which suggests that these syndromes represent a spectrum of severity of the same disease with additional genetic factors contributing. Some familial febrile seizures may also be caused by monogenetic channelopathies (Mantegazza et al 2005).

Benign myoclonic epilepsy in infancy

Benign myoclonic epilepsy in infancy (Dravet and Bureau 1981, Dravet et al 1992) is very rare and occurs from age 4 months to 2 years, presenting with single generalised myoclonic seizures, or brief bursts of myoclonus, but not serial seizures. There is often a family history of febrile convulsions or epilepsy. No other cause or underlying pathology is known. Tonic–clonic seizures may appear in adolescence and, without early treatment, there may be some slowing of intellectual development.

The EEG in the awake state shows normal background activity for age. Clusters of spike or multiple spike-and-slow-wave complexes at about 3/s accompany the myoclonic jerks, which themselves are generally multiple. Few discharges occur without accompanying myoclonus during the waking state. The discharges and myoclonias increase with drowsiness and early sleep, but the myoclonias usually disappear and the discharges become fewer in slow wave sleep.

This syndrome should be distinguished from 'benign myoclonia in early infancy' (Lombroso and Fejerman 1977, Dravet et al 1986), which starts between 4 and 9 months of age with a series of myoclonic contractions of the head and neck, together with a jerky tremor of the upper limbs, and disappears by the age of 2 years. It is not epileptic in origin and the EEG is always normal. The EEG may also be helpful in the differential diagnosis from benign neonatal sleep myoclonus (normal record), infantile spasms (hypsarrhythmia) and the Lennox–Gastaut syndrome (see below).

Severe myoclonic epilepsy in infancy (Dravet's syndrome)

In contrast with the above, severe myoclonic epilepsy in infancy (Dravet 1978, 2000), presenting at 2–12 months of age, is a slightly less rare condition. Several hetero-geneous mutations of SCN1A (encoding for the α1 subunit of the voltage-gated sodium channel) have been reported in children with severe myoclonic epilepsy in infancy, including missense, nonsense and insertion/deletion alleles (Baulac et al 2004). There may be a family history of febrile seizures, GEFS+ or other idiopathic epilepsy. Onset is often associated with a febrile illness. Initially, there are generalised or unilateral clonic seizures; later, myoclonias, complex partial seizures and atypical absences appear. Outcome is poor, with early death in up to 20%, learning difficulties, ataxia and interictal myoclonias.

The EEG changes progressively; it is normal initially or shows only slight postictal slowing, but photosensitivity may appear early, even in the first year of life. Rhythmic theta may be seen in the centroparietal areas and at the vertex. Later, with the advent of myoclonus, the background activity becomes abnormal and fast generalised spike-and-slow-wave, or multiple spike-and-slow-wave, discharges appear, together with focal abnormalities (*Fig. 6.8(a)*). Photosensitivity is found in 40%, and the myoclonic seizures may be precipitated by changes in lighting, eye closure or viewing patterns (*Fig. 6.8(b)*). The generalised myoclonias are accompanied by EEG discharges, but segmental myoclonias may occur without EEG changes. Absences, where they occur, are often accompanied by myoclonus or atonia; the EEG shows irregular spike-and-slow-wave activity (*Fig. 6.8(c)*).

Lennox–Gastaut syndrome

There are various, possibly related, syndromes, characterised by myoclonic seizures in childhood, the classification of which is controversial. At present, it remains questionable whether these syndromes represent distinct nosological entities. Here, the terminology of the proposed classification of the Commission on Classification and Terminology of the ILAE (1989) is followed.

Perhaps least in dispute is the Lennox–Gastaut syndrome, although even this term is not used with the same meaning in every country. In some, Lennox–Gastaut syndrome includes myoclonic astatic epilepsy of early childhood (here considered as a separate entity; see below).

The Lennox–Gastaut syndrome (Roger et al 1989, Aicardi and Levy 1992) consists of a triad of symptoms:
- a symptomatic generalised epilepsy, which includes atonic and axial tonic seizures of focal origin, atypical absences, myoclonic jerks and generalised tonic–clonic as well as partial seizures
- EEG abnormalities with diffuse slow spike-and-wave discharges and bursts of fast activity during sleep
- slow mental development and behavioural problems.

This syndrome represents some 2–3% of childhood epilepsies. It presents at 2–10 years of age, with a male predominance. The causes are varied. The syndrome is a common sequel to infantile spasms; about 25% of cases are cryptogenic.

241

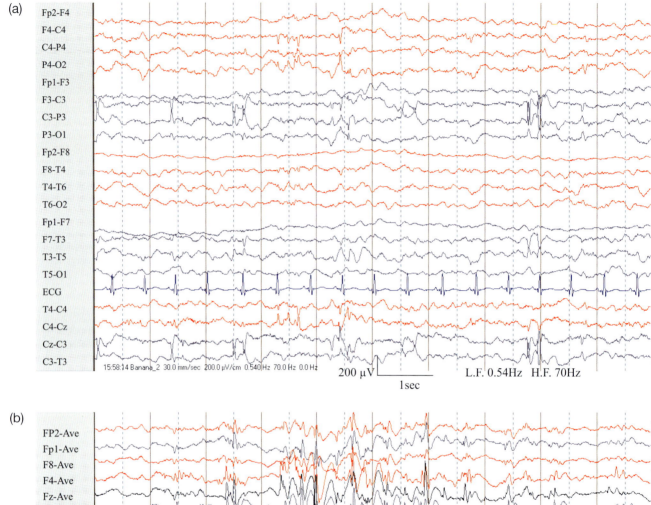

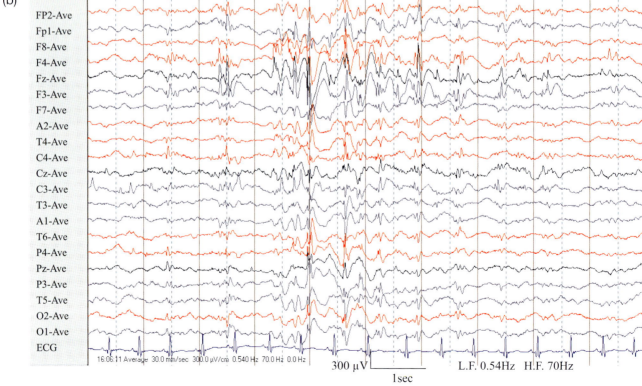

Fig. 6.8   Severe myoclonic epilepsy in infancy. A 4-year-old boy presented with febrile convulsions and a normal EEG, then photogenic seizures, learning disability and severe attention paedicit hyperactivity disorder. He now has absences, and atonic and complex partial seizures. (a) Multifocal spikes at the C3 and C4–P4 electrodes. (b) Generalised irregular multiple spike-and-slow-wave discharges, maximal anteriorly and of greater amplitude on the left. (From Binnie et al (2003), by permission.)

The background EEG activity is shifted to the theta and delta ranges and is often poorly structured for the patient's age. Interictal 2–2.5 Hz sharp-and-slow-wave or spike-and-slow-wave discharges are a prominent feature, tending to involve all brain regions. Focal, as well as multifocal, spikes are often seen, predominantly over the temporal and frontal regions (Hughes and Patil 2002). When myoclonic jerks predominate clinically, a multiple spike-and-slow-wave pattern (at < 2.5 Hz) may be seen continuously for long periods. The individual spikes are slow (~ 150 ms), often blunted, bi- or triphasic, and each is followed by an irregular slow wave (the whole complex lasting half a second or more).

The sharp-and-slow-wave activity occurs in clusters or in brief or longer runs, sometimes with initial spikes or spike-and-slow-wave complexes in a pseudo-rhythmic sequence. They may be symmetrical and bilaterally synchronous, or sometimes asynchronous and/or asymmetrical in amplitude, and sometimes vary in their distribution (Aicardi and Levy 1992). Their inter- and intraindividual variability is considerable. The discharges are not influenced by eye opening, but their incidence may be reduced by mental activity; they are sometimes activated by hyperventilation, but not by photic stimulation. They are, however, activated during sleep, especially in stages I and II, but only occasionally in REM sleep. Rhythmic bilaterally synchronous 10–20 Hz

discharges of high amplitude with frontal predominance can be seen during sleep (the 'grand mal discharges' of Gibbs and Gibbs; the 'rhythmic spikes' of Gastaut). They may be accompanied by tonic seizures. Both tonic seizures and the fast discharges are facilitated by sleep, particularly slow wave sleep.

The ictal EEG is characterised by the following patterns:

- During atypical absences, irregular, diffuse, slow spike-and-slow-wave complexes are seen; these are more or less symmetrical and are sometimes difficult to differentiate from interictal discharges (*Fig. 6.9(a)*).
- During tonic seizures, bursts of bilateral rhythmic 10–20 Hz activity over the anterior areas and the vertex occur (*Fig. 6.9(b)*). Prior to the burst an electrodecremental event or generalised discharges can sometimes be observed.
- During atonic and myoclonic–atonic seizures the EEG shows multiple spike-and-slow-wave or diffuse spike-and-slow-wave activity, or fast rhythms with an anterior predominance.

Non-convulsive status epilepticus has been reported in more than two-thirds of patients; this can last for long periods (up to several months) and is difficult to treat. The EEG shows continuous widespread spike-and-slow-wave or sharp-slow-wave complexes most often between 1.5 and 2.5/s (*Fig. 6.10*).

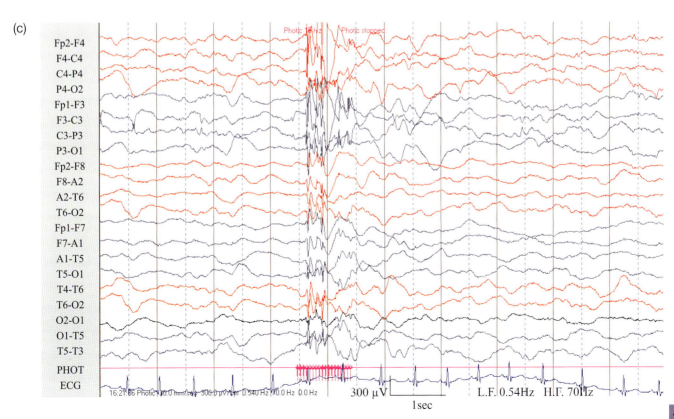

Fig. 6.8 *(continued)* (c) Photoparoxysmal response, grade 4. (From Binnie et al (2003), by permission.)

(a)

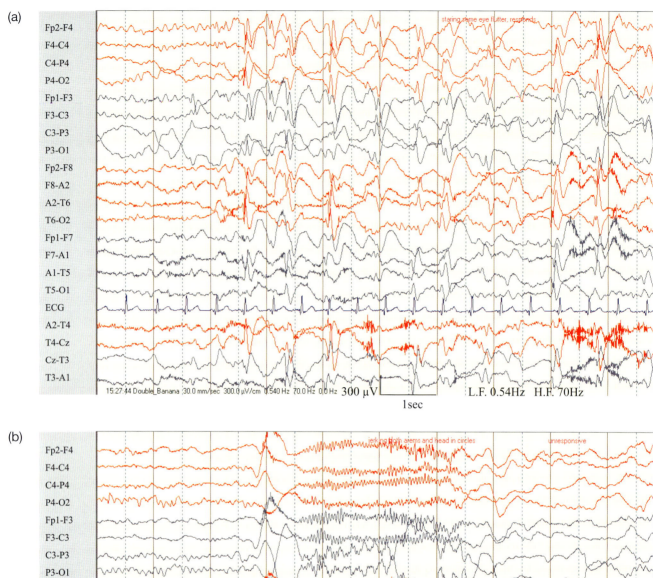

(b)

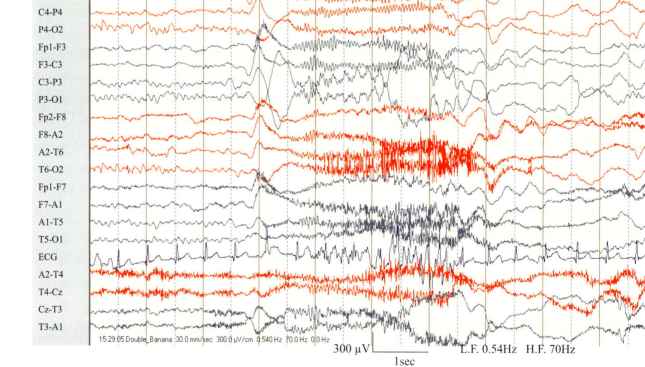

Fig. 6.9   Lennox–Gastaut syndrome: an 11-year-old boy, with long-standing learning disability and tonic–clonic, complex partial, atonic, axial tonic and atypical absence seizures. (a) Slow generalised spike-and-slow-wave activity during atypical absence; the boy exhibits staring and fluttering eyelids, but can be brought out of the attack and responds when called. The background activity preceding the discharge is relatively normal in this cryptogenic case. (b) A tonic seizure with generalised spiky fast activity preceded by 10 Hz activity. (From Binnie et al (2003), by permission.)

244

Myoclonic astatic epilepsy of early childhood (Doose's syndrome)

This syndrome (already briefly mentioned above) is considered to be idiopathic, starting before the age of 5 years with a male predominance. In 25% of cases onset is in the first year of life. It represents some 1% of childhood epilepsies. There is a high familial incidence of epilepsy. Onset in two-thirds is with a febrile or afebrile generalised tonic–clonic seizure. Children later develop myoclonic, astatic and myoclonic–astatic seizures, in some cases also short absences, absence status and tonic seizures. In the myoclonic–astatic seizures the myoclonias precede the atonic drop. Development prior to seizure onset is usually normal, but 50% of children develop learning difficulties in the long term.

The EEG may initially be normal for a long period; most children, however, gradually develop prominent abnormal theta rhythms at 4–7 Hz, especially in the parieto-occipital regions (*Fig. 6.11*) (Doose and Baier 1988). This EEG feature is also found in 16% of the siblings and in 6% of their parents. It represents a genetically determined EEG pattern and has to be distinguished from theta waves due to drowsiness.

Once myoclonic jerks and astatic seizures occur, irregular fast spike-and-slow-wave and multiple spike-and-slow-wave complexes appear (*Fig. 6.12*). With status epilepticus, 2–3 Hz discharges become prominent, but often remain very irregular. Sometimes the pattern resembles hypsarrhythmia. In the same patients, spike-and-slow-wave discharges may be elicited by photic stimulation. In cases with an unfavourable outcome, the alpha rhythm fails to develop and rhythmic slow waves at 5–7 Hz remain prominent over many years. A third of patients have episodes of non-convulsive status epilepticus, associated with 2–3/s spike-and-slow-wave activity. During nocturnal tonic seizures 10–15/s spikes occur.

Epileptic syndromes of childhood

The following syndromes are considered here:
- childhood absence epilepsy
- epilepsy with myoclonic absences
- epilepsy with eyelid myoclonia
- benign partial epilepsies in children:
  - benign childhood epilepsy with centrotemporal spikes (BECTS)
  - childhood epilepsy with occipital paroxysms
  - benign partial epilepsy with affective symptoms
  - benign partial epilepsy with extreme SEPs
  - atypical benign partial epilepsy
  - other forms of benign partial epilepsies

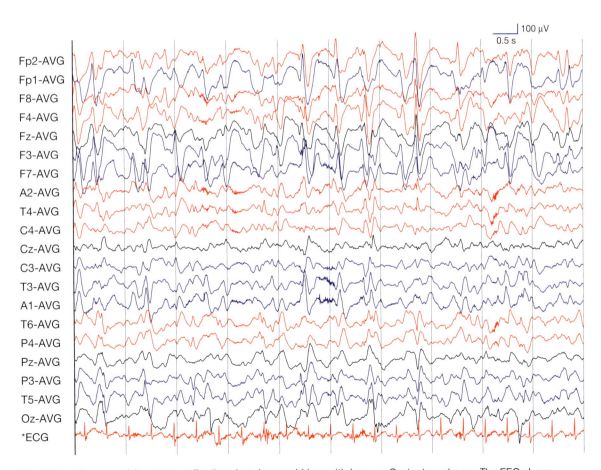

Fig. 6.10   Non-convulsive status epilepticus in a 4-year-old boy with Lennox–Gastaut syndrome. The EEG shows continuous, high amplitude 1–2/s sharp slow wave complexes with an anterior emphasis.

245

- epilepsy with continuous spike and wave during slow wave sleep (CSWS) (or electrical status epilepticus during slow wave sleep (ESES))
- acquired epileptic aphasia (Landau–Kleffner syndrome)
- chronic progressive epilepsia continua of childhood.

Childhood absence epilepsy

Absence seizures are a feature of idiopathic generalised epilepsy and present as various syndromes. Childhood absence epilepsy (pyknolepsy) occurs mainly in girls (4:1), with a peak prevalence at 6–7 years. The incidence is about 6/100,000 children, comprising 8% of epilepsies during school age. Childhood absence epilepsy appears to have a polygenic pattern of inheritance (Honovar and Meldrum 1997). All types of typical absence may occur, with the exception of those characterised by myoclonus. Absences are very frequent (hundreds per day) and can be simple, or include atonic or tonic components, with automatism or autonomic components. The seizure frequency is often grossly underestimated by patients and carers. Both routine EEG and long term monitoring have a role in determining the frequency of seizures. They are very brief (5–40 s, usually around 10 s) and end abruptly with resumption of the previous activity. Absences can be triggered by hyperventilation, hypoglycaemia, emotional upset, boredom and inactivity. Photosensitivity is thought to be incompatible with childhood absence epilepsy. Infrequent generalised tonic–clonic seizures develop in the further course in around 40% of patients. Absence status is rare. Intelligence is normal, but some cognitive deficits and behavioural problems may be found, partly due to frequent seizures, medication and expectations of parents and teachers, and partly a reflection of the underlying epilepsy syndrome (Pavone et al 2001). Prognosis is good, with 80% becoming seizure-free on medication, but generalised tonic–clonic seizures may occur 5–10 years after onset of childhood absence epilepsy (Loiseau et al 1983). Absences not infrequently persist into adulthood.

The ictal EEG shows classical 3/s spike-and-slow-wave activity (*Figs. 6.13–6.15*). The spike-and-wave frequency can vary during the discharge: from 3.5 to 4.5/s at the onset to 2.5 to 3.5/s at the end. The background is normal, but mild slowing has been described in some children (Holmes et al 1987). Cognitive assessment during EEG recording suggests that most of these discharges are accompanied by clinical events (Delgado-Escueta 1979). Simple tests may be employed to detect clinical changes during EEG recording: presenting

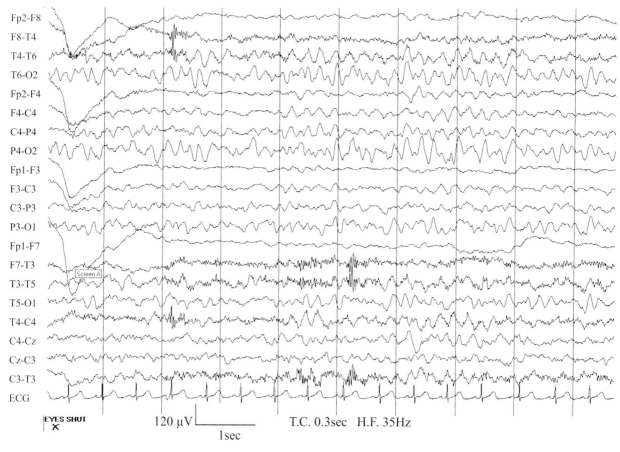

Fig. 6.11 Myoclonic astatic epilepsy: rhythmic theta activity in a 3.5-year-old boy. Excerpt starts on eye closure; the patient was alert but the normal 7–8 Hz postcentral dominant rhythm is virtually absent.

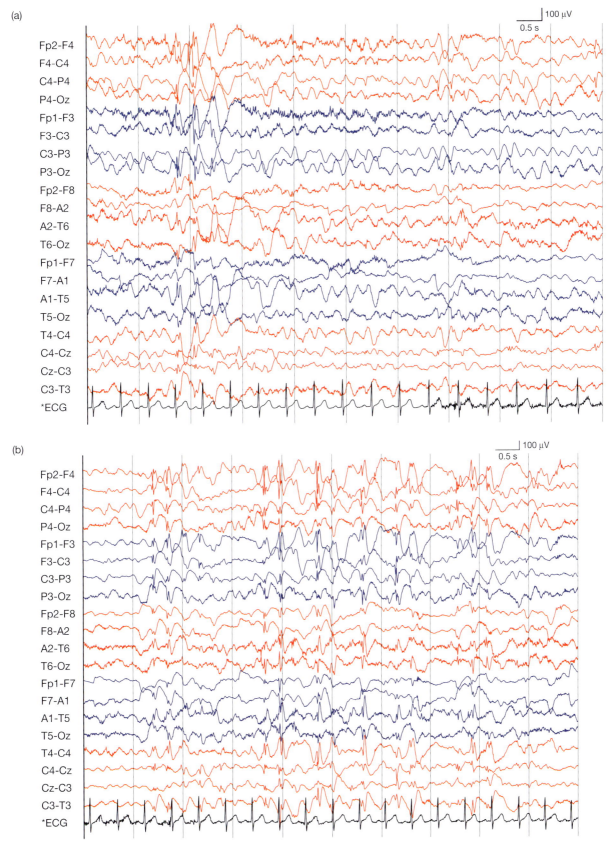

Fig. 6.12 Myoclonic astatic epilepsy in a 5-year-old boy. Onset of seizures was 12 months previously, but overt seizures have now ceased with treatment with sodium valproate and clobazam. (a) The interictal EEG is dominated by rhythmic theta activity interrupted by frequent brief bursts of posterior dominant, irregular spike-and-wave complexes. (b) Several episodes of unresponsiveness and staring (previously not noted by parents) occurred during the EEG, which were accompanied by runs of generalised irregular 2–4/s spike-and-wave activity.

numbers or phrases during the discharge to test recall, asking the patient to count on the fingers, or to raise the arms so that momentary loss of tone becomes apparent. Minor asymmetries of the discharges are common and should not lead to a suspicion of partial epilepsy. Isolated single spike-and-slow-wave complexes are not uncommon and may be asymmetrical.

Intermittent high voltage rhythmic delta activity may occur as a variant of physiological ongoing activities, but is *not* an ictal phenomenon (see *Fig. 6.14(b)*). This activity is often seen in children with childhood absence epilepsy aged between 6 and 10 years old, but is rare after the age of 15 years. Moreover, at seizure onset, this activity may become notched and then evolves into a generalised spike-and-slow-wave pattern. Reproducible induction by hyperventilation is usual if not universal, individual critical pCO_2 levels varying between 19 and 28 mmHg in the study by Wirrell et al (1996).

Occasionally, periods of unusual behaviour may occur in children with absence epilepsy and the EEG confirms non-convulsive status epilepticus (*Fig. 6.16*). Both may be reversed by slow intravenous injection of a small amount of diazepam, given cautiously under EEG control because of the rare precipitation of tonic status (Prior et al 1972, Tassinari et al 1972).

Other idiopathic generalised epilepsy syndromes in childhood with absences

Myoclonic absence epilepsy   Those children (mostly male) whose absences are accompanied by marked myoclonus have a less favourable prognosis than do those with childhood absence epilepsy (Tassinari et al 1992). The age of onset of myoclonic absence epilepsy is between 1 and 12 years, and it seems to combine features of idiopathic and symptomatic/cryptogenic generalised epilepsy. Children with learning difficulties before the onset of seizures are probably symptomatic cases, who are not considered part of the syndrome by the new ILAE classification. However, half of those with an idiopathic cause will develop learning difficulties and behavioural problems. Frequent absences are accompanied by severe bilateral rhythmical myoclonias of the shoulders, arms and, rarely, legs, often associated with tonic contractions of shoulders and arms. Other seizures (generalised tonic–clonic seizures, atonic seizures, absences without severe myoclonias) may be seen, but are usually infrequent.

The EEG findings are closely similar to those in patients with childhood absence epilepsy. Ongoing activity is normal, and posterior rhythmic slow variants may occur. In contrast to childhood absence epilepsy, interictal discharges are not uncommon with generalised

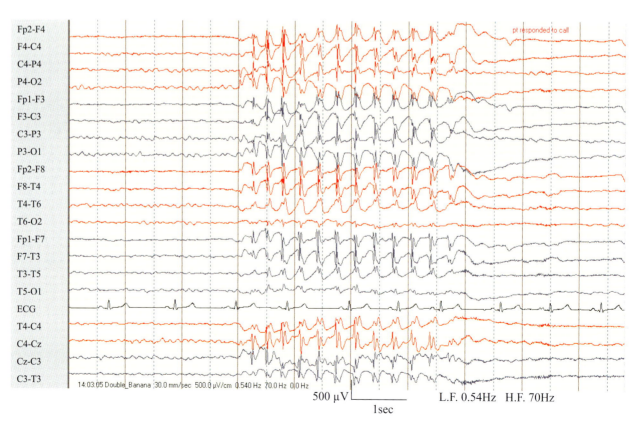

Fig. 6.13   Childhood absence epilepsy in a 10-year-old girl with a 3-year history of typical absences: generalised 3 Hz spike-and-slow-wave activity with frontal emphasis. (From Binnie et al (2003), by permission.)

spike-and-slow-wave activity, and focal or multifocal discharges may be seen. Ictal discharges comprise typical 3/s spike-and-slow-wave activity or multiple spike-and-slow-wave activity. Here too, asymmetries or even some

(a)

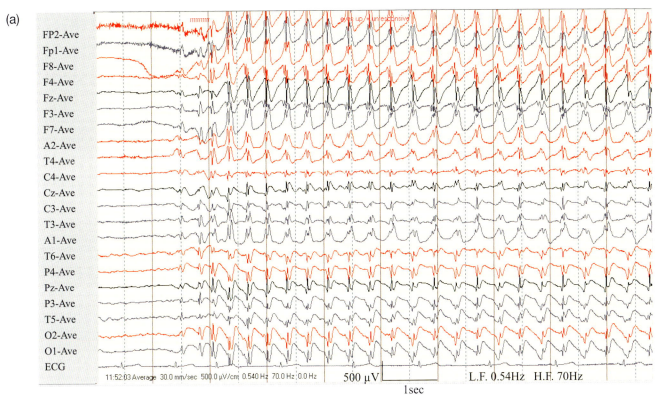

(b)

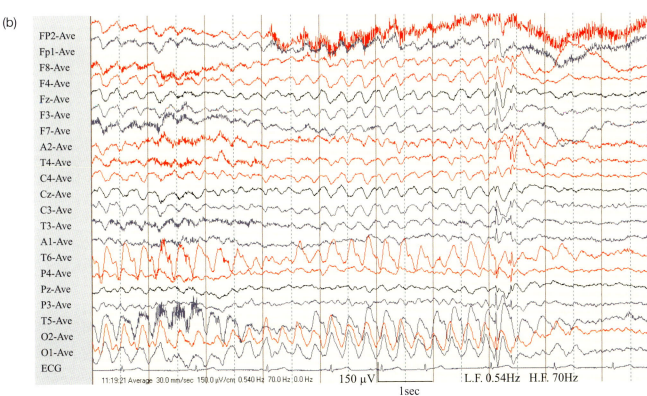

Fig. 6.14 Childhood absence epilepsy in a 6-year-old boy with a 2-year history of 'blank spells': (a) generalised 3 Hz spike-and-slow-wave activity. Spikes maximally negative at the back of the head – an unusual topography. (b) Notched rhythmic posterior slow activity with intermixed spikes, terminating in a frank spike-and-slow-wave discharge. (From Binnie et al (2003), by permission.)

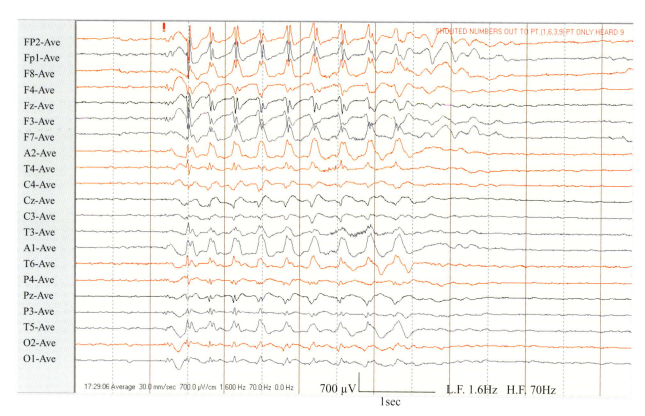

FP2-Ave		
Fp1-Ave		
F8-Ave		
F4-Ave		
Fz-Ave		
F3-Ave		
F7-Ave		
A2-Ave		
T4-Ave		
C4-Ave		
Cz-Ave		
C3-Ave		
T3-Ave		
A1-Ave		
T6-Ave		
P4-Ave		
Pz-Ave		
P3-Ave		
T5-Ave		
O2-Ave		
O1-Ave		

SHOUTED NUMBERS OUT TO PT. (1,6,3,9) PT ONLY HEARD 9

17:29:06 Average 30.0 mm/sec 700.0 μV/cm 1.600 Hz 70.0 Hz 0.0 Hz   700 μV   L.F. 1.6Hz   H.F. 70Hz
1 sec

Fig. 6.15 Childhood absence epilepsy in an 11-year-old girl. There was no overt seizure during generalised spike-and-slow-wave activity, but the technologist called out a series of numbers, of which only the last was recalled. (From Binnie et al (2003), by permission.)

focal discharges may be found. However, the initial spike may be positive, and in most cases synchronous with the myoclonus. EEG investigation may contribute to diagnosis, if simultaneous EEG and EMG recording is undertaken in any patient with therapy-resistant absences.

About half of patients are not responsive to therapy and other types of seizure often develop later, including features of Lennox–Gastaut syndrome, but these may be the symptomatic cases.

Eyelid myoclonia with absences (Jeavons' syndrome) This syndrome is not recognised by the ILAE classification. It has been postulated as a genetic, reflex, idiopathic, generalised epilepsy by some (Giannakodimos and Panayiotopoulos 1996) with the following characteristics: eyelid myoclonia with and without absences, eye-closure-induced seizures and/or EEG paroxysms and photosensitivity. However, it is considered as self-induced photosensitive seizures by others (Binnie 1996).

Eyelid myoclonia consists of jerking of the eyelids, often associated with eye deviation upwards and retropulsation of the head. This may be associated with a brief impairment of consciousness (eyelid myoclonia with absences). These seizures happen mainly after eye closure. Other seizures include generalised tonic–clonic seizures (spontaneous or light induced), myoclonia of the limbs and eyelid myoclonic status epilepticus.

The EEG shows normal ongoing activity and a normal sleep pattern. There are frequent high amplitude 3–6/s spike-and-wave discharges or multiple spike-and-wave discharges lasting 1–6 s. These are typically related to eye closure (within 0.5–2 s) and often occur in association with rhythmic eyelid myoclonia. Epileptiform activity is enhanced by hyperventilation and sleep, although a reduction may be observed in some. All patients are photosensitive. EEG abnormalities usually persist when patients are treated with antiepileptic drugs and even when they are seizure-free.

Benign partial epilepsies of childhood

The benign partial epilepsies are common (about one-third of all childhood epilepsies) and share several electro-clinical features, such as: idiopathic aetiology; age dependence; usually infrequent seizures that are easy to treat; benign prognosis of the epilepsy, with seizures disappearing at around puberty; activation of epileptiform discharges during sleep; and both focal and generalised EEG discharges (Dalla Bernadina et al 1992). The Commission on Classification and Terminology of the ILAE (1989) currently recognises two syndromes: BECTS and childhood epilepsy with occipital paroxysms (early-onset and late-onset variants). However, it has been suggested that there are other forms of benign partial epilepsies of childhood, which are less well defined: atypical benign partial

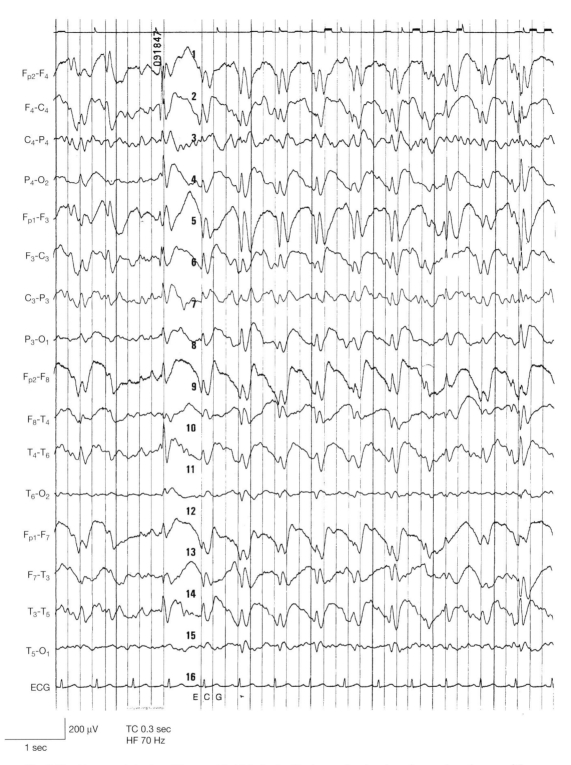

Fig. 6.16 Absence status in a 10-year-old child who had had occasional major seizures since the age of 5 years. Attacks had increased to every morning in the last 3 weeks and the child was forgetful at school. Runs of high voltage (note calibration), 1.5 Hz sharp-and-slow-wave activity dominate the EEG, with occasional spikes slightly more marked on the right side anteriorly. Discharges decreased, or were separated by longer intervals, when the child was asked to recite multiplication tables. (From Binnie et al (2003), by permission.)

epilepsy (Aicardi and Chevrie 1982, Aicardi 2000), benign partial epilepsy with affective symptoms and benign partial epilepsy with extreme SEPs. Atypical evolution has been found in 7% of children who initially presented with typical signs of BECTS and subsequently developed atypical benign partial epilepsy, Landau–Kleffner syndrome, status of BECTS, mixed atypical evolution or epilepsy with CSWS (Fejerman et al 2000).

(a)

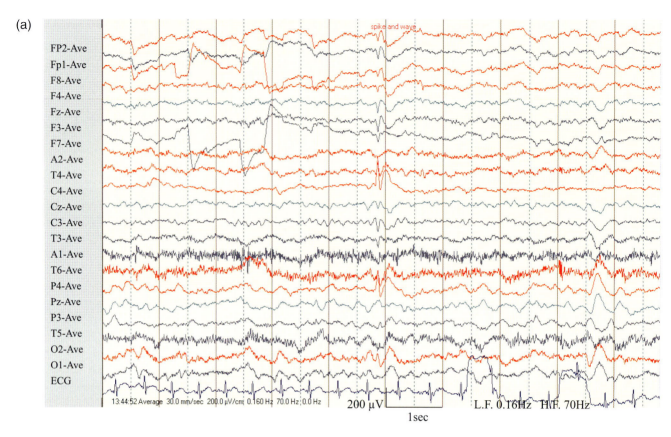

(b)

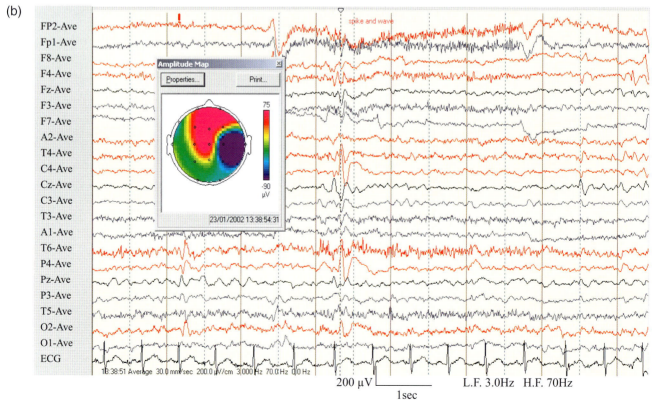

Fig. 6.17   Benign childhood epilepsy with centrotemporal spikes: a 3-year-old girl with onset at 2 years of simple partial seizures involving the left arm, and later generalised tonic–clonic seizures. (a) A typical Rolandic spike, maximally negative at the C4 and T4 electrodes with positivity in the midfrontal region. (b) Minor topographical variation in the same subject, this time with P4 and C4 maximum; the brain map also shows dipolar topography. (From Binnie et al (2003), by permission.)

The benign partial epilepsies of childhood may appear to form a rather well-defined group of syndromes with distinctive electroclinical findings. In practice, many patients presenting with a typical clinical picture show inappropriate EEG findings, or vice versa, while others show features of more than one syndrome, or an evolution from one syndrome to another. The electroencephalographer must therefore be prepared to be confronted by a classical EEG picture that in a clinical context is hardly reconcilable with benign partial epilepsy, a typical clinical presentation with an atypical EEG, or an electroclinical picture that changes over the years from one syndrome to another.

Benign childhood epilepsy with centrotemporal spikes (BECTS; benign Rolandic epilepsy) This is a common condition (up to 15–20% of all childhood epilepsies) presenting between 3 and 12 years of age and characterised by infrequent, mainly nocturnal partial (somatosensory and/or motor) seizures of orofacial onset with speech arrest. Secondary generalisation usually occurs only in nocturnal seizures. Absences can occur. The EEG is characterised by Rolandic spikes (see below). The aetiology is thought to be genetic: a family history of febrile convulsions, subclinical EEG discharges or idiopathic epilepsy is not uncommon, but a genetic aetiology has not been confirmed in twin studies. Asymptomatic siblings may show the EEG abnormality. In most cases seizures are infrequent, but where antiepileptic drugs are indicated seizure control is easily achieved. Prognosis is excellent, with remission of seizures by about 16 years of age. However, the benign nature of this syndrome has been exaggerated. Some degree of transient learning difficulties and behavioural problems are common (Weglage et al 1997, Massa et al 2001), and may be caused by maturation disorder (Doose and Baier 1989) or subclinical EEG discharges (Binnie et al 1992, Deonna et al 2000). Continuous spike-and-wave discharges during sleep are also commonly associated with cognitive problems; sleep EEG recording is mandatory in any child showing educational difficulties in association with BECTS.

The interictal EEG shows usually diphasic centrotemporal spikes or sharp waves with a high amplitude, electronegative component (sometimes exceeding 300 μV) often followed by a slow wave (*Fig. 6.17(a)*). Approximately 60% of cases show a unilateral focus, but in 40% it is bilateral, either synchronous or asynchronous. The spikes are generally of maximum amplitude in the central or Sylvian region, and less commonly (up to 20%) so at the parietal, vertex, occipital or mid-temporal electrodes (Wirrell et al 1995). The number of sharp waves and the degree of abnormality may vary from one recording to another, and the focus may shift within one hemisphere or to the opposite side. Typically, there is a simultaneous midfrontal electropositivity producing a dipolar configuration in most of these children (*Fig. 6.17(b)*). High resolution EEG studies suggest that

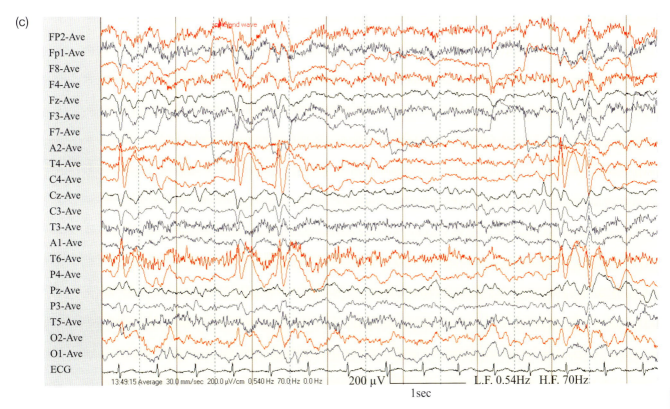

Fig. 6.17 (*continued*) (c) Discharge rate increases in drowsiness. (From Binnie et al (2003), by permission.)

two distinct dipole sources are involved to produce the typical interictal spikes (van der Meij et al 2001). The spikes often appear in clusters or short runs. Dramatic activation during drowsiness and sleep is characteristic and can be a helpful diagnostic tool in cases with infrequent or dubious discharges in the awake EEG (*Fig. 6.17(c)*). Twenty per cent of patients also show generalised spike-and-wave discharges or multiple spike-and-wave discharges, usually during slow wave and REM sleep. Apart from the epileptiform discharges the EEG is usually normal, although mild slowing is noted in some cases. Some structural brain lesions may mimic Rolandic spikes, e.g. low-grade astrocytoma or unilateral polymicrogyria (see pp. 283–287) (Ambrosetto 1993). Where the EEG topography is atypical, the clinical course of the seizure disorder is less likely to be benign (van der Meij et al 1992).

Deterioration of academic performance and socio-familial adjustment have recently been correlated with distinctive EEG patterns such as an intermittent slow wave focus, multiple asynchronous foci (*Fig. 6.18*), long sharp-wave clusters of more than 6 s, generalised 3/s spike-and-slow-wave complexes, conjunction of interictal paroxysms with negative or positive myoclonia, and abundance of interictal abnormalities during wakefulness and sleep, but not to seizures or drug treatment (Massa et al 2001). Sleep recording is important in all children with cognitive deficits to exclude Landau–Kleffner syndrome and epilepsy with CSWS.

Half of children with centrotemporal sharp waves never experience overt clinical seizures (Doose and Baier 1989). Centrotemporal sharp waves are found in 1–3% of healthy children (Eeg-Olofsson et al 1971, Cavazzuti et al 1980) and in up to 30% of siblings of children with BECTS (Doose and Baier 1989). It is assumed that the EEG trait follows an autosomal dominant mode of inheritance, with incomplete and age dependent penetrance.

A centrotemporal spike focus may also occur coincidentally in children with learning difficulties and neurological symptoms who have abnormal ongoing activity in the EEG. Neuroimaging is essential in all children with atypical electroclinical signs. The foci may disappear during treatment, but their persistence or reappearance does not necessarily indicate that seizures will recur.

Childhood epilepsy with occipital paroxysms

In its typical form this condition presents at 1–17 years of age with brief visual symptoms (amaurosis, elementary or complex hallucinations), which may progress to a partial or generalised motor seizure. Nocturnal adversive seizures with lateral tonic eye deviation associated with vomiting are typical in younger children and may last for > 30 min. Some children may experience complex paertial seizures. Migraine-like headache, nausea and vomiting often develop after the seizure. Episodes variously described as 'collapse' or 'going floppy' may occur, with or without vomiting.

A distinction is claimed between two variants (Panayiotopoulos 2000). The more common early-onset form (Panayiotopoulos' syndrome) appears around age 4 years (range 1–8 years), usually with nocturnal seizures with vomiting, eye deviation and impaired consciousness; there is secondary generalisation and hemiconvulsions. The late-onset variant (Gastaut's syndrome) peaks between age 7 and 9 years, with diurnal seizures with visual symptoms followed by postictal headache. In practice, many children show at different times features of both syndromes and a rigorous distinction between the two variants cannot always be maintained. In addition, it has been suggested that Panayiotopoulos' syndrome should be classified as an autonomic rather than an occipital epilepsy (Ferrie et al 2006).

A family history suggestive of migraine is common. This syndrome represents some 10–20% of benign partial epilepsy. Misdiagnosis is common, due to symptoms suggestive of basilar migraine and a high incidence of atypical presentations (e.g. episodes of collapse associated with vomiting). Spontaneous remission usually occurs by the age of 12 years in the early-onset variant; in the late-onset variant seizures often persist into late adolescence or adulthood and the prognosis is more uncertain.

The EEG shows normal background activity and interictal paroxysms of high amplitude (200–300 µV) spike-and-wave activity or sharp waves in one or both occipital and posterotemporal regions (*Fig. 6.19*). These features are typically diphasic with a high amplitude negative spike followed by a smaller positive peak and a negative slow wave. Discharges may cluster in a semi-rhythmic pattern at 1–4 Hz. Like Rolandic discharges they may not be present in every EEG recording; in some children they are found only in the postictal phase. Some children may exhibit additional or exclusive extra-occipital foci (Panayiotopoulos 2000). Usually discharges can be blocked by eye opening (*Fig. 6.20*). Fixation-off sensitivity (see pp. 69–70) is common (*Fig. 6.21*), whereas photosensitivity is not. During the seizures continuous occipital discharges spread to the central or temporal regions. Evolution to or from one of the other benign partial epilepsy syndromes may occur.

Benign partial epilepsy with affective symptoms (benign psychomotor epilepsy)

This rare syndrome is arguably a variant of BECTS. The distinctive feature is sudden fear or terror, e.g. screaming, yelling or calling for parents, sometimes associated with oral automatisms, laughing, speech arrest or autonomic symptoms. This has to be distinguished from gelastic seizures (see pp. 288–291). Consciousness is usually impaired, but not lost completely. The duration is usually short (1–2 min). Onset is at 2–10 years of age. In up to 40% of cases there is a positive family history. There may be a previous history

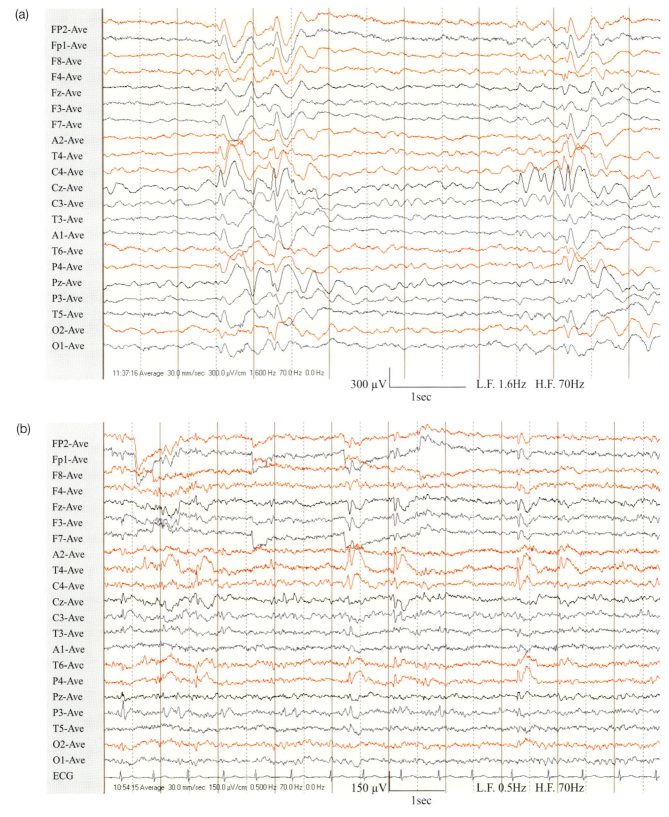

Fig. 6.18 Atypical EEG and clinical evolution of BECTS: this boy developed blank episodes aged 5 years when the EEG (a) showed episodic generalised slow waves with a possible sharp component at the vertex. Then a cluster of nocturnal tonic–clonic seizures was followed by episodes of head nodding, atonic attack and learning difficulties; EEGs at ages 6 and 7 years (not shown) were little changed but spikes were more obvious among the slow activity. At age 7 years there was onset of simple partial motor seizures involving the right arm. (b) The EEG at age 9 years shows bilaterally independent Rolandic spikes. (From Binnie et al (2003), by permission.)

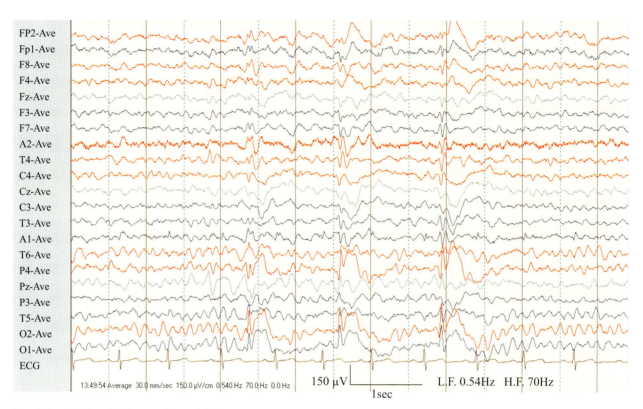

Fig. 6.19 Childhood epilepsy with occipital paroxysms: a 3-year-old girl was twice found beside her bed at night, floppy, unable to stand, vomiting and staring to one side. Further episodes of vomiting and abdominal pain followed over the next 6 months. The EEG shows occipital spike-and-wave activity of greater amplitude on the right. (From Binnie et al (2003), by permission.)

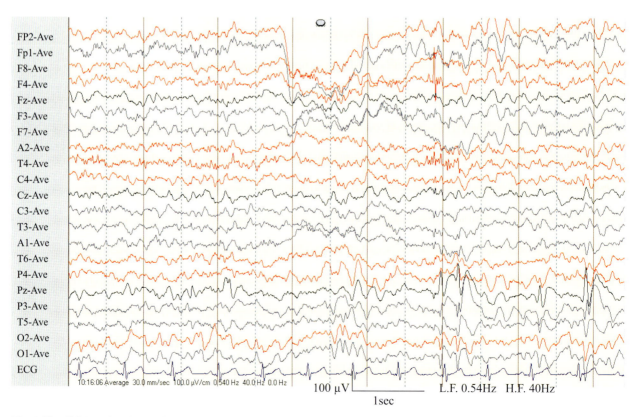

Fig. 6.20 Childhood epilepsy with occipital paroxysms: a 3.5-year-old girl suffered two episodes of unconsciousness, with cyanosis, some twitching but not a full convulsion. The EEG was misinterpreted as indicating an occipital lesion. Note the typical increase in occipital paroxysms on eye closure. (From Binnie et al (2003), by permission.)

of febrile seizures. Neurological and mental development is normal, but behavioural problems may be present. Untreated, the seizures may be rare or can occur many times throughout a 24-h period.

The background EEG and sleep organisation are normal. Interictally, there are frequent slow spikes followed by a slow wave (similar to the discharges seen in BECTS), with spikes over the frontotemporal or parietotemporal regions, which are activated by sleep. In about 50% of cases generalised discharges are seen, particularly during drowsiness. Ictal discharges are usually focal, but may be generalised.

Benign partial epilepsy with extreme SEPs　This rare condition is probably not an independent epilepsy but a variant of BECTS. Tactile stimuli, mainly tapping the soles or heels, elicit high voltage EPs over the contralateral somatosensory area. Onset is at age 1–13 years, with a peak between 4 and 6 years of age. Only about a third of children with extreme SEPs develop seizures. Usually the seizures are rare and of a simple partial motor type, occasionally with secondary generalisation. Mental and neurological development is normal and outcome is favourable.

The EEG shows centrotemporoparietal spikes similar to those seen in BECTS, with the added feature of unusually large SEPs.

Atypical benign partial epilepsy　It is debatable whether atypical benign partial epilepsy should be recognised as distinct from BECTS. It is claimed as an entity presenting between 2 and 8 years of age, with simple partial motor seizures, atypical absences, myoclonic seizures, generalised tonic–clonic seizures (often nocturnal) and atonic/astatic seizures (Aicardi and Chevrie 1982, Deonna et al 1986). Cognitive deficits and behavioural problems may be seen before onset, and often develop during the course of the epilepsy. In contrast to Lennox–Gastaut syndrome or myoclonic– astatic epilepsy, the prognosis regarding the seizures is excellent; however, permanent learning difficulties are common (Hahn et al 2001). Atypical benign partial epilepsy broadly overlaps with BECTS, electrical status epilepticus during sleep and Landau–Kleffner syndrome.

The awake EEG shows focal or multifocal sharp waves and spikes similar to Rolandic sharp waves (see *Fig. 6.17*), but with exceptional pronounced activation and generalisation during sleep, which can be nearly continuous,

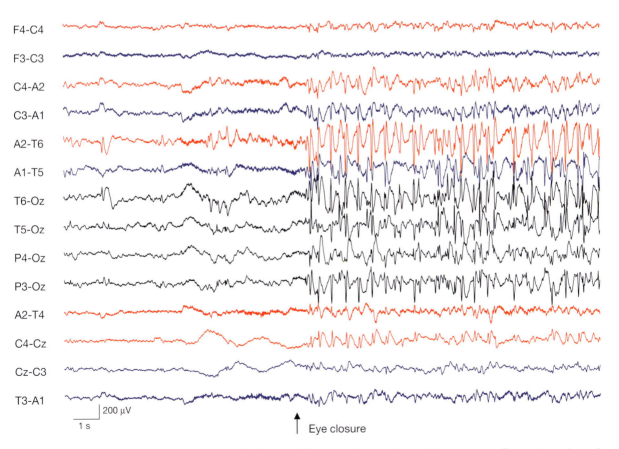

Fig. 6.21　Fixation-off sensitivity in an 8-year-old girl with childhood epilepsy with occipital paroxysms. She had experienced repeated attacks with eye deviation and vomiting since the age of 5 years. Eye closure is accompanied by bilateral 3/s spike-and-wave discharges over the posterior half of the head.

similar to Landau–Kleffner syndrome or CSWS (see below). There is a high incidence of centrotemporal sharp waves in siblings (Doose et al 2001).

Other forms of benign partial epilepsy Other forms have been described, such as benign partial epilepsy with frontal spikes (Beaumanoir and Nahory 1983), benign partial epilepsy with parietal spikes (Fonseca and Tedrus 2000), idiopathic photosensitive occipital lobe epilepsy (Guerrini et al 1995) and benign partial epilepsies in adolescence (Loiseau and Orgogozo 1978), but these may be rare.

Epilepsy with CSWS and ESES

ESES is an EEG phenomenon describing the occurrence of epileptiform discharges, the majority of the time in non-REM sleep, of continuous, bilateral diffuse slow spike-and-wave activity, which abates during REM periods. It occurs in the following four epileptic syndromes:

- CSWS
- Landau–Kleffner syndrome
- unilateral polymicrogyria
- some cases of BECTS and atypical benign partial epilepsy.

These syndromes have additionally in common: onset in early to middle childhood; a self-remitting course; steroid sensitivity of ESES in a proportion of cases; and continuing cognitive and language disability in the CSWS and Landau–Kleffner syndrome and in some cases of BECTS and atypical benign partial epilepsy. The symptoms displayed probably depend on the site of origin of the electrical status. For the unilateral polymicrogyria and Landau–Kleffner syndromes this is the anterior and posterior ends, respectively, of the Sylvian fissure, for BECTS (Rolandic epilepsy) it is towards the caudal end of the Rolandic fissure, and for CSWS it is presumably frontal, although this remains to be demonstrated. The aetiology in most cases is unknown, although occasional cases with demonstrable pathology at the site of origin of the epileptic discharge occur. It has been suggested that BECTS, Landau–Kleffner syndrome and CSWS represent a spectrum, with BECTS forming the benign common and Landau–Kleffner syndrome and CSWS the severe uncommon end. Detailed routine and sleep EEG recordings are essential in children where any of these syndromes is suspected.

CSWS is an uncommon syndrome seen in approximately 0.5% of all childhood epilepsies and is characterised by mild epilepsy, behavioural problems and ESES (Jayakar and Seshia 1991). It may present at 1–12 years of age with a slight male predominance. Seizures are usually infrequent; they may be generalised (myoclonic, convulsive or absences) or partial (simple or complex). Nocturnal seizures are common, but in contrast to the Lennox–Gastaut syndrome no tonic seizures occur. Seizures usually respond well to treatment and disappear during

adolescence as in BECTS. Symptomatic CSWS has been linked to early thalamic injury such as hypoxic ischaemic encephalopathy, periventricular leucomalacia or neonatal stroke (Guzetta et al 2005). Seizures are often severe and not all become seizure free. The aetiology in other cases is unclear. Family history is positive in a minority of cases.

Most children (about 70%) have normal development before onset. Nearly all children with symptomatic CSWS have neurological disability and learning difficulties before the onset of ESES. Behavioural disturbances occur in nearly all cases. Learning disability is common; some authors demand cognitive deficits as a mandatory feature, others do not (Aicardi and Chevrie 1982, Jayakar and Seshia 1991). Cognitive deficits include aphasia, apraxia, memory, attention problems and temporospatial disorientation. Often a widespread multisystem cognitive decline leads to moderate to severe learning difficulties. It is thought that long-lasting persistence of discharges during sleep is responsible for the cognitive decline. Other neurological abnormalities may be present. The progressive neuropsychological deterioration is usually of greater concern than the seizures. Although the condition is age dependent and self-limiting, the EEG phenomena are resistant to medication, and the eventual prognosis for psychological function has to be guarded.

The diagnostic criterion of this condition is that bilateral and diffuse slow spike-and-slow-wave discharges are present during at least 85% of slow wave sleep (*Fig. 6.22*) (Dalla Bernardina et al 1978, Tassinari et al 2002). However, this condition can occur with ESES less frequent than this, and often there is some fluctuation between EEGs. The figure of 85% should not be applied rigidly to this syndrome. Discharges are often predominant over the frontal areas. In REM sleep the continuous discharges are replaced by intermittent bursts of diffuse, rhythmic, synchronous and symmetrical spike-and-slow-wave discharges, often with frontal emphasis. Sometimes subclinical focal seizures over the frontocentral areas can be observed at the end of REM sleep (Dalla Bernardina et al 1978). The awake EEG usually shows focal discharges over the centrotemporal (see *Fig. 6.22(a)*) or frontotemporal regions; diffuse or generalised discharges may also be seen. In patients with symptomatic CSWS the morphology of the epileptiform activity may be more variable and asynchronous, or even unilateral (Guzetta et al 2005).

Acquired epileptic aphasia (Landau–Kleffner syndrome)

The Landau–Kleffner syndrome is characterised by a progressive loss of previously acquired language skills associated with epileptiform EEG activity during slow wave sleep (Deonna 1991). Onset is at 3–7 years of age. Overt epileptic seizures are present in not more than 80% of patients and, in about half of these, are infrequent and easily controlled. The essential clinical feature is an auditory verbal agnosia, which may show marked fluctu-

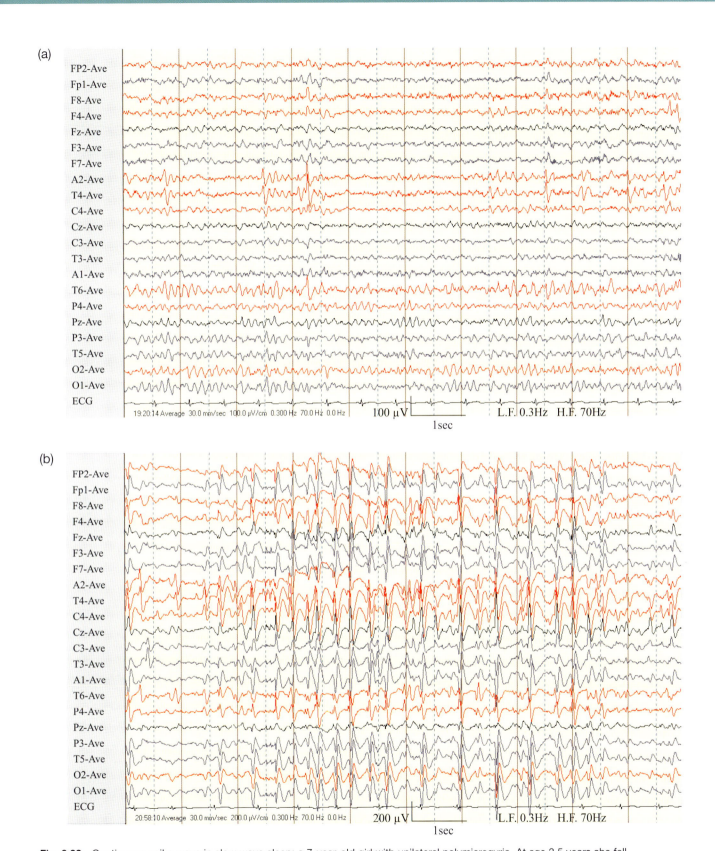

Fig. 6.22 Continuous spike wave in slow wave sleep: a 7-year-old girl with unilateral polymicrogyria. At age 3.5 years she fell unconscious, with left-arm twitching. She is now subject to continuous head nodding, exacerbated by voluntary movement. A seizure at age 5 years with version to the left: left facial twitching and vomiting. The girl has global developmental delay, can barely form sentences, is clumsy and has reduced swing in left arm. Recently, a new seizure type has developed, with pallor, dilated pupils and spitting, gulping or dribbling, followed often by writhing of limbs. (a) The waking EEG shows centrotemporal spikes (note that the field is not dipolar as in typical BECTS). (b) In slow wave sleep generalised spike-and-slow-wave activity occupies 85% of the recording. (From Binnie et al (2003), by permission.)

ations, from a loss of verbal understanding to a loss of understanding of all environmental sounds. In mild or early cases some communication is preserved through gesture or signing, but the patient may become entirely mute with amimia. There may be a loss of all social interaction and, occasionally, psychosis. The condition may be mistaken for autism or elective mutism, leading to the omission of sleep EEG investigation, which is essential in establishing the diagnosis. In Landau–Kleffner syndrome, as with the CSWS syndrome, seizures and EEG abnormalities resolve during adolescence; language impairment improves in many of the children, but if aphasia persists for more than 3 years complete recovery is unusual (Morrell et al 1995, Robinson et al 2001). Most cases are idiopathic, but in a few focal pathology is found.

In the awake state the discharges may be focal, focal with secondary generalisation or generalised. Focal discharges are typically located over the centrotemporal and parietal areas, similar to Rolandic discharges (*Fig. 6.23(a)*). Frontal lobe discharges in the awake EEG are particularly associated with disturbed behaviour (Robinson et al 2001). Sleep activates discharges at a frequency of 1.5–5 Hz, which become diffuse and nearly continuous, but not always occupying > 85% of slow wave sleep (*Fig. 6.23(b)*). During REM sleep the discharges fragment or disappear. Typically the focus in the awake and sleeping patient is not stable but shifting in pattern, topography and abundance from one day to another (Hirsch et al 1990). A distinctive feature of the Landau–Kleffner syndrome is that the discharges are predominant in the posterotemporal region. Close approximation of spikes to the primary auditory cortex has been demonstrated on magnetoencephalography (Paetau et al 1991, Morrell et al 1995). Unilateral discharges at or near the auditory cortex may disrupt auditory discrimination in the affected hemisphere and lead to suppression of auditory information from the opposite hemisphere, thereby accounting for two main aspects of the Landau–Kleffner syndrome. This can involve either side, since the contralateral propagation of paroxysms creates bilateral dysfunction that disrupts normal language development, regardless of the side of origin. Such localised abnormalities have led to surgical treatment by multiple subpial transection in selected patients (Morrell et al 1995). Presurgical assessment demands detailed EEG investigation to demonstrate a surgically treatable focal site of origin of the secondarily generalised discharges. The thiopental suppression test, carotid amytal test and electrocorticography may be required.

Epilepsia partialis continua of childhood
The original definition of epilepsia partialis continua given by Kojewnikov (1895) includes an association of localised epileptic jerks with more or less continuous Jacksonian seizures. Two forms are found in childhood. The first is due to a demonstrable focal lesion (due to cortical dysplasia, perinatal asphyxia or trauma) giving rise

to the seizures. The other, more frequent, type is a feature of Rasmussen's encephalitis (see p. 280). The children develop a progressive hemiparesis and semicontinuous or continuous focal motor or somato-motor seizures. Other seizure types may occur.

In the first group the EEG usually shows focal central EEG discharges against a normal background. In the second, ongoing activity is usually abnormal with widespread delta activity that is greater in amount contralateral to the hemiparesis. The epileptiform discharges are multifocal, uni- or bilateral, including generalised spike-and-slow-wave activity. The ictal discharges during partial seizures may be poorly localised. Myoclonic jerks occur without obvious EEG concomitants. At an advanced stage of the condition abnormal activities, both slow waves and epileptiform patterns, may, paradoxically, be more prominent over the less affected hemisphere.

Epileptic syndromes of late childhood and adolescence
The following syndromes are discussed here:
- juvenile absence epilepsy
- juvenile myoclonic epilepsy
- epilepsy with generalised tonic–clonic seizures on awakening
- primary reading epilepsy
- photosensitive epilepsy
- autosomal dominant nocturnal frontal lobe epilepsy
- other familial partial epilepsy syndromes
- progressive myoclonic epilepsies of childhood and adolescence
- ring chromosome 20 syndrome.

Juvenile absence epilepsy
Juvenile absence epilepsy is an idiopathic generalised syndrome accounting for some 2–8% of epilepsy and presents at age 7–16 years with a peak at 10–12 years of age. Absence seizures are generally much less frequent (1–10/day) than in childhood absence epilepsy. Remission is less likely, although seizures are usually easily controlled. A majority of patients also develop tonic–clonic seizures. The duration of absences is between 4 and 30 s (most around 15 s) and clinical recovery may occur before the end of the spike-and-wave activity. All variants of the absence may be found, but simple absences predominate. Myoclonic jerks occur in up to half of patients. Absence status may occur. One-third of patients have a first-degree relative with epilepsy. Intelligence is normal. The electroclinical picture overlaps with those of childhood absence epilepsy, juvenile myoclonic epilepsy and epilepsy with generalised tonic–clonic seizures on awakening.

Background EEG activity is normal. The ictal EEG shows generalised symmetrical 3–4/s spike-and-slow-wave activity with a frontal accentuation (*Figs 6.24* and *6.25*), which is often less regular than that in childhood absence epilepsy. Mild asymmetries at the onset of the

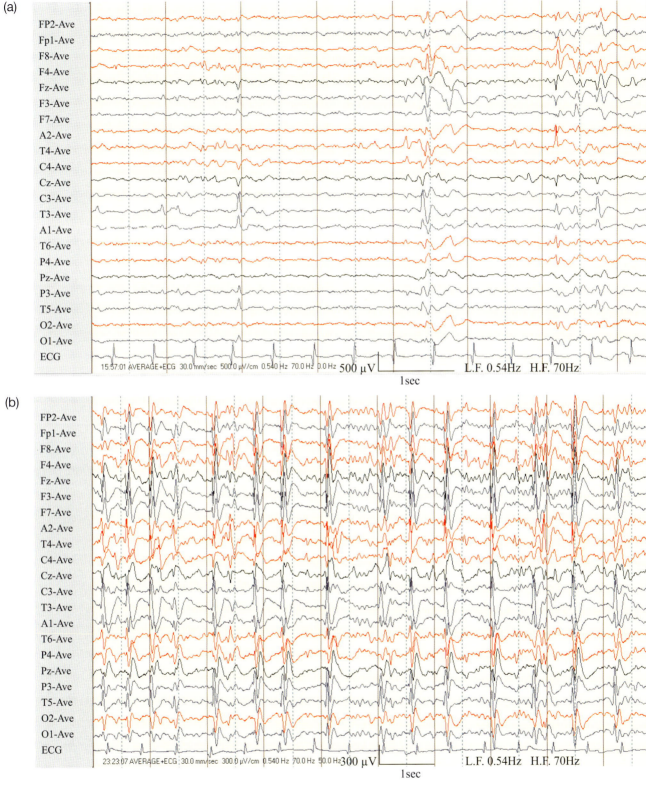

Fig. 6.23 Landau–Kleffner syndrome: a 7-year-old girl with deterioration of language skills from 2.5 years of age. Right handed. From 3 years old she was thought to be deaf, which was contested between her school and the ENT service. She was also said to be 'dreamy' at school. The first febrile seizure was at age 2.5 years: she vomited, lost consciousness and suffered left hemiconvulsion by day. The seizure was repeated after 3 months, also with pyrexia. She then developed left-sided nocturnal seizures. The seizures improved but her behaviour deteriorated on treatment with valproate. The EEG showed bilateral centrotemporal spikes – no sleep record. At age 6 years her condition was similar, but a more severe left-sided nocturnal seizure was followed by abrupt deterioration of language and behaviour. She is now mute, disoriented in place and incontinent. (a) The waking EEG shows Rolandic spikes, independent or synchronous on both sides, left more than right. (b) In slow wave sleep almost continuous spike-and-slow-wave activity is present. (From Binnie et al (2003), by permission.)

261

discharge may be seen (see *Fig. 6.24*). Multiple spike-and-slow-wave complexes can occur and are more common in those patients who also have myoclonias. The frequency at the start of the discharge is sometimes even faster. The discharges tend to be longer than in childhood absence epilepsy, yet loss of consciousness is often less complete (see *Fig. 6.25*) (Panayiotopoulos et al 1989). Ten to twenty per cent of patients are photosensitive and discharges can be triggered by hyperventilation and sleep.

Juvenile myoclonic epilepsy (Janz' syndrome)

This form of idiopathic generalised epilepsy represents 5–10% of all cases of epilepsy, and is characterised by irregular bilateral myoclonic jerks, which may cause the patient to fall. Sometimes the myoclonus is quite subtle and confined to the head, shoulders and arms. Onset is typically in the mid-teens, but sometimes much earlier, with myoclonic jerks and, less commonly, absences; later tonic–clonic seizures may appear. All seizure types tend to occur most frequently soon after wakening. The syndrome is often familial with a polygenic inheritance (Honovar and Meldrum 1997). Intelligence is normal. Up to 90% of patients with juvenile myoclonic epilepsy

become seizure-free with optimal medication, but notably carbamazepine exacerbates the condition. The relapse rate after withdrawing medication is very high, even if the patient has been seizure-free for several years.

The EEG shows generalised, fast, multiple spike-and-slow-wave discharges at 3–6/s in most patients; in some, 3/s or irregular spike-and-slow-wave discharges occur (*Fig. 6.26(a)*). The multiple spikes may be only two or three in number per slow wave, or may occur in a sustained burst of up to 20 spikes before the onset of spike-and-wave activity. Bursts are usually short (1–5 s) and there may be intradischarge fragmentation and unstable interdischarge frequency. The spike-and-slow-wave activity is often asymmetrical and there may be brief abortive focal discharges (*Fig. 6.26(b)*) in 20–30% of patients (Lancman et al 1994), mainly over the frontal regions. These may be misinterpreted as evidence of a partial epilepsy, with adverse consequences if this leads to use of carbamazepine. Background activity is usually normal, unless modified by medication. However, high voltage alpha activity is more common in patients with juvenile myoclonic epilepsy than in controls (Delgado-Escueta and Enrile-Bacsal 1984). Discharges are pro-

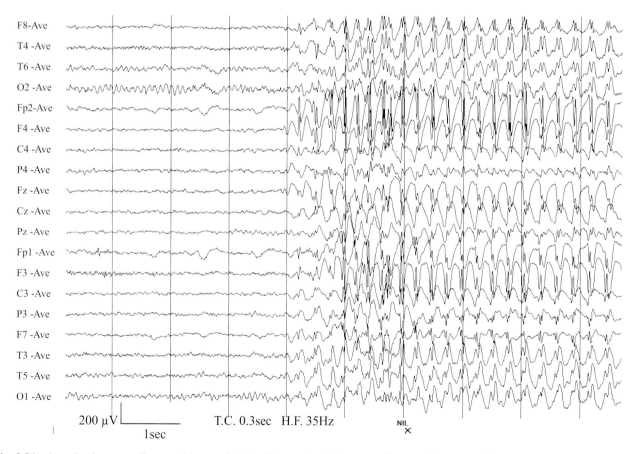

Fig. 6.24 Juvenile absence epilepsy: a 14-year-old girl, with occasional absences since age 12 years and two recent tonic–clonic seizures on awakening. The EEG shows a burst of generalised 4 Hz spike-and-slow-wave activity, without overt clinical signs. At onset the discharge is of greatest amplitude in the right frontal region, which was misinterpreted as evidence of partial epilepsy with secondary generalisation. (From Binnie et al (2003), by permission.)

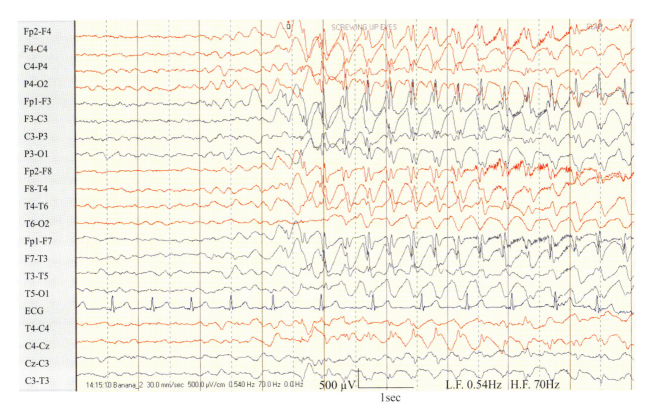

Fig. 6.25 Juvenile absence epilepsy with atypical features: a 10-year old girl with a 3-year history of staring and inattention, and sometimes incontinent of urine. She was counting aloud during the recording; during the generalised spike-and-slow-wave discharge illustrated, she continued to count but made an error. Note the atypical onset, with diffuse slow waves and slight asymmetry. (From Binnie et al (2003), by permission.)

voked by sleep, hyperventilation, sometimes by eye closure, and in 30–50% by photic stimulation (*Fig. 6.27*). Myoclonic jerks always correspond to bursts of generalised spike-and-slow-wave or multiple spike-and-slow-wave complexes.

The syndrome is underdiagnosed, partly because it appears to be less well known than it should be, but also because the myoclonus often goes unreported, the patient seeking medical advice only after the first tonic–clonic seizure. As noted above, correct diagnosis is essential to avoid inappropriate medication and prognostication; often this is achieved through recognition of the characteristic EEG findings.

Epilepsy with generalised tonic–clonic seizures on awakening

This syndrome accounts for up to 20% of epilepsies with generalised tonic–clonic seizures, presenting usually between 9 and 25 years of age (rarely as early as 4 years), with a slight male predominance. The seizures occur without an aura (this is an idiopathic generalised epilepsy) typically within 2 h of waking from sleep. Some patients show other diurnal patterns, and may for instance have seizures late in the evening. The convulsion may be preceded by a cluster of absences or myoclonic jerks.

Seizures are readily precipitated by deviation from habitual sleep patterns, retiring or rising late, or sleep deprivation. Unsurprisingly, given the age group, the first seizure often occurs after a late-night party.

The interictal EEG is often abnormal and various types of generalised spike-and-slow-wave discharge occur; they can be regular at 3 Hz, but more commonly are irregular at 2.5–4 Hz or with multiple spike-and-slow-wave activity. Asymmetrical or focal discharges are rather less common than in juvenile myoclonic epilepsy. Some 10–15% of patients are photosensitive. It will be noted that both the clinical picture and the EEG findings have several features in common with juvenile myoclonic epilepsy, and it is debatable whether all three are part of a spectrum.

Primary reading epilepsy

This rare syndrome presents between 12 and 25 years of age with a male preponderance. It is characterised by the precipitation through reading and other language-related activities (e.g. talking and writing) of myoclonic jerks of the throat, jaw or face, which may progress to a generalised tonic–clonic seizure if the patient continues to read. Half the patients have a family history of seizures, often of reading epilepsy. Similar symptoms can occur in the

(a)

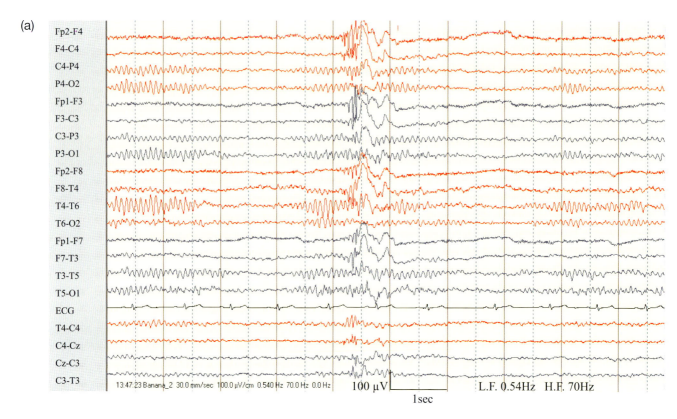

(b)

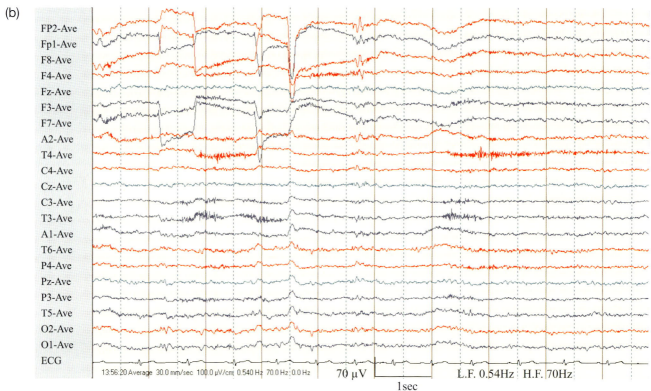

Fig. 6.26   Juvenile myoclonic epilepsy: a 17-year-old girl with early morning myoclonus since the age of 5 years that is exacerbated by bright lights. At age 16 years a myoclonic episode evolved to a tonic–clonic seizure. (a) Brief interictal burst of generalised fast multiple spike-and-slow-wave activity. (b) Focal spike in the right frontal region. (From Binnie et al (2003), by permission.)

(a)

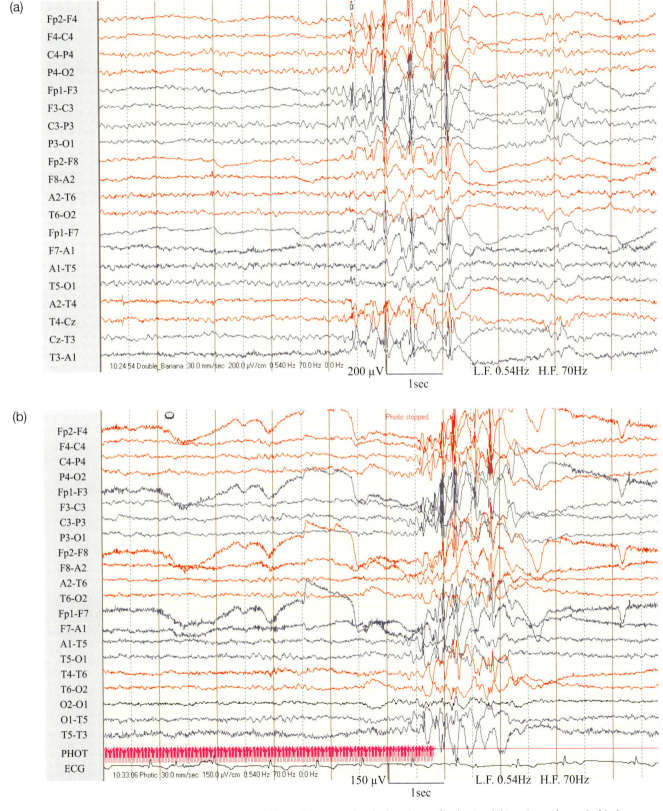

Fig. 6.27 Juvenile myoclonic epilepsy: an 11-year-old boy with one tonic–clonic seizure after la ate-night party, and reported to be clumsy in the early morning and to have been 'daydreaming' in recent months. His uncle has epilepsy. (a) The EEG shows multiple spike-and-slow-wave bursts that are more sustained than those of the patient in *Fig. 6.26(a)*. Discharge was apparently subclinical, but is followed by movement artefacts that suggest it was not. (b) Self-sustaining generalised photoparoxysmal response. (From Binnie et al (2003), by permission.)

context of partial or symptomatic generalised epilepsies, but these are regarded as different syndromes.

The interictal EEG is usually normal; reading and language-related activities precipitate epileptiform discharges over the parietotemporal regions, either bilaterally or left-sided, and rarely right-sided only (*Fig. 6.28*). Initially isolated spikes occur, with complaints of the jaw jerking or the text moving. Continued reading may lead to a build up of discharges and a tonic–clonic seizure. Reading-provoked paroxysmal alexia without motor symptoms, associated with prolonged focal ictal EEG abnormalities, has also been observed (Koutroumanidis et al 1998). As is common in reflex epilepsies with cognitive precipitants, the emotional context of the cognitive activity may be important and prolonged exposure may be required to elicit a seizure. To demonstrate reading epilepsy in the EEG laboratory it is important to use whatever type of text the patient finds most provocative, to make the experience somewhat stressful (e.g. reading aloud) and to allow sufficient time (at least 30 min of reading) for an effect to be seen. Photosensitivity is uncommon and does not appear to be related to the mechanism of epileptogenesis.

Photosensitive epilepsy

It is disputed whether epilepsies with reflex precipitation, and specifically photosensitive epilepsies, should be regarded as a syndrome. Although photosensitivity is associated with idiopathic generalised epilepsy, it can occur in partial, cryptogenic and symptomatic epilepsies, in patients who clearly have another recognised syndrome. It may be noted here that 1/4,000 of the general population, but 5% of people with epilepsy, exhibit a self-sustaining generalised photoparoxysmal response, which may be regarded as the hallmark of photosensitivity (see p. 66). On the other hand, up to 95% of people with such a self-sustaining generalised photoparoxysmal response attending an EEG department for diagnostic investigation (see pp. 64 ff.) have epilepsy. Photosensitivity is most often detected at age 7–19 years, with a peak at 12–14 years; there is a female predominance (2:1). However, this may be an artefact of referral patterns, as the syndromes associated with photosensitivity present typically about puberty, leading to EEG investigation at this time. Prolonged follow-up of photosensitive subjects shows that most retain their abnormal responses to intermittent photic stimulation into middle age, unless on antiepileptic medication.

Seizures include generalised tonic–clonic seizures, myoclonic jerks and, more often than is generally realised, partial seizures. Attacks may be precipitated by television, computer animation, discotheque lighting, linear patterns and sunlight flickering, and in 40% of patients these are apparently the only seizures, there being no spontaneous attacks.

The phenomenology of photic responses is considered on page 66. In the present context, it may merely be noted that paediatric syndromes commonly associated with photosensitivity are listed in *Table 6.2*.

Autosomal dominant, nocturnal, frontal lobe epilepsy

Autosomal dominant, nocturnal, frontal lobe epilepsy is a familial partial epilepsy with incomplete penetrance first described by Scheffer et al (1995). Various mutations of genes have been found on chromosome 20q, 15q and 1p

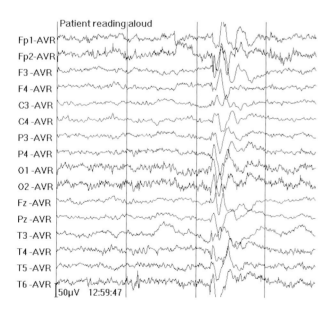

Fig. 6.28   Reading epilepsy in an 18-year-old adolescent. Reading and language had induced orofacial myoclonic jerks since the age of 14 years. A burst of generalised sharpened slow waves induced by reading aloud; this discharge was accompanied by a mild oral myoclonus. Vertical bars are at 1-s intervals. (From Binnie et al (2003), by permission.)

Table 6.2   Syndromes associated with photosensitivity

Syndrome	Percentage of patients being photosensitive
Juvenile myoclonic epilepsy	30–50%
Juvenile absence epilepsy	20%
Epilepsy with generalised tonic–clonic seizures on awakening	13%
Severe myoclonic epilepsy in infancy	70%
Epilepsy with myoclonic–astatic seizures	> 50%
Idiopathic photosensitive occipital lobe epilepsy	Up to 100%
Progressive myoclonic epilepsies	Variable

coding for subunits of the neuronal acetylcholine receptor comprising the sodium/potassium ion channel, but many families do not carry the described mutations (George 2004, Gourfinkel-An et al 2004). Onset is in childhood or adolescence, but misdiagnosis is frequent, especially confusion with nightmares or parasomnias (see p. 274). Seizures are brief and stereotypical, with a sudden, often explosive onset, and may include screaming, agitation, stiffening, turning and kicking or cycling of the legs. They usually occur in clusters during non-REM sleep, but some patients may have occasional diurnal seizures. Ongoing EEG activity is normal and interictal discharges are uncommon. The ictal EEG is usually unhelpful, as hyperkinetic movements create many muscle artefacts. In some, diffuse flattening, bifrontal or unilateral abnormalities may be seen (Picard et al 2000). These abnormalities consist of rhythmic theta, sharp waves or spikes. Video–EEG is helpful to verify the stereotypical movements.

Other familial partial epilepsy syndromes

Other familial partial epilepsy syndromes have recently been described (Gourfinkel-An et al 2004), such as:
- familial temporal lobe epilepsy
- autosomal dominant, lateral, temporal lobe epilepsy
- autosomal dominant partial epilepsy with auditory features
- familial partial epilepsy with variable foci.

Genetically determined temporal lobe epilepsy syndromes are generally milder than symptomatic forms. Interictal EEG abnormalities are found in about half of patients with familial temporal lobe epilepsy but are very common in familial partial epilepsy with variable foci (Berkovic and Steinlein 1999, Picard et al 2000). They consist of focal theta or delta slow waves and/or sharp waves or sparse spikes. The location of the interictal EEG abnormalities is usually concordant with the clinically presumed epileptic focus. Similar EEG abnormalities may be found in asymptomatic carriers (Picard et al 2000). The ictal EEG is characterised by a focal recruiting pattern, with increasing amplitude and decreasing frequency. The EEG is generally not helpful in distinguishing idiopathic from symptomatic partial epilepsy.

Progressive myoclonic epilepsies of childhood and adolescence

This term includes a heterogeneous group of unrelated disorders that share some clinical features (Roger et al 1992, Shahwan et al 2005):
- myoclonias usually including a combination of segmental, arrhythmic, asynchronous, asymmetrical myoclonias and massive myoclonias
- epilepsy usually with generalised tonic–clonic seizures and often other seizures
- dementia
- variable neurological symptoms always including cerebellar manifestations.

These disorders account for about 1% of all epilepsy occurring in childhood. Many of these conditions are recognised as distinct syndromes, and some are considered elsewhere. They include: degenerative progressive myoclonic epilepsy, including the Unvericht–Lundbourg Baltic type and Mediterranean type; Lafora disease; neuronal ceroid lipofuscinosis (see p. 303); mitochondrial encephalomyopathy with myoclonus epilepsy with ragged red fibres (MERRF) (see p. 304); sialidosis type 1 and type 2 (see p. 302); Gaucher's disease (see p. 303); childhood Huntington's disease (Jervis 1963); dentatorubropallidoluysian atrophy (Naito and Oyanagi 1982); neuroaxonal dystrophy (Dorfman et al 1978); action myoclonus–renal insufficiency syndrome (Andermann et al 1986); and juvenile GM2 gangliosidosis type 3 (Suzuki et al 1970). For an overview of the clinical and electrographic signs see *Table 6.3*.

In the early stages the EEG features of some forms of progressive myoclonus epilepsy can be mistaken for idiopathic generalised epilepsy. These features include generalised multiple spike-and-slow-wave discharges, normal background and photosensitivity. Subsequently, the EEG gradually begins to change and generalised slowing of the background activity, progressive disorganisation of sleep patterns, bursts of spikes, spike-and-slow-wave and multiple spike-and-slow-wave discharges occur, in relation to the overall clinical deterioration. The only progressive myoclonic epilepsy with characteristic EEG signs is the late infantile type of ceroid lipofuscinosis; photic stimulation at low flicker frequencies (1–5 Hz) evokes giant abnormal VEPs over both occipital areas, and possibly irregular generalised spike-and-wave discharges (see p. 303). There are also high amplitude VEPs and SEPs. For a full review see Berkovic et al (1991) and Honovar and Meldrum (1997).

Ring chromosome 20 syndrome

This is a rare chromosomal disorder due to mosaic r(20) characterised by intractable seizures, mental retardation to a varying degree, absence of dysmorphic features and a typical evolution. Pre- and postnatal growth are normal and initial psychomotor development is age appropriate. Onset of seizures is usually between 1 and 6 years of age, but may be earlier or later (Ville et al 2006). Seizures are usually frequent and drug resistant, and include complex partial seizures, nocturnal frontal lobe seizures, generalised tonic–clonic seizures and non-convulsive status epilepticus. Status epilepticus presents mainly in two forms: absence status epilepticus and complex partial status epilepticus (Inoue et al 1997). Learning difficulties and behavioural problems develop after the onset of seizures and may be associated with periods of non-convulsive status.

The interictal EEG may show irregular high amplitude slow waves, which are often bilateral and frontal but may be asynchronous or lateralised, as well as interictal spikes, sharp waves or spike-and-wave complexes bilater-

Table 6.3   Overview of progressive myoclonic epilepsies

Epilepsy type	Onset	Inheritances,	Clinical features chromosome, protein	Seizures	EEG
Unverricht–Lundborg disease	6–18 years	AR, 21q22.3, cystatin B	Ataxia, tremor, dysarthria, mild cognitive decline, all mild and late	Stimulus-sensitive myoclonus, GTCS, often on awakening, absences. Myoclonus becomes incapacitating	At onset normal ongoing activity, generalised spike-and-wave discharges, becoming more frequent; normal sleep pattern; photosensitivity. In long-established cases abnormality of ongoing activity, less epileptiform discharges
Lafora disease	10–19 years	AR, 6q24, laforin (EPMA2 gene); AR, 6p22, malin (NHLRC1 gene)	Learning difficulties, psychiatric symptoms, ataxia, cortical blindness, poor prognosis	Myoclonus, occipital seizures with visual symptoms, GTCS, complex partial seizures	At onset normal ongoing activity and generalised 3/s discharges; photosensitivity at slow IPS (1–6 Hz); later slowing of ongoing activity and fast (6–12 Hz) spike-and-wave activity; no normal sleep pattern
Neuronal ceroid lipofuscinosis NCL 1 (infantile)	0–1 years	AR, 1p32, palmitoyl protein thioesterase (PPT1)	Visual failure early, ataxia, rapid mental deterioration	Massive bilateral myoclonus	Progressive slowing and loss of amplitude: 'vanishing EEG'. VEP abolished by 4 years, ERG extinguished
NCL 2 (late infantile)	2–4 years	AR, 11p15, protein tripeptidyl peptidase 1 (TPP1)	Ataxia, psychomotor regression, extrapyramidal signs, visual failure late	Massive myoclonus, atypical absences, atonic seizures, GTCS	Slow ongoing activity, irregular bursts of spikes or multiple spike-and-wave discharges, multifocal discharges. Slow IPS: giant 'spikes' biocciptal. Flash VEP enlarged. SEP enlarged. ERG extinguished
NCL 3 (juvenile/Batten's disease)	4–10 years	AR, 16p21 (CLN 3: protein with unclear function)	Visual failure early, ataxia, behavioural and psychiatric problems, regression, extrapyramidal signs	GTCS and myoclonus, may be subtle	Slow ongoing activity, generalised spike-and-wave discharges, increase in sleep. VEP gradually extinguished. ERG extinguished early
NCL 4 (adult/Kuf's disease)	Variable, up to third decade	AR, AD, sporadic	Ataxia, dementia, extrapyramidal signs, no visual failure	Myoclonus (may develop late)	Generalised fast spike-and-wave discharges; photosensitivity
NCL 5 (late infantile Finnish variant)	4–6 years	AR, 13q21-q32 (CLN5)	Clumsiness, hypotonia, visual impairment, later ataxia	Myoclonus, GTCS	Similar to NCL 2, but response to slow IPS develops later at 7–8 years
NCL 6 (intermediate/late infantile variant/early juvenile)	5–7 years	AR, 15q21-23 (CLN6)	Marked ataxia, mental deterioration, visual failure, extrapyramidal signs	Myoclonus may occur, often other seizure types	Slow ongoing activity, irregular bursts of spikes or multiple spike-and-wave discharges. Slow IPS. Giant VEP. ERG extinguished

	Age	Genetics/enzyme	Clinical features	Seizure types	EEG/investigations
Myoclonus epilepsy and ragged red fibres (MERRF)	3–65 years	Sporadic or maternal, MtDNA point mutation	Myopathy, neuropathy, regression, ataxia, optic atrophy, sensorineural hearing loss, pyramidal signs	Myoclonus, generalised or partial seizures	Abnormal background activity in 80%, generalised spikes and multiple spike-and-wave discharges, diffuse delta, focal abnormalities (2–5/s); no normal sleep pattern; photosensitivity
Sialidoses Type 1	8–15 years	AR, 6p21.3, α-neuraminidase	Visual failure, cherry-red spot, ataxia	Action and intention myoclonus, facial myoclonus persisting in sleep, GTCS	Normal to fast, low voltage background, generalised spike-wave discharges during massive myoclonus. VEP decreased. SEP enlarged
Type 2	Variable, neonatal period to second decade	AR, 20, α-neuraminidase and β-galactosidase	Dysmorphic features, hepatosplenomegaly, progressive learning disability, renal failure, poor prognosis	Myoclonus	Low voltage fast activity, may show slowing at later stage, generalised spike-wave discharges
Gaucher's disease type III	–	β-Glucocerebrosidase	Ataxia, horizontal supranuclear gaze palsy, slow intellectual deterioration, anaemia, hepatosplenomegaly	Myoclonus, generalised or partial seizures	Normal to slow background activity, predominantly posterior or multifocal multiple spike-and-wave; photosensitivity
Juvenile Huntington's disease	Teens (may occur > 3 years)	AD with anticipation (juvenile usually paternal inheritance), 4p16.3, huntingtin	Arrest of psychomotor development, rigidity and dystonia, death 4–6 years after onset	Massive myoclonus, GTCS, absences, status epilepticus	Photosensitive, often before clinical seizures occur
Dentatorubropallidoluysian atrophy, PME type	6–70 years	AD with anticipation,1 2p13.31, atrophin 1	Choreoathetosis, ataxia, regression/dementia, psychiatric symptoms	Myoclonus, GTCS	Abnormal background, atypical spike-wave complexes, slow wave bursts
Juvenile GM2 gangliosidosis type 3	5–15 years	AR, β-N-acetylhexose aminidase A	Progressive ataxia, pyramidal signs	Myoclonus, other seizure types	–

ERG, electroretinogram; GTCS, generalised tonic–clonic seizures; IPS, intermittent photic stimulation; VEP, visual evoked potential.

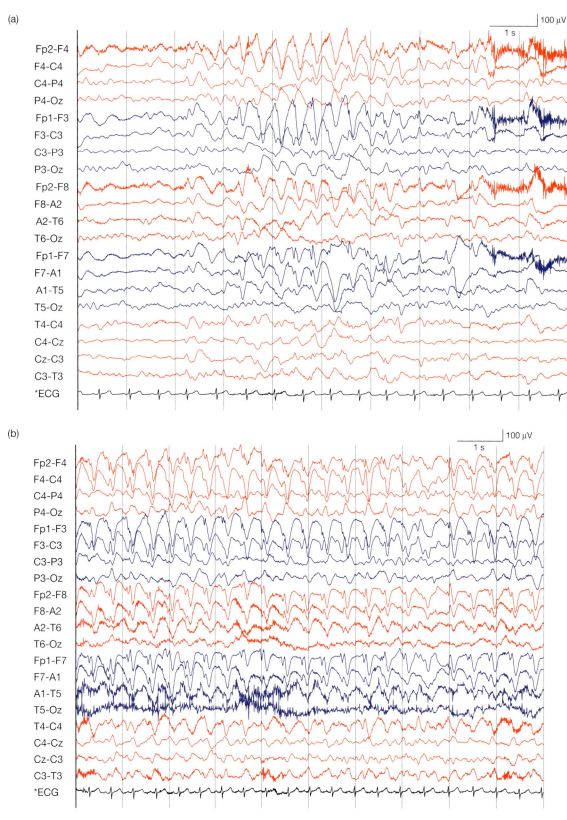

Fig. 6.29   Ring chromosome 20 in a 9-year-old girl with frequent seizures, including generalised tonic–clonic and complex partial seizures, as well as episodes of non-convulsive status. (a) The interictal EEG shows frequent bursts of high amplitude, generalised 2–3/s slow wave activity with some spikes intermixed, maximal over the anterior regions, often ending in a run of bifrontal sharp waves. Note the abnormal ongoing activity, which is dominated by 6–8/s activity. (b) During an episode of prolonged unresponsiveness (20 min) continuous bilateral 2–3/s spike-and-wave discharges are seen with a frontal emphasis.

ally over the frontal regions. In infants the interictal abnormalities are more diffuse and seizures are shorter (Ville et al 2006).

The ictal EEG shows prolonged periods of high amplitude slow waves with occasional spikes predominantly in the frontal or frontopolar areas (*Fig. 6.29(a)*). Initially this activity can be lateralised, but usually becomes bilateral. The duration of seizures is prolonged, lasting usually 10–50 min. Prolonged periods of continuous epileptiform activity are often associated with changes in awareness (*Fig. 6.29(b)*) and responsiveness (non-convulsive status epilepticus) that may last for days or even weeks.

NON-EPILEPTIC PAROXYSMAL DISORDERS

Some 40% of children presenting to epilepsy clinics do not have epilepsy (Robinson 1984, Uldall et al 2006). There are several conditions in childhood that mimic epilepsy. Most of them are benign and a detailed history from a person who has witnessed the attack is crucial for making the correct diagnosis (*Table 6.4*). Since some of these disorders are common, obviously they may occur in healthy children with a genetically determined EEG anomaly. This chance concurrence does not justify the assumption of a causal relationship, always providing that the parent's description clearly points, for instance, to simple breathholding.

In all paediatric work it has to be remembered that certain epileptiform EEG abnormalities may occur by chance, merely representing a genetic trait, and not relating to any symptoms. Focal spikes occur in about 2% of healthy children between the ages of 6 and 14 years (Eeg-Olofsson et al 1971, Cavazzuti et al 1980, Doose and Baier 1989) and generalised spike-and-wave paroxysms in about 1%.

Breathholding attacks

Breathholding attacks (cyanotic spells) are a common problem in toddlers and occur in about 4% of all children under 5 years of age. They are harmless and should be diagnosed by obtaining a precise description from the parents. The attack is usually provoked when the child is upset; he cries vigorously, holds his breath in expiration, becomes cyanotic and occasionally loses consciousness. Usually the limbs become flaccid and, rarely, this can be followed by a generalised stiffness lasting until the child resumes breathing. In some children prolonged sobbing results in breathholding spells.

In between breathholding attacks the EEG is normal. During the attacks – depending on the depth and duration of the breathholding – generalised slowing of the EEG may develop due to deepening hypoxia. The electrocardiogram (ECG), which should be recorded together with the EEG in all patients, generally shows bradycardia.

Reflex anoxic seizures

Reflex anoxic seizures or 'pallid breathholding spells' also occur in infants and toddlers but are less common than cyanotic breathholding attacks. They may be provoked by emotion or unexpected minor trauma, especially to the head. Placing the child in a bath is another common precipitant. The child stops breathing, loses consciousness, and becomes pale and hypotonic. The seizure is caused by cardiac asystole from vagal inhibition. The subsequent hypoxia may induce an epileptic seizure. In between attacks the EEG is normal.

Both types of breathholding, the pallid and the cyanotic, may be found in the same child on different occasions. Again the child's age and a full description of the event enables differentiation from reflex epilepsy.

Syncopes

Syncopal reactions are not infrequent events in children, especially in girls of school age. Fainting as a vasovagal reaction, whether or not followed by asystole, usually lasts some time before the child falls. However, it can occur quite abruptly and lead to tonic stiffness of the body, and even occasionally to some clonic movements and incontinence (convulsive syncope) during a period of isoelectric EEG (*Fig. 6.30*). The interictal EEG in such children may show some augmentation of slower components, especially over the occipital regions; the composition of the ongoing activity may thus consist of a broader spectrum of frequencies than usual. Overbreathing to the point at which high voltage slow waves occur can elicit a typical syncopal reaction. It is advisable to make routine use of a simultaneous lead I ECG for recording during hyperventilation. Tilt-table testing may confirm the diagnosis.

Syncopes of cardiac origin

Cardiac arrhythmias may be diagnosed by recording the ECG together with the EEG. Prolongation of the QT interval occurs in the 'sick sinus' syndrome, causing loss of consciousness during exercise or stress. It may be autosomal dominant (Romano–Ward syndrome), or autosomal recessive, and accompanied by deafness (Jervell–Lange–Nielsen syndrome). Depending on the duration of asystole, hypoxic slowing of the EEG may be seen during the attack. When hypoxic attacks are long, severe and frequent enough, the brain may be left impaired, with the EEG showing persistent abnormalities. Supraventricular tachycardia, such as Wolff–Parkinson–White syndrome can present with syncope.

In cyanotic heart failure, with thromboembolic complications or with metastatic bacterial embolisation of the brain, and in children with malformation of the great vessels, the EEG may show focal or generalised slowing, usually without spikes or spike-and-wave discharges.

Near-miss sudden infant death syndrome (SIDS) consists of episodes of apnoea and bradycardia, probably as a consequence of immature central regulatory control. The

Table 6.4 Episodic conditions that mimic epilepsy

Syndrome	Age	Aetiology	Attack	EEG	Prognosis
Breathholding attacks (cyanotic spells)	< 5 years (4% incidence)	–	Triggered by frustration, cries, holds breath in expiration: blue ± unconscious ± stiff	Normal, except when slowed due to hypoxia during attack	Harmless
Reflex anoxic seizures (pallid spells)	Infants and toddlers	Due to cardiac asystole from vagal inhibition	Triggered by pain, fever, minor trauma, etc.: falls, pale, apnoeic; hypoxia may induce a seizure	Normal, except when slowed due to hypoxia during attack	Harmless
Syncope	School children (girls > boys)	Vasovagal reaction ± asystole	Faint in hot environment: from fear or long standing ± tonic stiffness + rare clonic movements	Possible excess posterior slow waves during attack. Hyperventilation: high voltage slow waves	–
Migraine	2 years to puberty	Neurovascular instability of genetic origin?	Paroxysmal headache, unsteadiness, visual or gastrointestinal symptoms	Normal between attacks, except in hemiplegic migraine (slow waves). During attacks: mild (focal) slowing	–
Benign paroxysmal vertigo	1–5 years	–	Episodes of vertigo, lasting one to several minutes, associated with nystagmus and ataxia	Normal	–
Transient tic disorder	3–10 years (prevalence 5%)	Environmental	One or more simple motor tics, phonic tics	Normal	Self-limiting
Tourette's syndrome	3–10 years	Genetic	Motor tics, quasi-purposive movements, echopraxia, gutteral sounds, echolalia, coprolalia	Normal	Often improves after puberty
Masturbation/gratification phenomena	Infants to toddlers	–	Movements of masturbation, stereotypical movements while appearing withdrawn	Normal	–

Disorder	Age	Aetiology	Clinical features	EEG	
Psychogenic non-epileptic seizures	6 years to puberty	Model for symptoms, based on own experience or close relative	Performance graded according to audience, semi-purposeful activities, non-stereotypical, lack of injuries	Movement artefacts	–
Munchausen-by-proxy	Any age	Gain for carer	Protean manifestations	May be rendered abnormal by poisoning, asphyxia, etc.	Risk of harm
Paroxysmal choreoathetosis	1 year to puberty	Familial or acquired	Episodes of extrapyramidal symptoms, lasting one to a few minutes	Normal	–
Startle disease (hyperekplexia)	6 months to puberty	Genetic	Excessive startle, falling	Normal	–
Pavor nocturnus (see p. 274)	1–5 years	–	Sudden arousal from non-REM sleep (stage III–IV) during the first half of the night	Normal	–
Benign neonatal sleep myoclonus	0–3 months	–	Myoclonias confined to sleep in a healthy infant	Normal	Excellent
Cardiac arrhythmias	–	Long QT syndromes (e.g. Romano–Ward syndrome, Lange-Nielsen syndrome) Supraventricular tachycardia (e.g. Wolff–Parkinson–White syndrome)	–	Hypoxic slowing during prolonged attack	–
Other syncopes of cardiac origin	–	Cyanotic cardiopathies, valvular disorders, cardiomyopathy, etc.	–	Hypoxic slowing during prolonged attack	–
Near-miss sudden infant death syndrome	Peak 1–6 months	–	Apnoea, hypo- or hypertonia, cyanosis, etc.	Permanent EEG abnormalities may occur	–

EEG may recover after such attacks or remain permanently impaired.

Other episodic disorders

The diagnosis of migrainous attacks in children requires a detailed history from the parent and child. Attacks of complicated migraine may be mistaken for focal epileptic seizures, particularly in epilepsy of childhood with occipital paroxysms (see p. 254). During the attack, however, the EEG shows no spikes or spike-and-wave activity, but rather focal slowing; this may be confined to a small area of one hemisphere, but not infrequently spreads to adjacent areas. The EEG between attacks may be misleading, since in up to one-third of children with migraine the ongoing activity shows intermittent non-specific abnormalities, which at times occur paroxysmally (Gamstorp 1985). Photic responses, in contrast to those in adults, are not more prominent in children with migraine than in normals.

Benign paroxysmal vertigo is characterised by recurrent attacks of vertigo, lasting from one to several minutes. It may be related to migrainous attacks. It is seen in toddlers and in children up to 5 years of age, and has not been associated with EEG abnormality.

In paroxysmal choreoathetosis, paroxysmal kinesigenic choreoathetosis and in startle disease (hyperekplexia), which are genetically transmitted, the EEG is almost always normal. There are reports of epileptiform activity – which is, of course, difficult to differentiate from artefact, but this finding is exceptional.

The parasomnias (for a review see Mahowald and Ettinger 1990) may present with diagnostic problems in children, particularly if a question of epilepsy is raised. A useful tabulation of the various childhood problems according to sleep stage is given on page 36. Night terrors (pavor nocturnus) (*Fig. 6.31*) and sleepwalking (somnambulism) occur in non-REM sleep stages III–IV, nightmares in REM sleep, tooth-grinding (bruxism) and sleep-talking (somniloquy) in sleep stages I–II and REM sleep, whilst body rocking and bed wetting (sleep-related enuresis) may each occur in any sleep stage. Night terrors are common in toddlers (1–3% incidence) and are

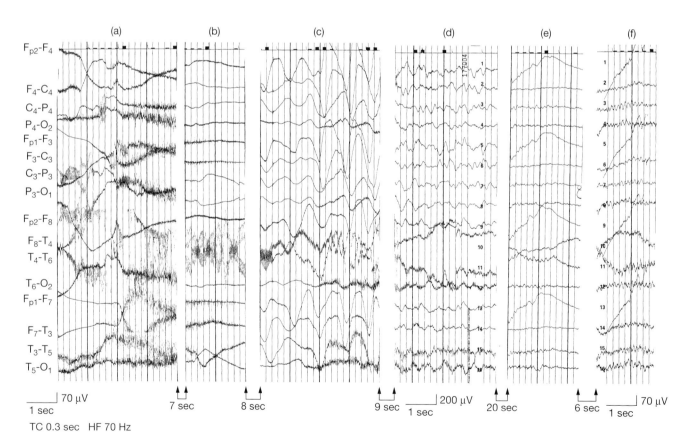

Fig. 6.30 Syncope in a 13-year-old boy who said that he did not like hospitals'. He felt 'an attack coming on' during electrode impedance testing before the start of the EEG and before the ECG electrodes were in place. Six extracts from EEG at documented intervals. (a) Patient slides down in chair, his body becomes rigid and his eyes are wide open; record shows almost no EEG activity, but increasing scalp muscle potentials. (b) Flaccid, still no apparent EEG activity. (c) More rigid, gurgling noises and dribbling; 50 Hz notch filters applied because of detachment of right midtemporal electrode T4; EEG shows high voltage, substantially symmetrical, slow delta waves. Sensitivity reduced in (d) and (e) as the patient starts to recover and increasing amounts of normal EEG activity appear, initially mixed slow waves, giving way in (f) to an 8 Hz alpha rhythm 70 s after the onset of syncope. (From Binnie et al (2003), by permission.)

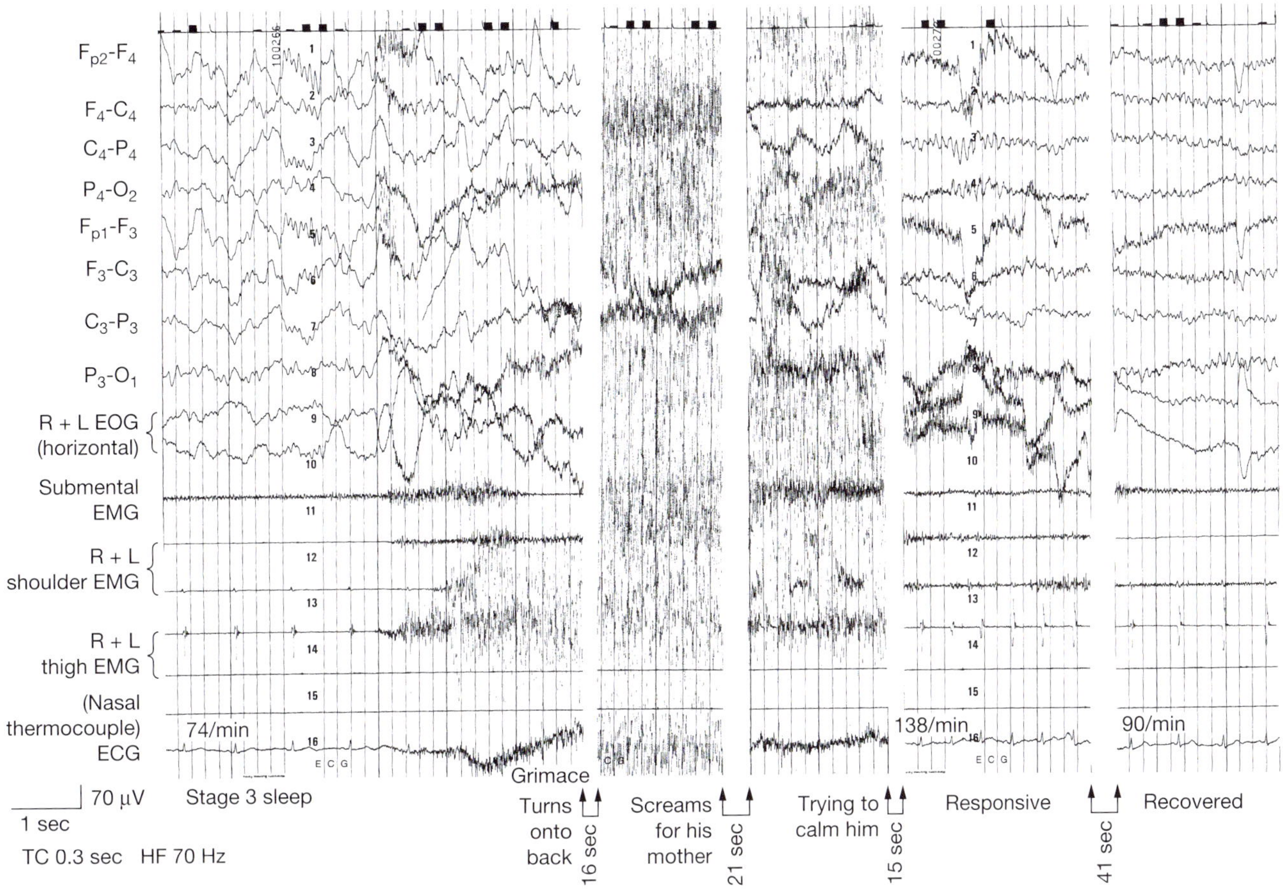

Fp2-F4
F4-C4
C4-P4
P4-O2
Fp1-F3
F3-C3
C3-P3
P3-O1
R + L EOG (horizontal)
Submental EMG
R + L shoulder EMG
R + L thigh EMG
(Nasal thermocouple)
ECG
74/min
138/min
90/min
70 μV
1 sec
TC 0.3 sec HF 70 Hz
Stage 3 sleep
Grimace
Turns onto back
16 sec
Screams for his mother
21 sec
Trying to calm him
15 sec
Responsive
41 sec
Recovered

Fig. 6.31 Pavor nocturnus during an afternoon sleep in a 13-year-old boy. He had a history of frequent night terrors every night for the previous 3 years (note: pavor nocturnus is usually seen in younger age groups). These had caused complaints from neighbours because of the associated ear-piercing screams. Phenytoin and diazepam given at bedtime had no beneficial effect. Video and polygraphic recording (nasal thermocouple not functioning) of a clinical attack during stage III sleep. The patient, who was on his side, suddenly grimaced and turned on to his back with teeth clenched. His arms flexed as he rolled over, crying out for his mother and screaming loudly for 20–30 s. He was comforted and gradually became more relaxed and responsive, his tachycardia settling in 2–3 min from the onset of the attack. (From Binnie et al (2003), by permission.)

characterised by sudden arousal from slow wave sleep with a piercing scream or cry, accompanied by autonomic and behavioural manifestations of intense fear. Night terrors, sleepwalking and sleep-talking tend to occur during the first half of the night.

There has been relatively little investigation of events during sleep, such as sleepwalking, nightmares and night terrors, in children but the absence of epileptiform activity on video–EEG monitoring, together with a clear history of the circumstances of the attacks are helpful in differentiation from frontal lobe epileptic seizures.

Myoclonic jerks during sleep (hypnagogic jerks), especially while falling asleep, have no abnormal EEG correlate (movement artefact can occur), but may be associated with arousal phenomena such as K-complexes.

Tics are usually clinically obvious, but occasionally need polygraphic recording to exclude an epileptic origin (for Tourette's syndrome see p. 320).

INFECTIONS OF THE NERVOUS SYSTEM

Meningitis and encephalitis

Bacterial meningitis

Meningitis may lead to generalised synchronous or asynchronous slow waves in children; only rarely can it cause focal slow waves. The slowing often begins posteriorly and may be focally accentuated in diffuse viral encephalitis. This does not necessarily imply the presence of a discrete lesion of the brain, since these foci move during the course of the disease. Even in bacterial meningitis such a slow wave focus does not necessarily indicate the development of a brain abscess. In neonates and infants focal spikes may occur. Meningitis can reduce the amplitude of the background activity during the active phase of the disease.

Fever, metabolic abnormalities and electrolyte imbalance associated with infection can cause generalised synchronous or asynchronous slow waves in children, even if the brain is not infected.

The possibility of hearing loss can be monitored with brainstem auditory evoked potentials (BAEPs) in young or neurologically compromised children unable to cooperate with standard clinical testing.

Encephalitis

In viral encephalitis the EEG reflects a diffuse involvement of cortical activity, showing generalised synchronous or asynchronous slow waves (rarely just brainstem dysfunction). Focal slow waves are seen when there is more localised brain damage, as in herpes simplex encephalitis. The amplitude can be reduced locally in the area of maximal involvement or, more often, overall. In congenital cytomegalovirus infection BAEPs are used to detect and monitor possible hearing loss; VEPs are used when chorioretinitis is present and affecting the central retina, so as to determine the involvement of afferent input to visual cortex.

Acute disseminated encephalomyelitis (ADEM) is a demyelinating response to a viral infection elsewhere, notably measles, rubella or chickenpox, predominantly affecting the white matter. The EEG shows moderate to severe, diffuse, high voltage theta–delta activity, reflecting white matter involvement (*Fig. 6.32*). In some patients there may also be abnormalities of EPs, suggesting demyelination. VEPs may be used to rule out the possibility of a coexisting optic neuritis, which when present is bilateral in these patients, unlike what is observed in multiple sclerosis.

Encephalitis of any form in children, if it is sufficiently severe, may, during its course, show periodic complexes similar to the type described by Radermecker (1956) in subacute sclerosing panencephalitis.

Herpes simplex encephalitis

Herpes simplex encephalitis (*Fig. 6.33*) in childhood may present in two different forms; infection with type II herpes virus, acquired during birth from the mother, or with type I herpes virus as in adults. No specific EEG patterns are pathognomonic for herpes simplex encephalitis, but a focal or lateralised EEG abnormality in the presence of encephalitis is highly suspicious of herpes simplex encephalitis. The EEG initially shows high amplitude slow waves with changing maxima but, in

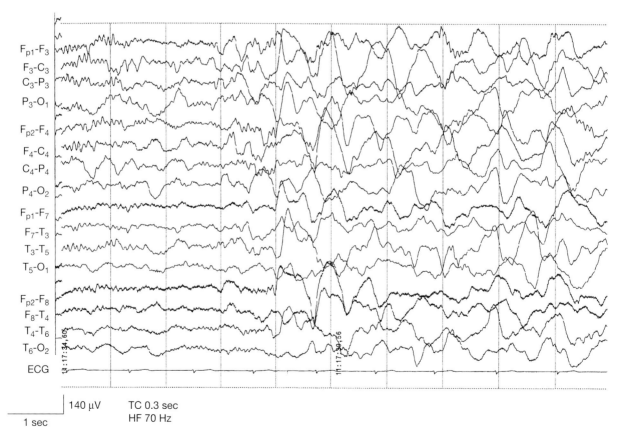

Fig. 6.32 Acute disseminated encephalomyelitis (ADEM) in a 15-year-old girl who had been ill for 5 days. She fluctuated between sleep and agitation during the EEG, which showed preservation of alpha-frequency components on both sides anteriorly but frequent bursts of irregular, high voltage generalised delta activity, increasing during periods of arousal and usually beginning in the left frontal region. A CT scan showed a left, frontal, low-density, white matter lesion, and on MRI there was widespread and massive loss of myelin. (From Binnie et al (2003), by permission.)

(a)

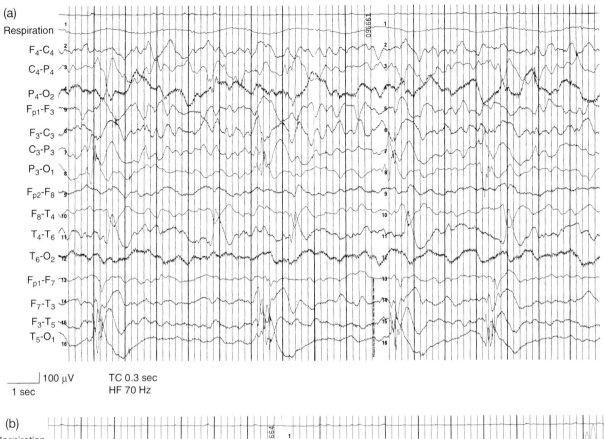

100 µV	TC 0.3 sec
1 sec	HF 70 Hz

(b)

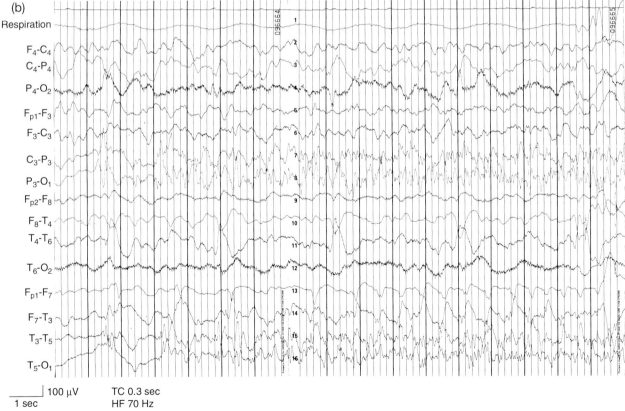

100 µV	TC 0.3 sec
1 sec	HF 70 Hz

Fig. 6.33 (a) Herpes simplex encephalitis in an 11-month-old boy who presented after a 48-h febrile illness and a single convulsion on the previous evening. He was unresponsive, but moved all limbs and followed fingers briefly before falling asleep. Treatment with aciclovir had just begun. The EEG shows stereotypical, repetitive complexes (emerging PLEDs) arising independently in each temporal region and spreading parietally, particularly on the left side. Ongoing background activity is depleted over the temporal areas, especially on the left. (b) Continuation of the previous EEG after a 16-s interval. Note the development during sleep of a localised ictal discharge from the left temporal lobe, spreading parietally. The outcome was poor. (From Binnie et al (2003), by permission.)

contrast to adults, focal EEG signs are not always seen. Constant focal abnormalities, especially with spikes or sharp waves, usually persist for weeks or months. The amplitude may be relatively reduced over an area of haemorrhagic necrotic inflammation. Periodic lateralised epileptiform discharges (PLEDs), as described by Upton and Gumpert (1970) in adults, may also be seen in children, even in infancy, but are relatively rare (see p. 58). In serial EEGs, PLEDs may be seen for only a short period at about the end of the first week from onset of symptoms (i.e. rather later than in adults). Their disappearance does not necessarily signal a good prognosis (Eggers et al 1976). Fleeting subclinical epileptiform discharges are uncommon in infants. When given sufficiently early, the antiviral agent aciclovir halts these stages of EEG evolution along with its effect on the clinical state.

Tuberculous meningoencephalitis

This is rarely seen nowadays in children in developed countries, but remains common in some parts of the world (Chaptal et al 1954, Patwari et al 1996). Generalised slowing of ongoing activity, especially with a subacute course, may occur and either be minimal or develop frontally predominant delta bursts (frontal intermittent rhythmic delta activity (FIRDA)). The latter characterises the severe, untreated case and may be associated with ventricular dilatation from the effects of a glutinous basal meningitis. Localised abnormalities may be seen in association with tuberculomata. It is also reported that diffuse beta activity may be prominent and persist for years. Seizures are associated with generalised or focal discharges in the EEG (Patwari et al 1996) and are attributed to cerebral oedema (57%), a syndrome of inappropriate antidiuretic hormone (ADH) secretion (32%), tuberculoma (27%), abnormal electrical foci (25%) and cerebral infarction (13%).

Cerebral thrombophlebitis

This is not uncommon in severe inflammatory processes in children, usually presenting with marked asymmetry, generalised slowing and often focal depression of the ongoing EEG. Focal spikes and sharp waves occur over the area concerned.

Brain abscess

Brain abscess may develop by direct spread from a mastoid or paranasal sinus infection or from a distant infection through septic emboli. The most common causative organisms are *Staphylococcus* and *Streptococcus* infections; infections caused by fungi, mycobacteria and protozoans occur considerably less frequently. As a result of focal suppuration, the EEG may show changes comparable to those occurring in brain tumours. Slow local polymorphic delta activity and FIRDA (see p. 48), as well as suppression of the local physiological activities, such as mu rhythm, alpha rhythm and beta activity, may occur (Michel et al 1979). Sometimes the delta activity is

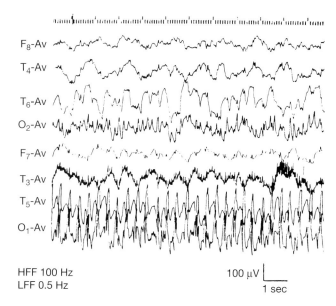

HFF 100 Hz
LFF 0.5 Hz

100 μV

1 sec

Fig. 6.34 Child with cerebral malaria. Continuous sharp-and-slow-wave discharge, maximal over the left posterior temporo-occipital region, during a partial seizure comprising nystagmus and deviation of the eyes to the right. (From Crawley et al (1996), by permission of Oxford University Press.)

generalised, not permitting localisation of the focus. Imaging (CT or MRI) is superior in detecting and localising chronic and infratentorial abscesses. In the early stages of supratentorial abscess formation, when there is still a localised acute cerebritis and not yet an encapsulated abscess, EEG abnormality may precede CT/MRI scan abnormality for several days, and thus provides a useful clue to the need for further imaging studies and timing of surgical intervention. The EEG together with CT/MRI scans are sensitive methods for monitoring antibiotic and surgical treatment. Whereas epileptiform activity is a rare occurrence in the acute stages, after successful surgical treatment many patients (up to 70%) will develop focal epileptiform activity and subsequent seizures (Legg et al 1973). However, these need not be related, because epileptiform activity may exist without seizures.

Cerebral malaria

The cause of coma in cerebral malaria remains unclear. The cerebral microcirculation is impaired by sequestration of red blood cells containing the parasite *Plasmodium falciparum*; neuronal damage may occur due to various factors including:

- hypoxic–ischaemic changes resulting from reduced blood flow
- raised intracranial pressure due to increased intracranial blood volume
- local toxins.

Cerebral malaria has a mortality of 10–40%, with neurological sequelae (hemiplegia, speech problems, cortical blindness, epilepsy) occurring in 5–15% of survivors. Almost two-thirds of 65 children admitted to a Kenyan hospital between January and September 1994 with cerebral malaria were found to have seizures, half with periods of status epilepticus (Crawley et al 1996, 2001). Other common seizure types are partial motor seizures and subtle or electrographic seizures. The EEG background activity was dominated by high amplitude delta activity. Ictal discharges always arose from the posterior parieto-temporal region, a border zone area lying between territories supplied by the carotid and vertebrobasilar circulations, often spreading more widely and lasting longer than the clinical accompaniments suggested (*Fig. 6.34*). During prolonged coma, both postictal depression or slowing, and subclinical seizure discharges, were often found and thus may play an important part in the pathogenesis of the coma in cerebral malaria (Crawley et al 1996). Initial EEG recordings of very slow frequency, particularly when delta activity appears in bursts (variants of FIRDA and occipital intermittent rhythmic delta activity (OIRDA), see p. 51) in relation to opisthonic posturing in those with intracranial hypertension (*Fig. 6.35*), or with background asymmetry, burst-suppression pattern or interictal discharges, were associated with an adverse outcome (Crawley et al 2001).

Slow virus infections
Subacute sclerosing panencephalitis

Subacute sclerosing panencephalitis (SSPE) is a chronic encephalitis with high levels of measles antibodies in the cerebrospinal fluid (CSF) which occurs several years after measles infection. It presents in late childhood or adolescence, with regression, personality changes and periodic, often subtle, abnormal movements. It has become a rare condition since the widespread introduction of measles vaccine, but the incidence may increase in the future with the fall in vaccination rate due to parental anxiety over the measles–mumps–rubella (MMR) triple vaccination. The 'characteristic' EEG pattern of periodic generalised complexes consists of stereotypical, repetitive, high voltage (300–1,500 μV) polyphasic complexes of 0.5–3 Hz that reoccur every 3–20 s, usually bilaterally or generalised, but not necessarily synchronous, on a relatively low voltage background (Cobb and Hill 1950) (see *Figs 2.49* and *6.36*). The characteristic EEG signs generally precede these clinical manifestations and are thus of great diagnostic significance. The complexes persist during sleep, although in a less distinct way. Complexes become more synchronous and symmetrical as the disease progresses. These are accompanied by synchronous myoclonus or loss of tone. In the final stages of the

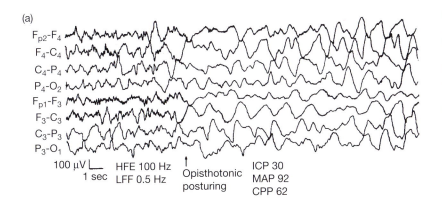

(a)

Fp2-F4, F4-C4, C4-P4, P4-O2, Fp1-F3, F3-C3, C3-P3, P3-O1

100 μV, 1 sec, HFE 100 Hz, LFF 0.5 Hz, Opisthotonic posturing, ICP 30, MAP 92, CPP 62

Fig. 6.35 Child with cerebral malaria presenting with coma and intracranial hypertension. (a) Initial EEG shows bursts of delta activity accompanying opisthotonic posturing. (b) 10 h later there is asymmetry with localised reduction of activity in the right parietotemporal watershed region. (From Crawley et al (2001), by permission of the BMJ Publishing Group.)

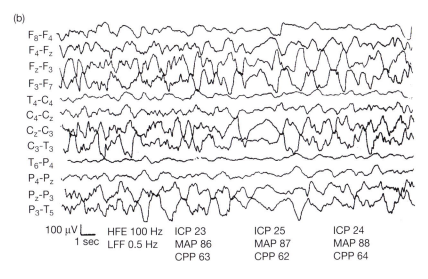

(b)

F8-F4, F4-Fz, Fz-F3, F3-F7, T4-C4, C4-Cz, Cz-C3, C3-T3, T6-P4, P4-Pz, Pz-P3, P3-T5

100 μV, 1 sec, HFE 100 Hz, LFF 0.5 Hz, ICP 23, MAP 86, CPP 63, ICP 25, MAP 87, CPP 62, ICP 24, MAP 88, CPP 64

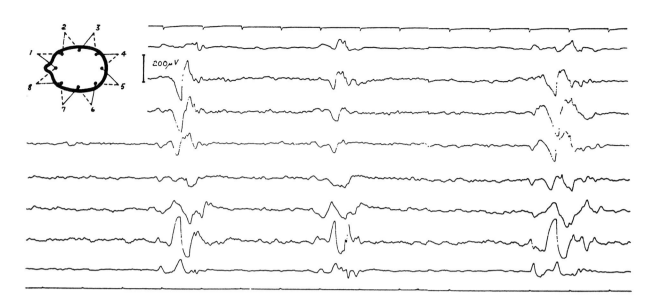

Fig. 6.36 Subacute sclerosing panencephalitis in a 7-year-old girl with a 4.5 month history of progressive organic deterioration, apraxia and dysphasia, latterly with generalised myoclonus. EEG (recorded in 1960) shows generalised, high voltage, slow wave complexes occurring periodically on a relatively low voltage background. (From Binnie et al (2003), by permission.)

illness the EEG progressively diminishes in amplitude to an almost isoelectric pattern. The periodic pattern with regular intervals of more than 4 s is considered to be one of the most specific of all EEG findings (Cobb 1966). When associated with the characteristic clinical signs it is pathognomonic. To demonstrate the complexes it may be necessary to carry out repeated EEG examinations, in both waking and sleep.

Although periodic complexes are almost universal, albeit varying in morphology (Yaqub 1996), atypical features have been reported, including frontal rhythmic delta activity, electrodecremental periods, diffuse sharp waves and sharp-and-slow-wave complexes over frontal regions, and multifocal epileptiform discharges, usually in frontal, central and temporal regions (Dogulu et al 1995, Gurses et al 2000). Pre-existing childhood pathology, such as cerebral palsy, may affect the presentation (Santoshkumar and Radhakrishnan 1996).

It should be noted that periodic complexes are also seen in other infections, such as herpes encephalitis and progressive rubella panencephalitis. These complexes should also be distinguished from PLEDS (see p. 58). The complexes of SSPE are, however, generally distinctive on account of their marked stereotypy, despite the complex waveform and the low repetition rate.

HIV infections and AIDS

The EEG and BAEP reveal non-specific abnormalities in children with acquired immune-deficiency syndrome (AIDS). It seems that EEG changes parallel disease progression; in two-thirds of children with human immuno-deficiency virus (HIV) encephalopathy the EEG was abnormal, but the EEG was abnormal in only 20–50% of children in earlier stages of AIDS (Schmitt et al 1992,

Vigliano et al 2000). Abnormalities are non-specific and epileptiform discharges are rare. In addition, some BAEP interpeak interval (IPI) abnormalities were found independently of the progression of AIDS (Vigliano et al 2000).

Other neurological disorders with presumed infectious aetiology
Rasmussen's syndrome (chronic focal encephalitis)

Rasmussen's encephalitis is characterised by uncontrollable focal seizures, associated with focal inflammation in cerebral tissue (Rasmussen et al 1958, Rasmussen 1978). Typically onset is before age 10 years, with simple focal seizures; in most cases epilepsia partialis continua develops. In addition, complex partial seizures or secondary generalised seizures may occur (Cockerell et al 1996). A progressive hemiparesis develops, accompanied by progressive focal cortical atrophy on imaging. Histologically, this is a progressive active microglial encephalitis with appearances strongly suggestive of a viral origin. However, no definite agent has been proven, except in the original cases due to a Siberian tick-borne virus (Kojewnikov 1895). More recently, an autoimmune reaction involving the glutamate receptor 3 has been proposed (Rogers et al 1994), but this remains controversial. Granata et al (2003) found that the association of partial seizures with focal EEG and neuroimaging changes allows a tentative diagnosis of Rasmussen's encephalitis at an early stage.

The EEG background activity is abnormal in all cases, showing asymmetry and bursts of slow waves, which are asymmetrical but usually bilateral with lateralised predominance (*Fig. 6.37*). Interictal discharges are often

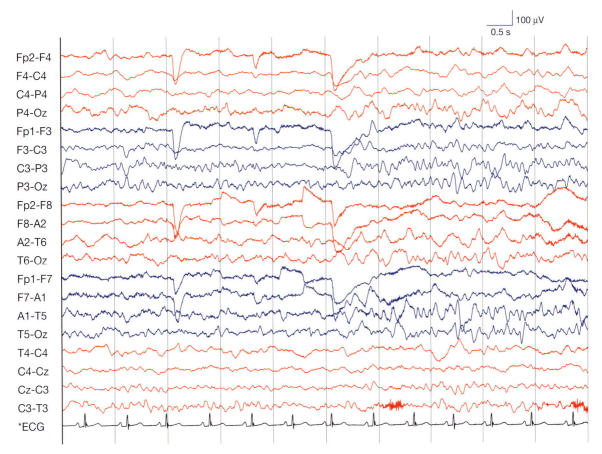

100 µV
0.5 s

Fp2-F4
F4-C4
C4-P4
P4-Oz
Fp1-F3
F3-C3
C3-P3
P3-Oz
Fp2-F8
F8-A2
A2-T6
T6-Oz
Fp1-F7
F7-A1
A1-T5
T5-Oz
T4-C4
C4-Cz
Cz-C3
C3-T3
*ECG

Fig. 6.37 Rasmussen's encephalitis in a 6-year-old girl. She presented with frequent partial seizures of the left side of the body. The MRI showed atrophy and signal change of the right hemisphere. The initial EEG is asymmetric with normal ongoing activity over the left and excess of slowing maximal over the right posterior region. She subsequently had functional hemispherectomy and pathology confirmed the diagnosis.

multifocal, either unilateral or bilateral, but always with a unilateral predominance. Bilateral synchronous, irregular spike-and-slow-waves or sharp-and-slow-waves are seen in half of patients. Despite the existence of diffuse abnormalities in many patients, it is usually possible to determine the hemisphere involved from EEG records (So and Gloor 1991). Often there is less activity, both normal and abnormal, over the affected hemisphere, presumably due to the atrophy seen on MRI; the discharges may therefore, paradoxically, predominate ipsilateral to the hemiparesis and motor seizures. The EEG eventually becomes of low voltage, which is in keeping with a burned-out encephalitis (Rasmussen 1978).

Reye's syndrome

This increasingly rare syndrome is a non-inflammatory encephalopathy of uncertain cause with fatty degeneration of the liver presenting with vomiting and coma 3–4 days after the onset of a mild viral illness (Reye et al 1963). An association with aspirin ingestion in children has led to prohibition of this drug for minor illnesses in children under 12 years of age. This may account for the increasing rarity of the syndrome. Many disorders of

metabolism may present with a 'Reye-like illness'. EEG features include non-specific changes seen in other types of coma, such as severe diffuse slowing and high voltage delta activity. Improvement in the EEG correlates closely with clinical improvement (Trauner et al 1977, Tasker and Cole 1997). A high incidence of 14 and 6 Hz positive spike bursts has been reported (Yamada et al 1977), as well as other patterns such as hypsarrhythmia and alpha coma (Yamada et al 1979).

NEURODEVELOPMENTAL DISORDERS

Neural tube defects and related conditions

Around 20% of children with myelomeningocoele may develop seizures at some time during the course of their illness (Talwar et al 1995). Most of them have shunted hydrocephalus. EEG abnormalities are related to additional CNS pathology.

An Arnold–Chiari malformation is found in 80% of children with myelomeningocoele. This consists of a protrusion of the cerebellar tonsils and brainstem through the foramen magnum, together with elongation of the fourth ventricle into the cervical canal. Other anomalies, such as stenosis of the aqueduct, grey matter heterotopias or absence of brainstem nuclei, may occur. Patients can present with additional symptoms of hind brain, lower cranial nerves and cervical cord compression, possibly leading to death. EPs have been proven helpful in the clinical evaluation of brainstem function, for monitoring the natural history and the surgical decompression procedure, as well as for predicting the neurological outcome (Taylor et al 1996).

BAEP abnormalities, consisting of poorly formed and delayed waves III and V with prolonged I–V IPI, have been found in about 70% of patients, but the relationship to the clinical picture remains controversial. The latency of wave V and the I–V IPI of such children are significantly longer than those of normal subjects (Lütschg et al 1985). The longest latencies were found in those with hydrocephalus or cranial nerve defects, such as fasciculation or atrophy of the tongue, stridor or cerebellar signs. The BAEP abnormalities reflect the severity of the Arnold–Chiari malformation. In myelomeningocoele posterior tibial nerve stimulation SEPs (PTN-SEPs)

show various abnormalities from normal responses in lesions below the L5 level, to absent or greatly attenuated responses recorded at the level of the conus medullaris and at the more distal locations, corresponding to the spinal cord anomaly. In this condition it can be of value to record from T12 and L5 sites to determine whether the attenuation of the response at T12 is due to spinal cord dysfunction or to a 'caudally displaced' conus medullaris of a tethered cord syndrome not associated with spinal cord dysfunction (i.e. attenuated response at T12 but unusually high amplitude at L5). In most cases there is a positive correlation with the child's sensory defect in the lower limb. The role in the evaluation of retethering following myelomeningocoele repair is debated, but is probably related to the degree of the SEP N22 component impairment.

Hydrocephalus

EEG abnormalities in children with hydrocephalus include generalised slow waves, more often asynchronous than bisynchronous, focal slow waves and focal spikes. Focal slow waves and spike-and-wave discharges are more common on the side of the shunt (*Fig. 6.38*). Generalised

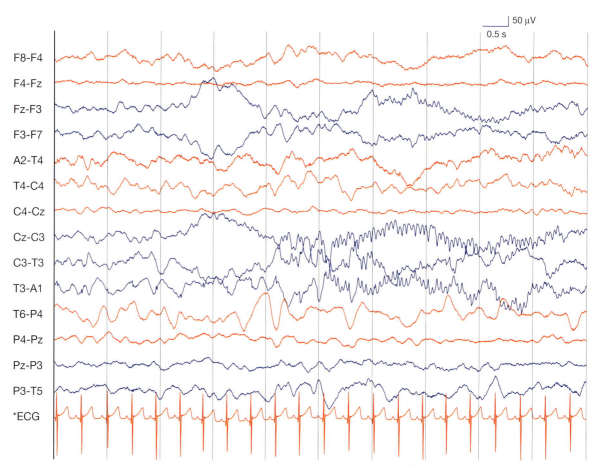

Fig. 6.38 Hydrocephalus with ventriculoperitonial shunt on the right in a 3-month-old infant. The right hemisphere is dominated by 2–4/s activity, with only a small amount of faster components intermixed, while on the left age-appropriate sleep spindles are seen.

or focal epileptiform discharges are often seen before the first shunt insertion (Suakkonen et al 1990). The incidence of seizures is higher in shunted than in unshunted hydrocephalus, ranging from 20% to 50%, and was found to be significantly affected by the original cause of the hydrocephalus (Ines and Markand 1977, Bourgeois et al 1999). There may be EEG evidence of an epileptogenic focus even in the absence of clinical attacks.

Disorders of neuronal migration

Development of the cerebral cortex involves three distinct but overlapping processes consisting of neuronal, and later glial, proliferation, neuronal migration and cortical organisation. Abnormal cortical development can be caused by genetic or acquired factors affecting neuronal proliferation and migration. The peak time for the latter is between the third and fifth month of gestation. The clinical spectrum of neuronal migration disorders is wide, ranging from early-onset epileptic encephalopathy to mild epilepsy. *Table 6.5* gives an overview of the more common neuronal migration disorders. Many other conditions are associated with abnormalities of gyrus and sulcus formation (Barth 1987), such as metabolic diseases (Zellweger disease, neonatal adrenoleucodystrophy, GM2 gangliosidosis, mitochondrial disorders, Menkes' disease), chromosomal abnormalities (trisomy 13, trisomy 18, trisomy 21, Miller–Dieker syndrome), neuromuscular syndromes (Walker–Warburg syndrome, Fukuyama muscular dystrophy, myotonic dystrophy, congenital myotonic dystrophy), neurocutaneous diseases (incontinentia pigmenti, neurofibromatosis, hypomelanosis of Ito, tuberous sclerosis, epidermal nevus syndrome), multiple congenital anomalies syndromes (Potter syndrome, Cornelia de Lange syndrome, Smith–Lemli–Opitz syndrome), and maternal and environmental causes (cytomegalovirus infection, fetal alcohol syndrome, carbon monoxide intoxication, ionising radiation). For reviews on migration disorders see Barth (1987) and Kuzniecky and Barkovich (2001).

Abnormal neuronal proliferation and differentiation

Hemimegalencephaly   See page 239.

Dysplasia of the cerebral cortex   In focal dysplasia, histological abnormalities involve a smaller area, which may be undetectable by current radiology (Taylor et al 1971). Patients usually develop intractable partial epilepsy. Focal status epilepticus or epilepsia partialis continua are relatively frequent. The interictal EEG often shows focal rhythmic epileptiform discharges, sometimes of high amplitude.

Neuronal migration

Classical lissencephaly   This is histologically characterised by a four-layered cortex. Several syndromes have been described, of which the X-linked, recessive Miller–

Dieker syndrome is the best known. It consists of lissencephaly, with profound learning disability, hypotonia, seizures and facial dysmorphia. Seizure types include infantile spasms, myoclonic, tonic and tonic–clonic seizures. A point mutation and microdeletion of the LIS1 gene occurs at the 17p1 3.3 locus. The EEG is characterised by rhythmic, high amplitude (around 300 μV) 10–18 Hz (typically 14 Hz) activity (*Fig. 6.39*). It has been suggested that, below the age of 1 year, this activity may be specific for this malformation and may be related to anomalous orientation of the cortical neurons (Gastaut et al 1987). In older children diffuse slow wave activity or high amplitude rhythmic 5–11 Hz activity of more than 300 μV may be prominent. Very high amplitude sharp-and-slow-wave complexes are common. About 80% of children present with infantile spasms, most of whom have atypical hypsarrhythmia and rhythmic fast activity (see *Fig. 6.7*).

Subcortical band heterotopia   This occurs in carrier females of the classic X-linked type 1 lissencephaly and may be an anatomical demonstration of Lyonisation. A thin band of white matter separates the cortex from the heterotopic grey matter. Epilepsy may be absent, mild or severe, and mild to moderate cognitive impairment may occur. Infantile spasms occur but less often than in lissencephaly. There appears to be a close correlation between the thickness of the heterotopic band, the likelihood of developing Lennox–Gastaut syndrome and the degree of cognitive deficit (Barkovich et al 1994). The EEG may be normal, but usually shows widespread theta, focal or multifocal epileptiform discharges, and sometimes generalised spike-and-slow-wave discharges (Barkovich et al 1994).

Lissencephaly and subcortical band heterotopia comprise a single malformation spectrum, often referred to as the agyria–pachygyria band spectrum (Kuzniecky and Barkovich 2001).

Aicardi's syndrome   See page 239.

Cortical organisation

Congenital bilateral peri-Sylvian syndrome (bilateral peri-Sylvian polymicrogyria)   This presents with congenital faciopharyngomasticatory diplegia, learning disability, variable speech impairment and seizures (Kuzniecky et al 1993). Epilepsy, which occurs in 90% of cases, usually begins between 6 and 12 years of age and may comprise atypical absences, tonic–atonic and tonic–clonic seizures. Infantile spasms have been described. Typical interictal EEG findings are bursts of bilateral synchronous spike and sharp-and-slow-wave, or multiple spike-and-slow-wave complexes at 2.5–3 Hz, with predominance over the frontal areas. In some patients multifocal abnormalities are seen. In patients without epilepsy the EEG may be normal.

Table 6.5 Overview of migration disorders

Disorder	Aetiology	Clinical features	Seizures	EEG
Abnormal neuronal and glial proliferation				
Hemimegalencephaly	Isolated or associated with neurocutaneous syndromes	Often severe learning disability, various degrees of hemiparesis	Very common, early epileptic encephalopathy, infantile spasms, epilepsia partialis continua, partial seizures	Asymmetrical burst-suppression pattern, abnormal fast activity, frequent epileptiform activity, asymmetric background activity with higher amplitudes over abnormal side
Focal cortical dysplasia	Sporadic, environmental, genetic predisposition in some	Developmental delay, hemiplegia	Very common, early onset, refractory, partial seizures	High amplitude rhythmic activity, focal or multifocal epileptiform activity, sometimes widespread, may be normal
Abnormal neuronal migration				
Classical lissencephaly type I	Chromosome 17-linked (Miller–Dieker syndrome), X-linked, autosomal recessive, sporadic	Learning disability, hypotonia, dysmorphic features	Infantile spasms (in 80%), Lennox–Gastaut syndrome, myoclonic, tonic, GTCS	Atypical hypsarrhythmia with rhythmic fast activity, high amplitude, rhythmic fast activity, diffuse slow wave activity
Cobblestone dysplasia (type II lissencephaly)	May be associated with congenital muscular dystrophy, Walker–Warburg syndrome, muscle–eye brain disease	Depending on syndrome: mental retardation, muscular dystrophy, eye malformations, hydrocephalus	Depending on syndrome	Depending on syndrome
Focal subcortical heterotopia	Often sporadic	Variable degree of motor and cognitive deficits	Almost all develop focal epilepsy	Regional rather than focal areas of epileptogenesis

Subcortical band heterotopia	Sporadic or X-linked recessive (band heterotopia in females and lissencephaly in males)	Mild to moderate developmental delay, neurological deficits	Very common, sometimes infantile spasms, Lennox–Gastaut syndrome, multiple seizure types	May be normal, widespread theta, focal or multifocal epileptiform discharges; sometimes generalised spike-and-slow-wave discharges
Subependymal or periventricular nodular heterotopia	Isolated, X-linked dominant or metabolic (e.g. Zellweger syndrome, congenital adrenoleucodystrophy)	Learning disability relatively uncommon unless secondary to metabolic disorder	Common, focal, multifocal and generalised seizures; infantile spasms and Lennox–Gastaut syndrome may be seen	May be normal, focal slow waves, focal or bilateral epileptiform discharges, may mimic classical 3/s spike-and-slow-wave activity
Abnormal cortical organisation				
Diffuse polymicrogyria	Environmental (e.g. intrauterine cytomegalovirus infection)	Learning disability, neurological deficits	Variable	Generalised or multifocal epileptiform discharges
Unilateral polymicrogyria	Sporadic or genetic	Hemiparesis, mild learning disability	Partial, atypical absences, GTCS	May mimic benign partial epilepsy with centrotemporal spikes or CSWS, focal abnormalities, but more often widespread or generalised
Congenital bilateral perisylvian syndrome	X-linked, autosomal dominant or sporadic	Faciopharyngomasticatory diplegia, learning disability, language impairment	In 50–85% Lennox–Gastaut-syndrome-like picture with atypical absences, tonic–atonic and GTCS	Bursts of bilateral synchronous spike and sharp-and-slow-wave, or multiple spike-and-slow-wave complexes with frontal predominance, sometimes multifocal abnormalities
Schizencephaly (cleft brain)	Sporadic, genetic in some bilateral cases	Related to type of defect, hemiplegia, developmental delay	Very common, severe, usually focal	Widespread areas of epileptiform activity

CSWS, continuous spike and wave during slow wave sleep; GTCS, generalised tonic–clonic seizures.

Unilateral polymicrogyria   This is characterised by hemiparesis, mild learning disability and epilepsy. Almost all seizure types and several epileptic syndromes may be observed, the most common being infantile spasms, partial seizures, atypical absences and generalised tonic–clonic seizures. Patients with the typical clinical and EEG picture of benign partial epilepsy with centrotemporal spikes (see *Fig. 6.22(a)*) have been described (Ambrosetto 1993). Unilateral polymicrogyria has been described as a cause of CSWS (see *Fig. 6.22(b)*), evolving towards spontaneous remission (Guerrini et al 1998).

Schizencephaly (cleft brain)   The spectrum of anatomical and clinical presentations is wide. Seizures are often the presenting symptom and severe epilepsy is frequent. No specific electroclinical findings have been observed.

Other   Several forms of malformation of the brain, including agenesis of the corpus callosum, holoprosencephaly (an anomaly of hemisphere cleavage with a single forebrain ventricle), and lissencephaly (absence of cortical gyri) may be associated with infantile spasms (Aicardi 1994) (see pp. 231 ff.).

Neurocutaneous syndromes

This group of developmental disorders affects both the skin and the nervous system and has a high association with tumours and congenital malformations. The more common syndromes are:
- linear sebaceous naevus
- Klippel–Trenaunay–Weber syndrome
- Proteus syndrome
- Ito's hypomelanosis
- incontinentia pigmenti
- tuberous sclerosis
- neurofibromatosis.

The neurological deficit, degree of learning difficulty and incidence of epilepsy depend on the extent of cerebral malformation (Cross 2005). The EEG can show an asymmetrical burst-suppression pattern, abnormal fast activity, frequent paroxysmal activity that may become continuous, and asymmetric background activity with higher amplitudes over the damaged side (Paladin et al 1989, Vigevano et al 1996). The abnormalities may be uni- or bilateral.

Tuberous sclerosis   This is an autosomal dominantly inherited neurocutaneous syndrome, characterised by

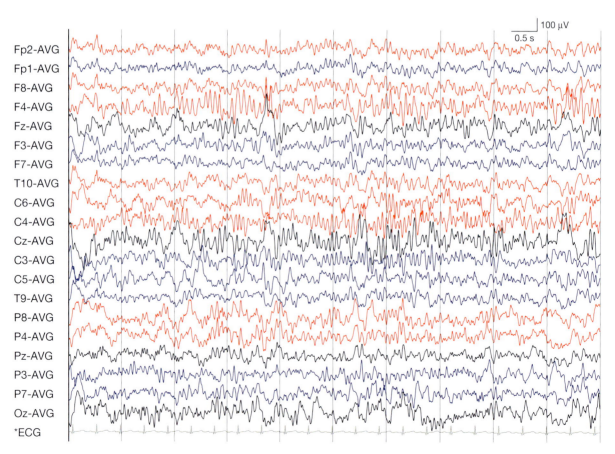

Fig. 6.39   Lissencephaly in a 10-month-old infant. The EEG shows widespread, continuous, high amplitude 10–14/s activity present when awake and asleep.

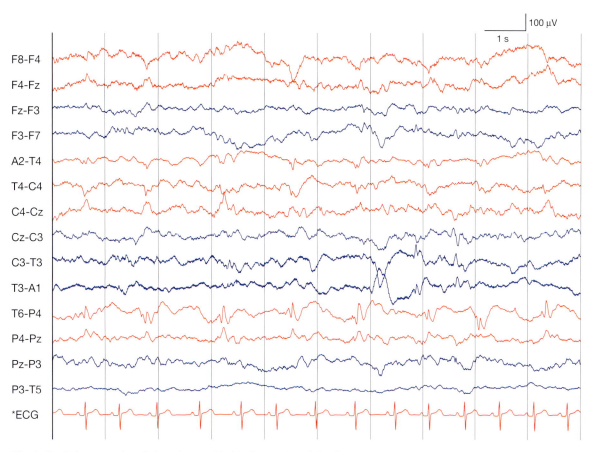

Fig. 6.40 Tuberous sclerosis in a 4-year-old girl with previous infantile spasms. Now she has nocturnal seizures during which she is distressed, runs around the room and becomes 'shaky'. The interictal EEG shows long runs of sharp-and-slow-wave complexes over the right parietotemporal region and, independently, less frequent discharges over the left temporal region, corresponding to cortical tubers on MRI.

learning difficulties, characteristic skin changes and epilepsy occurring in 80–90% of cases in a strongly age-related pattern (Cross 2005, Curatolo et al 2005). Tuberous sclerosis is observed in around 10% of children with infantile spasms (Aicardi 1994). A third of spasms are preceded by partial seizures, and the spasms may be asymmetrical. The EEG may show typical hypsarrhythmia, but more often modified hypsarrhythmia or focal and multifocal paroxysmal discharges, in addition to single or multiple slow wave foci (Curatolo et al 2001). EEG foci often correspond to cortical tubers seen on MRI (*Fig. 6.40*). Although spasms are almost always responsive to treatment in children with tuberous sclerosis, subsequent epilepsy, learning disability and behavioural problems are common. Later, Lennox–Gastaut syndrome, myoclonic, atonic, tonic, partial or generalised seizures may develop.

Sturge–Weber syndrome This is characterised by facial naevus (port wine stain) involving the first division of the trigeminal nerve and a meningeal angioma affecting the same side that is overlying the ischaemic atrophic cortex.

The major clinical problems consist of partial epilepsy, with or without secondary generalisation, progressive hemiparesis, cognitive decline and also glaucoma, which may be congenital or develop subsequently (unless intra-ocular pressure is monitored regularly). The interictal EEG is normal at birth and for most of the first year. A typical and characteristic evolution follows, with a poverty of activity and suppression of amplitude developing over the anterior half of the head on the affected side (*Fig. 6.41*). When calcifications become visible, interictal spikes and sharp waves appear over the posterior half of the affected hemisphere. The degree of abnormality is variable, and bilateral changes may be seen.

SPINAL CORD DISORDERS

Methods of investigating the level of spinal cord compression syndromes and spinal cord tumours by SEPs are similar to those used in adults; they are discussed in detail by Smith and Prior (2003). Some paediatric conditions such as tethering of the cord in spinal dysraphysm are discussed on page 326.

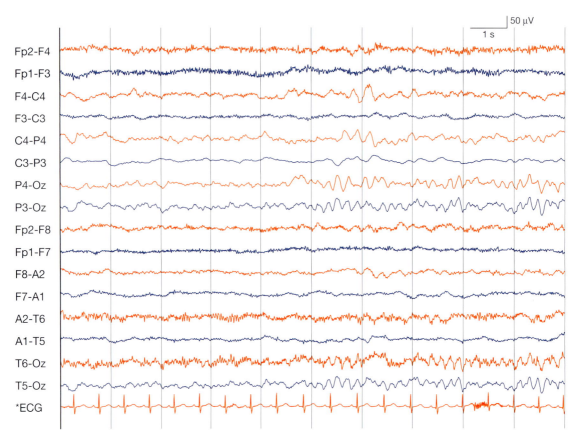

| | | 50 μV |
| 1 s | | |

Fp2-F4
Fp1-F3
F4-C4
F3-C3
C4-P4
C3-P3
P4-Oz
P3-Oz
Fp2-F8
Fp1-F7
F8-A2
F7-A1
A2-T6
A1-T5
T6-Oz
T5-Oz
*ECG

Fig. 6.41 Sturge–Weber syndrome in a girl aged 20 months with an angioma over the left hemisphere and right-sided weakness. She presented with right-sided seizures, often with secondary generalisation. The EEG shows amplitude suppression, with lack of normal activity over the left frontal region.

BRAIN TUMOURS

EEG findings in children with supratentorial brain tumours do not differ from those in adults (van Huffelen 2003). A high proportion of intracranial tumours in childhood, however, arise in the posterior fossa. Thus they often give rise to increased intracranial pressure. With acute changes in pressure, large amplitude bursts of FIRDA or OIRDA occur, usually attenuating on eye opening. When the pressure increase is slow, irregular very slow waves are found over the parietal or occipitoparietal regions more often than are rhythmical ones. After surgical removal of the tumour, the EEG may take months to normalise. During chemotherapy, progress towards EEG normalisation is usually continuous. Very occasionally epileptiform activity occurs in the absence of seizures (Lipinski et al 1975).

Dysembryoplastic neuroepithelial tumours   These are well-recognised glial–neuronal neoplasms associated with cortical dysplasia. They often cause focal, drug-resistant epilepsy with onset in childhood. Neurological signs are either absent, or minor and unchanging. Outcome after epilepsy surgery is often favourable. The EEG may show localised slow activity (*Fig. 6.42*) and interictal focal or multifocal epileptiform activity. Discharges are often discordant to the tumour.

Gelastic seizures (giggling attacks)   These are often associated with hypothalamic hamartoma or low-grade astrocytoma. Onset of seizures is usually in the first year of life and seizures are resistant to treatment. Other seizure types include atonic seizures, myoclonias, complex partial seizures and generalised tonic–clonic seizures. Precocious puberty is a common feature with pedunculated tumours. The interictal EEG may be normal or show focal but usually bilateral epileptiform changes over the frontal or anterotemporal regions. Studies with depth electrodes demonstrate that the epileptogenic discharges originate from the hypothalamic hamartoma and adjacent structures (Kuzniecky et al 1997).

EPs do not have a major role in the diagnosis of intracranial tumours in children. In lesions affecting the visual system, VEPs can complement clinical estimation of the visual deficit. Both pattern and flash stimulation may elicit abnormal potentials, the former with half-field stimulation. Findings are much the same as in adults, showing at an early stage a different pattern of latency and/or amplitude pointing to compression or infiltration. In infants and children with optic pathway gliomata,

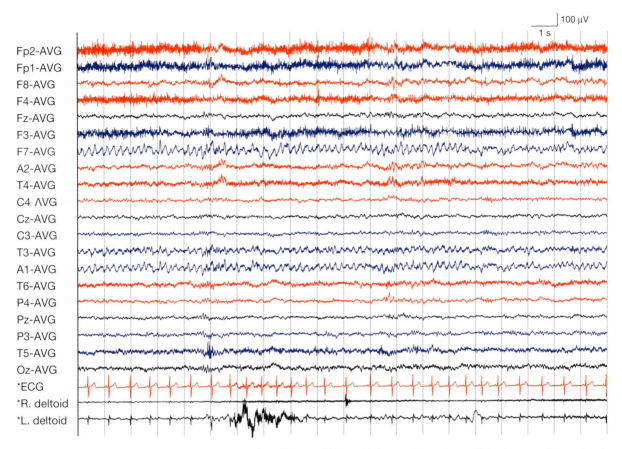

Fig. 6.42 Dysembryogenic neuroepithelial tumour in a 10-year-old boy with frequent complex partial seizures and occasional secondary generalised tonic–clonic seizures. The interictal EEG shows continuous 4–5/s activity (up to 100 μV) over the left anterior temporal region, at times with a sharp appearance concordant with the location of the tumour.

particularly when associated with neurofibromatosis type I, a regression of the tumour may be seen, allowing a non-aggressive expectant approach, which requires close monitoring. In this context VEPs are a useful adjunct to clinical and neuroimaging evaluation. In optic gliomas anatomically confined to one optic nerve, full-field, pattern-reversal VEPs from the other eye can demonstrate an occipital asymmetry, indicating tumour extension to nasal fibres crossing at the chiasm (Kriss and Thompson 1997). The BAEP may be of some diagnostic and prognostic help in tumours of the brainstem and cerebellum – the most frequent locations for brain tumours in infancy and childhood. Various patterns of abnormal conduction may be seen when tumours are both infiltrating and compressing the brainstem (*Fig. 6.43*). Follow-up examination of the BAEP may be useful in the evaluation of the course of brainstem and cerebellar tumours.

Moyamoya disease This is a rare progressive cerebral arterial occlusion disease in children and young adults, which has been found after radiation of intracranial tumours, in craniopharyngioma, neurofibromatosis and Down's syndrome, but is usually sporadic. The patho-genesis is unclear. The common presentation is with

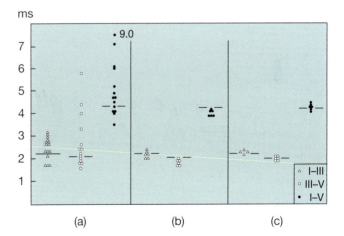

Fig. 6.43 BAEP IPIs (I–III, III–V and I–V) in patients with a lesion (a) infiltrating the brainstem and (b) compressing the brainstem, and (c) with tumours of the pineal body. The horizontal line represents the upper limit in normal children more than 3 years old. (From Binnie et al (2003), by permission.)

acute hemiplegia, dysphasia, acute neurological deficits and/or seizures. Children may recover completely or partially, but in a few cases repeated episodes may lead to

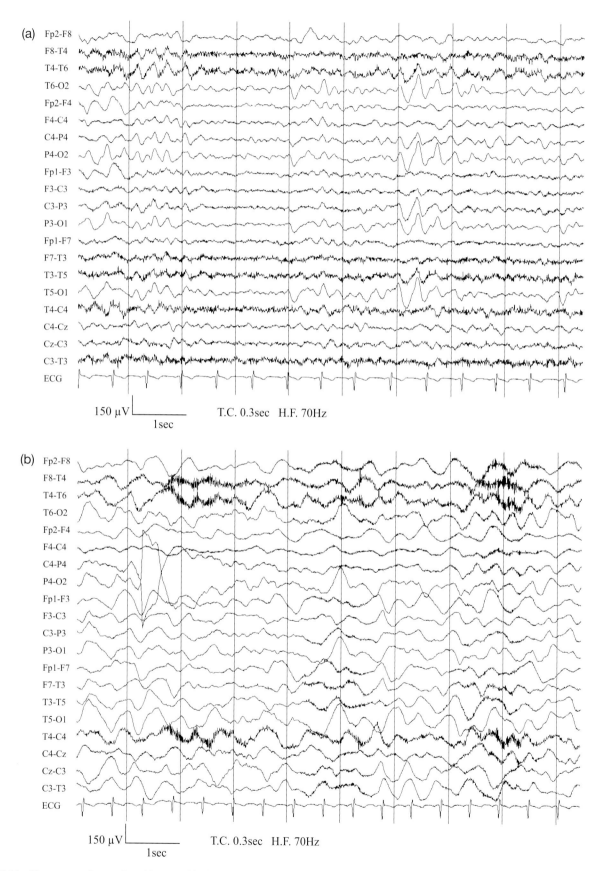

Fig. 6.44 Moyamoya disease in a 10-year-old girl, referred with a single episode of hemiparesis. (a) In the resting state, poorly formed alpha rhythm and prominent rhythmic posterior slow activity. (b) After hyperventilation, generalised slowing persists, leading to respiratory arrest requiring resuscitation. *Note*: in patients with confirmed or suspected moyamoya disease, hyperventilation should not be performed. (From Binnie et al (2003), by permission.)

death. The EEG is characteristic, with localised or lateralised slow activity during hyperventilation or persistence of marked slow activity for more than a minute after hyperventilation (*Fig. 6.44*). High amplitude, slow activity of 2–3 Hz may persist for several minutes and be accompanied by headache and/or motor deficits. It is important to be aware of this syndrome since neurological sequelae have been reported after voluntary hyperventilation, and thus this procedure should be avoided in patients with known moyamoya or sickle cell disease.

TRAUMA

Head injury

Overall, the EEG features do not differ greatly from those in adults (Prior 2003). The specific features of interest soon after head injury include asymmetries of amplitude, frequency, abundance and reactivity of ongoing activity, localised or diffuse depression of activity (including burst suppression and electrocerebral silence), localised or diffuse polymorphic delta activity, FIRDA or OIRDA, abnormal arousal patterns, epileptiform discharges, PLEDs, alpha coma, spindle coma and beta coma.

In contrast to adults, prominent slow wave arousal responses (Schwartz and Scott 1978, Bricolo and Turella 1990) are probably the most common finding in childhood injuries of moderate severity (classic examples are given in *Figs 6.45–6.47*).

Simple EEG reactivity to stimuli has been shown by Gutling et al (1995) to be superior to both the initial Glasgow Coma Scale (GCS) score and the somatosensory central conduction time (CCT) in prognostic accuracy. Fifty comatose patients were studied within 48–72 h of their severe head injury and the findings compared with outcome at 1.5 years using discriminant analysis. EEG reactivity correctly classified 92%, CCT classified 82%, and both measures together classified 98% of the patients into globally good or bad outcome groups, outperforming any other reported method. The GCS score allowed a correct classification in only 72% and, combined with either of the two electrophysiological measures, did not further increase predictability. In a study of 29 comatose patients (not exclusively with head injuries), the response to a sound stimulus (increase in EMG, change in EEG frequency and appearance of sharp waves or K-complex) was the single best predictor for outcome, with significant response rates for the good, deficit, vegetative and death outcomes at 83%, 57%, 37% and 18%, respectively (Alster et al 1993). Others have emphasised the superiority of SEPs over EEG measures, including simple reactivity, in predicting outcome in the greatest number of patients (e.g. Hutchinson et al 1991).

EEG abnormalities after head injury in children are generally more severe than in adults. Even a minor localised head injury may induce generalised EEG slowing, which can persist for days or even weeks. Conversely, a concussion may present with a focal slow wave abnormality in the EEG without indicating a contusional lesion. However, children react to widespread damage with focal epileptiform discharges more often than do adults. In spite of the fact that children with seizure disorders may consequently sustain injuries, there is only a weak association between the coincidental finding of epileptiform activity and accidental brain trauma. Changes after mild head injuries may be seen only during drowsiness or light sleep (Enomoto et al 1986).

Non-accidental injury (child abuse, see p. 295) has to be considered when infants present with head injury. A careful record should be made of any unusual findings, such as bruising or abrasions, noted during electrode application; this should be included in the factual report together with the description of the state of the patient at the time of recording.

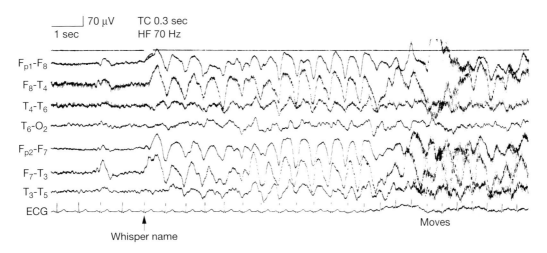

Fig. 6.45 Paradoxical delta arousal response to whispering the patient's name in a 19-year-old man after blunt head injury. Note the very low frequency of the delta waves (1.5–2 Hz), and the increase in heart rate and onset of muscle potentials on the right side of the sample. (From Binnie et al (2003), by permission.)

Occasionally, clinical events in children appear to be delayed after a latent period varying from minutes to hours, but are followed by recovery in less than 48 h. Of the 28 children described by Vohanka and Zouhar (1990), clinical events included deterioration in level of consciousness in 17, cortical blindness in seven and epileptic seizures in two. Their EEGs were either normal or showed OIRDA. The authors suggested that previous problems, such as perinatal abnormality, minimal brain dysfunction or migraine, may have played a part in the 'post-traumatic encephalopathy syndrome' in over 70% of the children, although none showed any residual difficulties in a 3-year follow-up period. Other causes for apparent late deterioration are subclinical epileptiform discharges (Snoek et al 1984, Verduyn et al 1992).

There is evidence that EEG signs allow prediction of outcome in children after head injury (Snoek et al 1984, Dusser et al 1989) on the basis of features similar to those seen in adults. Dusser et al (1989) studied the short and long term prognostic values of the EEG in 24 children with severe head injury, followed for a period of 8–36 months. During coma, four EEG patterns were found, comparable to those reported by Bricolo and Turella (1990):

- borderline
- sleep-like
- changeable (*Fig. 6.48*)
- slow monotonous.

The children had been given anticonvulsant prophylactic doses of phenobarbital, but were not sedated. The slow monotonous pattern (found in 12/24 patients) indicated a poor prognosis. It was associated with a longer duration of coma and slower awakening than other EEG patterns, and was observed in the three patients who died. Of those surviving, only half achieved as good an intellectual and motor outcome as the survivors who displayed other EEG patterns. Arousal patterns (either a paradoxical slow wave response or 'attenuation') tended to be absent or inconstant in those with poor outcome and more often (but not universally) present in those who did well (see *Fig. 6.48(b, c)*). These authors also described the value of finding a 'pre-awake' pattern of clear regional differentiation with monomorphous posterior slow waves and low voltage fast activity anteriorly, which heralded the onset of a 'complete' awakening; this did not occur in those who died or suffered residual damage.

Coma in childhood

Acute encephalopathies of any type that lead to coma are of serious import in children. Many are evaluated neurophysiologically for both diagnostic and management purposes. Although the general principles of such applications are similar to those pertaining to the adult, certain aspects merit discussion here.

Causes of acute encephalopathy in childhood have been reviewed by Tasker and Cole (1997), who found encephalitis, meningitis and hypoxic–ischaemic insults to be the most common causes in a series of 82 children (*Box 6.1*).

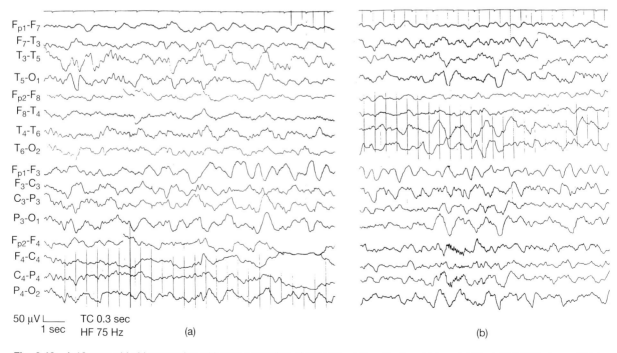

50 µV   TC 0.3 sec
1 sec   HF 75 Hz   (a)

(b)

Fig. 6.46 A 10-year-old girl presenting with some clouding of consciousness after head injury. (a) Her initial EEG showed high amplitude polymorphic, left-sided slow activity. (b) During sleep beta spindles (channels 13–15) were depleted on the left side. A large left frontotemporal extradural haematoma was removed soon afterwards and she made an uneventful clinical recovery. Her EEG a year later showed only slight localised changes. (From Binnie et al (2003), by permission.)

EEG monitoring has proved of considerable importance in the management of non-traumatic coma in children and in status epilepticus (Tasker et al 1988, 1989). Methods require careful assessment and purpose-built devices, such as the cerebral function monitor (CFM) or Cerebral Function Analysing Monitor (CFAM) (RDM Consultants, Uckfield, Sussex, UK; http://www.cfams.com), as described on page 175, should be used in conjunction with intermittent conventional EEG recording (Tasker et al 1990).

Conventional EEG techniques are of value in identifying several patterns of clinical relevance and in guiding management in the comatose child, especially when ventilation and muscle relaxants make neurological assessment difficult. Interpretation of neurophysiological recordings may be complex because of both the multifactorial nature of the clinical problems and the influence of medication (*Fig. 6.49*). Indeed, when severe depression

of the EEG occurs during heavy sedation, it may be necessary to record BAEPs and short latency SEPs (which are largely unaffected by CNS-depressant drugs) to check the viability of sensory pathways.

The EEG may indicate the occurrence and nature of subclinical seizure discharges (e.g. in the acute phase of haemorrhagic shock and encephalopathy syndrome (Harden et al 1991)), the functional effects of impaired cerebral perfusion indicative of cerebral damage with a poor outcome, and depth of coma. Intravenous anaesthetic agents, such as thiopental, used to control seizures in status epilepticus, will of course lead to patterns of burst suppression or electrocerebral silence by virtue of their depressant effect on cortical metabolism. In the absence of cortical damage, these drug-induced changes are reversible. However, when a question of brain death arises, matters are more complex. Sustained electrocerebral silence – once serum levels of drugs have fallen to zero, and providing there is normothermia and neither metabolic derangement nor any other potentially reversible cause – may provide one indicator of irreversible cortical damage. The issues relating to the usefulness of EEG in brain death in children has been much discussed (Task Force on Brain Death in Children 1987, Moshe 1989, Schneider 1989, British Paediatric Association 1991, Lynch and Eldadah 1992). The consensus is that the determination of brain death in children is a clinical decision with well-defined criteria in which the EEG has no place. However, the role of the EEG in the diagnosis of brain death in neonates and young infants remains to be defined.

Several groups have used EPs in the evaluation of comatose patients. These measures have been considered as a useful aid in obtaining information about the severity, the course and the localisation of a brain lesion. SEPs seem more reliable predictors of outcome than do VEPs (Taylor and Farrell 1989). The bilateral loss of the N20 cortical component of the SEP is considered the best predictor of a pessimistic outcome (Wohlrab et al 2001). In children with severe brain injury, a 5-year follow-up revealed a positive predictive value of 85.4% (sensitivity 62.5%, specificity 87.8%) for a normal SEP, and a positive predictive value of 90.9% (sensitivity 61.2%, specificity 94.6%) for a bilaterally absent cortical SEP (Carter et al 1999). However, it has to be appreciated that the EP results depend not only on the severity of the brain damage but also on the aetiology of the coma. In children, as in adults, the prognostic reliability of EPs is higher in postanoxic compared with post-traumatic and postinfectious aetiologies (Carter et al 1999, Wohlrab et al 2001).

In patients suffering from coma due to a severe head injury, a craniocaudal loss of BAEP central components was seen in patients who either survived with severe permanent defects or who died soon after the accident. Severe latency prolongation was also an unfavourable sign, whereas patients with normal IPIs had a good

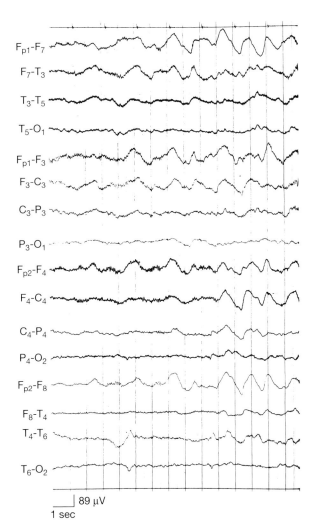

Fig. 6.47 FIRDA in a 16-year-old boy who remained comatose 2 months after a severe head injury. There is fluctuating, low voltage, beta activity at 20 Hz in both frontal regions. (From Binnie et al (2003), by permission.)

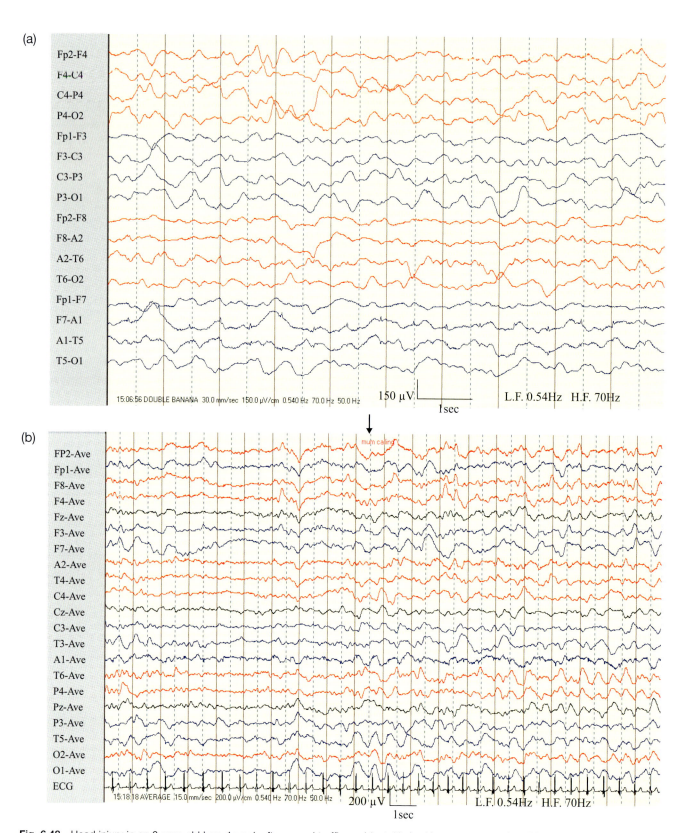

Fig. 6.48 Head injury in an 8-year-old boy, 1 week after a road traffic accident. He had been comatose for 48 h, and was now distressed, unsettled and confused. (a) Generalised, high voltage slow waves, slowest over the left posterior region. (b) During drowsiness, EEG activity was not responsive to painful stimuli, but was to his mother's voice (arrow). When she called his name there was an increase in amplitude. Note the slower paper speed. (From Binnie et al (2003), by permission.)

outcome. Care must be taken in BAEP interpretation when there is a skull fracture, as the finding of absence of all BAEP waves may be due to VIIIth cranial nerve or cochlear damage. Similar results have been described for SEPs. A unilateral or a bilateral loss of the N20 component of the SEP was always a prognostically bad sign, whereas normal latencies or latency prolongation of N20 was seen in patients with neurological sequelae and in children surviving without any defects (*Fig. 6.50*). The prognostic accuracy is better when SEP and BAEP results are evaluated together (Lütschg et al 1983, Lütschg 1985). SEPs seem to be a better predictor of an unfavourable than of a favourable outcome (Schalamon et al 2005).

In children who are comatose due to hypoxic–ischaemic encephalopathy, a latency prolongation or a loss of BAEP components, as well as persistent bilateral absence of the N20 component of the SEP, is an indicator of irreversible brain damage. A good prognosis was seen only when the BAEP IPIs and SEP N13–N20 central latencies were normal (*Fig. 6.51*).

In comatose patients with encephalitis the initial BAEP and SEP results are not precise indicators of the prognosis. However, a loss of components is a bad prognostic sign. Normal survivors showed either normal or rapidly normalising values; the values in handicapped survivors tended to normalise more slowly or showed permanent abnormalities.

Initial EPs of patients, comatose due to intoxication, should be interpreted with caution. Most of these data have no prognostic validity as a normalisation of the EP results may be seen within a few days, if the patient has no additional brain damage (e.g. due to a concomitant hypoxic–ischaemic encephalopathy).

Flash VEPs are generally not useful in comatose children because they depend on the level of consciousness and are modified by sedatives and anaesthetics, but can be helpful in the recovery period if visual function cannot be assessed clinically (Boyd and Harden 1991).

The role of BAEPs, VEPs and SEPs as early indicators of neurological outcome following extracorporeal membrane oxygenation in neonates and children suffering from cardiopulmonary failure has been recently evaluated and compared with other neurodiagnostic tests (Amigoni et al 2005). SEPs were the only useful EP predictor, having the advantage over the EEG of being less affected by sedation and over neuroimaging of being available at the bedside, a relevant issue considering the difficulty of transporting such patients for neuroimaging.

Child abuse: drugs, physical trauma

There has been a growing awareness of the frequent occurrence of physical violence against children. In 1972, Caffey described a constellation of clinical findings (retinal haemorrhages, intracranial injury with subdural

(c)

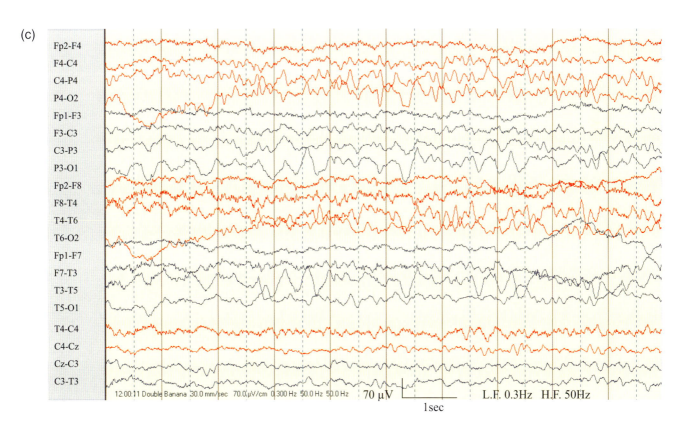

Fig. 6.48 *(continued)*   (c) Three weeks after the accident, now awake and talking, but still unsteady on his feet. The EEG activity has improved, but is still slow for his age (7 Hz). Focal slow waves over the left posterior region correspond to a brain contusion seen on MRI. (From Binnie et al (2003), by permission.)

and/or subarachnoid haemorrhage with little or no evidence of head trauma) that he called the 'whiplash shaken baby syndrome'. A recent US estimate of non-accidental injury in children gives an incidence of 10/1,000, with 50% of cases involving children under the age of 3 years and 25% of those being aged under 1 year (Kinney and Armstrong 1997). These authors point out that, in their experience, in the USA non-accidental injury is now second only to SIDS (see p. 271) as a cause of postneonatal death in infants under 1 year old. It is now routine for casualty officers to consider non-accidental causes when children are brought to the department with physical injuries or poisoning. As part of the clinical work-up, EEG or EP investigations may be requested. The results of such investigation may show brain damage, which is more severe than would be expected for the type of injury that the child's parents claim to have occurred. Thus, for example, with 'acci-

dental' poisoning, an EEG may show either typical drug-induced fast activity (as it occurs following ingestion of barbiturates, benzodiazepines and several other drugs) when the parents claim that the child took only aspirin, or the child's level of consciousness may be such as to suggest non-accidental poisoning. Such findings would lend support to a suspicion that the injury was non-accidental. However, some unmedicated children do exhibit a large amount of fast activity; the finding of 'excess' beta activity can therefore never provide more than corroboration of a clinical suspicion of drug overdose. It is the responsibility of the consultant neurophysiologist to make clear to the referring physician both the relevance and the limitations of the EEG findings in an appropriate manner. Occasionally, the EEG findings are a major factor in the case, and the police may request access to the investigations as evidence so that a prosecution can be brought.

Box 6.1　Causes of childhood encephalopathy*

1. Trauma:
   (a) accidental
   (b) non-accidental injury

2. Hypoxic–ischaemic injury:
   (a) cardiorespiratory arrest
   (b) near-miss SIDS
   (c) near drowning
   (d) smoke inhalation
   (e) shock syndrome

3. Intracranial infection:
   (a) meningitis
   (b) encephalitis
   (c) postinfectious (acute disseminated encephalomyelitis)
   (d) AIDS – *Toxoplasma, Cryptococcus*

4. Mass lesions:
   (a) haematoma
   (b) abscess
   (c) tumour

5. Fluid, electrolyte, acid–base disorders:
   (a) hypernatraemia
   (c) syndrome of inappropriate secretion of ADH
   (d) water intoxication
   (e) acidosis
   (f) alkalosis

6. Acute ventricular obstruction:
   (a) Blocked arteriovenous shunt

7. Seizure disorders

8. Complications of malignancy (disease or treatment)

9. Systemic infection:
   (a) Gram- negative septicaemia

10. Poisoning

11. Vascular:
    (a) arteriovenous malformation
    (b) hypertensive encephalopathy
    (c) embolism
    (d) aneurysm
    (e) migraine (hemiplegic, basilar)
    (f) venous thrombosis
    (g) arteritis
    (h) homocystinuria
    (i) coagulopathy

12. Endocrine dysfunction:
    (a) hypoglycaemia
    (b) diabetes mellitus
    (c) diabetes insipidus

13. Respiratory failure

14. Renal failure

15. Hepatic failure

16. Reye's syndrome

17. Inherited metabolic disorder:
    (a) lactic acidosis
    (b) urea cycle disorders
    (c) amino acidopathies
    (d) organic acidaemias
    (e) mitochondrial disorders

18. Hypo/hyperthermia

19. Haemorrhagic shock and encephalopathy syndrome

20. Iatrogenic:
    (a) rapid overcorrection of dehydration, acidosis
    (b) drug overdose

*Modified from Tasker and Cole (1997).
ADH, antidiuretic hormone.

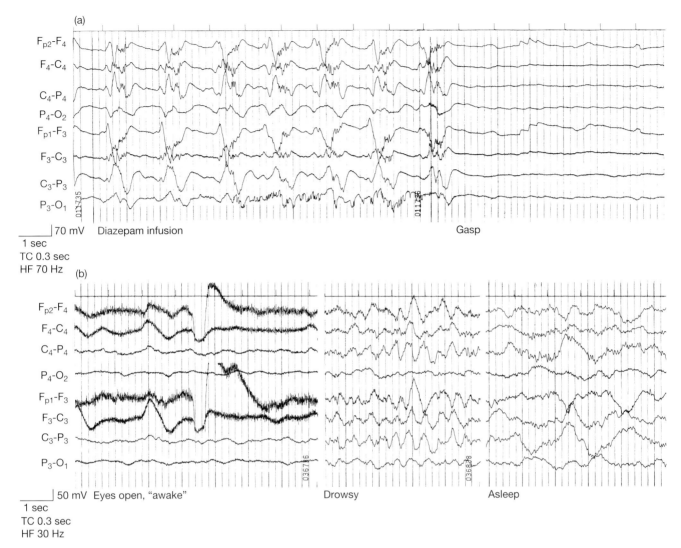

Fig. 6.49 Coma: complexities of EEG interpretation in the intensive care unit. Recording 4 days after postoperative cardiac arrest in a 1-year-old boy with pelvic neuroblastoma. He exhibited episodes of 'sunsetting' eye movements, associated with intracranial pressure rise. Diazepam infusion was being given when an 8-min episode of repetitive complex discharges occurred, terminating with eye opening and a gasp, when the EEG became isoelectric. These findings suggested that his episodes of repetitive abnormal eye movements were epileptiform in nature. His clinical recovery was limited, and 2 weeks later (b) he was in a vegetative state. Whilst he lay with his eyes open the EEG was featureless, containing only low voltage, slow components and muscle potentials anteriorly (lower left-hand recording). However, when his eyes were closed, there was relative preservation of the features of drowsiness and sleep (lower centre and right-hand recordings). (From Binnie et al (2003), by permission.)

NEUROLOGICAL MANIFESTATIONS OF SYSTEMIC DISEASES

If the systemic disorders described below are severe and remain uncorrected, they will render the EEG strikingly abnormal. Most of the changes are non-specific; they consist of generalised slowing occurring at the time when psychomotor retardation, or progressive loss of physical or mental abilities cause the patient to present. Spikes and spike-and-slow-wave activity, which may be generalised or focal, may appear, particularly if the disorder provokes epileptic seizures. Overall, the EEG changes do not differ from those seen in adults.

Endocrine disorders

As in adults, generalised synchronous or asynchronous slow waves occur in many endocrine disorders and are commonly associated with slowing of alpha activity and, less often, with focal or generalised epileptiform discharges. A detailed discussion of the EEG features in endocrine disorders in the adult is given by MacGillivray (2003).

In children, in terms of the EEG, the most important endocrine disorder is hypothyroidism. Diffuse slowing of the EEG and low voltage records are characteristic (Harris et al 1965), especially in infants with a complete absence of the thyroid gland. Sleep spindles are often reduced or absent. Supplementation with thyroid hormone restores a rhythmic pattern normal for the child's age. In an athyroid patient (where the diagnosis is confirmed by hormone studies), the development of a normal EEG and full mental capacity are often not achieved by hormone replacement.

BAEP

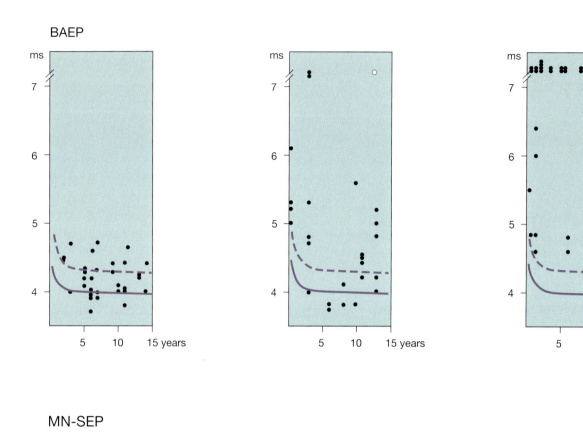

MN-SEP

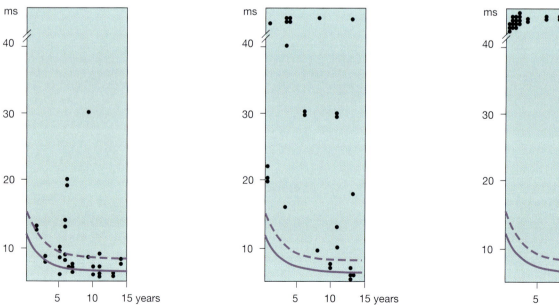

Fig. 6.50 BAEP I–V IPIs (upper part) and median nerve stimulation SEP (MN-SEP) N20 latencies (lower part) in children of different ages with head injuries. (a) Patients surviving without defects; (b) patients surviving with permanent defects; (c) children who died. The symbols above breaks in the *y* axes represent unmeasurable I–V IPIs due to loss of BAEP components, or an undetectable N20 wave due to loss of SEP components. (From Binnie et al (2003), by permission.)

Disturbances of carbohydrate metabolism

Hypoglycaemia causes slowing of the ongoing EEG activity. The threshold varies individually. The first sign is usually an exaggerated EEG response to overbreathing, which does not cause clinical symptoms. Repeating the EEG 20–60 min after glucose ingestion demonstrates a normal response to overbreathing. Slow waves are usually generalised, but focal slow waves may be seen. As in other endocrine encephalopathies, epileptiform discharges may occur. Hypoglycaemic coma results in high amplitude delta activity, but the EEG slowing in diabetic coma may

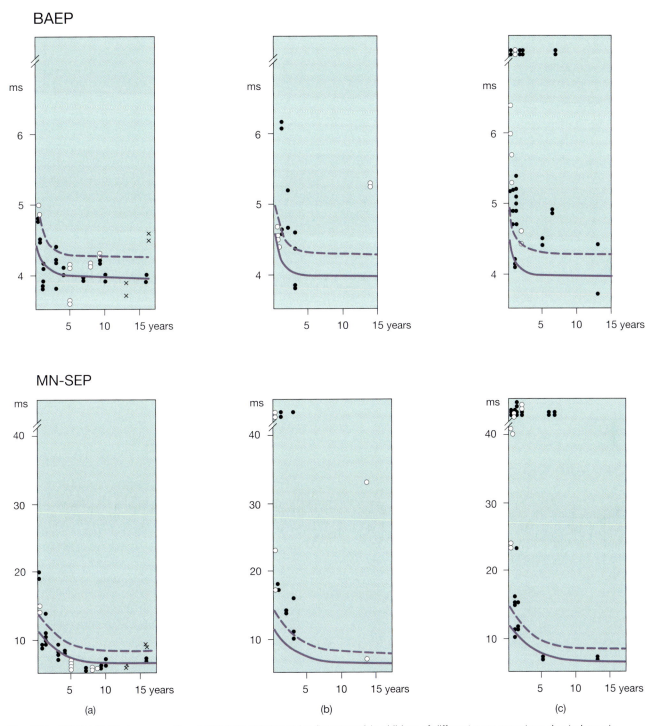

Fig. 6.51 BAEP I–V IPIs (upper part) and MN-SEP N20 latencies (lower part) in children of different ages comatose due to hypoxic–ischaemic encephalopathy (●), encephalitis (O), and intoxications (×). (a) Patients surviving without defects; (b) patients surviving with permanent defects; (c) children who died. The symbols above the broken line represent unmeasurable I–V IPIs due to loss of BAEP components, or an undetectable N20 wave due to loss of SEP components. (From Binnie et al (2003), by permission.)

299

be due to hyperglycaemia, electrolyte abnormalities or a combination of the two. Repeated severe hypoglycaemic episodes in diabetic children are followed by EEG changes such as slowing of the EEG, at times accompanied by spikes or spike-and-slow-wave activity, focal or generalised seizures and disturbances of cognitive function, probably due to neuronal damage (Doose 1959).

Disturbances of electrolyte balance

Hypocalcaemia, hypoparathyroidism and hypomagnesaemia may each present with non-specific slowing of the ongoing activity, which does not necessarily parallel the degree of electrolyte disturbance. When seizures occur, epileptiform activity may be seen in the EEG; it may take some time to normalise after the electrolyte abnormality has been corrected.

Disturbances in serum sodium and potassium levels, especially if they are associated with water overload or dehydration, are particularly important in infancy. It remains difficult in the individual patient to attribute the EEG changes to one or other electrolytic factor and to separate their influence from that of a causal infection and its consequences. In infancy, acidosis is much better tolerated than alkalosis. Almost any type of EEG abnormality can be encountered – slowing of every severity, with or without focal preponderance, and, depending on the degree and duration of the disorder, reduction of amplitude, even complete electrical silence. Complete normalisation of the EEG is possible in children, and especially in infants, even after major voltage depression or a silent trace, provided that the metabolic disorder and related encephalopathy per se were the cause of the EEG abnormality.

Renal and hepatic disorders in children do not differ significantly in their EEG effects from those described in adults (MacGillivray 2003).

INBORN ERRORS OF METABOLISM

This group of conditions has traditionally been divided according to whether white matter or neurons are predominantly affected. The leucodystrophies fall into the former group, while disorders such as Batten's disease and Tay–Sachs disease, with material stored in neurons, fall into the latter. At present, there is no universally agreed system of classification, and terminology has been somewhat controversial. The discovery of other groups of disorders, such as the mitochondrial diseases and peroxisomal disorders, has complicated the matter still further. However, recent developments in molecular biology herald a much more fundamental understanding of these conditions. In the interim, we have therefore adopted the approach of tabulating the various groups of disorders to be discussed, as far as is possible, in relation to the underlying pathological process, and limiting textual description to those where a fairly clear picture has emerged and clinical neurophysiology has a role. The

area is one of rapidly increasing knowledge, particularly in relation to the specific chromosomal lesions responsible.

Peroxisomal disorders

Certain enzymes that would digest the cell's cytoplasm remain confined within peroxisomes. Peroxisomes play an important role in cellular metabolism. Diseases due to inborn metabolic errors may affect peroxisomal functions such as beta-oxidation, whether due to virtual absence of peroxisomes or, more commonly, to single or multiple loss of peroxisomal beta-oxidation enzyme activities (see the reviews by Raymond 2001, Baumgartner and Saudubray 2002).

In all inborn errors of peroxisomal beta-oxidation known at present, the clinical picture is one of multiple abnormalities, especially involving the nervous system; these usually lead to death in the first decade of life. The diseases can be grouped into those with early onset (infantile Refsum's disease, neonatal adrenoleucodystrophy and Zellweger's syndrome), all of which have multiple gene deletions, and those with later onset (such as X-linked adrenoleucodystrophy and Refsum's disease), which are single gene disorders. Most are associated with neuronal migration disorders (see p. 283).

Many of these disorders, particularly infantile Refsum's syndrome and even a few cases of Zellweger's cerebrohepatorenal syndrome, may be associated with normal EEGs. The EP findings may be quite a useful pointer to these conditions, often with loss of the ERG.

Neonatal adrenoleucodystrophy This is a neurodegenerative disorder characterised by a variable degree of adrenal insufficiency and a progressive white matter degeneration associated with abnormal facial features, moderate to severe hypotonia, neonatal convulsions, hepatomegaly and retinitis pigmentosa (Verma et al 1985). The EEG shows slowing, and often spikes, which may be asymmetrical or lateralised (Aubourg et al 1986), and hypsarrhythmia has also been described (Verma et al 1985) (*Fig. 6.52*). The EP abnormalities suggest peripheral involvement due to photoreceptor degeneration and optic nerve, cochlear and/or auditory nerve involvement (Verma et al 1985). The MN-SEP N9 component (Erb's point potential) can be markedly delayed and the SEP CCT (N13–N20) prolonged. Depending on the duration of the disease, the BAEP may show either normal values, latency prolongation or even a loss of some components (Markand et al 1982). The EMG may show polyphasic motor units, suggesting a neurogenic process; nerve conduction velocity is normal or only moderately slowed, the lowest value in one series being 31 m/s in the tibial nerve (Domagk et al 1975). Neuropathological examination has demonstrated severe degeneration of cerebral and cerebellar white matter with demyelination, and gliosis of cerebellar white matter with heterotopic Purkinje cells. Sural nerve biopsy may show decreased density of myelinated sheaths (Aubourg et al 1986).

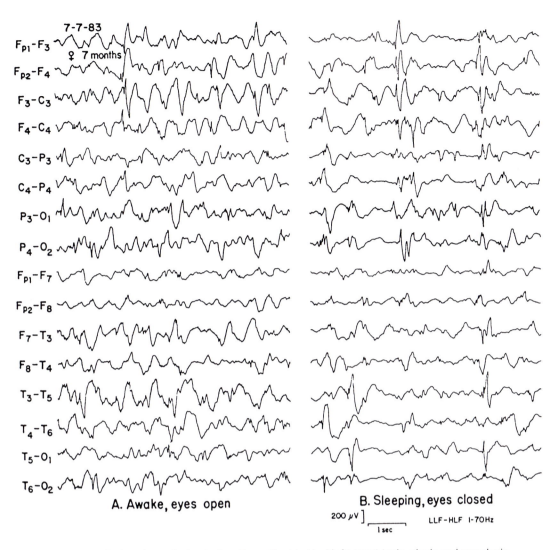

7-7-83
♀ 7 months

$F_{P1}-F_3$
$F_{P2}-F_4$
F_3-C_3
F_4-C_4
C_3-P_3
C_4-P_4
P_3-O_1
P_4-O_2
$F_{P1}-F_7$
$F_{P2}-F_8$
F_7-T_3
F_8-T_4
T_3-T_5
T_4-T_6
T_5-O_1
T_6-O_2

A. Awake, eyes open

B. Sleeping, eyes closed

200 μV
1 sec
LLF–HLF 1–70Hz

Fig. 6.52 Neonatal adrenoleucodystrophy in a 7-month-old girl with frequent tonic, clonic and myoclonic seizures. The EEG shows high voltage slow waves and epileptiform discharges when she is awake; these become pseudo-periodic when she falls asleep. (From Verma et al (1985), by permission.)

Infantile Refsum's disease This is characterised by learning disability, minor facial dysmorphia, chorioretinopathy, sensorineural hearing defects, hepatomegaly, failure to thrive and hypocholesterolaemia. The EEG typically shows generalised fast activity. ERGs are absent, but flash VEPs may be present, albeit 'feeble', but absent with pattern stimulation. BAEPs may be delayed or absent with stimuli up to 70 or 80 dB normal hearing level (nHL), and SEPs may show delayed central conduction. The EMG and sensory and motor nerve conduction may be normal.

Zellweger's cerebrohepatorenal syndrome This is an autosomal recessive inherited disorder (Harding and Copp 1997) with very consistent clinical features, including severe hypotonia, hepatomegaly, developmental delay, facial dysmorphia and seizures. The condition is generally considered to be lethal during infancy (Govaerts et al 1985). Neuropathological find-

ings comprise cortical dysplasia, widespread brain and spinal cord demyelination, and accumulation of neutral fat in histiocytes in white and grey matter of the cerebrum, cerebellum and basal ganglia. The EEG shows progressive abnormality of ongoing activities, diffuse epileptiform discharges persisting in sleep, with some focal features at the vertex in both awake and sleeping states, and an absence of photic responses. BAEPs are generally absent (in relation to hearing loss and brainstem dysfunction) and SEPs delayed or reduced at the scalp, showing little spread beyond the appropriate somatosensory cortex (Govaerts et al 1985).

Lysosomal enzyme disorders and other leucodystrophies

Lysosome dysfunction results in storage of intermediary material within the lysosomes, sometimes resulting in leucodystrophies. In this section the neurophysiological findings in some lysosomal disorders are discussed.

These disorders are due either to an insufficient myelin production or to an accumulation of abnormal myelin degradation products. Usually both the CNS and the peripheral nervous system are affected. Leucodystrophies were previously known under the generic term of 'Schilder's disease', a term now used only to describe a localised acute leucodystrophy of uncertain aetiology.

Krabbe's globoid body leucodystrophy This condition starts in infancy and usually leads to death in the second year of life. There is a deficiency of galactocerebrosidase activity. From the outset there is progressive slowing of ongoing EEG activity, most prominently in the third stage of the disease, with high amplitude slowing when the child has reached a decerebrate state (Kliemann et al 1969). Even at an early stage small multifocal discharges are seen. Multifocal spikes and spike-and-slow-wave complexes become increasingly widespread and may lead to hypsarrhythmia. Organisation of sleep activity and differentiation between sleeping and resting records may be disturbed. Abnormalities of nerve conduction studies indicate an involvement of peripheral myelin.

In contrast to other neurodegenerative diseases in infancy and early childhood, in Krabbe's disease VEPs are present until death in the juvenile form, while in the infantile form, although better preserved, they become delayed and then progressively reduced in amplitude.

Abnormal BAEPs are also described in Krabbe's disease: here, in adjunct to the disruption of central components, there is an abnormally delayed wave I in the absence of middle-ear and cochlear abnormality, which seems to be an indication of VIIIth nerve demyelination. This finding is observed only in Krabbe's and metachromatic leucodystrophies and, curiously, in those who show also a peripheral myelin involvement (A. Suppiej, personal observation, 2001). BAEP central abnormalities are seen in many other metabolic disorders that probably cause significant disruption of cerebral myelin (e.g. Leigh's disease, maple syrup urine disease, pyruvate dehydrogenase deficiency, phenylketonuria, propionic acidaemia and Menkes' kinky hair disease).

Metachromatic leucodystrophy Metachromatic leucodystrophy (MLD) is associated with deficiency of the enzyme arylsulphatase A. MLD is separated into three subgroups (the infantile, juvenile and adult forms) depending on the age of onset of symptoms. Early in the disease there is rhythmic, high voltage, slow wave activity at 4 Hz, sometimes with superimposed fast activity at 16–24 Hz. Sleep organisation is normal initially but, as the disease progresses, it becomes disturbed. In the later stages irregular spikes and spike-and-slow-wave or sharp-and-slow-wave activity may be seen, but are less frequent than in Krabbe's disease.

The most marked abnormalities in EPs are found in the infantile form of MLD. There is marked slowing of the nerve conduction velocity, especially in the late infantile form where values of 0–30 m/s have been reported (Lütschg 1984); correspondingly, the latency of the N20 component of the SEP as well as the CCT are prolonged. In the juvenile form the nerve conduction velocity is normal or subnormal during the first years of the disease and the latency of the SEP N20 component is moderately prolonged. The BAEP I–V IPI is increased in the infantile form and is normal or only moderately increased in the juvenile form. The VEP P100 latency elicited by checkerboard stimulation as well as by flashes is prolonged in the juvenile form. In the infantile form, or late in the course of juvenile forms, the P1 component is often no longer detectable.

GM2-gangliosidosis (Tay–Sachs disease) Caused by deficiency of hexosaminidase A, this is the most common type of leucodystrophy. From the first months of life acoustic startle or myoclonus without EEG manifestations is prominent. In the second year of life spontaneous and induced myoclonic jerks, as well as erratic partial seizures, develop. At the onset, under 1 year of age, there may be some augmentation of slow waves but sleep organisation is normal. The ERG and VEPs are also normal. In the following year the increasing amplitude and slowing of the ongoing EEG is progressive, and spikes or sharp waves, sometimes focal, may be found. Sleep organisation becomes less clear and VEPs may gradually become impaired. In the third year, low voltage slowing, pronounced focal sharp waves and spike-and-slow-wave activity occur but are not prominent. Sleep organisation is lost, the ERG becomes impaired and VEPs are absent (Pampiglione and Harden 1984).

Juvenile GM2-gangliosidosis type 3 This disorder is considered on page 267.

Cherry-red spot myoclonus syndrome (sialidosis type I) This is a rare chronic neuronal storage disorder with onset in later childhood and is typified by a cherry-red spot at the macula. Type II sialidosis is associated with dysmorphic features and has a more progressive course. Progressive myoclonus (see p. 267) and easily controlled epileptic seizures occur, with relative preservation of cognitive function. At different times, myoclonus may be confined to the lower face or consist of massive bilateral jerks. The EEG shows rhythmical runs of positive spikes over the vertex, culminating in a slow wave. Myoclonus occurs during the spike bursts (*Fig. 6.53*), and both increase during the early stages of sleep. The ERG is normal but VEPs may be of reduced amplitude. Peripheral nerve conduction is normal but EMG studies demonstrate denervation. SEPs have early scalp components that are about five times the normal amplitude (Engel et al 1977).

Niemann–Pick disease This disease is associated with a deficiency of sphingomyelinase. No consistent EEG

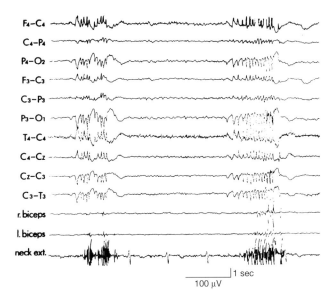

Fig. 6.53 Sialidosis type I (cherry-red spot myoclonus syndrome). Myoclonic jerks from age 11 years, cherry-red spots found at age 12 years, and vision had deteriorated to finger counting by age 28 years, when this combined EEG and surface EMG trace was recorded. It shows prominent rhythmical runs of electropositive spikes over the vertex, culminating in a slow wave. Myoclonus develops during the spike bursts. (From Engel et al (1977), by permission of Blackwell Publishing.)

abnormalities have been described in the different types of this disease.

Gaucher's disease   This is the result of a deficiency in glucocerebrosidase. Epileptic seizures occur in the juvenile and infantile forms. The infantile form can present as progressive myoclonus epilepsy (see p. 267). It is associated with various EEG findings, including normal to slow background activity, bursts of multiple spike-and-slow-wave complexes and rhythmical sharp waves at 6–10 Hz (Green 1974). Most patients are photosensitive. The VEPs are normal but SEPs are abnormally enlarged in some patients. The EEG is unaffected in the juvenile and adult forms of the disease.

Pelizaeus–Merzbacher disease   This is not a lysosomal disorder, but the pathological picture is similar. Abnormality of the myelination component of the proteolipid protein (PLP) gene is considered responsible (Harding and Copp 1997). The BAEPs of these cases are somewhat different from those in adrenoleucodystrophy or metachromatic leucodystrophy. There is a loss of central BAEP waves from birth onwards (Ochs et al 1979, Markand et al 1982), as well as a prolongation of major positivity of flash VEPs and of the N20 of upper limb SEPs, with an abnormal CCT. In contrast to other leucodystrophies, the EP abnormalities remain stable. These findings are in agreement with the hypomyelinating nature of the disorder. The clinical findings

associated with PLP gene mutations show a wide spectrum, extending from Pelizaeus–Merzbacher disease to relatively mild, late-onset spastic paraplegia. In the latter a progressive multifocal, predominantly axonal, peripheral neuropathy can often be detected by EMG (Schiffmann and Boespflüg-Tanguy 2001).

Neuronal ceroid lipofuscinosis (Batten's disease)

This group of neuronal ceroid lipofuscinosis (NCL) disorders can be described as a deficiency of a specific lysosomal hydrolase or as polyunsaturated fatty acid lipidosis; they should be separately regarded as disorders of peroxidation of polyunsaturated fatty acids.

There are seven types of NCL:
- infantile NCL (type 1)
- classic late infantile NCL (type 2), Jansky–Bielschowsky disease
- juvenile NCL (type 3), Spielmeyer–Vogt–Sjogren or Batten's disease
- adult NCL (type 4), Kuf's disease
- late infantile Finnish variant NCL (type 5)
- late infantile variant NCL (type 6)
- progressive epilepsy with mental retardation or northern epilepsy (type 7).

Each form of the disease is genetically distinct (Mitchison and Mole 2001), with an autosomal recessive inheritance in all except for the adult form, which may have autosomal dominant inheritance (see *Table 6.3*).

Variations between the clinical syndromes with, for example, different ages of onset or atypical ultrastructural inclusion bodies, are described and will require further characterisation to be included in this scheme. It represents a major advance for the purposes of diagnosis by DNA analysis, both in the child and in the fetus. The neurophysiological findings in NCL are summarised in *Table 6.3*.

Infantile NCL (type 1) starts after normal development at between 8 and 18 months of age as an acute encephalopathy. The EEG shows irregular, high voltage slowing and irregular, generalised spikes and spike-and-wave activity associated with myoclonic jerks. Normal sleep patterns are preserved but the ERG may be lost and is an index of progressive blindness. During the course of the disease the frequency of the generalised slow waves, which may be grouped, gradually becomes lower. The amplitude remains high over the central and frontal areas, but diminishes occipitally. In the third year of life sleep organisation and the EEG differentiation between the waking and sleeping states are lost. After the third year of life the EEG becomes electrically silent.

Classic late infantile NCL (type 2) starts after 2 years of age with developmental delay and myoclonic or motor seizures; later, ataxia, dementia and progressive visual impairment develop. The EEG is characterised by a peculiar response to slow photic stimulation at a rate of 1–5 Hz, first observed by Carels (1960); each flash produces a single bioccipital spike or polyspike, which are in fact giant VEPs (*Fig. 6.54*). Irregular slow activity and spikes and polyspike and wave discharges can be seen in the resting EEG. The VEP and SEP are very large, although the VEP only transiently as it may become abolished at a later stage. The enlarged SEP is a more constant feature, being observed in the different stages of the disease. The ERG is extinguished.

Juvenile onset NCL (type 3), or Batten's disease, is the is the most common form in Scandinavia. With onset between 4 and 14 years of age it is characterised by early visual impairment, mental deficiency, ataxia and seizures, mainly absences and tonic–clonic seizures. Progressive myoclonus epilepsy occurs 1–4 years after onset. In the early stages the background activity slows and bursts of paroxysmal activity, sometimes long lasting, appear. The background activity may be low and featureless. Photic stimulation does not affect the EEG. The ERG becomes abolished early in the course and VEPs decrease as the disease progresses.

The various degrees of ERG abnormality in all types of NCL is an important feature as it helps to distinguish this group of disorders from other neurodegenerative disorders with clinical features of grey matter involve-

ment, such as mental deterioration, epilepsy and progressive cortical atrophy.

Mitochondrial disorders (mitochondrial encephalomyopathies)

The mitochondria are exclusively maternally transmitted and have their own DNA (mt-DNA) encoding for 13 proteins of the respiratory chain, two ribosomal RNAs and 22 transfer RNAs. The nuclear genome encodes for the ~50 remaining proteins of the respiratory chain. Mitochondrial cytopathies are caused by alterations of nuclear- or mitochondrial-encoded genes. Respiratory chain disorders are associated with a wide range of clinical presentations, with conditions ranging from infantile lactic acidosis, ocular myopathy, fatiguable limb weakness with or without retinopathy, to multisystem syndromes. The multisystem syndromes mainly affect the CNS and include Kearns–Sayre disease, MERRF and mitochondrial encephalopathy with lactic acidosis and stroke-like episodes (MELAS). For a classification and reviews, see Sue et al (1999) and Gillis and Kaye (2002).

Conditions where respiratory chain disorders have been identified (Holt et al 1989, Moraes et al 1989)

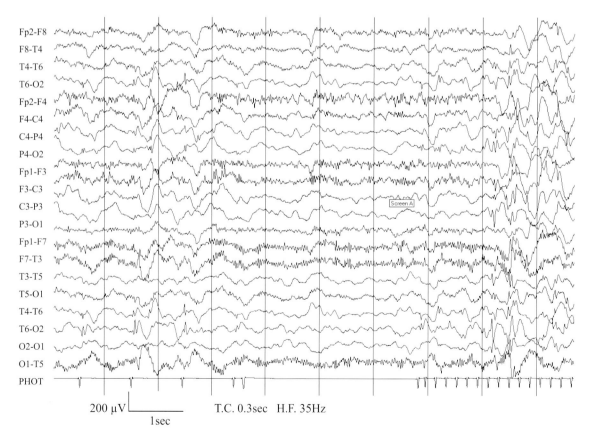

Fig. 6.54 Late infantile NCL in a 4-year-old boy with a 2-year history of global delay, ataxia, multiple seizure types (tonic–clonic, absences, tonic, atonic, complex partial) and recently deteriorating vision. The CT scan shows diffuse atrophy. Intermittent photic stimulation at 1 flash/s (left) elicits giant VEPs; at 5/s (right) a generalised photoparoxysmal response occurs. (From Binnie et al (2003), by permission.)

include MERRF, MELAS, Kearns–Sayre syndrome, Leber's hereditary optic neuropathy, Leigh's disease (sub-acute necrotising encephalomyelopathy, with ataxic encephalopathy, ophthalmoplegia, hypotonia, myopathy and neuropathy), Alpers' disease (progressive polio dystrophy), chronic progressive external ophthalmoplegia, diabetes mellitus with deafness, Pearson's syndrome and others.

Clinically, a variety of myopathic and encephalopathic symptoms occur. Diagnosis often requires a comprehensive laboratory screen (Di Mauro et al 1985). Epileptic seizures are the presenting symptom of many mitochondrial disorders. Generalised seizures typically occur in MERRF, whereas partial seizures are frequent in MELAS and in children with undefined mitochondrial encephalopathy (Canafoglia et al 2001).

The electrodiagnostic investigations show abnormalities that may be summarised as follows. The ECG shows ST-segment or T-wave abnormality and conduction defects (partial heart block, complete heart block, intraventricular conduction defects, right bundle branch block). The EEG shows diffuse, slow wave abnormality and focal or multifocal EEG epileptiform activity. Photo-paroxysmal responses are typical of MERRF, but can also be seen in MELAS, or overlapping phenotypes, and

Leigh's syndrome. The effect of the strokes in MELAS leads to variable topography of the resulting EEG foci. The EMG is myopathic in 60%, shows denervation in 20% and decreased sensory nerve action potentials (SNAPs) in 30%. ERGs are abnormal in a minority, being most likely in Kearns–Sayre syndrome and NARP (neuropathy, ataxia and retinitis pigmentosa), where retinal degeneration occurs.

The detailed neurophysiological findings differ widely in this group of diseases according to whether a chronic encephalopathy or epileptic seizures are the more prominent cerebral component, whether a retinopathy is present, and according to the nature of associated muscle disorders.

Alpers' disease   This is an autosomal recessive disorder causing progressive neuronal degeneration in association with liver disease starting in early childhood. The genetic basis has been recently described as a defect in POLG1. Intractable epilepsy and regression are characteristic. Epilepsia partialis continua may be seen. The EEG is abnormal with asymmetrical, high voltage, 1–3/s slow waves and small multiple spikes superimposed, occurring in runs of rather periodic, often multifocal, discharges (*Fig. 6.55*). These discharges often have a side emphasis,

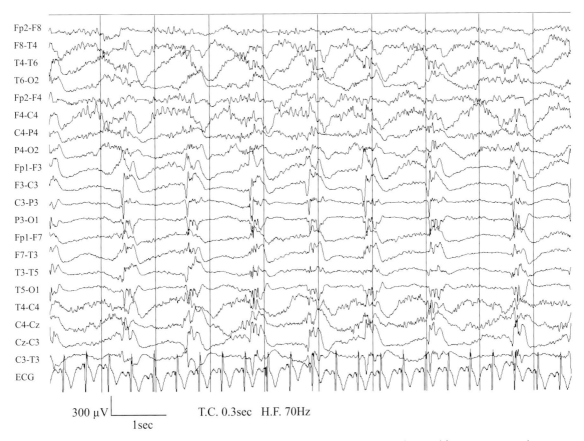

Fig. 6.55   Alpers' disease: a 3.5-year-old boy with intractable epilepsy and regression, and frequent motor seizures involving the right arm for the previous month. Thalamic lesions were seen on MRI. The EEG shows periodic, multifocal discharges maximal posteriorly on the left. (From Binnie et al (2003), by permission.)

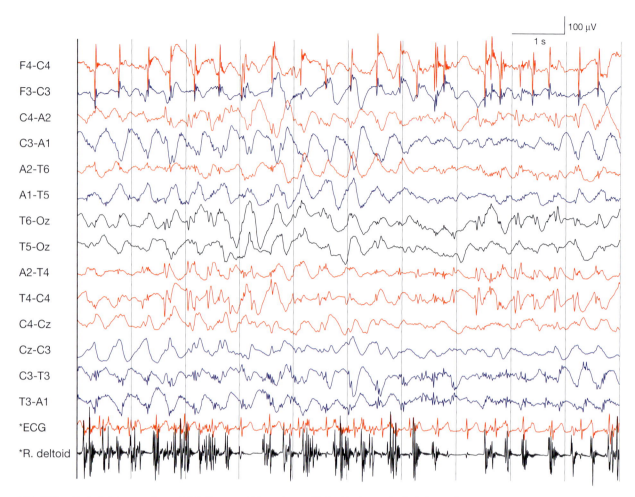

100 µV

1 s

F4-C4

F3-C3

C4-A2

C3-A1

A2-T6

A1-T5

T6-Oz

T5-Oz

A2-T4

T4-C4

C4-Cz

Cz-C3

C3-T3

T3-A1

*ECG

*R. deltoid

Fig. 6.56   Alpers' disease: at onset this 3-year-old boy showed left-sided myoclonic jerks and a loss of motor skills and speech. Now presenting with right-sided epilepsia partialis continua. The EEG shows 1–2/s slow activity on the left associated with low amplitude multiple spikes over the left temporal region. These spikes are also seen independently over the right posterior region. Note the EMG evidence of epilepsia partialis continua over the right deltoid and pick-up of right facial jerking over F4–C4.

which may change over time (*Fig. 6.56*). There is progressive slowing of background activity. The ERG remains normal, but the VEPs are abnormal at an early stage, with degraded waveforms that are often asymmetrical in distribution, with rapid progression until extinction (Boyd et al 1986).

Myoclonus epilepsy with ragged red fibres (MERRF)   The 'ragged red fibres' are due to the appearance of sarcolemmal accumulation of mitochondria stained red by modified Gomori trichrome. The clinical features are generalised and sometimes partial seizures, myoclonus, myopathy, progressive dementia, sensorineural hearing loss, ataxia and usually a positive family history. The most notable EEG findings are: background abnormalities (80%); irregular generalised spike-and-slow-wave discharges (70%); diffuse slow delta bursts (30%); focal epileptiform discharges (40%), most commonly over the occipital regions; photosensitivity (25%); and giant VEPs (80–100%) (So et al 1989).

MELAS (mitochondrial myopathy, encephalopathy, and lactic acidosis, with stroke-like episodes)   This is characterised by vomiting, cortical blindness and hemiparesis or hemiplegia, myopathy, dementia, short stature, sensorineural hearing loss, lactic acidosis, and usually a positive family history. In the acute stage after a stroke-like episode, the EEG often shows focal high voltage delta waves with multiple spikes (Fujimoto et al 1999) (see *Fig. 2.41*). Later focal spikes or sharp waves and 14 and 6 Hz positive bursts may be seen. Photosensitivity has been described (Canafoglia et al 2001).

Leigh's syndrome   This syndrome is frequently associated with respiratory chain disorders and is characterised by hypotonia, vomiting, weight loss, psychomotor regression, movement disorders and deterioration following infections. The EEG shows no specific abnormalities. Data suggest that different patterns of BAEP abnormalities point to different biochemical defects (Taylor and Robinson 1992). Patients with pyruvate dehydrogenase deficiency show poor morphology and reproducibility of waveforms, while patients with complex-4 or cytochrome oxidase deficiency demonstrate increased I–V IPI and low amplitude or absent waves IV/V; in complex 1 deficiency, the BAEPs may be normal or abnormal, depending on the form and severity of disease.

Amino acid disorders and organic acidurias

Phenylketonuria (PKU)   This is the best known and most frequent of the amino acid disorders. Classical PKU with a hepatic phenylalanine hydroxylase defect occurs in about 1/10,000 newborns and if untreated results in delayed psychomotor development and learning disability. The EEG is abnormal in 80% of untreated patients, with generalised slowing of the ongoing activity, pronounced beta waves, sometimes theta bursts, abnormal sleep features and usually generalised spikes and spike-and-wave discharges (Metcalf 1972). However, only 25% of children develop epilepsy, either infantile spasms or generalised tonic–clonic seizures. Classical hypsarrhythmia with multifocal spikes is usually seen in infantile spasms. Metabolic screening of newborns leads to early discovery and dietary treatment of this condition. Children who are treated adequately and sufficiently early show normal development of alpha activity with age but enhanced beta activity. The frequency of EEG abnormalities increases with age despite an early and strict dietary control. Generalised slowing and paroxysmal activity, with or without spikes, are more frequent than in controls (Scheffner et al 1978, Pietz et al 1988). There is no strong association between the EEG findings and blood levels of phenylalanine, although patients with normal levels will have plentiful alpha activity, whereas in those with high levels of phenylalanine theta waves are predominant.

The PKU variants (e.g. dehydropteridine reductase deficiency) are rare and information about specific EEG alterations is not available. Other amino acid disorders, such as homocystinuria, present with mainly non-specific EEG changes. However, serial recordings may demonstrate that more striking EEG changes occur during periods of acute metabolic decompensation.

Maple syrup urine disease   This is an autosomal recessive disease caused by a defect in the branched chain α-keto acid dehydrogenase complex (19q13.1-13.2), leading to an accumulation of valin, leucin and isoleucin and their keto acids. Shortly after birth neonates become lethargic and develop feeding difficulties. Subsequently, symptoms progress to stupor, apnoea, opisthotonus, and epilepsy with myoclonia, partial and generalised seizures. The characteristic odour may only be detectable a few weeks after birth. This disease, if left untreated, is fatal. The EEG shows a characteristic comb- or mitten-like rhythm during the first few weeks of life (Clow et al 1981, Tharp 1992). Bursts consist of primarily monophasic negative

5–7/s activity in the central and parasagittal regions (*Fig. 6.57*) and are seen during wakefulness and sleep, with the most abundant bursts occurring during quiet (non-REM) sleep. This pattern usually disappears with treatment. Subsequently, the EEG shows non-specific abnormalities, often with severe slowing, spikes, multiple spikes and spike-and-wave complexes and triphasic waves (Sonksen et al 1971).

Homocystinuria   Around 20% of untreated patients develop generalised tonic–clonic seizures, but the EEG is abnormal in most patients, even those without seizures (*Fig. 6.58*).

Glutaric aciduria type I   This is an autosomal recessive disorder of organic acid metabolism secondary to glutaryl coenzyme A (CoA) dehydrogenase deficiency (Gordon 2006). It presents in infancy with acute encephalopathic episodes that are often precipitated by infections and result in acute striatal necrosis, focused on the putamen. Damage to the basal ganglia leads to a dystonic–dyskinetic disorder, with choreoathetosis, dystonia and rigidity, sometimes mixed with spastic signs. Peculiar serial EEG findings have been reported during an episode of acute encephalopathy. The day following the onset of acute encephalopathy, the EEG showed suppression bursts including continuous 14–15 Hz rhythmic waves. Then, periodic synchronous discharges appeared and lasted for about 40 min. The periodic discharges finally disappeared and nearly total electrical silence continued (Fujimoto et al 2000).

Urea cycle disorders   Disorders of this type, such as ornithine carbamyl transferase deficiency (OCTD) and arginosuccinate synthetase deficiency (ASSD), develop from early infancy. The EEG is abnormal, showing multifocal spikes, spike-and-slow-wave activity and generalised slowing of the ongoing EEG.

Disorders of neurotransmitter metabolism

Non-ketotic hyperglycinaemia (glycine encephalopathy) This is an autosomal recessive disorder of glycine metabolism. The commonest variant presents in the early neonatal period with lethargy, seizures and respiratory failure or periodic respiration. It is a cause of early myoclonic encephalopathy and infantile spasms. Normal EEG background activity is absent. The EEG in both wake and sleep consists of complex, high voltage bursts of spikes and irregular sharp waves lasting for 1–5 s, alternating with periods of suppression lasting 3–10 s (*Figs 6.59* and *6.60*), similar to the burst-suppression pattern seen in early infantile myoclonic encephalopathy (Aicardi and Goutières 1978, Scher et al 1986). Superimposed on this pattern are electrographic seizures and spike discharges that may be localised to the midline. The vertex spikes may be elicited by tactile stimulation. The

(a)

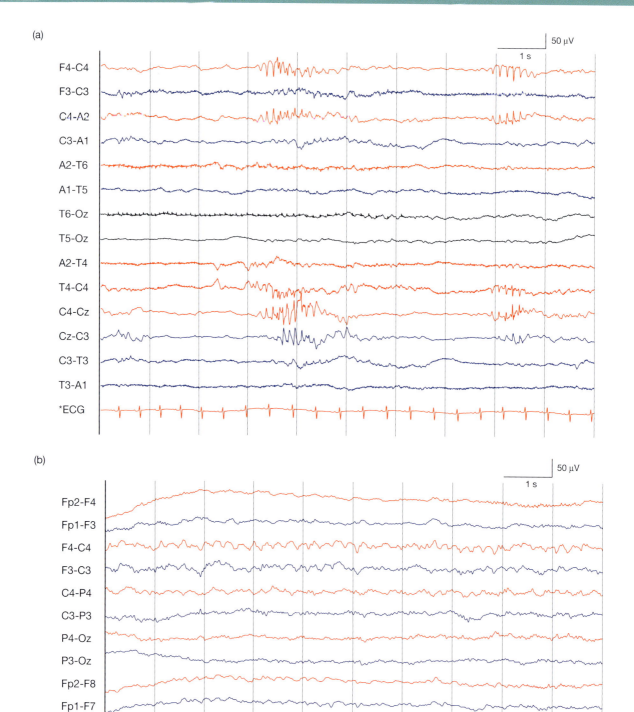

Fig. 6.57 Maple syrup urine disease in a 3-week-old neonate who presented with lethargy and weakness. (a) The EEG shows comb-like bursts at 7/s in central regions. (b) In the same child, now 4 months old, the comb-like bursts have disappeared and are replaced by non-specific runs of 4–5/s activity in the central regions.

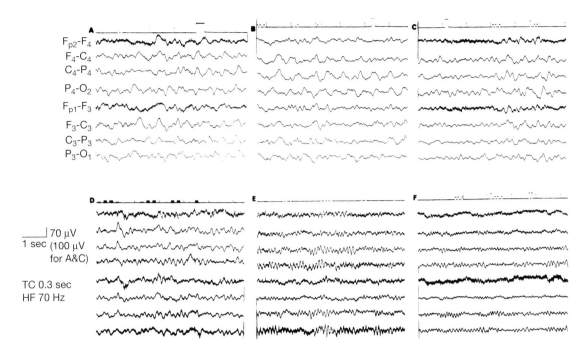

Fig. 6.58 Serial EEGs in a boy with homocystinuria. At the age of 6 years he had developed absence attacks; his EEG was described as showing normal background rhythms, generalised spike-and-wave discharges and photosensitivity. He experienced increasing learning difficulties, and when he was 14 years old truncal ataxia and generalised tonic–clonic seizures were reported. Phenytoin was prescribed. His sister had died at 15 years of age after a similar illness. A diagnosis was made of homocystinuria secondary to 5,10-methylene-tetrahydrofolate-reductase deficiency with superimposed phenytoin intoxication due to related folate metabolism abnormality. He was treated with valproate, methionine, folinic acid and hydroxycobalamine. The six EEG samples were obtained over a 3.5-month period and show gradual improvement of the severe encephalopathic abnormality. Sporadic epileptiform discharges and mild photosensitivity were present throughout and persisted in follow-up EEGs over the subsequent 10 years. (From Binnie et al (2003), by permission.)

EEG later evolves towards atypical hypsarrhythmia or multifocal epileptiform abnormalities.

BAEPs may show prolonged latencies for waves III and V. The neuropathology is a spongy leucodystrophy involving all myelinated tracts, especially in the reticular activating system, cerebellar peduncles and optic tracts (Scher et al 1986). These authors comment that the evident brainstem involvement 'emphasises the role of the non-specific diffuse somatosensory projection system in the generation of myoclonus and stimulus-evoked seizures' in these patients.

Atypical forms with a later onset (during the first year of life, and very rarely during childhood) present after febrile illness with mental deterioration and seizures that are difficult to control. The EEG is characterised by high amplitude slow waves of 2–3 Hz.

Pyridoxine-dependent seizures Seizures that respond only to pyridoxine may present 'typically' in the neonatal period (or even as intrauterine seizures) or 'atypically' later in early childhood, although upper age limits for the diagnosis have not been clearly defined. Seizures are commonly generalised tonic–clonic, focal, myoclonic or spasms, and are refractory to conventional antiepileptic drugs. The condition is (probably) rare, and currently there is no routine screening test available, although raised pipecolic acid levels have been described and this or similar metabolic tests may become available in the near future. Evidence is emerging that mutations in the antiquitin gene, which codes for an aldehyde dehydrogenase in lysine metabolism, may be responsible for many of the cases.

Rare cases that do not respond to pyridoxine, but only to its metabolite, pyridoxal-5-phosphate, have been found to have mutations in the PNPO gene (Mills et al 2005).

Both the clinical seizures and the EEG findings may be very variable, but Nabbout et al (1999) described a suggestive pattern in those presenting in status epilepticus. They found burst-suppression patterns in 2/5 cases and a mixed continuous and discontinuous picture in a further two. Four neonates showed periods of high amplitude, rhythmic delta activity (often mixed with sharp waves). The most characteristic feature was a rapid change from runs of this activity, accompanied by apparently generalised seizures, to focal/multifocal and burst-suppression patterns, the latter being accompanied by massive myoclonic jerks or longer attacks resembling infantile spasms (*Fig. 6.61(a)*). Treatment with intravenous pyridoxine was associated with cardiovascular collapse. The EEG concomitant of such a collapse is loss

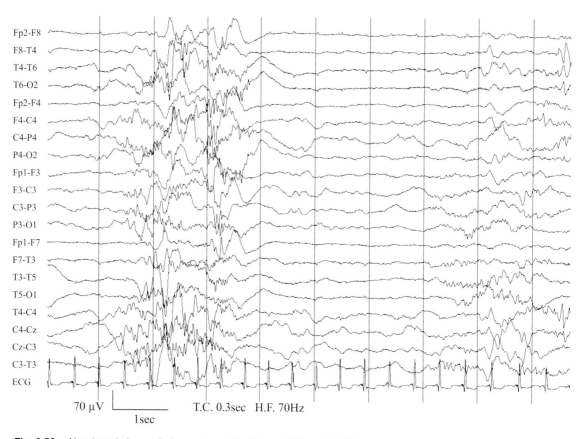

Fp2-F8
F8-T4
T4-T6
T6-O2
Fp2-F4
F4-C4
C4-P4
P4-O2
Fp1-F3
F3-C3
C3-P3
P3-O1
Fp1-F7
F7-T3
T3-T5
T5-O1
T4-C4
C4-Cz
Cz-C3
C3-T3
ECG

70 μV ⌊_____⌋ T.C. 0.3sec H.F. 70Hz
        1sec

Fig. 6.59 Non-ketotic hyperglycinaemia: an infant born at 38 weeks' GA, now 2 weeks old. She presented with hypotonia, poor feeding and frequent myoclonic seizures. There were raised glycine levels in the CSF and plasma. The EEG shows high voltage bursts of irregular sharp waves and spikes lasting 2–3 s, alternating with periods of suppression lasting 4–5 s. The bursts are different from the discontinuous pattern seen in preterm infants or term babies with hypoxic–ischaemic encephalopathy, where less fast components or no multifocal spikes are seen. (From Binnie et al (2003), by permission.)

of all EEG activity lasting for periods of several hours, followed by the return of near-normal continuous activity, a phenomenon that remains poorly understood, but is only seen in association with this condition (*Fig. 6.61(b, c)*).

In most cases, when pyridoxine is given intravenously under EEG monitoring the seizures will stop abruptly and the EEG will normalise during the next few hours. Resuscitation equipment should always be immediately available when a trial of pyridoxine is undertaken. Oral administration of pyridoxine diminishes, but does not entirely remove, the risk. It seems likely that the presence of this EEG pattern may relate to the severity of the condition, and no EEG findings could be said to definitely exclude it. A subgroup of affected neonates responds only to very high doses of pyridoxine given for a longer period, and this should be assessed by serial EEGs.

Disorders of metal metabolism

Wilson's disease (hepatolenticular degeneration) This is a chronic hepatocerebral degeneration caused by an autosomal recessive mutation of ATP7B on chromosome 13q14-21. It is associated with a copper-transport defect,

and copper therefore accumulates in various organs, such as the liver, brain, kidney and eye (Kayser–Fleischer ring), producing haemolysis, cirrhosis, lenticular degeneration and cortical atrophy. Onset in childhood commonly presents as hepatic dysfunction (hepatitis, cirrhosis, hepatomegaly) with dysarthria and parkinsonian-like clinical features, usually later.

In the early subclinical stages of the disease the EEG is usually normal. When the free copper content (total copper – ceruloplasmin bound copper) becomes elevated and neurological and psychiatric signs ensue, the EEG also becomes abnormal. It shows a variable degree of slowing of the ongoing activity, slow wave paroxysms and, occasionally, sharp transients (Heller and Kooi 1962). Frank epileptiform EEG phenomena and seizures may occur (Dening et al 1988). When hepatic dysfunction progresses, triphasic waves (see p. 58) may develop. EEG abnormalities are related to the severity of the clinical presentation; minor abnormalities are also found in clinically unaffected siblings (Hansotia et al 1969). The EEG has been used to monitor the effect of copper-removing therapy with zinc sulphate or penicillamine (Nevšimalová et al 1986, Hoogenraad et al 1987).

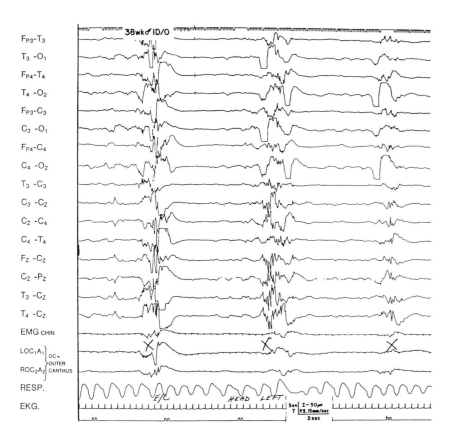

Fig. 6.60 Glycine encephalopathy (neonatal non-ketotic hyperglycinaemia). This infant, born after 38 weeks' gestation, was lethargic, with poor respiratory effort, a weak cry and inability to suck. Occasional myoclonias were induced by tactile and painful stimuli. At 2 h of age the EEG shows a near-silent background, with complex paroxysmal discharges accompanying whole-body myoclonic jerks (marked X). (From Scher et al (1986), by permission).

Menkes' 'kinky hair' disease This is an X-linked (Xq13.3) metabolic disease with a maldistribution of copper. Infants present with seizures, hypothermia and failure to thrive. Their hair shows a characteristic appearance, variously described as 'kinky' or 'steely'. In the early stages the EEG may show runs of rhythmic activity of various frequencies (*Fig. 6.62*). Later, hypsarrhythmia or multi-focal epileptiform discharges occur, but these may be quite variable in severity (White et al 1993). The ERG is normal but of low amplitude; the VEPs are of low amplitude or absent (Friedman et al 1978).

SPINOCEREBELLAR DEGENERATIONS AND RELATED DISORDERS

Friedreich's ataxia

Abnormal EPs are found in Friedreich's ataxia (see also p. 333). Abnormal BAEPs in this disease help to rule out other hereditary ataxias in which BAEPs are normal (see p. 325). Neuropathological studies confirm cell loss and gliosis in the brainstem of Friedreich's patients, but not in other ataxias. The BAEPs have a characteristic rostro-caudal loss of waves beginning very early in the course of the disease; they may reflect both the rate of progression and the severity of the condition.

The MN-SEP cervical response (N13 component) is frequently absent, but is only mildly delayed if it is present. The latency of the cortical N20 component is delayed, indicating a slowed CCT. The N20 component is broadened or split into subcomponents.

The VEP P100 component may be small, delayed or absent, despite normal or only mildly impaired visual acuity. In contrast to multiple sclerosis, however, the VEP changes in Friedreich's ataxia are usually similar for each eye.

The changes found in nerve conduction and EMG studies are detailed on p. 329.

Infantile neuroaxonal dystrophy

Infantile neuroaxonal dystrophy is a progressive disorder, which presents with a combination of increasingly severe upper and lower motor neuron signs. The child shows progressive motor and intellectual regression, early visual disturbance and progressive sensorimotor neuropathy. Seizures are rare. Hypotonia and optic atrophy are evident on examination. The histological finding of spheroid bodies in axonal endings is characteristic and can be documented in various tissues, including biopsy of skin. Neurophysiological features make an important contribution to the diagnosis, with characteristic EEG abnormalities and denervation seen on the EMG. The EEG shows high voltage, non-reactive, fast activity at 16–24/s (*Fig. 6.63*). The fast activity may persist during sleep and K-complexes are absent. These signs appear over the first year or two of the disease (Aicardi and Castelein 1979). The ERG is normal, but the VEPs to flash stimulation and cortical MN-SEPs are of low amplitude or absent.

311

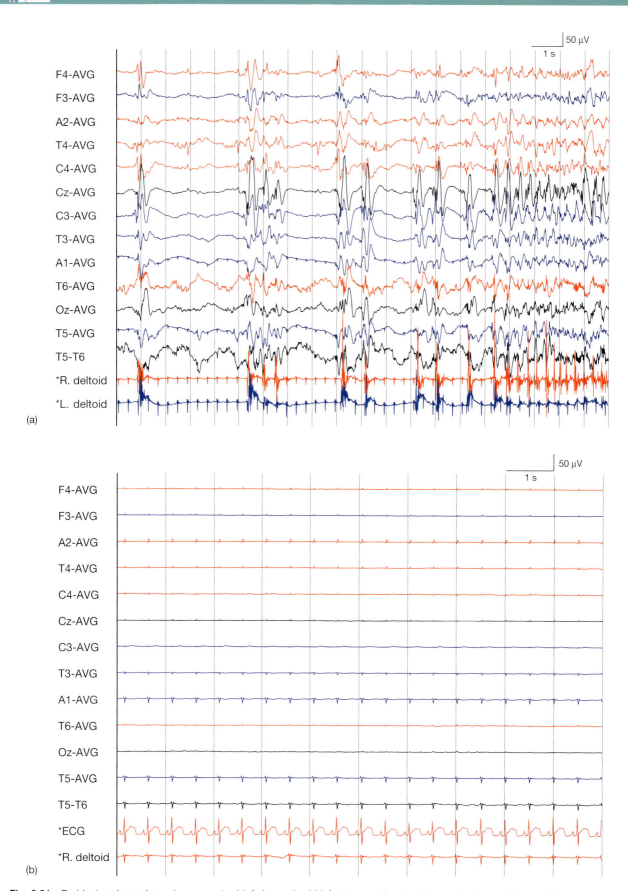

Fig. 6.61   Pyridoxine dependency in a neonate. (a) A 4-month-old infant presenting in status epilepticus, with onset of seizures at 1 month. The seizures did not respond to conventional antiepileptic treatment. The EEG shows an initial burst-suppression pattern accompanied by jerks of the upper limbs (see EMG channels). This is followed by a run of continuous high amplitude, spike-and-wave complexes predominantly over the left. (b) The EEG following a loading dose of intravenous pyridoxine.

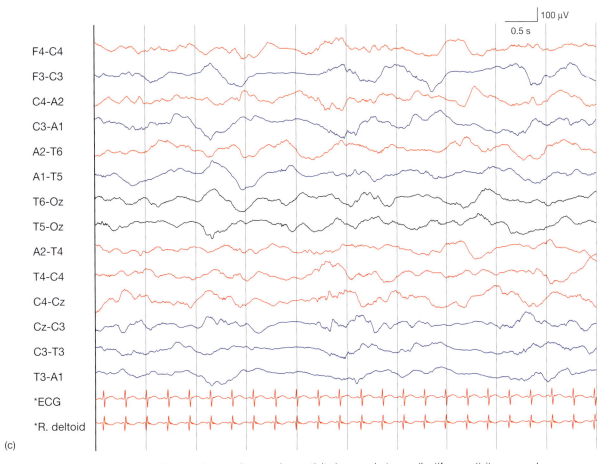

(c)

Fig. 6.61 *(continued)*   (c) The following day continuous slow activity is seen, but no epileptiform activity recurred.

Peripheral nerve conduction velocities are normal, but there is evidence of anterior horn cell disease on electro-neuromyography (Ramaekers et al 1987).

Demyelinating disease

Multiple sclerosis (MS)   This disease is uncommon in childhood, but patients have been reported with onset between 3 and 14 years of age, in whom the EEG and VEPs were studied (Bye et al 1985). Seizures of focal onset occurred in three of the five patients studied, and the EEGs showed appropriate localised changes in the context of abnormalities, which were classified as moderate to severe. CT scans demonstrated low-density, white matter lesions in all the children. Latency prolong-ation of the P100 component of the VEP to pattern reversal was generally marked. Useful clinical information was obtained from the EEG in 83% and from the EPs in 80% of the 16 children studied by Selcen et al (1996); they commented on the differences in presentation compared with adults, particularly noting the absence of a female preponderance, the frequency of EEG abnor-malities and the lower yield from CSF analysis.

The EP investigation of patients with demyelinating diseases, especially MS, is comparable to that in adults. Acute disseminated encephalomyelitis (ADEM) (see *Fig. 6.32*), a demyelinating response to a viral infection elsewhere, is discussed on page 276.

CHROMOSOMAL ABNORMALITIES

A few chromosomal aberrations associated with learning disability have neurophysiological findings sufficiently distinctive and consistently associated as to suggest the diagnosis. Unusual neurophysiological features shared by similarly affected siblings can also be used to help delineate a recognisable syndrome. Epilepsy occurs in 20–30% of children with learning disability. However, unless one of the specific syndromes is suspected or the child is thought to be having seizures, the EEG has little part to play in the clinical diagnosis and management of the child with learning disability. Where the learning disability is acquired after previous normal cognitive function, consideration should be given to recording a sleep EEG to exclude continuous spike-and-wave during slow wave sleep (see p. 258).

Trisomy 21 (Down's syndrome)

Infantile spasms are not uncommon in Down's syndrome and the EEG features are similar to those described on pages 231–239. Epilepsy, with the exception of reflex

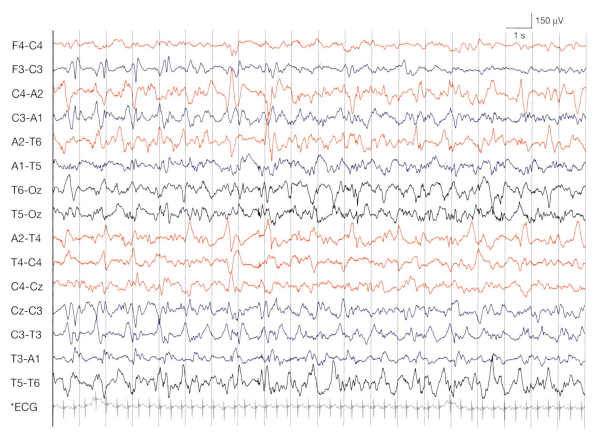

Fig. 6.62 A 4-month-old infant with Menkes' disease presenting with failure to thrive and seizures. The EEG shows runs of epileptiform activities that are independent over the right and left hemispheres.

seizures, is less frequent than in other patients with learning disability. The risks of neurological deficit due to cervicomedullary compression seem to be less than originally feared, but in those cases where further investigation seems warranted, SEPs are a useful component of the assessment (Pueschel et al 1987).

Fragile X syndrome

This is the second most common genetic cause for severe learning difficulty in males, but up to one-third of obligate female carriers have some degree of learning difficulty. About 20% of affected males develop mild, often age-related epilepsy, with onset between 2 and 9 years of age (Musumeci et al 1999). Typical EEG findings are centrotemporal sharp waves activated by sleep that are similar to benign Rolandic sharp waves. Other abnormalities include bisynchronous multiple spike-and-slow-wave complexes, rhythmic theta activity, and slowing of background activity. Female carriers only occasionally have epilepsy or EEG abnormalities, which may be non-specific or similar to those described above (Singh et al 1999).

Angelman's (happy puppet) syndrome

Angelman's syndrome is caused by a deletion or mutation of the maternally inherited 15q11-13 region and is characterised by severe developmental delay, ataxia, happy personality, characteristic facial appearance, frequent and sometimes inappropriate laughter, and epilepsy.

Distinctive patterns of EEG abnormality are found from infancy through to later childhood (Boyd et al 1988, Rubin et al 1997). These consist of the following (Boyd et al 1988):

- persistent rhythmic 4–6 Hz activity exceeding 200 μV in the absence of drowsiness
- prolonged runs of 2–3 Hz activity at 200–500 μV that are often more prominent anteriorly and sometimes associated with ill-defined spike-and-slow-wave complexes (*Fig. 6.64*)
- spikes mixed with 3–4 Hz components reaching 200 μV, especially posteriorly, facilitated by eye closure (*Fig. 6.65*).

Two or three of these features may be present in the same recording. The EEG changes are sufficiently characteristic to distinguish Angelman's syndrome from other disorders with retardation and to help establish the diagnosis (Boyd et al 1988, Dan and Boyd 2003). Burst-locked averaging in children with myoclonus of the hands and face, accompanied by rhythmic theta bursts, showed a component preceding the myoclonus by 19 ± 5 ms (Guerrini et al 1996). The laughter, which is distinct from gelastic seizures, is not associated with specific EEG phenomena.

(a)

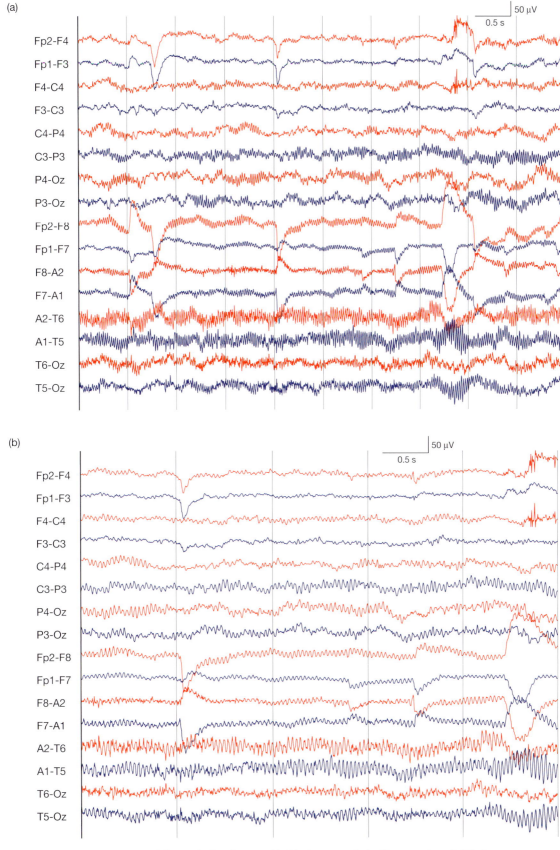

(b)

Fig. 6.63 Infantile neuroaxonal dystrophy in a 3-year-old girl who presented with regression of skills from age 15 months. Now there is increased tone and decreased reflexes. (a) The EEG shows an excess of high amplitude (50–100 μV), fast activity at around 24/s, seen particularly over the posterior region (no history of benzodiazepine treatment). (b) The fast activity is more clearly seen when the paper speed is increased.

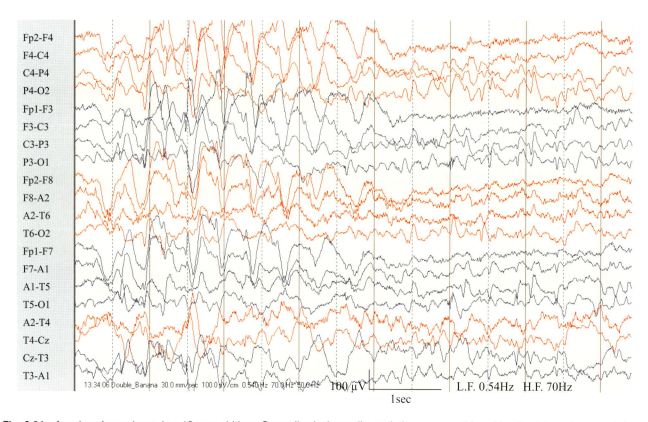

Fig. 6.64   Angelman's syndrome in a 10-year-old boy. Generalised, slow spike-and-slow-wave activity with a frontal maximum and sharp posterior theta activity. (From Binnie et al (2003), by permission.)

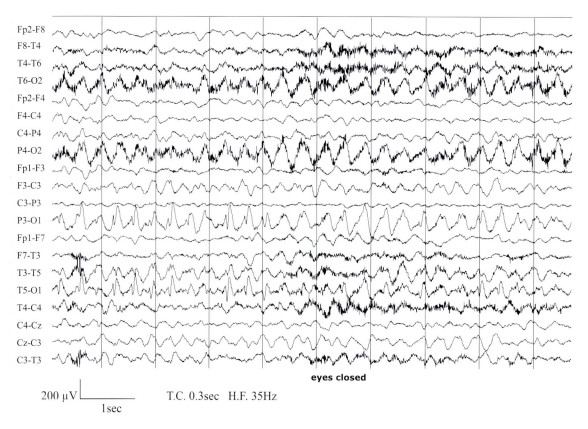

Fig. 6.65   Angelman's syndrome in a 6-year-old boy. Atonic absences from age 2 years; later, myoclonic jerks with absence. Subtle jerks make him unstable walking. Posterior spike-and-slow-wave activity (contrast with *Fig. 6.64*). (From Binnie et al (2003), by permission.)

By contrast, no striking EEG changes are seen in Prader–Willi syndrome, the other condition associated with learning disability and the same chromosome abnormality, but of the paternal inherited 15q region. Inversion duplication of the 15q region can be associated with autism and epilepsy, but these children do not have the typical EEG changes seen in Angelman's syndrome.

Rett's syndrome

Rett's syndrome is a X-linked, dominant, inherited progressive disorder that occurs predominantly in girls and is characterised by severe learning difficulties with autistic features, stereotypical hand-wringing movements, episodes of hyperventilation and epilepsy. Abnormalities of the MECP2 gene are found in a majority of affected girls. The gene abnormality has also been described in girls with less severe symptoms and in boys with a severe neonatal encephalopathy.

The EEG features evolve with the clinical stage of the disease (Glaze et al 1985). The EEG development is normal until about 18 months of age, when there is often loss of sleep spindles and normal rhythmic activities. The background activity becomes too slow for the child's age (see *Fig. 6.66(a)*), but may be normal until the age of 3 years (Laan et al 1998). First, epileptiform discharges appear during sleep (Hagberg et al 1993), consisting of often high amplitude, focal or multifocal sharp waves over the parasagittal (see *Fig. 6.66(b)*) and centrotemporal regions. Discharges may be seen before clinical seizures occur (Robb et al 1989). In the later stages, repetitive bursts of, usually, high amplitude spikes or sharp waves over the same regions are seen, which are enhanced by or confined to light sleep (*Fig. 6.67*) (Hagne et al 1989). In a third of the cases seen by Robb et al (1989) the discharges could be elicited by passive finger tapping by the EEG technologist during recording. These EEG discharges appear to be giant SEPs, which originate from a hyperexcitable sensorimotor cortex.

Apnoeic episodes are associated with bursts of high amplitude slow waves and episodic hyperventilation or normal breathing with faster activities. Although the breathing pattern is normal during sleep, Nomura et al (1984) found abnormalities of the tonic and phasic components of sleep and an incremental increase in REM sleep with increasing age.

The EPs showed variable abnormalities in the nine girls aged 10–12 years investigated by Bader et al (1989a, b).

Wolf–Hirschhorn syndrome

Wolf–Hirschhorn syndrome is caused by a variably sized deletion of the distal portion of the short arm of chromosome 4 involving band 4p16. It is characterised by typical craniofacial malformations (including 'Greek warrior helmet appearance'), intrauterine and postnatal growth retardation, developmental delay of variable degree, skeletal anomalies, hearing loss and structural brain abnormalities. Epilepsy occurs in up to 90% of affected children, with early onset and predisposition for febrile seizures, generalised seizures and status epilepticus. Typical EEG abnormalities include high voltage delta activity with a notched appearance or superimposed sharp waves or spikes in the centroparietal and parieto-occipital regions, and bursts of diffuse spike-and-wave discharges.

CHILD PSYCHIATRY

Autism spectrum disorder and related disorders
Autism

'Autism spectrum disorder' is an umbrella term for children with disordered social and language development who exhibit specific, often ritualistic, behaviours. Like cerebral palsy it has been associated with a wide variety of conditions, including the fragile X syndrome, phenylketonuria, and Duchenne muscular dystrophy. However, in the majority a single cause is not identified, although a genetic contribution is likely in most.

Clinical neurophysiological abnormalities have been reported more often in autism than in any other psychiatric condition in childhood (Small 1975), and the EEG abnormalities described have all varied widely (Tuchman and Rapin 2002). This complicates interpretation of the findings, but slow activity or discharges, generalised or in a variable focal distribution, will be found in around 30–45% in single recordings (Small 1975, Tsai et al 1985). The percentage was much higher when multiple recordings and sleep recordings were used. Small (1975) has reviewed the EEG studies of autism and other developmental disorders. In 14 studies involving 800 patients there was a mean incidence of abnormalities in 50%, with a large range (10–83%) of abnormal EEGs depending on the criteria used for diagnosis and the criteria for EEG abnormality. This compared to an incidence of 44% abnormal EEGs in the children with other psychiatric diagnoses in these studies and 6% abnormal EEGs in the normal controls. Both non-specific slow activity as well as epileptiform activity have been reported. In a retrospective review of 24-h ambulatory digital EEG data collected from 889 children with autistic spectrum disorders 60.7% of children had epileptiform activity in sleep, with no difference being found based on clinical regression (Chez et al 2006).

Up to 35% of children with autism will at some stage develop epilepsy (Olsson et al 1988). Complex partial seizures seem to be the most common seizure type. These can affect behaviour, which may even lead to a misdiagnosis of autism. Identification of persistent EEG foci with changes during episodic behaviour disturbances should identify children who may respond to anticonvulsant treatment. However, there is no good evidence that treating occasional discharges with antiepileptic drugs alleviates the symptoms of autism, and there is no justification at this time for considering more drastic measures, such as multiple subpial transection, to normalise the EEG.

There is a superficial similarity between autism and the Landau–Kleffner syndrome in that both share language difficulties and some children with Landau–Kleffner

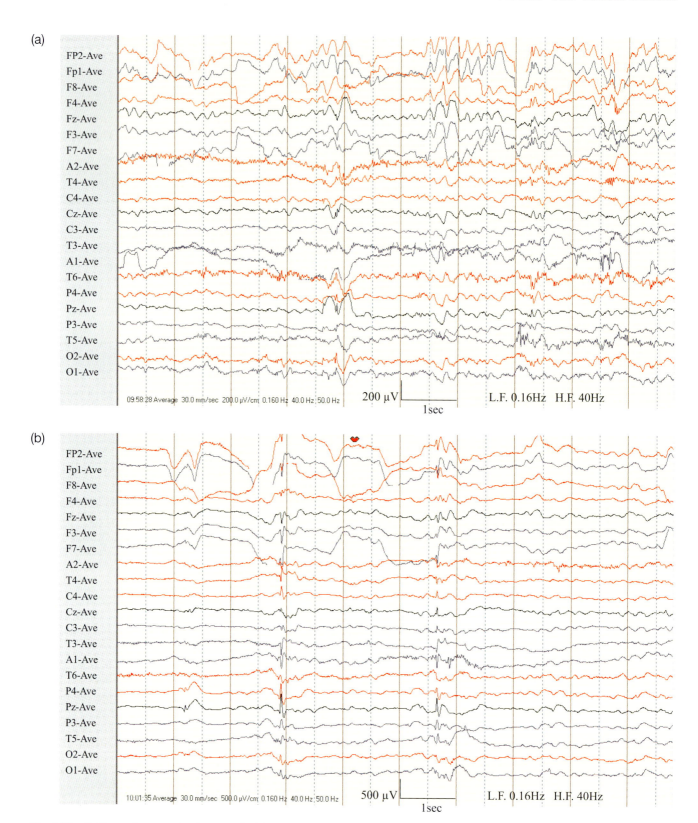

Fig. 6.66   Rett's syndrome: a developmentally delayed 4-year-old girl, with frequent seizures, little speech, cries and screams. (a) The EEG shows a slowed background, with frontal delta and midparietal spike-and-slow-wave discharges. (b) More prominent midparietal spike-and-slow-wave discharges. (From Binnie et al (2003), by permission.)

syndrome have behavioural problems. However, clinically the two are usually distinguishable on closer examination. The onset of Landau–Kleffner syndrome is usually later than that of acquired autism. The language disorder of Landau–Kleffner syndrome is specifically an auditory agnosia. Once the child grasps what is required of him, he can respond appropriately, and hence, in contra-distinction to autism, performance IQ is unimpaired. Finally, the sleep EEG showing ESES in Landau–Kleffner syndrome is diagnostic. Discharges in a sleep EEG can vary during the night and from night to night, so that a single negative recording cannot definitely exclude ESES (Deonna 1991, Tuchman et al 1991).

EP studies, particularly of the auditory system, have been widely used to test various hypotheses concerning putative defects in acquisition and/or processing of sensory information. Between a third and a half of autistic patients show impairment of function in the auditory pathways, particularly through the upper brainstem (Thivierge et al 1990). However, it remains difficult to assess the clinical relevance of this finding and routine BAEP studies are not indicated, although BAEPs have been used to rule out unrecognised hearing loss, a cause of autistic behaviour that needs to be differentiated from autism.

Grillon et al (1989) examined middle latency auditory evoked potentials (MLAEPs) (and BAEPs) in eight non-retarded autistic subjects and found no alterations.

Buchwald et al (1992) believe that the reductions found in the P1 component in drowsiness and sleep in 11 young autistic adults represent alterations in the reticular activating system. It appears that all MLAEPs in children are rather inconstantly recordable under the age of 8 years (Kraus et al 1985), which would complicate interpretation of findings in young autistic children. All these data should be interpreted very cautiously in view of the different structural brain abnormalities that have been associated with autism (Rapin and Katzman 1998).

Developmental dysphasia
The prevalence of EEG abnormalities in children with developmental dysphasia is around 60% of those with seizures and 20% of those without. Both the proportion and the type of EEG changes were similar to those seen in a comparison group of 314 children with autism (Tuchman et al 1991). However, the condition is heterogeneous and children are much more likely to have an EEG recorded if seizures are suspected. Although early reports emphasised the occurrence of BAEP abnormalities, these are much less apparent when clearly organic disorders are excluded.

Attention-deficit/hyperactivity disorder
The interpretation of neurophysiological findings in hyperactivity is complicated by differences in both clinical classification and methodology. EEG abnormal-

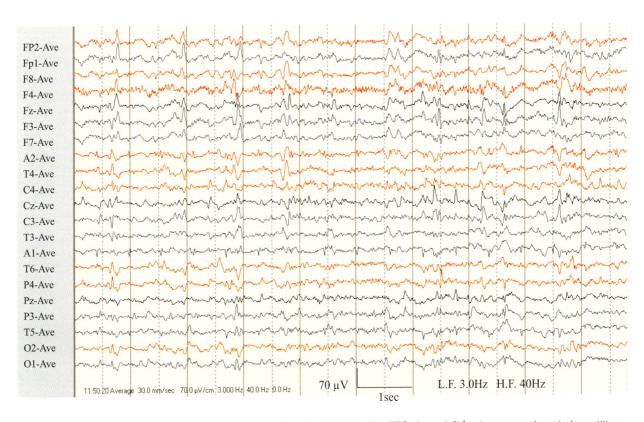

Fig. 6.67 Rett's syndrome in a 12-year-old girl with severe learning disability. The EEG shows left frontal–temporal–central repetitive 1–2/s spikes. (From Binnie et al (2003), by permission.)

ities have been found in about 35% of children with symptoms of hyperactivity, impulsivity and inattention (Cantwell 1980). These include diffuse non-specific changes, excessive slow activity, epileptiform activity and reduced amounts of well-organised alpha waves (Shetty 1973, Satterfield et al 1974). Most of the EEG abnormalities tend to disappear with increasing age.

Satterfield et al (1988) conducted a long term study of 110 boys with attention-deficit disorder and hyperactivity, comparing them with 75 controls studied with EEGs, auditory evoked potentials (AEPs) and EEG spectral data between the ages of 6 and 12 years. These patients were followed up between the ages of 14 and 20 years, by which time 31 (28%) had become delinquent with multiple arrests. The patients who became delinquent in their teens had normal EEGs and spectral analysis, as well as AEPs. The non-delinquent group, however, had more clinical EEG abnormalities, with more activity in the alpha and beta bands and more total power on spectral analysis as well as lower amplitude AEPs. This was taken to indicate that the non-delinquent children had a higher incidence of underlying brain dysfunction than the delinquent group, who may have been subject to more adverse environmental, family or genetic influences. For a recent review, see Barry et al (2003a, b).

Behavioural problems

Children with epilepsy are at a higher risk of developing behavioural problems and psychiatric disorders than are their healthy peers, particularly inattention, hyperactivity and conduct problems (Rutter et al 1970) and this may in some cases be related to EEG abnormalities (Stores et al 1978, Weglage et al 1997). In a large longitudinal study, epileptiform discharges were found in 3.5% of normal children; half of these had behavioural problems such as hyperactivity and emotional disturbances (Cavazzuti et al 1980).

Interictal or subclinical epileptiform discharges occur in up to 3% of children without neurological disease and may cause a short impairment of cognitive performance in about 50% of patients investigated (Aarts et al 1984, Binnie et al 1990). This so-called transitory cognitive impairment (TCI) may adversely affect behaviour and school performance (Kasteleijn–Nolst Trenité et al 1988). However, this should not be taken as a licence to administer antiepileptic drugs to every underachieving child with an abnormal EEG. TCI has to be demonstrated by using specific techniques of EEG-monitored psychological testing, which are not generally available. A diagnosis of epileptic absences similarly requires demonstration of unawareness during a discharge. If a trial of treatment aimed at abolishing TCI (and presumptively improving academic performance) is undertaken, then its effectiveness should be monitored with sequential neurophysiological and psychometric assessments (Binnie 2003, Pressler et al 2005).

The EEG is also useful in the differential diagnosis of absence seizures and attention-deficit/hyperactivity disorder, as these may also present with similar clinical symptoms (see the discussion in McConnell and Duncan 1998). TCI and absence seizures should both be considered in the differential diagnosis of behavioural problems and attention-deficit/hyperactivity disorder, and the EEG is thus useful clinically in evaluating children with intermittent attention or behavioural problems.

In practice, an EEG showing non-specific abnormalities usually serves only to sow confusion in the minds of parents and physicians.

Anorexia nervosa

In a controlled study, Crisp et al (1968) found a high (59%) incidence of EEG background abnormalities in patients with anorexia nervosa, with 31% showing an abnormal response to hyperventilation and 12% showing paroxysmal abnormalities. Similarly, Bridgers (1987) found epileptiform activity in 10%. These abnormalities were felt likely to be secondary to the effects of starvation, although the high incidence of soft neurological signs in binge eaters and the response of some patients with eating disorders to antiepileptic drugs has been interpreted as evidence that such EEG abnormalities may also have aetiological significance in some patients. They may also reflect the effect of psychotropic drugs.

Tourette's syndrome

It had been suggested that non-specific EEG abnormalities, such as sharp waves and diffuse slowing, occur in patients with Tourette's syndrome. However, more recent investigations could not confirm a consistent difference between patients with Tourette's syndrome and normal controls (Peterson 1995), while sleep EEG studies found immature arousal patterns (Glaze et al 1983).

Depression and schizophrenia

Whereas earlier studies reported no consistent sleep EEG features in children and adolescents with depression, more recently, reduced REM latency and increased REM time have been found, as in adults (Emslie et al 1990). It has, however, been suggested that this may be accounted for, in part, by methodological problems and the effects of treatment (Thaker et al 1990). The abnormalities associated with depression seem to occur less frequently in prepubertal patients, but may be expressed at a younger age when there is familial evidence for depression and abnormal sleep in a parent (Dahl et al 1991, Giles et al 1992).

In children with, or at high risk of, schizophrenia, non-specific EEG changes (Waldo et al 1978), excessive fast activity and lack of alpha waves (Itil et al 1976) and abnormalities in event related potentials (Strandburg et al 1991) have been reported.

CEREBRAL PALSY

Cerebral palsy is defined as a group of disorders of movement and posture due to a defect or lesion of the developing brain (Bax 1964). It occurs in 2.0–2.5/1,000 live births (*Lancet* 1989). The causes of cerebral palsy include genetic disorders, cerebral malformations, prematurity, perinatal asphyxia and postnatal factors such as sepsis, cerebral haemorrhage and kernicterus, but in the majority of cases the cause is unknown (*Lancet* 1989). The clinical features are summarised in *Table 6.6*.

The range of underlying pathologies makes it clear that a unitary pattern of neurophysiological abnormalities is unlikely, except changes that reflect the severity and distribution of any cortical damage. The EEG background activity is often disorganised and asymmetries, asynchrony and diffuse or focal slow activities may be found. The incidence of EEG abnormalities is high and was found to be greater in children with spasticity than in those with athetosis (Aird and Cohen 1950). Others found a relatively high incidence (44%) of epileptiform discharges, even in children without clinical evidence of seizures, peaking between the ages of 4 and 6 years (Perlstein et al 1955). The risk of developing epilepsy is much higher in patients with epileptiform discharges (Gibbs et al 1963). Focal, multifocal epileptiform and generalised epileptiform activity may be seen (*Fig. 6.68*).

Occipital spikes (*Fig. 6.69*) are common in children with cerebral palsy and strabismus, congenital cataract, retrolental fibroplasia or other ocular disorders, even if they do not suffer from epilepsy (Stillerman et al 1952, Gibbs et al 1955, Smith and Kellaway 1964). Failure or abnormal development of the occipital rhythms occurs in children deprived of early visual input. Jeavons (1964) found that the alpha rhythm was absent in 60% of blind children, including those in whom blindness was due solely to lesions confined to the eye. In addition, both ERG and VEP studies may demonstrate abnormalities in children with retinal and ocular disorders.

CLINICAL APPLICATION OF EVOKED POTENTIALS

Electroretinography and visual evoked potentials

Combined ERG and VEP recordings have a wide range of clinical applications in both paediatric ophthalmology and paediatric neurology, ranging from pre-retinal and retinal disorders to optic nerve and postchiasmal diseases (for reviews see Harden and Pampiglione 1970, Kriss and Thompson 1997).

Retinal disorders in childhood are not infrequent; they can be congenital or acquired, more or less progressive, isolated or associated with systemic disorders, or may occur within the context of neurodegenerative or syndromic diseases. The ERG usually becomes abnormal before clinical symptoms and ophthalmoscopic signs (e.g. appearance of the fundus oculi), giving an early clue to diagnosis, often with implications for genetic counselling. Different basic responses of the ERG permit localisation of dysfunction within the retina. The patterns of ERG abnormalities and their evolution can assist the differential diagnosis in retinopathy (*Fig. 6.70*).

In children with poor visual behaviour, the question arises as to whether the visual loss is due to a retinal or a cortical problem. In this situation, it is useful to record the ERG and VEP simultaneously. Pathology that affects peripheral retinal areas has a greater effect on the ERG than on the VEP. If the ERG is normal and the VEP (to pattern or flash) is abnormal, blindness has to be attributed to pathology of the visual pathways or the cortex,

Table 6.6   Clinical features of cerebral palsy

Subtype	% of cerebral palsy	Associated features	Epilepsy/EEG abnormalities	Literature
Spastic tetraplegia (bilateral hemiplegia)	25%	Learning disability, amblyopia, dysphagia and dysarthria	Epilepsy in 60%, EEG abnormalities in 80%	Aird and Cohen (1950) Zafeiriou et al (1999)
Spastic diplegia	30%	Severe learning disability (30% of patients), seizures (30% of patients)	Epilepsy in 60%, EEG abnormalities in 80%	Aird and Cohen (1950) Zafeiriou et al (1999)
Spastic hemiplegia	30%	Right side more often affected, sensory loss in the affected limbs, hemianopia, hemiatrophy of the affected side, seizures	Epilepsy in 30–40%, EEG abnormalities in 80%	Aird and Cohen (1950) Sussova et al (1990)
Dyskinetic cerebral palsy	10–20%	Epilepsy, deafness, dysarthria, supranuclear gaze palsies, some degree of learning disability may occur	Epilepsy in 35%, EEG abnormalities in 60%	Aird and Cohen (1950)
Ataxic cerebral palsy	Rare	Cerebellar signs, learning disability, up to a half of these cases are genetic	–	–

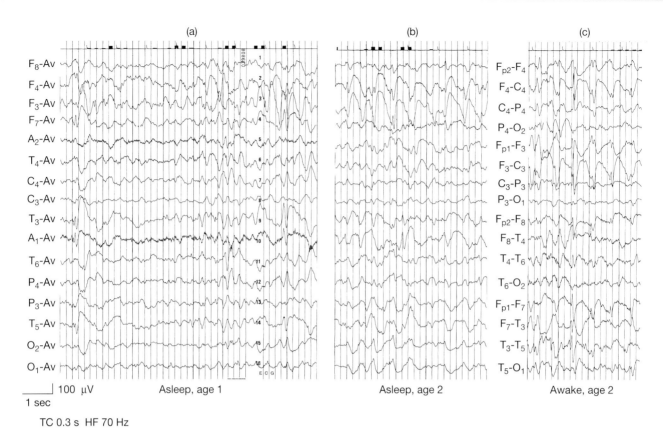

(a)   (b)   (c)

F_8-Av
F_4-Av
F_3-Av
F_7-Av
A_2-Av
T_4-Av
C_4-Av
C_3-Av
T_3-Av
A_1-Av
T_6-Av
P_4-Av
P_3-Av
T_5-Av
O_2-Av
O_1-Av

F_p2-F_4
F_4-C_4
C_4-P_4
P_4-O_2
F_p1-F_3
F_3-C_3
C_3-P_3
P_3-O_1
F_p2-F_8
F_8-T_4
T_4-T_6
T_6-O_2
F_p1-F_7
F_7-T_3
T_3-T_5
T_5-O_1

|____| 100 µV   Asleep, age 1   Asleep, age 2   Awake, age 2
1 sec

TC 0.3 s HF 70 Hz

Fig. 6.68   Severe cerebral palsy following near-miss cot death; generalised fits. (a) The EEG at 1 year shows complexes which are maximal in the left centrotemporal region during sleep. By the age of 2 the child could not crawl, the right limbs were weaker than the left, and he vocalised but did not speak. Major fits had ceased but brief twitches and jerks affecting the trunk and right limbs occurred 4–5 times a day. (b) The EEG during sleep shows increased sharp and slow wave discharges of wider distribution, and (c) when he was awake, profuse runs of bilateral spike and slow wave discharges with a slight left-sided preponderance (note change of montage). (From Binnie et al 2003, by permission.)

even though a restricted retinal portion in the macular or paramacular areas can cause an abnormal VEP.

The VEP gives information about the functional integrity of the visual pathways and visual cortical areas. The results in optic nerve abnormalities of various aetiologies are summarised in *Table 6.7*.

VEPs assist in the diagnosis of various optic nerve disorders, permit early identification of dysfunction, and can be used to monitor the evolution of both the underlying pathological processes and the beneficial or toxic effect of some therapeutic procedures; they can also assist in prognostication. An example of this type of application is shown in *Fig. 6.71*. In this patient, the first suspicion of optic nerve dysfunction due to compression by a tumour arose from the VEP findings; follow-up VEPs were used to monitor initially unsuccessful and subsequent successful therapy.

The VEP may be normal with cortical lesions outside area 17, although a normal VEP in patients with cortical blindness due to destruction of area 17 has also been reported (Kupfersmith and Nelson 1985). In infants with developmental visual agnosia, normal flash VEPs were reported before the restoration of vision. In two cases with basilar migraine and blindness, a VEP of very low amplitude was found that recovered completely a few weeks later.

Some children, particularly in the age range 5–10 years, present with psychogenic visual loss with a hysterical basis, mimicking acute visual pathway dysfunction. The finding of normal flash and pattern VEPs is helpful in distinguishing between the two conditions.

Unilateral amblyopia may be confirmed by the presence of an abnormal interocular amplitude difference of both the N70 and P100 components of the pattern VEP. In patients undergoing occlusion therapy, the improvement of vision can be monitored by EPs.

A specific pattern of the topographical distribution of VEPs from the occipital leads, called 'crossed asymmetry', has been demonstrated as being peculiar to albinism (Apkarian 1992, Kriss et al 1992). In this condition an excessive decussation of optic fibres at the chiasm is demonstrated by a maximal amplitude of the P2 component of the flash VEP from the occipital region contralateral to the stimulated eye, while a negativity is recorded over the ipsilateral scalp. This finding is helpful in the work-up of infants presenting with early-onset nystagmus, since the ocular and cutaneous stigmata are not always clinically obvious.

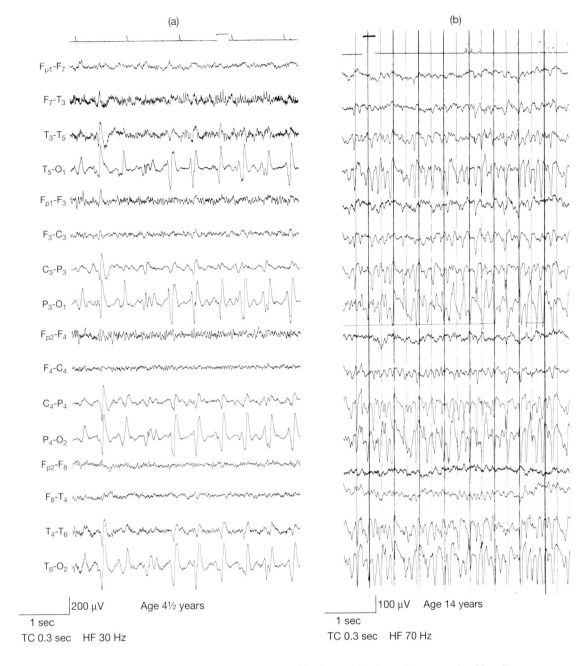

(a)   (b)

F_{p1}-F_7

F_7-T_3

T_3-T_5

T_5-O_1

F_{p1}-F_3

F_3-C_3

C_3-P_3

P_3-O_1

F_{p2}-F_4

F_4-C_4

C_4-P_4

P_4-O_2

F_{p2}-F_8

F_8-T_4

T_4-T_6

T_6-O_2

200 μV   Age 4½ years   100 μV   Age 14 years

1 sec   1 sec

TC 0.3 sec   HF 30 Hz   TC 0.3 sec   HF 70 Hz

Fig. 6.69 Occipital spikes at the ages of (a) 4.5 years and (b) 14 years, in a boy with severe visual handicap (congenital cataract and 'see-saw' nystagmus). Caesarian delivery at term. Febrile seizure aged 15 months; onset of seizures at 2 years 9 months. By 4.5 years he was experiencing 1–2 partial seizures per month consisting of transient loss of vision, which sometimes progressed to a fall or a brief period of unconsciousness. At age 14 years he had 1–3 nocturnal attacks per week consisting of shaking without loss of consciousness. There are alterations in morphology of the discharges in the two samples 10 years apart. Note: different sensitivities, but same chart speed. (From Binnie et al (2003), by permission.)

A combination of ERG and VEP integrated with information from EEG recordings can assist diagnosis of ophthalmologic and neurological disorders involving the visual system from the eye to the visual cortex. Extensive information on these disorders and the diagnostic, prognostic and monitoring role in ophthalmology can be found in textbooks such as that by Taylor (1997).

The prevalent or selective involvement of latencies and/or amplitudes of the main VEP components point to white or grey matter dysfunction, respectively, which is particularly relevant at an early stage in neurodegenerative disorders when the clinical picture may still be unclear. In some early-onset, hereditary leucoencephalopathies the VEP waveform is not yet compromised

scotopic ERG   1 Hz photopic ERG   30 Hz photopic ERG

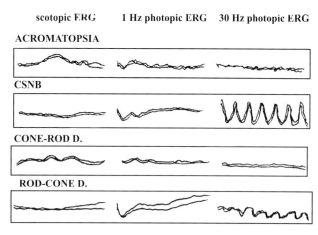

Fig. 6.70   Patterns of ERG abnormalities in differential diagnosis of retinopathy: note the involvement of the single flash ERG under photopic conditions and the 30 Hz response with sparing of scotopic ERG in acromatopsia, the involvement of the scotopic ERG with sparing of the photopic ERG in congenital stationary night blindness (CSNB) and involvement of both the photopic and scotopic responses with worse cone responses in cone–rod retinal dystrophy and worse rod responses in rod–cone retinal dystrophy.

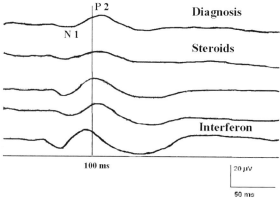

Fig. 6.71 VEP recordings from a 3-month-old girl suffering from haemangioma capillaris (strawberry naevus). In this disorder it is known that after the first few months the tumour undergoes spontaneous regression and there is no intervention unless excessive bleeding or optic nerve compression indicates the need for therapy. In this child the VEP raised suspicion of left optic nerve dysfunction, and steroids were started, but without success; after treatment with interferon the VEPs improved. The VEP in this child was the only way to evaluate the optic nerve function. Recordings from Oz–Fz following left-eye flash stimuli.

compared with other EP types, and therefore can be used to monitor evolution.

Auditory evoked potentials

The two main clinical uses of BAEPs in paediatrics are:
- the detection of hearing loss in children who, because of their young age or behavioural problems, are unable to cooperate sufficiently for conventional clinical audiological assessment of the auditory system
- the evaluation of the brainstem function in a variety of neurological disorders specific to the paediatric age group.

Table 6.7	VEP abnormalities in optic nerve disorders		
Disorder	**Amplitude**	**Latency**	**Morphology**
Optic neuritis	↓/Normal	↑↑	Normal
Optic nerve atrophy	↓↓	Normal/↑	Abnormal
Optic hypoplasia	↓↓	↑↑	Abnormal
Glaucoma	↓/Normal	Normal/↑	Normal
Papilloedema	Normal/↓	Normal/↑	Normal/abnormal
Tumours	↓	↓	Abnormal
Compression	↓	↑	Abnormal

↑, increased; ↓, decreased.

Detection of hearing loss

The most important clinical application of BAEPs in paediatric as well as neonatal medicine is in the assessment of auditory function, given the impossibility of performing formal audiometric testing in the very young or in those with behavioural problems; it is sometimes justified by the discovery of an unrecognised hearing deficit.

The protocol for estimation of hearing by means of BAEP testing in the first few months of life, centres on demonstrating a wave V at click levels of 30 dB above normal adult threshold (nHL). Each ear is tested separately. If wave V is not present at 30 dB nHL the intensity can be increased until a threshold is determined.

Auditory threshold estimation by means of BAEPs in children requires an understanding of the neural basis of the wave V generation and its maturation under normal and pathological conditions. It is important to implement the strict technical requirements for identification and demonstration of maximal amplitude of a still immature wave V during the first few months of life. Close cooperation between the audiologist and neurophysiologist is important.

It has to be appreciated that BAEPs do not really assess 'hearing'. There may be children with disorders of auditory perception who have normal BAEPs; some of them may have abnormalities of the later AEP components. A normal BAEP is helpful in that it implies that the child will probably not benefit from amplification therapy. There may also be children with normal hearing who have abnormal BAEPs, possibly because of dysfunction of the auditory pathway due to neurological damage. In these patients attention has to be paid to the latency and threshold of wave I.

The Joint Committee of Infant Hearing (2000) has recommended that early identification of hearing loss based on clinically assessed risk factors be abandoned because approximately 50% of infants with hearing loss will be missed, with serious consequences throughout life in speech, language and cognitive development. As a consequence, neonatal and paediatric units have started to adopt protocols for neonatal audiological screening that are based on the combination of otoacoustic emission (OAE) recordings and BAEPs or the automated auditory brainstem response (A-ABR) test. However, it has been recently shown (Johnson et al 2005) that the A-ABR test has the risk of false-negative results, and thus the possibility of the occurrence of permanent hearing loss in those who pass the test (see p. 215).

It has to be remembered that normal hearing at birth does not preclude delayed-onset or acquired hearing loss. Thus all infants at risk of 'postneonatal' hearing loss should undergo audiological monitoring recorded according to the criteria of the Joint Committee of Infant Hearing (2000); these are set out in *Box 6.2*.

Evaluation of brainstem function in neurological disorders particular to the paediatric age group

The BAEP evaluates the auditory pathways only at the brainstem level, and therefore can be normal in children with marked neurological abnormalities, in particular those with cerebral palsy. However, children with neuro-degenerative disorders often show abnormal BAEPs (see p. 135).

In the first months of life it is often difficult to identify signs of regression in a child presenting with psycho-motor delay, particularly if there are confounding factors that may provoke a clinical deterioration (e.g. undetected epileptic seizures, inappropriate use of antiepileptic drugs or evolving hydrocephalus). In such circumstances, distinguishing a progressive from a static encephalopathy can be a challenge. We endorse the view of Boyd and Harden (1991) that BAEPs, in combination with other neurophysiological techniques, are very useful in suggesting the possibility of a progressive encephalopathy involving white or grey matter, with particular patterns of abnormality being helpful in indicating specific types of biochemical and/or genetic testing.

At an early stage, grey matter diseases, such as infantile neuroaxonal dystrophy, Leigh's syndrome, Friedreich's ataxia and peroxisomal disorders, will show BAEPs with a general decrease in amplitude of all components but with either normal, or slightly abnormal, latencies. In contrast, white matter diseases, such as Krabbe's leucodystrophy, Canavan disease and Pelizaeus–Merzbacher disease, or disorders of lysosome function, such as metachromatic leucodystrophy, will show prolonged BAEP latencies and/or absent late components, with quite distinctive features according to age of onset and the degree of initial brainstem involvement (*Fig. 6.72*) (Markand et al 1982, De Meirleir et al 1988, Schiffmann and Boespflüg-Tanguy 2001).

Abnormal BAEPs are also described in many other metabolic disorders that probably cause significant disruption of cerebral myelin (e.g. maple syrup urine disease, pyruvate dehydrogenase deficiency, phenylketonuria, propionic acidaemia and Menkes' kinky hair disease). BAEPs also have an important role in paediatric neurology, where they provide a functional support to imaging diagnosis in brainstem structural pathology (see p. 324).

Box 6.2   Indicators of audiological risk in the age range 29 days to 2 years requiring audiological monitoring*

- Parental/caregiver concern about hearing/speech/anguage and development
- Stigmata of syndromes known to include sensorineural or conductive hearing loss
- Family history of childhood hearing loss
- In utero infections, such as cytomegalovirus, herpes, rubella, syphilis and toxoplasmosis
- Postnatal infections associated with sensorineural hearing loss, including bacterial meningitis
- Hyperbilirubinaemia requiring exchange transfusion, conditions requiring mechanical ventilation or extracorporeal membrane oxygenation
- Syndromes associated with progressive hearing loss (neurofibromatosis, osteopetrosis, Usher's syndrome)
- Neurodegenerative disorders (Hunter, Friedreich's ataxia, Charcot–Marie–Tooth)
- Head trauma
- Recurrent or persistent otitis media with effusion for at least 3 months

*Joint Committee on Infant Hearing (2000).

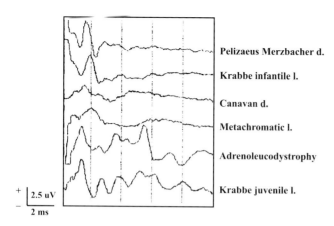

Fig. 6.72 Typical BAEP traces in children affected by different types of leucodystrophy. Note the degree of BAEP abnormalities in relation to the type of leucodystrophy and age of onset.

Somatosensory evoked potentials

Clinical examination of the sensory system can be difficult in young children. In infants and preverbal children, SEPs can provide information about sensory function and are, at present, the only way to uncover a sensory deficit. SEPs provide a non-invasive method of evaluating the entire neuraxis and have been accepted as a useful diagnostic tool in the assessment of paediatric neurological disorders (see also Chapter 5).

Here we briefly mention the use of the SEP that relates to the anatomical location of the disorders; for a more extensive review of the field the reader is referred to pages 136–150 in this book and Binnie et al (2004, Section 3.5.3).

Peripheral lesions

SEPs are useful for the evaluation of peripheral pathology in three circumstances:

- to determine the peripheral conduction velocity when SNAPs cannot be obtained at the periphery
- to explore proximal lesions of the peripheral sensory pathways that are not accessible to conventional EMG studies
- to investigate the whole somatosensory pathway up to the cortex in any pathology combining peripheral and central lesions.

Peripheral nerve, plexus and root lesions can be assessed by SEPs. In Erb's palsy, abnormal SEP findings may be observed. In the case of a root avulsion, the dorsal root ganglion is intact and the N9 component may be recorded but more rostral components are abnormal or absent. If there is a plexus lesion, the N9 component as well as the more proximal components are abnormal or absent.

Spinal cord lesions

In spinal cord disorders MN-SEPs are a valuable diagnostic tool for early indication of spinal-cord involvement in compressive lesions. These include cervical spinal

stenosis secondary to achondroplasia (Boor et al 1999), mucopolysaccharidoses (Boor et al 2000) or other bone disorders. In intrinsic abnormalities, such as Arnold–Chiari malformation (Nishimura and Mori 1996) or tumours, SEP findings correlate with the severity of clinical involvement and may provide useful documentation, but not always new information.

Tibial nerve SEPs have been extensively used in spinal dysraphism, especially to enhance clinical and neuroradiological diagnosis of tethered cord syndrome. In these patients imaging techniques often demonstrate a caudally displaced conus medullaris. With an appropriate montage this can be demonstrated in the PTN-SEP by means of an inverted amplitude gradient (maximal at the L5 lead instead of at the T12 lead) (*Fig. 6.73*). Excessive traction resulting in hypoxia of the lumbosacral spinal cord (Yamada et al 1981) is associated with a reduction in SEP amplitude of the postsynaptic lumbar component, N22 (Polo et al 1994). If surgical procedures to correct tethering are undertaken before irreversible damage has occurred, these findings are followed by neurophysiological improvement, often preceding clinical recovery (*Fig. 6.74*).

At an early stage of traumatic lesions the persistence of SEP on the scalp may suggest some residual spinal cord function when clinical examination would lead to a pessimistic conclusion; recovery of SEP may antedate clinical improvement (Rowed et al 1978).

In tumours of the spinal cord, SEPs can be used as a screening test for detecting spinal cord compression and to follow postoperative evolution; they do not compete with CT or MRI studies, which should be undertaken in

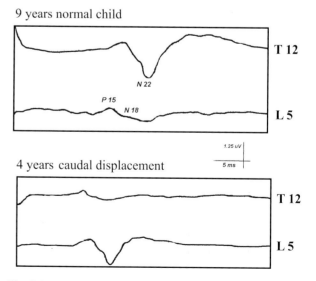

Fig. 6.73 Recordings at T12 and L5 vertebral apophyses following posterior tibial nerve stimulation in a normal child (upper traces) and in a child with asymptomatic caudal displacement of the conus medullaris (lower traces). Note the inverted amplitude gradient (maximal at the L5 lead instead of at the T12 lead).

Follow-up of tethered cord **Follow-up of surgical untethering of lumbo-sacral lipoma**

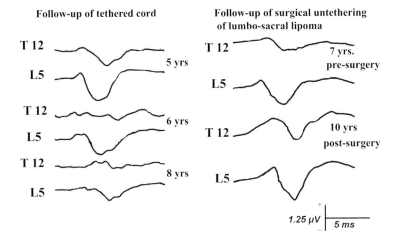

Fig. 6.74 PTN-SEP recorded at T12 and L5 vertebral locations. Follow-up of a patient with a tethered cord considered unsuitable for surgical treatment (left) and of another patient before and after surgery to remove tethering (right). Recordings from the spinous process of the T12 and L5 vertebrae referred to the umbilicus.

any patient with a suspected cord tumour. A good correlation has been shown between transient or persisting central intraoperative SEP conduction changes and the postoperative clinical and SEP outcome (Mauguière et al 1996).

Lesions of the cervicomedullary junction

When the volley of ascending impulses in the cervical dorsal columns is blocked or dispersed at the cervicomedullary junction, the segmental N13 potential to upper limb stimulation is normal, whereas far-field P14 and later components are absent or abnormal (Mauguière et al 1983, Yamada et al 1986). The lower limb P39 SEP is most often abnormal, but can be preserved in association with severe upper limb SEP abnormalities (Mauguière and Ibañez 1985); this occurs when the lesion compresses only the outer dorsal column fibres. When the P39 potential is abnormal, an absent or delayed P30 is consistently found (Tinazzi et al 1996). This 'cervicomedullary' pattern has no aetiological specificity; in children it is used in situations such as foramen magnum stenosis secondary to achondroplasia (Boor et al 1999), mucopolysaccharidoses (Boor et al 2000) or atlantoaxial subluxation (which can occur in Down's syndrome), where there is a risk of compression at the level of the cervicomedullary junction. Early neurophysiological signs of deterioration can assist in decisions concerning the role of decompressive surgery before structural irreversible damage has occurred, and also monitor postoperative functional recovery.

Brainstem lesions

The upper limb P14 potential and lower limb P30 are generated close to the dorsal column nuclei in the medulla oblongata. In lesions of the medulla oblongata affecting somatosensory input transmission, these components are reduced, delayed or absent.

More than 50% of patients with a lesion located in the pons or in the cerebral peduncles have a reduced or absent N20 to median nerve stimulation, and in less than half of them the P14 component is abnormal. Thus, apart from the unequivocal significance of the cervicomedullary pattern, no definite electroclinical correlations can be made between the appearance of the P14 potential and the lesion site in patients with brainstem lesions above the medulla oblongata.

Thalamic and capsulothalamic lesions

SEPs to median nerve stimulation are abnormal in more than 70% of capsulothalamic and in nearly 90% of posterior thalamic lesions (Mauguière and Ibañez 1990). After stimulation of the affected side the loss of both parietal N20–P27 and frontal components (P22–N30), with preserved scalp far-field positivities including P14, represent the most frequent SEP abnormality in thalamic or capsular lesions interrupting the somatosensory pathways (Nakanishi et al 1978, Mauguière et al 1982, 1983, Stöhr et al 1983, Yamada et al 1985, Graff-Radford et al 1985, Mauguière and Ibañez 1990). In thalamic lesions outside the ventroposterolateral nucleus, early cortical SEPs are normal.

Cortical lesions

In large lesions involving the central and parietal regions, abnormal SEPs are very similar to those encountered in posterior thalamic lesions deafferenting the hemisphere, showing a loss of all cortical SEPs but with a preserved P14 component. In lesions outside the centroparietal area, early cortical components are normal. When the lesion is close to the cortex it may damage separately pre- or post-Rolandic cortices or thalamocortical fibres in the corona radiata, and abolish selectively the parietal N20–P27 or frontal P22–N30 components.

There is a highly significant correlation between abnormal SEPs and the presence of sensory and/or motor deficits. When the N20 or P27 components are lost, patients with cortical lesions usually show no motor deficit but do show astereognosis, possibly combined with hypaesthesia for some elementary sensory modalities. Conversely, sensation is preserved and motor function impaired in patients with absent or grossly abnormal P22 or N30 components. Studies correlating SEP abnormalities and clinical evaluation of stereognosis and tactile and joint sensation in adult patients give clues to the somatosensory dysfunction in children unable to cooperate with clinical testing.

There are disorders where the pathological SEP finding is a 'giant' cortical response. We refer to progressive myoclonic epilepsies and focal motor seizures due to lesions of the peri-Rolandic area associated with cortical reflex myoclonus (Dawson 1947, Shibasaki and Kuroiwa 1975, Mauguière and Courjon 1980, Mauguière et al 1981, Rothwell et al 1984, Obeso et al 1985, Shibasaki et al 1985, Kakigi and Shibasaki 1987, Seyal 1991, Ikeda et al 1995, Valeriani et al 1997a, b). Giant SEPs mostly reflect an enhanced response of the N24–P24 generator in area 3b (Valeriani et al 1997a), while the parietal N20 potential and P14 far-field scalp positivity are normal in most patients. Cortical hyperresponses to either cutaneous inputs alone or to both cutaneous (area 3b) and proprioceptive (areas 3a or 2) inputs can occur in progressive myoclonic epilepsy with cortical reflex myoclonus.

Giant SEPs can also be observed over the damaged hemisphere in patients with supratentorial tumours, post-traumatic cortical atrophies or long after an ischaemic or haemorrhagic stroke, in the absence of myoclonus triggered by somatosensory stimuli, but the occurrence of focal motor seizures is frequent in the history of such patients (Laget et al 1967, Furlong et al 1993, Valeriani et al 1997b). Loss of inhibitory control and postlesional collateral sprouting of cortical afferents could be responsible for such SEP abnormalities.

In some children aged 3–13 years with normal neurological status, vertex and parietal EEG spikes, corresponding to high voltage SEPs (up to 400 μV), can be evoked by a single tactile stimulus (De Marco and Negrin 1973). The presence of these 'extreme SEPs' might forecast the possible occurrence of partial motor seizures with benign outcome (De Marco and Tassinari 1981). Median nerve cortical SEPs in these children are usually normal up to a latency of 60 ms and then show a large central negativity, which reaches its maximum only for low stimulation frequencies of 0.2–0.5 Hz (Plasmati et al 1989).

In central demyelination the main effect is to slow conduction and thus to increase the time dispersion of impulses in the dorsal columns, medial lemniscus and thalamocortical fibres, a finding observed in all types of childhood leucodystrophies and associated with abnormal latency of AEPs and VEPs. Abnormal SEPs are very similar whatever the cause of demyelination, and are frequently observed in the absence of any sensory symptoms. Peripheral SEP components, such as the Erb's point N9 and the scalp P9 or the popliteal N7 and lumbar P17, obtained after stimulation of the median and posterior tibial nerves, respectively, are normal in central demyelination. Their abnormality is often a useful clue

to involvement of the peripheral myelin, such as in Krabbe's globoid body leucodystrophy or Methachromatic leucodystrophies. The spinal N13 and N22 potentials evoked by stimulation of the median and posterior tibial nerves, respectively, are also most often normal in amplitude and latency in central demyelination. However, these components may be missing when the afferent fibres are demyelinated at the dorsal root entry zone before their synaptic contact with the dorsal horn neurons, which generate these spinal SEPs. Cortical N20 and P39 components may be delayed, delayed and reduced, or absent in central demyelination.

Clinical applications of SEPs in encephalopathies are described under specific disorders on page 326 ff.

SPECIAL EMG PROBLEMS: NERVE CONDUCTION STUDIES AND EMG IN THE INVESTIGATION OF CHILDHOOD WEAKNESS

Neurogenic causes of weakness
Spinal muscular atrophies
This group of genetic disorders is caused by abnormalities in the survival motor neuron (SMN) gene located on chromosome 5q13. Two copies of the SMN gene are contained at this locus, SMNt (telomeric SMN1) and SMNc (centromeric SMN2). Homozygous deletions of SMNt produce spinal muscular atrophy (SMA), but the phenotypic expression and severity of the disease are influenced by the activity of the SMNc gene (Brahe 2000). The phenotypic expression may also be influenced by other gene products, particularly the neuronal apoptosis inhibitory protein gene, which has been shown to be deleted in 68% of cases of SMA type I, but in only 2% of cases of SMA types II and III (Roy et al 1995). Diagnosis of SMA by DNA analysis for the SMN gene has been possible for some time (Lefebvre et al 1995).

SMA type I (Werdnig-Hoffmann disease) (Hausmanowa-Petrusewicz et al 1968), also referred to as proximal hereditary motor neuropathy (HMN) (Harding and Thomas 1984), is the most severe form of the disease, with onset before the age of 6 months, typically within the first few months of life. Babies have generalised hypotonia and proximal weakness, and sucking and swallowing difficulties, but the ocular muscles are spared and the babies often have an alert look. Respiratory difficulty with intercostal weakness is common. Prognosis is poor and death occurs generally during the first year of life.

SMA 0, an even more severe form of SMA, has been described. It presents prenatally with reduced fetal movements and death soon after birth (MacLeod et al 1999, Dubowitz 1999). These infants may have associated abnormalities, such as contractures or even fractures (Garcia-Cabezas et al 2004).

One of the original diagnostic criteria for SMA was that the diaphragm was spared (Rudnik-Schoneborn et al 1995). This produced a very characteristic chest shape and movement. However, involvement of the diaphragm can occur in SMA. SMA with diaphragmatic involvement is now classified under the SMA plus types, and has been shown not to be associated with the deletion of the SMN gene on chromosome 5 (Zerres and Davies 1999). It includes the condition SMARD1 (SMA with respiratory distress), also known as severe infantile neuropathy with respiratory involvement (Appleton et al 1994, Pitt et al 2003).

SMA type II   The intermediate form with onset usually between 6 and 12 months of age (HMN proximal type II) is characterised by delayed motor development with failure to sit independently by age 9–12 months, or failure to stand by age 1 year. Examination reveals atrophy of the proximal muscles, especially in the lower limbs. Fasciculations of the tongue are sometimes noted, and a postural tremor of the hands is a very frequent finding. However, the children are alert and cognitively normal (Moosa and Dubowitz 1973). Joint laxity and bone deformities are frequent, and progressive evolution towards immobilisation after 1–10 years of illness is the rule.

SMA type III (Kugelberg–Welander disease)   This is the mildest form of the disorder, sometimes referred to as the 'adult form'. Age of onset is from 2 years, through childhood and adolescence. The clinical features are pseudomyopathic, including proximal weakness and atrophy, particularly in the lower limbs, with pseudohypertrophy of the calves and gluteal muscles, fasciculations and absent knee jerks. The course of the disease is either non-progressive or only mildly progressive, and most individuals can walk unaided. long term survival depends on respiratory function, but may be compatible with a near-normal life span.

Before the DNA diagnosis became widely and quickly available an electromyographer might have been called to do EMG on the early-onset SMA (0 and 1). This was to lend diagnostic support to the clinical diagnosis in case the infant deteriorated seriously and ethical questions of whether to ventilate or not occurred. This still can occur. Paradoxically, in the most severe cases the classical EMG changes of denervation and reinnervation, which are the mainstay of the diagnosis, may not be found and the EMG may be equivocal. In our experience, fibrillations are only variably present. Similar problems occurred when muscle biopsy was used if done early. The pathologist was often not able to be certain that the changes seen were neurogenic.

The nerve conduction studies will undoubtedly point to significant motor loss, with reduced CMAPs and decreased motor conduction velocity (Krajewska and Hausmanowa-Petrusewicz 2002). The sensory responses in these infants can be absent, either a result of the associated involvement of the dorsal root ganglia that is known to occur (Rudnik-Schoneborn et al 2003) or because they can be technically very difficult to obtain. Theoretically, distinction from sensorimotor axonal neuropathy may be impossible, but practically this is seldom an issue.

The clinical diagnosis of the later forms of SMA types II and III is sometimes not straightforward. The neurophysiological signs of a widespread axonal degeneration of the motor nerves will be present, with or without concomitant sensory involvement. In the age group in which these conditions are seen, distinction from CMT type 2 may not be possible on EMG. A useful technique is to go immediately from the EMG demonstration of chronic denervation in a limb muscle to EMG of the tongue. Via the submental route this investigation is well tolerated, and demonstration of denervation will point towards the diagnosis of SMA. Genetic diagnosis may be possible.

Other forms of SMA   There are several other much rarer disorders involving the anterior horn cells, including Kennedy's disease or X-linked spinobulbar muscular atrophy. This disease is due to a defect in the androgen receptor gene (La Spada et al 1991). Onset is usually in adolescence or early adult life. Fasciculations of the lower face and tongue are often a prominent feature. Other findings include gynaecomastia, congenital fractures, joint contractures and a sensory neuronopathy.

Fazio–Londe disease primarily affects the lower cranial nerves. Starting in the first decade of life it often presents with dysphagia, nasal speech and recurrent respiratory infections (McShane et al 1992). The disorder is progressive, leading to death in 1–10 years. There may be a phenotypic overlap with Brown–Vialetto–van-Laere syndrome, which has a variable association with deafness (Dipti et al 2005).

There are also several rare forms of SMA with predominantly distal muscle involvement (Harding and Thomas 1980). Most are only slowly progressive, but they require neurophysiological studies to distinguish them from hereditary motor and sensory neuropathies (HMSNs) (see below).

Peripheral neuropathies

Hereditary and acquired peripheral neuropathies in childhood have been reviewed by Rocha and Escolar (2004) and Ryan and Ouvrier (2005).

Hereditary neuropathies   Children with hereditary neuropathy may be referred to the clinical neurophysiology department, either because of presenting symptoms and signs of peripheral neuropathy or because of a complex medical picture in which neuropathy constitutes a relatively minor feature. Patients are usually younger in this latter group. Some hereditary neuropathies are listed in *Table 6.8.*

Table 6.8 Hereditary neuropathies

Type of disorder	Disease
Neuropathy as part of generalised metabolic disorder	Leucodystrophy
	Refsum's disease
	Abetalipoproteinaemia
	Fabry's disease
	Hereditary amyloid neuropathies
	Hereditary hepatic porphyria
	Familial isolated vitamin E deficiency
Predominantly peripheral neuropathies	Hereditary motor and sensory neuropathies (HMSN):
	type I (Charcot–Marie–Tooth: demyelinating)
	type II (Charcot–Marie–Tooth: neuronal)
	type III (Déjerine–Sottas disease: hypomyelinating)
	Hereditary sensory and autonomic neuropathies (HSAN) types I–IV
	Hereditary neuropathy with liability to pressure palsy (HNPP)
	Neuroaxonal dystrophy
	Giant axonal neuropathy
	Friedreich's ataxia

HMSNs/Charcot–Marie–Tooth (CMT) disease The hereditary motor and sensory neuropathies (HMSNs) are the commonest forms of familial neuropathy. The group name HMSN is a non-specific term recommended by Thomas and colleagues (Thomas and Calne 1974, Dyck 1975). However, many authors, particularly those relating to the genetic aspects of these diseases, use the terms CMT disease (HMSN I), congenital hypomyelination and Déjerine–Sottas disease (HMSN III) to refer to these disorders. Major advances have been made in the genetic understanding of this group of disorders and the gene products involved in the different HMSN disease subtypes. However, the precise mechanism by which the protein products cause pathological changes in the nerves remains uncertain (Pareyson 1999).

HMSN types I and II. The most common form of HMSN is type 1A. It is caused by a 1.5-Mb duplication of the peripheral myelin protein 22 (PMP22, chromosome 17p11.2-p12) and has an autosomal dominant inheritance. The X-linked dominant form (HMSN IX) is caused by mutations in the connexin 32 gene. The clinical onset commonly occurs in the first or second decade of life. Patients most often present with pes cavus, toe walking and distal muscle wasting affecting the intrinsic foot muscles and the peronei – hence the alternative name of 'peroneal muscular atrophy'. Sometimes, hypotonia and foot deformity are present at birth. Many asymptomatic cases are detected by electrophysiological screening of the family when a member is affected.

HMSN type III (Déjerine–Sottas disease) (Dyck 1984) and *congenital hypomyelination*. These are both rare disorders, typically characterised by very slow motor conduction velocities below 12 m/s and absent sensory responses (Ouvrier et al 1987, 1990). Déjerine–Sottas disease can be inherited as an autosomal dominant disorder (PMP-22 point mutations, chromosome 17) (Ionasescu et al 1996), but many cases appear to be new mutations in heterozygotes. Onset is usually before 2 years of age, with a severe sensorimotor neuropathy frequently associated with sensory ataxia and scoliosis. Dysmorphic features, nerve enlargement and ocular symptoms (miosis, reduced response to light, ptosis and nystagmus) occur in some patients. Congenital hypomyelination is an autosomal recessive neuropathy, which usually presents at birth with either floppy infant syndrome or arthrogryposis (Guzetta et al 1982).

HMSN type IV (Refsum's disease) (Dyck 1984). This disorder results from a defect of phytanic acid metabolism; phytanic acid accumulates in the serum and tissues. Peripheral neuropathy, with a very variable degree of demyelination, usually develops in association with ataxia. Cardiomyopathy, night-blindness and other abnormalities may also be present. A few patients are asymptomatic until the age of 5 years. As in adults, nerve conduction velocities are markedly reduced.

In addition to the clinical picture, nerve conduction studies remain important, both to confirm the presence of a peripheral neuropathy and also to characterise the type of neuropathy and direct genetic testing. They can be used to divide affected individuals into 'demyelinating' (type I, X and III) or 'neuronal' (type II) on the basis of nerve conduction studies (Dyck and Lambert 1968a, b, Harding and Thomas 1980). Motor nerve conduction velocities in a large number of patients with HMSNs show a bimodal distribution (*Fig. 6.75*). The value of 38 m/s for conduction velocity in the median nerve has been found to discriminate between type I (conduction velocity < 38 m/s) and type II (conduction velocity > 38 m/s) in most affected individuals > 2 years old. Type III (Déjerine–Sottas disease) congenital hypomyelination typically shows conduction velocities below 12 m/s. It has been found that abnormalities remain virtually unchanged with ageing in longitudinal studies (Dyck et al 1989), and testing for demyelinating neuropathies can be undertaken in children after the age of 2 years, even before they become symptomatic.

Recent evidence has indicated that the degree of clinical progression and weakness is more closely related to progressive axonal loss rather than to the degree of conduction slowing or 'demyelination' (Krajewski et al 2000).

Motor conduction slowing in the 'demyelinating' HMSNs is generalised and diffuse, and conduction block and dispersion are rare. Therefore, nerve conduction studies demonstrating conduction block or dispersion should prompt consideration of acquired or inflammatory neuropathies. Spontaneous EMG activity, such as neuromyotonia, is particularly common in HMSN II.

In addition to the 'pure' forms of the HMSNs, there are some more complex forms, particularly of type I, in which the neuropathy is associated with optic atrophy, pigmentary retinopathy, deafness (Dyck 1984) or pyramidal signs (Harding and Thomas 1984).

Hereditary neuropathy with liability to pressure palsy (HNPP)   This quite common disorder is inherited as an autosomal dominant disorder, most commonly with a deletion or nonsense mutation of the PMP22 gene (this is the same gene responsible for HMSN type IA but results from a deletion rather than a duplication of the gene). Affected individuals can present at any age, but frequently do so in the first two decades of life, with recurrent mononeuropathies (mononeuritis simplex or multiplex) that are frequently related to minimal trauma. Significant slowing of motor and sensory nerve conduction velocities is found, particularly at the common sites of entrapment for the median, ulnar and peroneal nerves. Motor nerve conduction block at these sites is common. Motor and sensory conduction slowing can also be demonstrated in clinically unaffected nerves. Some degree of diffuse sensorimotor neuropathy is found in all individuals later in life. Morphological studies on nerve biopsy specimens show 'tomaculous' or sausage-like swellings of the myelin sheaths, and segmental demyelination, particularly on teased fibre preparations.

Hereditary sensory and autonomic neuropathies (HSANs)   The HSANs form a group of diseases in

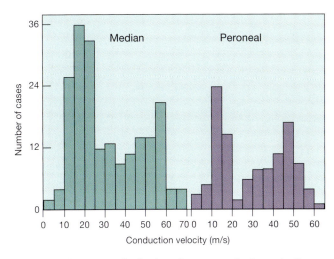

Fig. 6.75   Frequency distribution of motor conduction velocity in the median and peroneal nerves in patients with HMSN. (From Harding and Thomas (1980), by permission of Oxford University Press.)

which the lesions of sensory afferent fibres may involve large myelinated fibres to a variable extent, but always affect small myelinated or unmyelinated fibres, and hence autonomic fibres. These disorders, which are rare in childhood, have been described and classified by Dyck (1984). Classification is based on natural history and clinical characteristics, particularly the range of autonomic function disturbance. For example, anhidrosis is unique to HSAN type 4 and alacrima is unique to HSAN type 3, whereas orthostatic hypotension is most frequent in HSAN type 3. Nerve conduction and EMG findings can be helpful but they are often normal. Sural nerve biopsy is usually required to demonstrate the pathological abnormality. HSAN type I is the only type that is dominantly inherited; all the remaining disorders are thought to be recessive.

HSAN type I (hereditary sensory radicular neuropathy). This sensory neuropathy is characterised by loss of perception of pain and temperature. Spontaneous pain may also be troublesome. Insidious onset occurs in the first decade. Nerve biopsy shows a severe loss of unmyelinated axons and a reduction in the number of small myelinated fibres (Dyck 1984).

HSAN type II (congenital insensitivity to pain). This is a rare, recessively inherited sensory neuropathy that affects all sensory modalities from birth (Ohta et al 1973), with marked disturbance of sensory nerve conduction.

HSAN type III (familial dysautonomia or Riley–Day syndrome) is also rare. Almost all affected individuals come from an Ashkenazi Jewish background. The disorder is usually manifest from birth, with severe incapacity, autonomic impairment and failure to respond to painful stimuli. Orthostatic hypotension is often a severe feature. Sensory action potentials are severely reduced (Brown and Johns 1967), sometimes with a considerable temporal dispersion.

HSAN type IV (congenital insensitivity to pain with anhydrosis) has been described in very few patients, and there are no reports of nerve conduction studies (Thomas 1987).

HSAN type V (congenital insensitivity to pain with partial anhydrosis) has been described in two young children (Low et al 1978, Dyck et al 1983). There is a selective loss of small myelinated fibres. Motor and sensory nerve conduction velocities are normal using standard techniques.

Other hereditary neuropathies Other very rare disorders, including congenital autonomic dysfunction with universal pain loss, and progressive pandysautonomia, have been described.

Neuropathy as part of a generalised metabolic disorder

Mucopolysaccharidosis Carpal tunnel syndrome is exceedingly rare in paediatric patients, except in the mucopolysaccharidoses. Of these, mucopolysaccharidosis I and II (Hurler and Hunter syndromes) are commonly affected, but it is also found in mucopolysaccharidosis VI and mucolipidosis II and III (Haddad et al 1997). Any method to determine its presence is valid, but palm-to-wrist studies in the smallest infants may be impossible technically. Detection of a double peak when recording between the ulnar and median nerves at the wrist when stimulating digit IV is particularly valuable in these cases. Before the advent of bone marrow transplant it was remarkable how quickly the condition could deteriorate, going from minor sensory changes to an absence of any motor response over 12 months. This might have been due to infarction of the nerve following compression from the mucopolysaccharidosis material.

Leucodystrophies There is a known association between some of the leucodystrophies and peripheral neuropathy. The demonstration of a demyelinating neuropathy in a child who was showing regression was an important clue to the diagnosis. Now that specific diagnostic tests are available, such referrals are uncommon. It is of some continuing interest to review what is known for the atypical case that might still present.

In *metachromatic leucodystrophy* a demyelinating peripheral neuropathy is seen in all the major groups: late infantile (Aicardi 1993), early juvenile (Yatziv and Russell 1981), and juvenile (Clark et al 1979). At the upper end of the age range there have been instances where the unexpected discovery of a demyelinating neuropathy in a child with regression has led to the correct diagnosis (Percy et al 1977, Bosch and Hart 1978, Wulff and Trojaborg 1985).

Krabbe's leucodystrophy is caused by a deficiency of the lysosomal enzyme galactocerebrosidase β-galactosidase, and usually presents early (before 6 months of age; early infantile form). A peripheral neuropathy has been demonstrated at 7 weeks in an antenatally diagnosed case (Lieberman et al 1980). This is unusual, as nerve conduction studies, along with MRI of the brain, may be normal this early, becoming abnormal at beyond 3 months (Zafeirou et al 1997). The late infantile form presents between 6 and

18 months of age, with the floppiness being a mixture of central and peripheral demyelination (Aicardi 1993). The peripheral neuropathy is much less consistently seen in the late-onset or juvenile forms. It was seen in only one of the four cases reported (Phelps et al 1991), who demonstrated highly variable presentations, some being wheelchair-bound and others in active employment.

Cockayne syndrome is also associated with a demyelinating peripheral neuropathy (Moosa and Dubowitz 1970, Sasaki et al 1992). It is caused by a defective chromosome or DNA repair. There is a wide spectrum of severity, with an increasing severity in the peripheral and central demyelination with age (Smits et al 1982).

Adrenoleucodystrophy is diagnosed by the detection of long-chain fatty acid abnormalities and is variably associated with a demyelinating neuropathy. The neonatal form of the condition may present with signs and neurophysiological findings that would suggest infantile SMA (Paul et al 1993).

Pelizaeus–Merzbacker syndrome is characterised by abnormal eye movements, choreoathetotic movements, cerebellar ataxia and progressive spasticity. MRI shows deficient CNS myelination. The most common mutation is in the PLP gene, producing misfolding of the PLP (chromosome Xq22); spastic paraplegia 2 is an allelic disorder. Other mutations have been described. PLP is found mainly in the CNS, and the absence of a peripheral neuropathy used to be an important diagnostic pointer in this context. However, a recent family study (Garbern et al 1997) has demonstrated that a peripheral demyelinating neuropathy can be seen, albeit with a unique mutation that leads to the absence of PLP expression (PLP null syndrome).

Other metabolic conditions with associated neuropathy

Hypobetalipoproteinaemia. The defect in hypobetalipoproteinaemia is a deficiency of lipoprotein β, which plays a crucial role in lipid transport. Neurological defects manifest late in childhood with ataxia and weakness. Tendon reflexes, as well as joint and vibration sensation, are lost. Motor conduction velocities are normal but sensory action potentials can be reduced, and there is some evidence of chronic partial denervation on EMG sampling (Miller et al 1980).

Fabry's disease and the *hereditary amyloid neuropathies.* A selective small-fibre involvement, giving rise to severe neurogenic pain, is a particular feature of these disorders. Nerve conduction studies are normal. Thermal thresholds may become elevated during adulthood.

Hereditary hepatic porphyrias. These conditions can present with an acute onset of motor neuropathy and should be considered in the differential diagnosis of Guillain–Barré syndrome.

Neuropathy in spinocerebellar degeneration

Friedreich's ataxia This is an autosomal recessive disorder caused by a trinucleotide repeat in the frataxin

(X25) gene on chromosome 9q13 (Campuzano et al 1996). Onset is usually between the ages of 2 and 16 years, typically at 4–5 years of age, with clumsiness in walking and running often causing frequent falls. In early childhood the disorder may present as a delay in attaining motor developmental milestones. Areflexia in the lower limbs and dysarthria are usual, but only 50% of affected individuals develop pes cavus, scoliosis, distal wasting and cardiomyopathy. A minority may develop optic atrophy, nystagmus, sensorineural deafness, sphincter dysfunction and diabetes mellitus (De Michele et al 1996). Motor conduction velocities are usually normal or only slightly reduced, but sensory or mixed nerve action potentials are absent or significantly reduced (McLeod 1971). Abnormalities of sensory conduction are demonstrable in patients as young as 3 years of age (Ouvrier et al 1982). The EMG may be mildly neurogenic. These electrophysiological features may be detected in routine screening of the siblings.

Familial isolated vitamin E deficiency This is a rare disorder but can mimic Friedriech's ataxia, and present with cerebellar ataxia, dysarthria and peripheral neuropathy. Measurement of serum vitamin E levels is essential to distinguish between these two disorders (Kayden 1993). Acquired vitamin E deficiency is much more common and presents in the same way.

Giant axonal neuropathy This is also an autosomal recessive, inherited ataxia. Onset during the first 7 years of life with progressive weakness and clumsiness of gait associated with a slurring of speech is usual. Other symptoms, encephalopathy and tightly curled hair, are often present. Tendon reflexes are absent; sensation, particularly vibratory and positional, is impaired. Motor nerve conduction velocities are normal or mildly decreased; sensory nerve potentials are usually absent (Berg et al 1972). The EMG is consistent with a neurogenic disturbance. Pathological findings are those of axonal degenerative neuropathy with severe focal axonal swelling and accumulation of neurofilaments.

Acquired inflammatory polyneuropathies These may be acute, with symptoms appearing over days or a few weeks, or chronic, with progression of at least 8 weeks.

Acute inflammatory demyelinating polyneuropathy (Guillain–Barré syndrome) This represents a frequent form of polyneuropathy in adults and children. It may occur in toddlers but is exceptional during the first year of life. It is now well established as an immunologically mediated, acute neuropathy involving both cell-mediated and humoral mechanisms. Characteristically there is multifocal inflammatory demyelination and a variable degree of secondary axonal loss. Typically, the onset of weakness is preceded by gastroenteritis or infection of the upper respiratory tract a week or two before. Several studies have implicated recent infection with *Campylobacter jejuni*, cytomegalovirus, Epstein–Barr virus or *Mycoplasma pneumoniae* as being of probable significance.

Campylobacter jejuni is the most important agent associated with axonal degeneration, lack of sensory disturbances, severe disability, slow recovery and poor outcome (Govoni and Granieri 2001). Conduction block is the main cause of the muscle weakness.

The typical presenting symptom is weakness of the lower limbs and sometimes hands, progressing typically for 1 or 2 weeks, but this may vary from a few hours to some weeks. Other initial symptoms include back pain and paraesthesias. Irritability, headache, ataxia and bladder dysfunction are seen in many children. The cranial nerves may be involved. Miller–Fisher syndrome is rare in children.

Electrophysiological abnormalities represent changes resulting from patchy, but extensive, multifocal demyelinating peripheral nerve lesions, with a variable degree of secondary axonopathy. The EMG diagnosis rests on the accurate identification of patchy demyelination in two or more motor nerves. Motor abnormalities are consistently more severe than sensory changes. Common findings include prolonged distal motor latencies (particularly those of the median nerve at the wrist and the common peroneal nerve at the ankle) and temporal dispersion of motor responses. There may or may not be conduction block in the routinely tested elbow–wrist and knee–ankle segments. Unlike classical axonal neuropathies, lower limb studies are no more likely to be abnormal than are upper limb responses. Motor conduction velocities are usually reduced to a range of 30–40 m/s. Some cases have the most dramatic changes on distal stimulation, with distal motor latencies in the median nerve of up to 15 ms with severe dispersion of the compound nerve action potential. In our experience, increased distal motor latency, marked slowing of motor conduction and very polyphasic, long-duration CMAPs are almost as prominent as conduction block signs. In some infants the only demonstrable abnormality may be absence of F-wave responses, indicating proximal conduction block. Sensory action potential abnormalities are common but variable. They include some slowing and dispersion, with reduction of amplitude. Delanoe et al (1998) used sensory as well as motor nerve abnormalities to reach the diagnostic criteria. SNAPs of upper limbs are commonly reduced or absent, while sural SNAPs may be preserved.

Chronic inflammatory demyelinating polyneuropathy (CIDP) This is a rare form of demyelinating polyneuropathy in childhood. A few cases of early onset during the first year of life were reported by Sladky et al (1986). A congenital form has been described (Pearce et al 2005). Electrophysiological data are the same as in the adult, and should be distinguished from those observed in HMSN I. In CIDP there is a variable degree of slowing in different segments of the same nerve and from one nerve to another, a marked dispersion of CMAPs and frequent conduction block (Lewis and Sumner 1982). CIDP in children tends to fluctuate more rapidly and be associated with a better prognosis (Simmons et al 1997).

Uniform slowing of nerve conduction constitutes strong evidence in favour of a familial demyelinating neuropathy (including HMSN I and other genetically determined neuropathies). Conversely, multifocal slowing is an argument in favour of acute or chronic acquired demyelinating neuropathy (Miller et al 1985).

Drug-induced neuropathies   There are many drugs that cause neuropathies. Some of the more important with their most common presentation (axonal, demyelinating, etc.) are listed in *Table 6.9*. Chemotherapeutic agents such as vincristine are proportionally the most common causes in children. long term antibiotic therapy is perhaps a peculiarity in children, more so than in adults. Children with cystic fibrosis who take chloramphenicol for long periods may be at risk of developing a predominantly sensory axonal neuropathy, as may also occur if metronidazole is used for long periods.

What is remarkable is how infrequently children are referred for investigation of suspected peripheral neuropathy given the frequency with which they are exposed to these agents. There may be many and various reasons for this. It is certainly true that many of the drugs are needed for life-threatening conditions, and many clinicians knowing at what doses they might expect to see signs, and recognising them easily, will manage the patient clinically. This avoids the need for a further investigation that might be painful. In many cases the neuropathy can be reversed on stopping treatment.

Thalidomide is the one exception to this strategy, with clinicians unhappy to prescribe it without nerve conduction monitoring. Yet its neurotoxicity is much less than the other drugs already discussed (e.g. vincristine). This fact is difficult to explain, but may result from the past reputation of this drug. There were well-documented cases of peripheral neuropathy (Burley 1961, Clemensen et al 1984) with thalidomide usage, and in fact it was the early reports of peripheral neuritis that

delayed its approval by the US Food and Drug Administration in 1960, not the fetal abnormalities, which were first reported in the following year. The other reason is that it is often given for conditions that are not considered as life-threatening as some of the cancers.

The neuropathy associated with thalidomide was often restricted to the sensory nerves and was painful (Wulff et al 1985). Some report that there is no improvement when the drug is withdrawn (Peltier and Russell 2006), while others (Isoardo et al 2004) suggest that as many as 50% might recover, prompting an MRI search for a dorsal root ganglionopathy. Only 25% of the cases reported by Isoardo et al showed this.

As thalidomide caused a sensory neuropathy that was either irrecoverable or very slow to recover, EMG studies were asked for routinely when the drug was reintroduced for a variety of conditions. In our practice these are most often dermatological conditions, including Behçets disease, but it is also used for lupus erythematosus, graft-versus-host disease, ulcerative colitis and Crohn's disease, although it has yet to be demonstrated to have proven effectiveness in some of these.

Neurotoxicity is clearly a concern. The results of the few large retrospective clinical studies are difficult to interpret, as the reported incidence of neurotoxicity varies from 1% to 50% (Burley 1961). Lack of awareness of the test–retest variability for the measurement of the SNAPs (Pitt 1996) may in part explain this wide variation in incidence. All studies have used the amplitude of the compound SNAP as an indicator of axonal loss. In our practice, in order to minimise stress and to ensure compliance with monitoring, it has been our policy to study one or two sensory nerves in the legs every 6 months, informing the clinicians of any decreases that are felt to be significant. As an observation it is interesting how often the patients and their families will choose to continue with the drug, despite indications of a neuro-

Table 6.9   Toxic neuropathies: clinical and pathological features

Axonal			Demyelinating	Mixed
Sensory	**Sensory and motor**	**Motor**		
Chloramphenicol	Alcohol (ethanol)	Gangliosides	Chloroquine	Amiodarone
Digoxin	Arsenic	Lead	Procainamide	Gold
Doxorubicin	Cadmium			Sodium cyanate
Isoniazid	Colchicine			
Metronidazole	Dapsone			
Phenytoin	Lithium			
Pyridoxine	Nitrofurantoin			
Statins	Organophosphates			
Thalidomide	Vincristine			

pathy, because of its efficacy. Children do not seem to suffer the burning pain that is experienced by adults on thalidomide.

Myopathic causes of weakness

Acquired and genetic myopathies in childhood are summarised in *Table 6.10*.

Congenital myopathies

The congenital myopathies form a group of inherited muscle disorders characterised by specific structural changes within the muscle corresponding to distinctive histological patterns on muscle biopsy. Clinical presentation is non-specific. Hypotonia (floppy infant syndrome, see p. 219) and bony abnormalities, such as skeletal deformities at birth or in infancy, and muscle

weakness and delayed milestones at later ages, are the most frequent signs of congenital myopathies. Most are non-progressive and some may show gradual improvement. Since the motor unit potential abnormalities are mild and a child's cooperation with needle EMG limited, it is usually not possible to study isolated potentials. It is important to note that a normal needle EMG examination does not exclude a congenital myopathy.

Central core disease   This disease (Shy and Magee 1956) usually has an autosomal dominant transmission (chromosome 19q13.1, several mutations identified) and is characterised by the presence of cores in the central part of muscle fibres, in particular in type I fibres. Onset of proximal weakness is congenital or during childhood and is non- or slowly progressive. Malignant hyperthermia is

Table 6.10   Myopathies

Basis	Disease
Acquired disorders	
Inflammatory	Idiopathic polymyositis or dermatomyositis
	Polymyositis secondary to collagen disease or underlying malignancy
Endocrine and metabolic	Thyrotoxicosis
	Hypothyroidism
	Acromegaly
	Cushing's syndrome
	Abnormalities of Ca^+ metabolism
Drug induced or toxic	Steroids
	Chloroquine
	Clofibrate
	Zidovudine
	Alcohol
Genetic disorders	
Muscular dystrophies	Duchenne dystrophy
	Becker dystrophy
	Facioscapulohumeral dystrophy
	Scapulohumeral dystrophy
	Limb girdle dystrophy
	Welander's distal myopathy
	Ocular and oculopharyngeal muscular atrophy
Myotonic disorders	Myotonic dystrophy
	Myotonia congenita
	Paramyotonia congenita
Myopathies with periodic paralysis	Hypokalaemic periodic paralysis
	Hyperkalaemic periodic paralysis
	Normakalaemic periodic paralysis
Metabolic disorders	Mitochondrial myopathies
	Disorders of glycogen metabolism
	Acid maltase deficiency
	McArdle's disease
	Disorders of lipid metabolism
	Carnitine deficiency
Congenital myopathies	Central core disease
	Nemaline myopathy
	Centronuclear myopathy

associated in some cases. Needle EMG is either mildly 'myopathic' or normal.

Nemaline (rod) myopathy   This myopathy (Shy et al 1963) is inherited as either an autosomal dominant or autosomal recessive trait (several genetic types identified). It is characterised by rod-shaped bodies within the muscle fibres, with a predominance of small type I fibres. Clinical presentation is usually that of a floppy baby or infant, with hypotonia, weakness and, frequently, skeletal deformities such as a high arched palate, long face, pes cavus or kyphoscoliosis. Respiratory problems can cause early death in the neonatal period or in infancy. The EMG can be normal before the age of 3 years; later, it usually shows myopathic abnormalities in the proximal and distal muscles, often associated with fibrillation potentials at rest. With increasing age the EMG abnormalities progress and 'neuropathic' EMG features develop in the distal muscles, sometimes with long-duration motor unit potentials reflecting active 'degeneration– regeneration' processes (Norton et al 1983, Wallgren-Pettersson et al 1989).

Centronuclear or myotubular myopathy   This type of myopathy (Spiro et al 1966) is characterised by type I fibre atrophy with persistence of the fetal myotube aspect. Inheritance is either dominant (19p13.2: dynamin 2) or recessive, and the severe neonatal form is often X-linked (Xq27.3: myotubularin). Floppy-baby features are associated with facial weakness, ptosis, extraocular palsy, dysmorphic features and cardiorespiratory failure. These symptoms are usually ascribed to congenital (or infantile) myotonic dystrophy or acid maltase deficiency. Needle EMG shows membrane hyperexcitability with spontaneous muscle activity, fibrillation potentials, positive sharp waves, complex repetitive and myotonic discharges (Munsat et al 1969) and brief small polyphasic motor unit potentials. Nerve conduction velocities are normal.

Multicore (minicore) myopathy   The name refers to the characteristic changes seen on histological examination. Approximately half of the cases of minicore myopathy are caused by a genetic error in one of two genes – selenoprotein N1 (SEPN1, chromosome 1p35-36) and ryanodine receptor 1 (RYR1, chromosome 19q13.1). With the latter gene abnormality there is an increased incidence of malignant hyperthermia, making anaesthetics dangerous (Jungbluth et al 2005). This may complicate pregnancy as well (Osada et al 2004). Multicore myopathy has four clinical types. The most common is the classic form, which presents at birth with floppiness and delayed milestones. The other forms are described by their most prominent features: mild with hand involvement, antenatal with arthrogryposis multiplex congenita and ophthalmoplegic (Nucci et al 2004).

In all myopathies the importance of doing EMG is to find other causes of the presentation rather than to diagnose a myopathy. At best an EMG will demonstrate a myopathic process, and all such examinations, including those that are negative, need to be followed by a muscle biopsy for the diagnosis to be firmly established.

Carnitine deficiency

Carnitine plays an essential role in the transfer of long-chain fatty acids into the mitochondria for beta oxidation. Low carnitine levels may not cause clinical symptoms until they reach less than 10–20% of normal. The most common manifestations are progressive cardiomyopathy, myopathy, hypoketotic hypoglycaemic encephalopathy and sudden death. Carnitine deficiency may be primary or secondary. Primary carnitine deficiency is caused by deficient transport of carnitine into the cells (mutations in the OCTN2 gene) and can be systemic or, rarely, purely muscle carnitine deficiency. Secondary carnitine deficiencies are caused by other metabolic disorders (e.g. organic acidaemias and fatty acid oxidation disorders such as medium-chain acyl-CoA dehydrogenase (MCAD) deficiency). Electrophysiological features are multiple, either 'myopathic' or 'neurogenic', with nerve conduction velocity abnormalities (Di Mauro 1979). The EMG may be normal.

Myotonias and periodic paralysis

The pathological basis of these disorders is now recognised to be a disturbance of voltage-gated ion channels, and they form part of the increasing group of disorders known as channelopathies. Apart from the recessive form of myotonia congenita, originally described by Becker in 1965, all these disorders are inherited as autosomal dominant disorders, although some recent reports suggest that inheritance patterns may be more complex (Kubisch et al 1998, Plassart-Scheiss et al 1998). The myotonic disorders are dominated by muscle stiffness and muscle hypertrophy in myotonia congenita, whereas the periodic paralyses present with episodes of recurrent weakness precipitated by cold exposure or resting after exercise. Hyperkalaemic periodic paralysis and paramyotonia congenita present in infancy. Myotonia congenita presents in infancy or childhood and hypokalaemic periodic paralysis often presents around puberty. For a review see Surtees (2000).

EMG is an important investigation and can confirm the characteristic myotonic discharges found in the myotonias and some cases of hyperkalaemic periodic paralysis between attacks. In myotonia congenita, cooling increases the myotonic discharges, but in paramyotonia congenita cooling paradoxically reduces the myotonic discharges and can induce paralysis during which the CMAP cannot be elicited. *Table 6.11* summarises the clinical and EMG features of the channelopathies.

Congenital myotonic dystrophy

Myotonic dystrophy type 1 is an autosomal dominant genetic disorder associated with a CTG expansion in the

3' untranslated region of the DMPK gene on chromosome 19q13.3 (Bonifazi et al 2006). It may present at birth with poor sucking and swallowing, respiratory insufficiency, marked hypotonia and facial weakness. Associated features are bone and joint deformities, arched palate, clubbed feet and arthrogryphosis. The EMG in distal limb musculature may show very short myotonic bursts or increased insertional activity. Sometimes the EMG features are those of a myopathic process. However, it is more often the case that the EMG examination of the infant is inconclusive, perhaps because the muscles are wasted. The cases that present in this way always inherit the condition from the mother (Rutherford et al 1989), in whom the diagnosis may not have been made. Usually the diagnosis is fairly obvious clinically, but EMG may confirm it. The confirmation should be sought by demonstrating the CTG repeats, which are invariably present. Short term prognosis is bad but, if the infant survives, respiratory function improves. However, in the long term, learning difficulties and motor handicap are significant.

Inherited muscular dystrophies

This is one of the most common neuromuscular disorders in children. Although Duchenne muscular dystrophy is by far the commonest and one of the most serious forms, muscular dystrophy is a group of inherited disorders all characterised by variable degrees and distributions of muscle weakness and wasting. *Table 6.12* summarises the main clinical features and genetic features of the most common forms of inherited muscular dystrophies.

Congenital muscular dystrophy   This affects a heterogeneous group of neonates or infants presenting with clinical weakness and hypotonia associated with arthrogryphosis and sometimes respiratory difficulties. Before the molecular genetics were understood the classification relied on the clinical features and the country of origin. Now they can be divided into three main groups according to the biochemical defects that result from the specific gene defects (Emery 2002, Muntoni and Voit 2004). The location of the defects and the associated disease are:

- extracellular matrix protein (merosin-deficient congenital muscular dystrophy, Ullrich congenital muscular dystrophy)
- glycosyltransferases (abnormal *O*-glycosylation of α-dystroglycan) (Walker–Warburg syndrome, muscle–eye–brain disease, Fukuyama congenital muscular dystrophy, congenital muscular dystrophies with mental retardation and pachygyria)
- proteins of the endoplasmic reticulum (rigid-spine syndrome).

Merosin-deficient congenital muscular dystrophy (MDC1A) is the most common. It presents in the first few months of life. At best those affected will only achieve independent standing with limited ambulation, with later development of a scoliosis. As laminin α$_2$ is expressed in the basement membrane of blood vessels in the brain, there are associated white matter changes. Early on there is demyelinating neuropathy, which is first seen in motor nerves, but as the children get older involves sensory nerves as well (Shorer et al 1995).

The next most common is Ullrich congenital muscular dystrophy. This may be autosomal dominant or recessive. Clinical features are variable and include weakness, skin abnormalities, hyperlaxity and contractures. Unlike MDC1A, normal intelligence and respiratory function are present. Presenting in the neonatal period the development of contractures is one of the most prominent features.

Fukuyama congenital muscular dystrophy is particularly frequent in Japan, where it represents the second most common form of muscular dystrophy after Duchenne dystrophy. The classical picture of a child with Fukuyama congenital muscular dystrophy is the combination of severe hypotonia, generalised muscle weakness and wasting, severe brain involvement with developmental delay, seizures and abnormal eye function.

The five conditions that are characterised by disorders of the *O*-glycosylation of α-dystroglycan have involvement of not only skeletal muscle but also the brain and eye. The changes in the brain are collectively classified under the type II lissencephaly spectrum. Eye changes vary and may include cataracts, myopia and disorders of ocular movement.

In the last group, the rigid spine group, axial hypotonia is noted in the first year of life, proceeding to the development of stiffness of the spine due to the development of contractures in the extensor muscles between 3 and 12 years of age.

There are dystrophic changes in the muscles of all affected children. If EMG is asked for, it will show a myopathic picture. No EMG may be asked for if the clinical suspicion is high, in which case muscle biopsy and search for the gene defect may be undertaken first.

Dystrophinopathies   Duchenne muscular dystrophy is the most common form of muscular dystrophy, occurring in up to 1/3,500 males. It is characterised by progressive proximal weakness, muscle hypertrophy, contractures, mild learning difficulties and cardiopathy leading to death by respiratory or cardiac failure by age 15–25 years. Becker muscular dystrophy is a related disorder with a milder clinical course. Both are caused by mutations in the *dystrophin* gene on Xp21.1. Since genetic testing has become routine there is no role for EMG. In a clinically typical presentation a raised creatine kinase should be followed by a search for the appropriate gene marker and, if this is not found, by a muscle biopsy.

Congenital myasthenic syndromes and autoimmune myasthenias

Myasthenia of any type is rare in paediatrics. In the infant the congenital myasthenic syndrome is far more common than transient neonatal myasthenia (see below).

Table 6.11   Channelopathies

Type of channelopathy	Inheritance: gene and chromosomal location	Ion channel abnormality	Main clinical features	Precipitants of attacks	Investigations	Treatment
Inherited channelopathies presenting with myotonia or muscle stiffness						
Myotonia congenita	AR (Becker type) more common, more severe AD (Thomsen) mild form. CLCN1, 7q35	Defective inward chloride current	Onset in early (Thomsen) or late (Becker) childhood. Non-progressive. Generalized and painless muscle stiffness. Lower limbs often affected. Paralysis prominent after a period of rest. Improves with continued activity ('warm-up' phenomenon). Muscle hypertrophy common but wasting and some fixed weakness may also occur; no periodic paralysis. Some distal forearm weakness and atrophy in recessive form, may be confused with myotonic dystrophy. Heterozygous parents of Becker type may show myotonia	Prolonged rest maintaining the same posture. Not influenced by cold	No ECG changes. Repetitive stimulation may show a decremental response. EMG: prominent myotonia, normal motor units. Muscle biopsy: type 2B fibres may be absent. Cause not known. No myopathic changes	Exercise and avoidance of prolonged rest. Mexiletine (a sodium channel blocker) to prevent myotonia. In most cases, no treatment required
Myotonic dystrophy (Charlet et al 2002, Mankodi et al 2002)	Type 1: 19q13.3 Type 2: 3q abnormal splicing of CLC-1 RNA	Defective chloride channel function	Type 1: classical features including muscle wasting, myotonia, subcapsular cataracts, cardiac conduction abnormalities, hearing defects, gonadal atrophy and cognitive deficits. Expansion of an unstable CTG repeat in myotonic dystrophy protein kinase gene (DMPK). Type 2: more proximal pattern (PROMM), other features variable. Expansion of a CCTG tetranucleotide repeats in zinc finger protein 9 (ZNF9) gene	Not episodic	EMG: typical dive-bomber myotonic discharges. Genetic testing	Symptomatic
Potassium aggravated myotonia	AD: sodium channel alpha subunit SCN4A, 17q23-q25	Mildly reduced fast sodium channel inactivation	Includes myotonia fluctuans (mild form), acetazolamide responsive myotonia and myotonia permanens (severe form). Clinical severity depends on degree of channel function disruption. Onset in childhood. Face, especially eyelids, hands and paraspinal muscles mainly affected. Stiffness only, often painful but no weakness and not temperature dependent. Some cases worsened by cold or exercise. In myotonia fluctuans there is marked variability in the severity of muscle stiffness, fluctuating daily. Myotonia is constant in myotonia permanens. A painful form is also reported	Ingestion of K⁺ rich food, fasting, exertion, exposure to cold, infection	ECG changes of hyperkalaemia. EMG: myotonia	Mexiletine or acetazolamide. Monitor during surgery for rigidity and rhabdomyolysis

Condition	Genetics	Molecular defect	Clinical features	Precipitating factors	Investigations	Treatment
Paramyotonia congenita	AD: sodium channel alpha subunit SCN4A, 17q23-q25	Reduced fast sodium channel inactivation	Onset in early childhood. Myotonia most obvious in facial, eyelid, forearm and hand muscles. Muscle stiffness is usually painless, aggravated by cold or exercise (paradoxical myotonia). Further continued cooling or exercise cause attacks of weakness which may last hours. Condition often non-progressive. (There may also be coexisting hyperPP.) No hypertrophy or atrophy of muscles	Cold exposure followed by exercise causes focal paralysis. (When associated with hyperPP there are also other precipitating factors)	Myotonia at rest but prominent fibrillations with cooling; with further cooling fibrillations disappear. CMAP amplitude is reduced after cooling of the limb. Serum K^+ may be low, normal or raised (raised in the presence of coexisting hyperPP). CK slightly elevated. Muscle biopsy myopathic	Mexiletine prevents myotonia as well as cold-provoked paralysis. Acetazolamide often provokes weakness, but some patients may benefit
Malignant hyperthermia (MH) and central core disease (CCD) (Quane et al 1993, Greenberg 1999)	MH: commonly AD, RYR1 or CACNL1AS calcium channel. Others include CACNL2A and SCN4A. CCD: RYR1	MH: defective ryanodine receptor causing increased Ca^{2+} release from sarcoplastic reticulum. CCD: defective Ca^{2+} release	MH: attacks of skeletal muscle rigidity, pain and weakness and rhabdomolysis associated with fever, sympathetic hyperactivity and hypermetabolism. Myoglobinurea may cause renal shutdown. CCD: caused by different mutations of the same gene RYR1. Presents during infancy with hypotonia, proximal weakness and skeletal abnormalities. Both dominant and recessive forms exist. Susceptible to MH	MH: inhalational anaesthetic agents and succinylcholine. Less severe episodes by alcohol, exercise, neuroleptic drugs and infections	MH: high CK, acidosis and myoglobinurea during an episode. Halothane contracture test. Gene testing. CCD: typical muscle histochemistry with amorphous central cores in type 1 fibres	MH: during an episode dantrolene, cooling, bicarbonate and maintenance of urine output. Discontinuation of anaesthesia, avoidance of Ca^{2+}, Ca^{2+} antagonists and beta blockers
Channelopathies presenting with periodic paralysis						
Hyperkalaemic periodic paralysis with myotonia (hyperPP + myotonia)	AD, sodium alpha subunit SCN4A, 17q23-q25	Reduced inactivation fast and slow sodium channels	Symptoms first noticed during childhood, episodic flaccid weakness, rapid onset, affecting mainly proximal limb muscles, may be focal when followed by specific exercise or generalised following sleep. Respiratory and bulbar muscles spared. Attacks begin within minutes, usually in the morning, last 20 min to 2 h, and can be aborted or delayed by mild exercise. Weakness milder than with hypoPP. No residual weakness. Reflexes reduced or absent. Occasional myotonia between paralytic attacks affecting facial, eyelid, tongue and hand muscles. Frequency of attacks lessens with time but repeated attacks over years may lead to persistent proximal muscle weakness	Rest after exercise, fasting, exposure to cold, sleep, anaesthesia, K^+ rich food	ECG changes of hyperkalaemia, CK increased during attacks. EMG: fibrillations and myotonia often precede an attack, affected muscles inexcitable or less excitable during paralysis. During an attack CMAP amplitude is reduced. Muscle biopsy myopathic. Provocative tests using potassium potentially dangerous	During an attack hydrochloro-thiazide or beta adrenergic inhalation. Mexiletine to prevent further attacks of myotonia. Also avoidance of strenuous exercise, prolonged fasting and exposure to cold. CHO rich and low potassium diet beneficial

Continued

Table 6.11 *Continued*

Type of channelopathy	Inheritance: gene and chromosomal location	Ion channel abnormality	Main clinical features	Precipitants of attacks	Investigations	Treatment
Hypokalaemic periodic paralysis (Hypo)	AD, calcium alpha subunit *CACNL1AS*, 1q31–q32, poorly expressed in females. M:F = 3:1, commonest type of HypoPP	Calcium channel inactive, leading to defect in control of resting membrane potential	Onset in adolescence (or 5–35 years); M:F = 3:1; may spontaneously improve thereafter. Episodic proximal weakness of limbs, lasts hours or days, hypotonic and often areflexic limbs during attack, rapid recovery. Weakness may be severe. Frequency: several/week to one every few months. Hypokalaemia during an attack. Residual, often slowly progressive, proximal leg weakness present, occasionally permanent. Eyelid myotonia may be present	CHO or salt rich food, after sleep, rest following exercise, alcohol, emotional changes, etc. Light physical activity may prevent an attack	Hypokalemia during, but not between, attacks. CK increased during attack. EMG: no myotonia, muscles inexcitable during an attack. CMAP amplitude reduced during attack. Exercise test to demonstrate defective repolarisation. Biopsy myopathic, with vacuoles and tubular aggregates. Provocative tests: oral glucose with or without insulin	Oral potassium solution for an attack. Low CHO and low Na diet, acetazolamide and other K$^+$ sparing drugs for prevention
Thyrotoxic periodic paralysis	Sporadic or autosomal dominant, males more often affected	Not known	Affects particularly patients of Asian ethnic origin. Common in summer months. Onset 18–40 years. M:F = 20:1. Hypokalaemia during paralysis. Weakness lasts hours or days and involves proximal > distal, legs > arms. Respiratory and bulbar muscles may also be affected	CHO rich food, rest after exercise, muscle cooling, thyroxine ingestion	Hypokalemia during attack. CMAP reduced during attack. Exercise test to demonstrate defective repolarisation. Myopathy uncommon. No myotonia	As for HypoPP, careful potassium replacement and correction of thyroid function. Not acetazolamide
Other channelopathies causing periodic paralysis (PP)						
Hypokalaemic periodic paralysis (HypoPP2), AD (Sternberg et al 2001)	AD, sodium alpha subunit *SCN4A*, 17q23–q25	? Reduced number of excitable Na$^+$ channels	Most features similar to calcium channel HypoPP. Rare. Single family reported. Onset 9 years. Muscle pain common after attacks. Permanent muscle weakness in some cases	Rest following exercise, sleep	ECG-changes of hypokalaemia. EMG: usually no myotonia. Muscle biopsy: myopathic with tubular aggregates	Not acetazolamide, often deleterious

HypoPP3 (Abbott et al 2001)	Potassium channel, KCNE3, 11q13	Not known	Rare. Paralysis lasts hours to days. No myotonia, not precipitated by high CHO meals. Myopathy with tubular aggregates. Both hypo- and hyper-kalaemic forms reported	Strenuous exercise	Serum K^+ may be high, low or normal. ECG changes	CHO loading
Anderson syndrome (Plaster et al 2001, Tristani-Firouzi et al 2001)	AD, potassium channel KCNJ2 Kir 2.1, 17q23	Reduced inward rectifying K^+ current	Features other than PP include skeletal and facial dysmorphism, cardiac conduction abnormalities causing syncope and sudden death, prolonged QT interval. Onset 2–18 years. Paralytic episodes last 1–36 h	Exercise and K^+	Serum K^+ may be high, low or normal. ECG changes	Amiodorone. or acetazolamide
Distal renal tubular acidosis (Nicoletta and Schwartz 2004)	AS or AR, SLC4A1, 17q21	Defective chloride–bicarbonate ion exchange	Endemic in north-east Thailand. Paralysis lasts hours. Other features include osteomalacia and pathological fractures. No myotonia	Not known	Hypokalaemia and hyperchloraemic acidosis in the presence of alkaline urine	K^+ and bicarbonate. Not acetazolamide (worsens acidosis)
X-linked form of periodic paralysis (Ryan et al 1999)	Xp 22.3, recessive	Not known	Onset in infancy and childhood, respiratory and bulbar muscles may be involved. Severity variable. Episodic prolonged weakness lasting days to months. Myalgia and cramps may be present	Not known	EM: dilatation and proliferation of sarcoplasmic reticulum. Basal ganglia calcification	Not known

CHO, carbohydrate; hyperPP, hyperkalaemic periodic paralysis.

Table 6.12 Inherited muscular dystrophies

Disorder	Gene (protein)	Age of onset	Distribution of weakness	EMG	Creatine kinase	Course	Additional features
Congenital							
Fukuyama dystrophy (AR)	9q31-33 (fukutin)	Infancy	Face and limbs	Myopathic	++	Progressive	Seizures (50%); contractures; ocular involvement; severe learning difficulties
Ullrich 1 (AD) Ullrich 2 (AD or AR) Ullrich 3 (AD or AR)	21q22.3 (collagen su α1) 21q22.3 (collagen su α2) 2q37 (collagen su α3)	Neonatal or prenatal	Diffuse: distal > proximal; neck flexors; variable severity	Myopathic	Normal/+	Progressive	Hypotonia; (congenital) contractures; hypoventilation in first decade
Merosin deficiency (AR)	6q (laminin α2 chain)	Neonatal	Proximal + distal; face	Myopathic, mild neuropathy in some	++	Progressive	Contractures
Congenital muscular dystrophy with muscle hypertrophy	19q13 (fukitin related protein)	Neonatal	Diffuse: limbs, neck, trunk, face; hypotonia	Myopathic	+++	–	Contractures; calf hypertrophy and wasting of other muscles; scoliosis
Rigid spine syndrome (AR)	1p (selenoprotein N1)	Neonatal to first decade	Variable distribution and severity	Myopathic	Normal	Stable or slow decline	Rigid spine onset at 3–7 years; progressive scoliosis; other contractures; progressive respiratory failure
Walker–Warburg syndrome	9q34 (O-mannosyl-transferase 1) and others	Prenatal	Hypotonia	Myopathic. VEP: giant	Can be +++	Death in infancy	'Cobblestone' cortex; seizures; anomalies in brain, muscle and eye
Muscle–eye–brain disease (AR)	1p (glycosyltransferase)	Neonatal	Hypotonia	Myopathic. VEP: giant	+	Death in childhood	Anomalies in brain, muscle and eye

Dystrophinopathies

Disorder	Gene/locus	Age of onset	Distribution	EMG	CK	Course	Other features
Duchenne (XR)	Xp21 frameshift (dystrophin)	3–5 years	Symmetric, proximal > distal	Myopathic	+++	Failure to walk by 9–15 years	Calf hypertrophy; contractures; scoliosis; mild learning difficulties; cardiomyopathy
Becker (XR)	Xp21 in-frame (dystrophin)	> 7 years	Symmetric, proximal > distal	Myopathic	+++	Failure to walk by 16–80 years	Calf hypertrophy; contractures; scoliosis; cardiomyopathy

Other muscular dystrophies in children

Disorder	Gene/locus	Age of onset	Distribution	EMG	CK	Course	Other features
Facioscapulohumeral (AD)	4q 35	First to fifth decade	Face, scapular; often asymmetric; later, foot extensors and pelvic girdle	Myopathic	Normal/++	Slowly progressive, worse with younger onset	Retinal vascular disease; sensorineural hearing loss
Emery–Dreifuss (EMD1: XR. EMD 2: AD/AR)	Xq28 (emerin), 1q (lamin)	Neonatal to third decade	Proximal ± distal leg (humeroperoneal)	Myopathic; occasional fibrillations and high amplitude potentials	+/++	Slowly progressive	Contractures before weakness (elbows, posterior neck in extension); cardiac disease (second or third decade); wasting of biceps and triceps, later calf
Distal (AR)	2p (dysferlin)	First to fourth decade	Mild, distal (especially gastrocnemius and biceps)	Myopathic	+++	Slowly progressive	–
Limb-girdle (AD)	Several genes described	Variable	Proximal limb girdle	Myopathic, some have myotonia, fibrillations	Normal/++	Slowly progressive	Myalgia and stiffness in some patients; occasionally neuropathy
Limb-girdle (AR)	Several genes described (2A 15q Calpain-3 most common)	First to fourth decade	Symmetrical; proximal limb girdle and milder distally	–	++/+++	Varies, more rapid with earlier onset	Quadriceps may be selectively spared; contractures (toe walking may be presenting sign); mild mental retardation in some

AD, autosomal dominant; AR, autosomal recessive; XR, X-linked recessive.

+, + mildly elevated; ++, elevated to high; +++, very high.

Congenital myasthenic syndromes Patients with congenital myasthenic syndrome or autoimmune myasthenia can present at varying intervals from birth to late childhood and adolescence depending on the severity and type of neuromuscular transmission defect. Symptoms include ptosis, varying degrees of oculobulbar and limb weakness, and fatigue.

The congenital myasthenic syndromes are currently divided according to the location of the abnormality in relation to the synapse. The congenital myasthenic syndromes are summarised in *Table 6.13* together with their neurophysiological findings.

Numerically, the acetylcholine receptor (AChR) defects are the most common in the Mayo clinical experience (Engel et al 2003). Most of these are due to gene mutations in the epsilon subunit of the receptor. Survival is possible by the low-level expression of the residual gamma subunits, which are found in fetal AChRs. In normal neonates the fetal AChR has been replaced by the adult form where the gamma unit is replaced by an epsilon unit. The presynaptic congenital myasthenic syndromes are associated with defects that curtail the evoked release of acetylcholine (ACh) quanta or ACh resynthesis, the latter being due to mutations in choline acetyltransferase, and are found in the congenital myasthenic syndrome with episodic apnoea (also called 'familial infantile myasthenia'). In end-plate acetylcholine esterase (AChE) deficiency a mutation in the collagenic tail subunit (ColQ) prevents the enzyme being localised appropriately in relation to the end-plate. Slow-channel and fast-channel syndromes are caused by mutations that increase or decrease the synaptic response to ACh. The second most common syndrome is end-plate AChR deficiency, which is caused by mutations in rapsyn. This molecule is critical in concentrating AChR in the postsynaptic membrane. Plectin has a similar action but is much less common.

The neurophysiological testing of the neuromuscular junction has been discussed in Chapter 4 (see pp. 160–162 and 165). In summary, stimulated single fibre EMG of orbicularis oculi is preferred as the most sensitive test and one can be performed successfully in most subjects under local rather than general anaesthesia. Repetitive nerve stimulation can still be valuable and is used in our practice to screen all infants under 1 year of age. Mostly the routine nerve conduction studies are non-contributory except when a repetitive CMAP is evoked, which is only seen in slow-channel syndrome and end-plate AChE deficiency.

Certain of the conditions have unusual responses to repetitive nerve stimulation that are worth highlighting.

CMS with episodic apnoea (familial infantile myasthenia) can present in infancy or early childhood with episodic respiratory failure. In some cases repetitive nerve stimulation demonstrates a similar picture to acquired myasthenia gravis, with decrement at low frequencies with improvement after exercise or tetanic stimulation

and a subsequent post-tetanic increase in decrement. In others short-duration, repetitive stimulation is usually normal and detection requires prolonged exercise for several minutes or a continuous 10-Hz train for 5 min before low-frequency (2–3 Hz) stimulation demonstrates significant decrement. In normal infants under 3 months old, stimulation over 3 Hz will produce an apparent abnormality due to the immaturity of the neuromuscular junction, and thus should be used with caution.

Congenital Lambert–Eaton myasthenic syndrome has been described in two cases of severe hypotonia with areflexia in the newborn (Albers et al 1981, Bady et al 1987). Nerve conduction studies demonstrate a very low CMAP in the rested limb or facial muscle, but > 700% facilitation after contraction or repetitive stimulation at 20–50 Hz. The exact physiological presynaptic defect has not been demonstrated.

In addition to the clinical neurophysiological studies most patients will require more detailed in vitro neurophysiological testing and ultrastructural pathological studies for definitive diagnosis.

Acquired disorders of neuromuscular transmission Clinical features and neurophysiological findings in acquired disorders of neuromuscular transmission in childhood are summarised in *Table 6.14*.

Transient neonatal myasthenia can occur in infants born to mothers with acquired myasthenia gravis. In this condition maternal AChR antibodies transferred across the placenta during gestation induce weakness and hypotonia shortly after birth (Fenichal 1978). The diagnosis is usually made by a combination of findings from clinical examination, the edrophonium test and elevated serum titres of AChR antibodies. It is rare for the neurophysiologist to be referred such cases.

Autoimmune juvenile myasthenia gravis is rare in early childhood, but can occur in older children. It is more common in China and Japan. Clinical features are similar to those in adult patients, but can present with rapid onset and progression leading to ophthalmoplegia and marked ptosis. Familial immune myasthenia gravis is a rare autosomal recessive disorder, which has an earlier onset than non-familial autoimmune myasthenia gravis between 2 and 20 years. More commonly there are no anti-AChR antibodies. In adults it has been found that some AChR antibody negative patients have antibodies to the muscle-specific tyrosine kinase (MuSK) (Vincent et al 2004). These antibodies are directed against the extracellular domain of MuSK and inhibit agrin-induced AChR clustering in muscle myotubes. Often the patients are female with bulbar weakness. We have seen children who have a similar presentation, but they are very rare.

The results of nerve conduction studies and EMG are similar to those found in adults (see *Table 6.14*). Repetitive nerve stimulation of affected muscles will show a progressive decline in CMAP amplitudes of more than 10% with the first four or five stimuli (see p. 161). Exer-

Table 6.13 Congenital myasthenic syndromes: motor responses and responses to repetitive stimulation

Syndrome	Inheritance: gene (gene produce)	CMAP	Low-frequency stimulation (2–3 Hz)	High frequency stimulation (20–50 Hz)	Comment
Presynaptic defects					
CMS with episodic apnoea	AR: 10q11 (ChAT)	Normal at rest, reduced with 10 Hz stimulation for 5 min	Decrement in weak muscles	Decrement	Decrement increases after exercise. Corrected by rest and AChE inhibitors
Episodic ataxia 2	AR: 19p13 (CaCnα1a)	–	–	–	EMG and nerve conduction studies often normal
Congenital Lambert–Eaton syndrome	AR	Very low amplitude	No change or may decrement	Increment	Rare
Synaptic defects					
Congenital AChE deficiency	AR: 3p25 (ColQ)	Repetitive CMAP to single stimuli	Decrement	Marked decrement	No response to AChE inhibitors
Postsynaptic defects					
Slow-channel CMS	All AD: ε. Also AR: 2q24 (AChR α subunit) 17p13 (AChR β subunit) 2q33 (AChR δ subunit) 17p13 (AChR ε subunit)	Repetitive CMAP to single stimuli, increases in size after AChE inhibitors	Decrement	Decrement	Multiple mutations identified with different functional effects, all including slow or prolonged AChR ion channel closure
Fast-channel CMS	All AR: 2q24 (AChR α subunit) 2q33 (AChR δ subunit) 17p13 (AChR ε subunit)	Normal	Decrement	Decrement	Decreased affinity of AChR for ACh. Brief or reduced channel activation due to several mechanisms
Rapsyn deficiency	AR: 11p11 (rapsyn)	Normal	Decrement in 50–80%	Decrement	Decrement after exercise
Plectin deficiency	AR: 8q24 (plectin)	–	Decrement	–	EMG: myopathic
AChR deficiency	All AR: 2q24 (AChR α subunit) 17p13.1 (AChR β subunit) 2q33 (AChR δ subunit) 17p13 (AChR ε subunit)	Normal	Decrement	Decrement	Improves with AChE inhibitors. ε most common. Onset and clinical features variable. Response to AChE inhibitors

ACh, acetylcholine; AChE, acetylcholine esterase; AChR, acetylcholine receptor; CaCnα1a, calcium channel α_{1a}; ChAT, choline acetyl transferase; CMS, congenital myasthenic syndrome.

345

Table 6.14   Acquired disorders of neuromuscular transmission in childhood

Disorder	Pathophysiology	Clinical features	Neurophysiology
Myasthenia gravis	Antibodies to AChR or MuSK	Weakness, fatigability, ptosis, diplopia	Normal CMAP. RNS: decrement at 2–3 Hz (> 10%). SFEMG: increased jitter and blocking
Neonatal myasthenia gravis	Maternal antibodies, usually AChR	Transitory weakness and respiratory depression for 1–4 weeks	Normal CMAP. RNS: decrement at 2–3 Hz. SFEMG: increased jitter and blocking
Lambert–Eaton myasthenic syndrome	IgG antibodies to nerve terminal voltage-gated calcium channels	Generalised or proximal weakness. Dry mouth, impotence and/or other signs of autonomic dysfunction	RNS: decremental response at low rates, incremental response at high rates. Facilitation with exercise
Botulism	Botulinum toxin blocking cholinergic synapses	Constipation, weakness, bulbar involvement. Infants: weak cry, feeding difficulties, weakness, respiratory failure	Low CMAP amplitude, increases after rapid RNS or sustained exercise. RNS: little decrement at low rates, increment at high rates. EMG: fibrillation potentials

AChR, acetylcholine receptor; IgG, immunoglobulin G; MuSK, muscle-specific tyrosine kinase; RNS, repetitive nerve stimulation; SFEMG, single fibre EMG.

cising muscle before testing will increase the decremental response (postexercise exhaustion). SFEMG is the most sensitive test for generalised and ocular myasthenia gravis and usually shows increased jitter when the test site includes facial muscles.

Lambert–Eaton myasthenic syndrome (LEMS) is extremely rare in children, but has been described with or without associated neoplasm. Nerve conduction studies typically show small CMAPs and a marked (> 100%) increment after rapid (50 Hz) repetitive nerve stimulation or sustained muscle contraction for 10–15 s (see *Table 6.14*).

Botulism is caused by botulinum toxin produced by *Clostridium botulinum*, either due to ingestion of the toxin (food-borne botulism), wound botulism or ingestion of spores from soil or honey products, which germinate into bacteria, colonise the bowel and synthesise toxin (infant botulism). As the toxin is absorbed, it irreversibly binds to AChR at neuromuscular junctions. Constipation is often the first symptom of infant botulism, and can precede weakness by several weeks. Other symptoms include weak cry, poor feeding, acute bulbar dysfunction and respiratory failure. Botulinum toxin has a presynaptic action of reducing the number of quanta of ACh released by a depolarising potential. The electrophysiological findings therefore reflect blocked neuromuscular transmission with preserved nerve conduction (see *Table 6.14*). Sensory conduction studies are normal. CMAPs can be markedly diminished in amplitude, with normal or near-normal motor conduction velocities (Cherington 1982). With slow rates of repetitive nerve stimulation (2 Hz) a decremental response may be seen, whereas with rapid rates (50 Hz) an increase in the amplitude of the responses may occur. Single fibre EMG shows increased jitter and

blocking, which may improve with continued contraction of the muscle or with higher rates of stimulation. EMG sampling reveals small, short-duration motor units; spontaneous fibrillations are a variable finding (Cherington 1982).

REFERENCES

Aarts JHP, Binnie CD, Smit AM, et al 1984 Selective impairment during focal and generalized epileptiform EEG activity. Brain 107: 293–308.

Abbott GW, Butler MH, Bendahhou S, et al 2001 MiRP2 forms potassium channels in skeletal muscle with Kv 3.4 and is associated with periodic paralysis. Cell 104: 217–231.

Advances in Neurology 2005 Myoclonic epilepsies in childhood. Adv Neurol 95: 1–307.

Aicardi J 1993 The inherited leukodystrophies: a clinical overview. J Inherit Metab Dis 16: 733–743.

Aicardi J 1994 Infantile spasms and related conditions. In: *Epilepsy in Children*, 2nd edn (ed J Aicardi). Raven Press, New York, pp 18–43.

Aicardi J 2000 Atypical semiology of Rolandic epilepsy in some related syndromes. Epileptic Disord 2(Suppl 1): S5–S9.

Aicardi J 2005 Aicardi syndrome. Brain Dev 27:164–171.

Aicardi J, Castelein P 1979 Infantile neuroaxonal dystrophy. Brain 102: 727–748.

Aicardi J, Chevrie JJ 1982 Atypical benign partial epilepsy of childhood. Dev Med Child Neurol 24: 281–292.

Aicardi J, Goutiéres F 1978 Encephalopathie myoclonique neonatale. Rev Electroencéphalogr Néurophysiol Clin 8: 99–101.

Aicardi J, Levy GA 1992 Clinical and electroencephalographic symptomatology of the 'genuine' Lennox–Gastaut syndrome and its differentiation from other forms of epilepsy of early childhood. Epilepsy Res (Suppl 6): 185–193.

Aird RB, Cohen P 1950 Electroencephalography in cerebral palsy. J Pediatr 37: 448–454.

Albers JW, Faulkner JA, Darovini-Zis K, et al 1981 Abnormal neuromuscular transmission in an infantile myasthenic syndrome. Ann Neurol 16: 28–34.

Alster J, Pratt H, Feinsod M 1993 Density spectral array, evoked potentials, and temperature rhythms in the evaluation and prognosis of the comatose patient. Brain Inj 1: 191–208.

Ambrosetto G 1993 Treatable epilepsy and unilateral opercular neuronal migration disorder. Epilepsia 34: 604–608.

Amigoni A, Pettenazzo A, Biban P, et al 2005 Neurologic outcome in children after extracorporeal membrane oxygenation: prognostic value of diagnostic tests. Pediatr Neurol 32: 173–179.

Andermann E, Andermann F, Carpenter S, et al 1986 Action myoclonus–renal failure syndrome: a previously unrecognized neurological disorder unmasked by advances in nephrology. Adv Neurol 43: 87–103.

Apkarian P 1992 A practical approach to albino diagnosis. VEP misrouting across the age span. Paediatr Genet 13: 77–78.

Appleton R, Riordan A, Tedman B, et al 1994 Congenital peripheral neuropathy presenting as apnoea and respiratory insufficiency. Dev Med Child Neurol 36: 547–553.

Aubourg P, Scotto J, Rocchiccioli F, et al 1986 Neonatal adrenoleukodystrophy. J Neurol Neurosurg Psychiatry 49: 77–86.

Bader GG, Witt-Engerstrom I, Hagberg B 1989a Neurophysiological findings in the Rett syndrome, I: EMG, conduction velocity, EEG and somatosensory evoked potential studies. Brain Dev 11: 102–109.

Bader GG, Witt-Engerstrom I, Hagberg B 1989b Neurophysiological findings in the Rett syndrome, II: Visual and auditory brainstem, middle and late evoked responses. Brain Dev 11: 110–114.

Bady B, Chauplannaz G, Carrier H 1987 Congenital Lambert–Eaton myasthenic syndrome. J Neurol Neurosurg Psychiatry 50: 476–478.

Barkovich AJ, Guerrini R, Battaglia G, et al 1994 Band heterotopia: correlation of outcome with magnetic resonance imaging parameters. Ann Neurol 36: 609–617.

Barry RJ, Johnstone SJ, Clarke AJ 2003a A review of electrophysiology in attention-deficit/hyperactivity disorder. I. Qualitative and quantitative electroencephalography. Clin Neurophysiol 114: 171–183.

Barry RJ, Johnstone SJ, Clarke AJ 2003b A review of electrophysiology in attention-deficit/hyperactivity disorder: II. Event related potentials. Clin Neurophysiol 114: 184–198.

Barth PG 1987 Disorders of neuronal migration. Can J Neurol Sci 14: 1–16.

Baulac S, Gourfinkel-An I, Nabbout R, et al 2004 Fever, genes and epilepsy. Lancet Neurol 3: 421–430.

Baumgartner MR, Saudubray JM 2002 Peroxisomal disorders. Semin Neonatol 7: 85–94.

Bax MCO 1964 Terminology and classification of cerebral palsy. Dev Med Child Neurol 6: 295–307.

Beaumanoir A, Nahory A 1983 Benign partial epilepsies: 11 cases of frontal partial epilepsy with favorable prognosis [in French]. Rev Electroencephalogr Neurophysiol Clin 13: 207–211.

Becker PE 1965 Genetik der Myopathien. Verh Dtsch Ges Inn Med 71: 171–183.

Berg BO, Rosenberg GA, Asbury AK 1972 Giant axonal neuropathy. Pediatrics 49: 894–899.

Berkovic SF, Steinlein OK 1999 Genetics of partial epilepsies. Adv Neurol 79: 375–381.

Berkovic SF, So SK, Andermann F 1991 Progressive myoclonus epilepsies: clinical and neurophysiological diagnosis. J Clin Neurophysiol 8: 261–274.

Binnie CD 1996 Differential diagnosis of eyelid myoclonia with absence and self-induction by eye-closure. In: Eyelid Myoclonia with Absence (eds JS Duncan, CP Panayatopoulos). John Libby, London, pp 98–102.

Binnie CD 2003 Cognitive impairment during epileptiform discharges: is it ever justifiable to treat the EEG? Lancet Neurol 2: 725–730.

Binnie CD, Channon S, Marston D 1990 Learning disabilities in epilepsy: neurophysiological aspects. Epilepsia 31(Suppl 4): S2–S8.

Binnie CD, de Silva M, Hurst A 1992 Rolandic spikes and cognitive function. Epilepsy Res (Suppl 6): 71–73.

Binnie CD, Cooper R, Mauguière F, et al (eds) 2003 Clinical Neurophysiology, Vol. 2, EEG, Paediatric Neurophysiology, Special Techniques and Applications. Elsevier Science, Amsterdam.

Binnie CD, Cooper R, Maugière F, et al 2004 Clinical Neurophysiology, EMG, Nerve Conduction and Evoked Potentials, Vol. 1. Elsevier, Amsterdam.

Bonifazi E, Gullotta F, Vallo L, et al 2006 Use of RNA fluorescence in situ hybridization in the prenatal molecular diagnosis of myotonic dystrophy type I. Clin Chem 52: 319–322.

Boor R, Fricke G, Brühl K, et al 1999 Abnormal subcortical somatosensory evoked potentials indicate high cervical myelopathy in achondroplasia. Eur J Pediatr 158: 662–667.

Boor R, Miebach E, Brühl K, et al 2000 Abnormal somatosensory evoked potentials indicate compressive cervical myelopathy in mucopolysaccharidoses. Neuropediatrics 31: 122–127.

Bosch EP, Hart MN 1978 Late adult onset metachromatic leukodystrophy dementia and polyneuropathy in a 63-year-old man. Arch Neurol 35: 475–477.

Bourgeois M, Sainte-Rose C, Cinalli G, et al 1999 Epilepsy in children with shunted hydrocephalus. J Neurosurg 90: 274–281.

Boyd SG, Harden A 1991 The clinical neurophysiology of the central nervous system. In: Paediatric Neurology, 2nd edn (ed EM Brett). Churchill Livingstone, Edinburgh, pp 717–795.

Boyd SG, Harden A, Egger J, et al 1986 Progressive neuronal degeneration of childhood with liver disease ('Alpers' disease'): characteristic neurophysiological findings. Neuropediatrics 17: 75–80.

Boyd SG, Harden A, Fatten MA 1988 The EEG in early diagnosis of the Angelman (happy puppet) syndrome. Eur J Pediatr 147: 508–513.

Brahe C 2000 Copies of the survival motor neuron gene in spinal muscular atrophy: the more, the better. Neuromusc Disord 10: 274–275.

Bricolo AP, Turella GS 1990 Electrophysiology of head injury. In: Head Injury. Handbook of Clinical Neurology, Vol. 13(57) (ed R Braakman). Elsevier Science, Amsterdam, pp 181–206.

Bridgers SL 1987 Epileptiform abnormalities discovered on electroencephalographic screening of psychiatric inpatients. Arch Neurol 44: 312–316.

British Paediatric Association 1991 Diagnosis of Brain Stem Death in Infants and Children. A Working Party Report. British Paediatric Association, London, pp 1–6.

Brown JC, Johns RJ 1967 Nerve conduction in familial dysautonomia Riley–Day syndrome. J Am Med Assoc 201: 200–203.

Buchwald JS, Erwin R, van Lancker D, et al 1992 Midlatency auditory evoked responses: P1 abnormalities in adult autistic subjects. Electroencephalogr Clin Neurophysiol 84: 164–171.

Burley D 1961 Neuropathy after thalidomide ('Distaval'). BMJ 2: 1286–1287.

Bye AME, Kendall B, Wilson J 1985 Multiple sclerosis in childhood: a new look. Dev Med Child Neurol 27: 215–222.

Caffey J 1972 On the theory and practice of shaking infants. Its potential residual effects of permanent brain damage and mental retardation. Am J Dis Child 124: 161–169.

Campuzano V, Montermini L, Molto MD, et al 1996 Friedreich's ataxia: autosomal recessive disease caused by an intronic GAA triplet repeat expansion. Science 271: 1423–1427.

Canafoglia L, Franceschetti S, Antozzi C, et al 2001 Epileptic phenotypes associated with mitochondrial disorders. Neurology 56: 1340–1346.

Cantwell DP 1980 Drug and medical intervention. In: *Handbook of Minimal Brain Dysfunction. A Critical Review* (eds HE Rie, ED Rie). Wiley, New York.

Carels G 1960 Etude physiopathologique d'un syndrome myoclonique chez deux infants atteints d'une forme infantile tardive de l'idiotie amaurotique. Acta Neurol Psychiatr Belg 60: 435–464.

Carter BG, Taylor A, Butt W 1999 Severe brain injury in children: long-term outcome and its prediction using somatosensory evoked potentials (SEPs). Intensive Care Med 25: 722–728.

Cavazzuti GB, Cappella L, Nalin A 1980 Longitudinal study of epileptiform EEG patterns in normal children. Epilepsia 21: 43–55.

Chaptal J, Passouant P, Jean R, et al 1954 Signes électroencéphalographiques des séquelles de méningite tuberculeuse. Rev Neurol 90: 830–834.

Charlet B, Singh GN, Philips AV, et al 2002 Loss of the muscle-specific chloride channelin type 1 myotonic dystrophy due to misreguled alternative splicing. Mol Cell 10: 45–53.

Cherington M 1982 Electrophysiologic methods as an aid in diagnosis of botulism: a review. Muscle Nerve 9(Suppl): 28–29.

Chevrie JJ, Aicardi J 1986 The Aicardi syndrome. In: *Recent Advances in Epilepsy*, Vol. 3 (eds TA Pedley, BS Meldrum). Churchill Livingstone, Edinburgh, pp 189–210.

Chez MG, Chang M, Krasne V, et al 2006 Frequency of epileptiform EEG abnormalities in a sequential screening of autistic patients with no known clinical epilepsy from 1996 to 2005. Epilepsy Behav 8: 267–271.

Clark JR, Miller RG, Vidgoff JM 1979 Juvenile-onset leukodystrophy: biochemical and electrophysiological study. Neurology 29: 346–53.

Clemensen OJ, Olsen PZ, Andersen KE 1984 Thalidomide neurotoxicity. Arch Dermatol 120: 338–341.

Clow CL, Reade TM, Scriver CR 1981 Outcome of early and long-term management of classical maple syrup urine disease. Pediatrics 68: 856–862.

Cobb WA 1966 The periodic events of subacute sclerosing leucoencephalitis. Electroencephalogr Clin Neurophysiol 21: 278–294.

Cobb WA, Hill D 1950 Electroencephalogram in subacute progressive encephalitis. Brain 73: 392–397.

Cockerell OC, Rothwell J, Thompson PD, et al 1996 Clinical and physiological features of epilepsia partialis continua. Cases ascertained in the UK. Brain 119: 393–407.

Commission on Classification and Terminology of the International League Against Epilepsy 1981 Proposal for revised clinical and electroencephalo-graphic classification of epileptic seizures. Epilepsia 22: 489–501.

Commission on Classification and Terminology of the International League Against Epilepsy 1985 Proposal for classification of epilepsies and epileptic syndromes. Epilepsia 26: 268–278.

Commission on Classification and Terminology of the International League Against Epilepsy 1989 Proposal for revised classification of epilepsies and epileptic syndromes. Epilepsia 30: 389–399.

Crawley J, Smith S, Kirkham F, et al 1996 Seizures and status epilepticus in childhood cerebral malaria. Q J Med 89: 591–597.

Crawley J, Smith S, Muthinji P, et al 2001 Electroencephalographic and clinical features of cerebral malaria. Arch Dis Child 84: 247–253.

Crisp AH, Fenton GW, Scotton L 1968 A controlled study of the EEG in anorexia nervosa. Br J Psychiatry 114: 1149–1160.

Cross JH 2005 Neurocutaneous syndromes and epilepsy-issues in diagnosis and management. Epilepsia 46(Suppl): 17–23.

Curatolo P, Seri S, Verdecchia M, et al 2001 Infantile spasms in tuberous sclerosis complex. Brain Dev 23: 502–507.

Curatolo P, Bombardieri R, Verdecchia M, et al 2005 Intractable seizures in tuberous sclerosis complex: from molecular pathogenesis to the rationale for treatment. J Child Neurol 20: 318–325.

Dahl RE, Ryan ND, Birmaher B, et al 1991 Electroencephalographic sleep measures in prepubertal depression. Psychiatry Res 38: 201–214.

Dalla Bernardina B, Tassinari CA, Dravet C, et al 1978 Benign focal epilepsy and 'electrical status epilepticus' during sleep. Rev Electroencéphalogr Neurophysiol Clin 8: 350–353.

Dalla Bernardina B, Sgro V, Fontana E, et al 1992 Idiopathic partial epilepsies in children. In: *Epileptic Syndromes in Infancy, Childhood and Adolescence* (eds J Roger, M Bureau, C Dravet, et al). John Libbey, London, pp 173–188.

Dan B, Boyd SG 2003 Angelman syndrome reviewed from a neurophysiological perspective. The UBE3A-GABRB3 hypothesis. Neuropediatrics 34: 169–176.

Dawson GD 1947 Investigations on a patient subject to myoclonic seizures after sensory stimulation. J Neurol Neurosurg Psychiatry 10: 141–162.

Delanoe C, Sebire G, Landrieu P, et al 1998 Acute inflammatory demyelinating polyradiculopathy in children: clinical and electrodiagnostic studies. Ann Neurol 44: 350–356.

Delgado-Escueta AV 1979 Epileptogenic paroxysms: modern approaches and clinical correlations. Neurology 29: 1014–1022.

Delgado-Escueta AV, Enrile-Bacsal F 1984 Juvenile myoclonic epilepsy of Janz. Neurology 34: 285–294.

De Marco P, Negrin P 1973 Parietal focal spike evoked by contralateral tactile somatotopic stimulations in four non-epileptic subjects. Electroencephalogr Clin Neurophysiol 24: 308–312.

De Marco P, Tassinari CA 1981 Extreme somatosensory evoked potential (ESEP): an EEG sign forecasting a possible occurrence of seizures in children. Epilepsia 22: 569–575.

De Meirleir LJ, Taylor MJ, Logan WJ 1988 Multimodal evoked potential studies in leukodystrophies of children. Can J Neurol Sci 15: 26–31.

De Michele G, Di Maio L, Filla A, et al 1996 Childhood onset of Friedreich ataxia: a clinical and genetic study of 36 cases. Neuropediatrics 27: 3–7.

Dening TR, Berrios GE, Walsh JM 1988 Wilson's disease and epilepsy. Brain 111: 1139–1155.

Deonna TW 1991 Acquired epileptiform aphasia in children (Landau–Kleffner syndrome). J Clin Neurophysiol 8: 288–298.

Deonna T, Ziegler AL, Despland PA 1986 Combined myoclonic–astatic and 'benign' focal epilepsy of childhood ('atypical benign partial epilepsy of childhood'). A separate syndrome? Neuropediatrics 17: 144–151.

Deonna T, Zesinger P, Davidoff V, et al 2000 Benign partial epilepsy of childhood: a longitudinal neuropsychological and EEG study of cognitive function. Dev Med Child Neurol 42: 595–603.

348

Di Mauro S 1979 Metabolic myopathies. In: *Diseases of Muscle* (Pt. II). *Handbook of Clinical Neurology*, Vol. 41 (eds PJ Vinken, G W Bruyn). North Holland, Amsterdam, pp 175–234.

Di Mauro S, Bonlla E, Zeviani M, et al 1985 Mitochondrial myopathies. Ann Neurol 17: 521–538.

Dipti S, Childs AM, Livingston JH, et al 2005 Brown–Vialetto–Van Laere syndrome; variability in age at onset and disease progression highlighting the phenotypic overlap with Fazio–Londe disease. Brain Dev 27: 443–446.

Dogulu CF, Ciger A, Saygi S, et al 1995 Atypical EEG findings in subacute sclerosing panencephalitis. Clin Electroencephalogr 26: 193–199.

Domagk J, Linke I, Argyrakis A, et al 1975 Adrenoleucodystrophy. Neuropädiatrie 6: 41–64.

Doose H 1959 Die posthypoglykämische Enzephalopathie. Monatsschr Kinderheilkd 107: 438–443.

Doose H, Baier WK 1988 Theta rhythms in the EEG – a genetic trait. Brain Dev 10: 347–354.

Doose H, Baier WK 1989 Benign partial epilepsy and related conditions: multifactorial pathogenesis with hereditary impairment of brain maturation. Eur J Pediatr 149: 152–158.

Doose H, Hahn A, Neubauer BA, et al 2001 Atypical 'benign' partial epilepsy of childhood or pseudo-Lennox syndrome. Part II: family study. Neuropediatrics 32: 9–13.

Dorfman LJ, Pedley TA, Tharp BR, et al 1978 Juvenile neuroaxonal dystrophy: clinical, electrophysiological, and neuropathological features. Ann Neurol 3: 419–428.

Dravet C 1978 Les épilepsies graves de l'enfant. Vie Méd 8: 543–548.

Dravet C 2000 Severe myoclonic epilepsy in infants and its related syndromes. Epilepsia 41(Suppl 9): 7.

Dravet C, Bureau M 1981 L'epilepsie myoclonique benigne du nourrisson. Rev Electroencéphalogr Neurophysiol Clin 11: 438–444.

Dravet C, Giraud N, Bureau M, et al 1986 Benign myoclonus of early infancy or benign non-epileptic infantile spasms. Neuropediatrics 17: 33–38.

Dravet C, Bureau M, Genton P 1992 Benign myoclonic epilepsy of infancy: electroclinical symptomatology and differential diagnosis from the other types of generalized epilepsy of infancy. Epilepsy Res (Suppl 6): 131–135.

Dubowitz V 1999 Very severe spinal muscular atrophy (SMA type 0): an expanding clinical phenotype. Eur J Paediatr Neurol 3: 49–51.

Dusser A, Navelet Y, Devictor D, et al 1989 Short- and long-term prognostic value of the electroencephalogram in children with severe head injury. Electroencephalogr Clin Neurophysiol 73: 85–93.

Dyck PJ 1975 Inherited neuronal degeneration and atrophy affecting peripheral motor, sensory and autonomic neurons. In: *Peripheral Neuropathy*, Vol. 2 (eds PJ Dyck, PK Thomas, EH Lambert). Saunders, Philadelphia, PA, pp 825–867.

Dyck PJ 1984 Neuronal atrophy and degeneration predominantly affecting peripheral sensory and autonomic neurons. In: *Peripheral Neuropathy*, 2nd edn (eds PJ Dyck, PK Thomas, EH Lambert, et al). Saunders, Philadelphia, PA, pp 1600–1655.

Dyck PJ, Lambert EH 1968a Lower motor and primary sensory neuron disease with peroneal muscular atrophy. I. Neurologic, genetic and electrophysiological findings in hereditary polyneuropathy. Arch Neurol 18: 603–618.

Dyck PJ, Lambert EH 1968b Lower motor and primary sensory neuron disease with peroneal muscular atrophy. II. Neurologic, genetic and electrophysiological findings in various neuronal degenerations. Arch Neurol 18: 619–625.

Dyck PJ, Mellinger JF, Reagan TJ, et al 1983 Not 'indifference to pain' but varieties of hereditary sensory and autonomic neuropathy. Brain 106: 373–390.

Dyck PJ, Karnes JL, Lambert, EH 1989 Longitudinal study of neuropathic defects and nerve conduction abnormalities in hereditary motor and sensory neuropathy type I. Neurology 39: 1302–1308.

Eeg-Olofsson O, Petersen I, Sellden U 1971 The development of the electroencephalogram in normal children from the age of 1 through 15 years. Paroxysmal activity. Neuropädiatrie 2: 375–404.

Eggers Chr, Lederer H, Scheffner D 1976 EEG-Befunde im Verlauf progredienter Hirnerkrankungen im Kindesalter. Monatsschr Kinderheilkd 124: 1–8.

Emery AE 2002 The muscular dystrophies. Lancet 23(359): 687–695.

Emslie GJ, Rush AJ, Weinberg WA, et al 1990 Children with major depression show reduced rapid eye movement latencies. Arch Gen Psychiatry 47: 119–124.

Engel J 2001 A proposed diagnostic scheme for people with epileptic seizures and with epilepsy: report of the ILAE Task Force on Classification and Terminology. Epilepsia 42: 796–803.

Engel J Jr, Rapin I, Giblin DR 1977 Electrophysiological studies in two patients with cherry red spot myoclonus syndrome. Epilepsia 18: 73–87.

Engel AG, Ohno K, Sine SM 2003 Congenital myasthenic syndromes: progress over the past decade. Muscle Nerve 27: 4–25.

Enomoto T, Ono Y, Nose T, et al 1986 Electroencephalography in minor head injury in children. Child Nerv Syst 2: 72–79.

Epilepsia 2002 *Epilepsy Through the Life Cycle*. Epilepsia 43(Suppl 3).

Epilepsia 2004 *Catastrophic Epilepsies in Childhood*. Epilepsia 45(Suppl 5): 1–26.

Fejerman N, Caraballo R, Tenembaum SN 2000 Atypical evolutions of benign localization-related epilepsies in children: are they predictable? Epilepsia 41: 380–390.

Fenichal GM 1978 Clinical syndrome of myasthenia in infancy and childhood. Arch Neurol 35: 97–103.

Ferrie C, Caraballo R, Covanis A, et al 2006 Panayiotopoulos syndrome: a consensus view. Dev Med Child Neurol 48: 236–240.

Fonseca LC, Tedrus GM 2000 Somatosensory evoked spikes and epileptic seizures: a study of 385 cases. Clin Electroencephalogr 31: 71–75.

Frantzen E, Lennox-Buchthal M, Nygaard A 1968 Longitudinal EEG and clinical study of children with febrile convulsions. Electroencephalogr Clin Neurophysiol 24: 197–212.

Friedman E, Harden A, Koivikko M, et al 1978 Menkes' disease. Neurophysiological aspects. J Neurol Neurosurg Psychiatry 41: 505–510.

Fujimoto S, Mizuno K, Shibata H, et al 1999 Serial electroencephalographic findings in patients with MELAS. Pediatric Neurol 20: 43–48.

Fujimoto S, Shibata H, Sugiyama N, et al 2000 Unique electroencephalographic change of acute encephalopathy in glutaric aciduria type I. Tohoku. J Exp Med 191: 31–38.

Furlong PL, Wimalaratna S, Harding GFA 1993 Augmented P22–N31 SEP component in a patient with unilateral space occupying lesion. Electroencephalogr Clin Neurophysiol 88: 72–76.

Gamstorp I 1985 *Paediatric Neurology*, 2nd edn. Butterworths, London.

Garbern JY, Cambi F, Tang X-E, et al 1997 Proteolipid is necessary in peripheral as well as central myelin. Neuron 19: 205–218.

349

Garcia-Cabezas MA, Garcia-Alix A, Martin Y, et al 2004 Neonatal spinal muscular atrophy with multiple contractures, bone fractures, respiratory insufficiency and 5q13 deletion. Acta Neuropathol 107: 475–478.

Gastaut H, Pinsard N, Raybaud CL, et al 1987 Lissencephaly (agyria–pachygyria): clinical findings and serial EEG studies. Dev Med Child Neurol 29: 167–180.

George AL Jr 2004 Molecular basis of inherited epilepsy. Arch Neurol 6: 473–478.

Giannakodimos S, Panayiotopoulos CP 1996 Eyelid myoclonia with absences in adults: a clinical and video–EEG study. Epilepsia 37: 36–44.

Gibbs EL, Fois A, Gibbs FA 1955 The electroencephalogram in retrolental fibroplasia. N Engl J Med 253: 1102–1106.

Gibbs FA, Gibbs EL, Perlstein MA, et al 1963 Electro-encephalographic prediction of epilepsy as a complication of cerebral palsy. Neurology 13: 143–145.

Giles DE, Roffwarg HP, Dahl RE, et al 1992 Electroencephalographic sleep abnormalities in depressed children: a hypothesis. Psychiatry Res 41: 53–63.

Gillis L, Kaye E 2002 Diagnosis and management of mitochondrial diseases. Pediatr Clin North Am 49: 203–219.

Glaze DG, Frost JD, Jankovic J 1983 Sleep in Gilles de la Tourette's syndrome: disorder of arousal. Neurology 33: 586–592.

Glaze DG, Frost JD, el Hibri HY, et al 1985 Rett's syndrome: polygraphic electroencephalographic–video characterization of sleep and respiratory patterns during sleep and wakefulness. Ann Neurol 18: 417–418.

Gordon N 2006 Glutaric aciduria types I and II. Brain Dev 28: 136–140.

Gourfinkel-An I, Baulac S, Nabbout R, et al 2004 Monogenic idiopathic epilepsies. Lancet Neurol 3: 209–218.

Govaerts L, Colon E, Rotteveel J, et al 1985 A neurophysiological study of children with cerebro-hepato-renal syndrome of Zellweger. Neuropediatrics 16: 185–190.

Govoni V, Granieri E 2001 Epidemiology of the Guillain–Barré syndrome. Curr Opin Neurol 14: 605–613.

Graff-Radford NR, Damasio H, Yamada T, et al 1985 Nonhaemorrhagic thalamic infarction: clinical, neuropsychological and electrophysiological findings in four anatomical groups defined by computerized tomography. Brain 108: 485–516.

Granata T, Fusco L, Gobbi G, et al 2003 Experience with immunomodulatory in Rasmussen's encephalitis. Neurology 61: 1807–1810.

Green JB 1974 Cerebral lipidoses. In: The Epilepsies. Handbook of Clinical Neurology, Vol. 15 (eds O Magnus, AML de Haas). North-Holland, Amsterdam, pp 423.

Greenberg DA 1999 Neuromuscular disease and calcium channels. Muscle Nerve 22: 1341–1349.

Grillon C, Courchesne E, Akshoomoff N 1989 Brainstem and middle latency auditory evoked potentials in autism and developmental language disorder. J Autism Dev Disord 19: 255–269.

Guerrini R, Dravet C, Genton P, et al 1995 Idiopathic photosensitive occipital lobe epilepsy. Epilepsia 36: 883–891.

Guerrini R, De Lorey TM, Bonanni P, et al 1996 Cortical myoclonus in Angelman syndrome. Ann Neurol 40: 39–48.

Guerrini R, Genton P, Bureau M, et al 1998 Multilobar polymicrogyria, intractable drop attack seizures and sleep-related electrical status epilepticus. Neurology 51: 504–512.

Gurses C, Ozturk A, Baykan B, et al 2000 Correlation between clinical stages and EEG findings of subacute sclerosing pan-encephalitis. Clin Electroencephalogr 31: 201–6.

Gutling E, Gonser A, Imhof HG, et al 1995 EEG reactivity in the prognosis of severe head injury. Neurology 45: 915–918.

Guzetta F, Ferriere G, Lyon G 1982 Congenital hypomyelination polyneuropathy. Pathological findings compared with polyneuropathies starting later in life. Brain 105: 395–416.

Guzetta F, Battaglia D, Veredice C, et al 2005 Early thalamic injury associated with epilepsy and continuous spike-wave during slow sleep. Epilepsia 46: 889–900.

Haddad FS, Jones DH, Vellodi A, et al 1997 Carpal tunnel syndrome in the mucopolysaccharidoses and mucolipidoses. J Bone Joint Surg Br 79: 576–582.

Hagberg B, Wahlström J, Anvret M (eds) 1993 Rett Syndrome – Clinical and Biological Aspects. Clinics in Developmental Medicine, No. 127. MacKeith Press/Cambridge University Press, London.

Hagne I, Witt-Engerström I, Hagberg B 1989 EEG development in Rett syndrome. A study of 30 cases. Electroencephalogr Clin Neurophysiol 72: 1–6.

Hahn A, Pistohl J, Neubauer BA, et al 2001 Atypical 'benign' partial epilepsy or pseudo-Lennox syndrome. Part I: symptomatology and long-term prognosis. Neuropediatrics 32: 1–8.

Hansotia P, Harris R, Kennedy J 1969 EEG changes in Wilson's disease. Electroencephalogr Clin Neurophysiol 27: 523–528.

Harden A, Pampiglione G 1970 Neurophysiological approach to disorders of vision. Lancet 18: 805–809.

Harden A, Boyd SG, Cole G, et al 1991 EEG features and their evolution in the acute phase of haemorrhagic shock and encephalopathy syndrome. Neuropediatrics 22: 194–197.

Harding B, Copp AJ 1997 Malformations. In: Greenfield's Neuropathology, 6th edn (eds DI Graham, PL Lantos). Arnold, London, pp 397–533.

Harding AE, Thomas PK 1980 The clinical features of hereditary motor and sensory neuropathy, types I and II. Brain 103: 259–280.

Harding AE, Thomas PK 1984 Peroneal muscular atrophy with pyramidal features. J Neurol Neurosurg Psychiatry 47: 168–172.

Harris R, Delia Rovere M, Prior PF 1965 Electroencephalographic studies in infants and children with hypothyroidism. Arch Dis Child 40: 612–617.

Hausmanowa-Petrusewicz I, Askonas W, Beduska B 1968 Infantile and juvenile spinal muscular atrophy. J Neurol Sci 6: 269–273.

Heller GL, Kooi KA 1962 The electroencephalogram in hepatolenticular degeneration (Wilson's disease). Electroencephalogr Clin Neurophysiol 14: 520–526.

Hirsch E, Marescaux C, Maquet P, et al 1990 Landau–Kleffner syndrome: a clinical and EEG study of five cases. Epilepsia 31: 756–767.

Holmes GL, McKeever M, Adamson M 1987 Absence seizures in children: clinical and electroencephalographic features. Ann Neurol 21: 268–73.

Holt IJ, Harding AE, Cooper JM, et al 1989 Mitochondrial myopathies: clinical and biochemical features of 30 patients with major deletions of muscle mitochondrial DNA. Ann Neurol 26: 699–708.

Honovar M, Meldrum BS 1997 Epilepsy. In: Greenfield's Neuropathology, 6th edn (eds DI Graham, PL Lantos). Arnold, London, pp 931–971.

Hoogenraad TU, Van Hattum J, Van den Hamer CJA 1987 Management of Wilson's disease with zinc sulphate. J Neurol Sci 77: 137–146.

Hrachovy RA, Frost JD Jr 1984 Hypsarrhythmia: variation of the theme. Epilepsia 25: 317–325.

Hughes JR, Patil VK 2002 Long-term electro-clinical changes in the Lennox–Gastaut syndrome before, during, and after the slow spike-wave pattern. Clin Electroencephalogr 33: 1–7.

Hutchinson DO, Frith RW, Shaw NA, et al 1991 A comparison between electroencephalography and somatosensory evoked potentials for outcome prediction following severe head injury. Electroencephalogr Clin Neurophysiol 78: 228–233.

Ikeda A, Shibasaki H, Nagamine T, et al 1995 Peri-Rolandic and frontoparietal components of scalp recorded giant SEPs in cortical myoclonus. Electroencephalogr Clin Neurophysiol 96: 300–309.

Ines DF, Markand ON 1977 Epileptic seizures and abnormal electroencephalographic findings in hydrocephalus and their relation to the shunting procedures. Electroencephalogr Clin Neurophysiol 42: 761–768.

Inoue Y, Fujiwara T, Matsuds K, et al 1997 Ring chromosome 20 and nonconvulsive status epilepticus. Brain 120: 939–953.

Ionasescu VV, Searby C, Greenberg SA 1996 Dejerine–Sottas disease with sensorineural hearing loss, nystagmus, and peripheral facial nerve weakness: de novo dominant point mutation of the PMP22 gene. J Med Genet 33: 1048–1049.

Isoardo G, Bergui M, Durelli L, et al 2004 Thalidomide neuropathy: clinical, electrophysiological and neuroradiological features. Acta Neurol Scand 109: 188–193.

Itil TM, Simeon J, Coffin C 1976 Qualitative and quantitative EEG in psychotic children. Dis Nerv Syst 37: 247–252.

Jayakar PB, Seshia SS 1991 Electrical status epilepticus during slow-wave sleep: a review. J Clin Neurophysiol 8: 299–311.

Jeavons PM 1964 The electroencephalogram in blind children. Br J Ophthalmol 48: 83–101.

Jervis GA 1963 Huntington's chorea in childhood. Arch Neurol 9: 244–257.

Johnson JL, White KR, Widen JE, et al 2005 A multicenter evaluation of how many infants with permanent hearing loss pass a two-stage otoacoustic emissions/automated auditory brainstem response newborn hearing screening protocol. Pediatrics 116: 663–672.

Joint Committee of Infant Hearing 2000 Year 2000 Position statement: principles and guidelines for early hearing detection and intervention programs. Am J Audiol 9: 9–29.

Journal of Clinical Neurophysiology 2003 Epileptic Encephalopathies in Childhood.

Jungbluth H, Zhou H, Hartley L, et al 2005 Minicore myopathy with ophthalmoplegia caused by mutations in the ryanodine receptor type 1 gene. Neurology 65: 1930–1935.

Kakigi R, Shibasaki H 1987 Generator mechanisms of giant somatosensory evoked potentials in cortical reflex myoclonus. Brain 110: 1359–1373.

Kasteleijn-Nolst Trenité DG, Bakker DJ, Binnie CD, et al 1988 Psychological effects of subcortical epileptiform discharges. I: Scholastic skills. Epilepsy Res 2: 111–116.

Kayden HJ 1993 The neurologic syndrome of vitamin E deficiency. Neurology 43: 2167–2169.

Kinney HC, Armstrong DD 1997 Perinatal neuropathology. In: Greenfield's Neuropathology, 6th edn (eds DI Graham, PL Lantos). Arnold, London, pp 535–599.

Kliemann FAD, Harden A, Pampiglione G 1969 Some EEG observations in patients with Krabbe's disease. Dev Med Child Neurol 11: 475–484.

Kojewnikov L 1895 Eine besondere Form von corticaler Epilepsie. Neurologisch Centralblatt 14: 47–48.

Koutroumanidis M, Koepp MJ, Richardson MP, et al 1998 The variants of reading epilepsy. A clinical and video–EEG study of 17 patients with reading-induced seizures. Brain 121: 1409–1427.

Krajewska G, Hausmanowa-Petrusewicz I 2002 Abnormal nerve conduction velocity as a marker of immaturity in childhood muscle spinal atrophy. Folia Neuropathol 40: 67–74.

Krajewski K, Lewis RA, Fuerst DR, et al 2000 Neurological dysfunction and axonal degeneration in Charcot–Marie–Tooth disease type 1A. Brain 123: 1516–1527.

Kraus N, Smith DI, Reed NI, et al 1985 Auditory middle latency responses in children: effect of age and diagnostic category. Electroencephalogr Clin Neurophysiol 62: 343–359.

Kriss A, Thompson D 1997 Visual electrophysiology. In: Pediatric Ophthalmology, 2nd edn (ed D Taylor). Blackwell Science, Oxford, pp 93–121.

Kriss A, Russel-Eggitt I, Harris CM, et al 1992 Aspects of albinism. Ophthal Paediatr Genet 13: 89–100.

Kubisch C, Schmitt-Rose T, Fontaine B, et al 1998 CLC-1 chloride channel mutations in myotonia congenita: variable penetrance of mutations shifting the voltage dependence. Hum Mol Genet 7: 1753–1760.

Kupfersmith MJ, Nelson JI 1985 Preserved visual evoked potential in infancy cortical blindness. Relation to blindsight. Neuroophthalmology 6: 85–94.

Kuzniecky RI, Barkovich AJ 2001 Malformations of cortical development and epilepsy. Brain Dev 23: 2–11.

Kuzniecky R, Andermann F, Guerrini R 1993 Congenital bilateral perisylvian syndrome: study of 31 patients. The CBPS Multicenter Collaborative Study. Lancet 341: 608–612.

Kuzniecky R, Guthrie B, Mountz J, et al 1997 Intrinsic epileptogenesis of hypothalamic hamartomas in gelastic epilepsy. Ann Neurol 42: 60–67.

Laan LA, Brouwer OF, Begeer CH, et al 1998 The diagnostic value of the EEG in Angelman and Rett syndrome at a young age. Electroencephalogr Clin Neurophysiol 106: 404–408.

Laget P, Mamo H, Houdart R 1967 De l'intérêt des potentiels évoqués somesthésiques dans l'étude des lésions du lobe pariétal de l'homme. Étude préliminaire. Neurochirurgie 13: 841–853.

Lancet 1989 Cerebral palsy, intrapartum care and a shot in the foot. Lancet ii: 1251–1252.

Lancman ME, Asconape JJ, Penry JK 1994 Clinical and EEG asymmetries in juvenile myoclonic epilepsy. Epilepsia 35: 302–306.

La Spada AR, Wilson EM, Lubahn DB, et al 1991 Androgen receptor gene mutations in X linked spinal and bulbar muscular atrophy. Nature 352: 77–79.

Lefebvre S, Burglen L, Reboullet S, et al 1995 Identification and characterisation of a spinal muscular atrophy determining gene. Cell 80: 155–165.

Legg NJ, Gupta PC, Scott DF 1973 Epilepsy following cerebral abscess. A clinical and EEG study of 70 patients. Brain 96: 259–268.

Lewis RA, Sumner AJ 1982 The electrodiagnostic distinction between chronic familial and acquired demyelinative neuropathies. Neurology 32: 592–596.

Lieberman JS, Oshtory M, Taylar RG, et al 1980 Peripheral neuropathy as an early manifestation of Krabbe's disease. Arch Neurol 37: 446–447.

Lipinski Chr, Lorenz HM, Scheffner D 1975 EEG-Veranderungen wahrend der Therapie von Tumoren der hirteren Schadelgrube im Kindersalter. Z Elektroenzephalogr Elektromyogr Gebiete 6: 188–194.

Loiseau P, Orgogozo JM 1978 An unrecognized syndrome of benign focal epileptic seizures in teenagers? Lancet 2: 1070–1071.

Loiseau P, Pestre M, Dartigues JF, et al 1983 Long-term prognosis in two forms of childhood epilepsy: typical absence seizures and epilepsy with Rolandic (centrotemporal) EEG foci. Ann Neurol 13: 642–648.

Lombroso CT, Fejerman N 1977 Benign myoclonus of early infancy. Ann Neurol 1: 138–143.

Low PA, Burlse WJ, MacLeod JG 1978 Congenital sensory neuropathy with selective loss of small myelinated nerve fibres. Ann Neurol 3: 79–182.

Lütschg J 1984 Pathophysiological aspects of central and peripheral myelin lesions. Neuropaediatrics 15(Suppl): 24–27.

Lütschg J 1985 *Evozierte Potentiate bei komatösen Kindern*. Gustav Fischer, Stuttgart.

Lütschg J, Pfenninger J, Ludin HP, et al 1983 Brain stem auditory evoked potentials and early somatosensory evoked potentials in neurointensively treated comatose children. Am J Dis Child 137: 421–426.

Lütschg J, Meyer E, Jeanneret-Iseli C 1985 Brain-stem auditory evoked potentials in meningomyelocele. Neuropediatrics 16: 202–204.

Lynch J, Eldadah MK 1992 Brain-death criteria currently used by pediatric intensivists. Clin Pediatr 31: 457–460.

MacGillivray BB 2003 The EEG in systemic disorders: the encephalopathies. In: *Clinical Neurophysiology*, Vol. 2, *EEG, Paediatric Neurophysiology, Special Techniques and Applications* (eds CD Binnie, R Cooper, F Mauguière, et al). Elsevier Science, Amsterdam, pp 307–354.

MacLeod MJ, Taylor JE, Lunt PW, et al 1999 Prenatal onset spinal muscular atrophy. Eur J Paediatr Neurol 3: 65–72.

Maher J, McLachlan RS 1995 Febrile convulsions. Is seizure duration the most important predictor of temporal lobe epilepsy? Brain 118: 1521–1528.

Mahowald MW, Ettinger MG 1990 Things that go bump in the night: the parasomnias revisited. J Clin Neurophysiol 7: 119–143.

Mankodi A, Takahashi MP, Jiang H, et al 2002 Expanded CUG repeats trigger aberrant splicing of CIV-1 chloride channel pre-mRNA and hyperexcitability of skeletal muscle in myotonic dystrophy. Mol Cell 10: 35–44.

Mantegazza M, Gambardella A, Rusconi R, et al 2005 Identification of an Nav1.1 sodium channel (SCN1A) loss-of-function mutation associated with familial simple febrile seizures. Proc Natl Acad Sci USA 102: 18177–18182.

Markand ON, Garg BP, De Meyer WE, et al 1982 Brain stem auditory, visual and somatosensory evoked potentials in leukodystrophies. Electroencephalogr Clin Neurophysiol 54: 39–48.

Massa R, Saint-Martin A, Carcangiu R, et al 2001 EEG criteria predictive of complicated evolution in idiopathic Rolandic epilepsy. Neurology 57: 1071–1079.

Mauguière F, Courjon J 1980 Effects of intravenous clonazepam on cortical somatosensory evoked responses (SER) in dyssynergia cerebellaris myoclonica (Ramsay–Hunt syndrome). In: *EEG and Clinical Neurophysiology* (eds H Lechner, A Aranibar). Excerpta Medica, Amsterdam, pp 433–444.

Mauguière F, Ibañez V 1985 The dissociation of early SEP components in lesions of the cervico-medullary junction: a cue for routine interpretation of abnormal cervical responses to median nerve stimulation. Electroencephalogr Clin Neurophysiol 62: 406–420.

Mauguière F, Ibañez V 1990 Loss of parietal and frontal somatosensory evoked potentials in hemispheric deafferentation. In: *New Trends and Advanced Techniques in Clinical Neurophysiology* (eds PM Rossini, F Mauguière). *Electroencephalogr Clin Neurophysiol* (Suppl 41): 274–285.

Mauguière F, Bard J, Courjon J 1981 Les potentiels évoqués somesthésiques dans la dyssynergie cérébelleuse myoclonique progressive. Rev EEG. Neurophysiol Clin 11: 174–182.

Mauguière F, Brunon AM, Echallier JF, et al 1982 Early somatosensory evoked potentials in thalamocortical lesions of the lemniscal pathways in humans. In: *Clinical Applications of Evoked Potentials in Neurology*. Advances in Neurology, Vol. 32 (eds J Courjon, F Mauguière, M Revol). Raven Press, New York, pp 321–338.

Mauguière F, Schott B, Courjon J 1983 Dissociation of early SEP components in unilateral traumatic section of the lower medulla. Ann Neurol 13: 309–313.

Mauguière, F, Ibañez, V, Turano, G, et al 1996 Neurophysiology in intramedullary spinal cord tumors. In: *Intramedullary Spinal Cord Tumors* (eds G Fischer, J Brotchi), Thieme, Stuttgart. pp 24–33.

McConnell HW, Duncan D 1998 Behavioral effects of antiepileptic drugs. In: *Psychiatric Co-Morbidity in Epilepsy: Basic Mechanisms, Diagnosis and Treatment* (eds HW McConnell, PJ Snyder). American Psychiatric Press, Washington, DC, pp 205–244.

McLeod JG 1971 An electrophysiological and pathological study of peripheral nerves in Friedreich's ataxia. J Neurol Sci 12: 333–349.

McShane MA, Boyd S, Harding B, et al 1992 Progressive bulbar paralysis of childhood. A reappraisal of Fazio–Londe disease. Brain 115: 1889–1900.

Metcalf DR 1972 EEG in inborn errors of metabolism. In: *Handbook of Electroencephalography and Clinical Neurophysiology*, Vol. 15B (ed A Rémond), pp 14–18.

Michel B, Gastaut JL, Bianchi L 1979 Electroencephalographic cranial computerized tomographic correlations in brain abscess. Electroencephalogr Clin Neurophysiol 46: 256–273.

Miller RG, Davis CJF, Illingworth DR, et al 1980 The neuropathy of abetalipoproteinemia. Neurology 30: 1286–1291.

Miller RG, Gutmann L, Lewis RA, et al 1985 Acquired versus familial demyelinative neuropathies in children. Muscle Nerve 8: 205–210.

Mills PB, Surtees RAH, Champion MP, et al 2005 Neonatal epileptic encephalopathy caused by mutations in the PNPO gene encoding pyridox(am)ine 5-phosphate oxidase. Hum Mol Genet 14: 1077–1086.

Mitchison HM, Mole SE 2001 Neurodegenerative disease: the neuronal ceroid lipofuscinoses (Batten disease). Curr Opin Neurol 14(6): 795–803.

Moosa A, Dubowitz V 1970 Peripheral neuropathy in Cockayne's syndrome. Arch Dis Child 45: 674–677.

Moosa A, Dubowitz V 1973 Spinal muscular atrophy in childhood: two clues to clinical diagnosis. Arch Dis Child 48: 386–388.

Moraes CT, Di Mauro S, Zeviani M, et al 1989 Mitochondrial DNA deletions in progressive external ophthalmoplegia and Kearns–Sayre syndrome. N Engl J Med 320: 1293–1299.

Morrell F, Whisler WW, Smith MC, et al 1995 Landau–Kleffner syndrome. Treatment with subpial intracortical resection. Brain 118: 1529–1546.

Moshe SL 1989 Usefulness of EEG in the evaluation of brain death in children: the pros. Electroencephalogr Clin Neurophysiol 73: 272–275.

Munsat TL, Thompson LR, Coleman RF 1969 Centronuclear ('myotubular') myopathy. Arch Neurol 20: 120–131.

Muntoni F, Voit T 2004 The congenital muscular dystrophies in 2004: a century of exciting progress. Neuromusc Disord 14: 635–649.

Musumeci SA, Hagerman RJ, Ferri R, et al 1999 Epilepsy and EEG findings in males with fragile X syndrome. Epilepsia 40: 1092–1099.

Nabbout R, Soufflet C, Plouin P, et al 1999 Pyridoxine dependent epilepsy: a suggestive electroclinical pattern. Arch Dis Child 81: F125–F129.

Naito H, Oyanagi S 1982 Familial myoclonus epilepsy and choreoathetosis: hereditary dentatorubral–pallidoluysian atrophy. Neurology 32: 798–807.

Nakanishi T, Shimada Y, Sakuta M, et al 1978 The initial positive component of the scalp-recorded somatosensory evoked potential in normal subjects and in patients with neurological disorders. Electroencephalogr Clin Neurophysiol 45: 26–34.

Nevšimalová S, Marecek Z, Roth B 1986 An EEG study of Wilson's disease. Findings in patients and heterozygous relatives. Electroencephalogr Clin Neurophysiol 64: 191–198.

Nicoletta JA, Schwartz GJ 2004 Distal renal tubular acidosis. Curr Opin Pediatr 16(2): 194–198.

Nishimura T, Mori K 1996 Somatosensory evoked potentials to median nerve stimulation in meningomyelocele: what is occurring in the hindbrain and its connection during growth? Child Nerv Sys 12: 13–26.

Nomura Y, Segawa M, Hasegawa M 1984 Rett syndrome: clinical studies and pathophysiological consideration. Brain Dev 6: 475–486.

Norton P, Ellison P, Sulaiman AR, et al 1983 Nemaline myopathy in the neonate. Neurology 33: 351–356.

Nucci A, Queiroz LS, Zambelli HJ, et al 2004 Multi-minicore disease revisited. Arq Neuropsiquiatr 62: 935–939.

Obeso JA, Rothwell JC, Marsden CD 1985 The spectrum of cortical myoclonus: from focal reflex jerks to spontaneous motor epilepsy. Brain 108: 193–224.

Ochs R, Markand ON, De Meyer WE 1979 Brainstem auditory evoked responses in leukodystrophies. Neurology 29: 1089–1093.

Ohta M, Ellefson RD, Lambert EH, et al 1973 Hereditary sensory neuropathy, type II. Clinical, electrophysiologic, histologic and biochemical studies of a Quebec kinship. Arch Neurol 29: 23–37.

Olsson I, Steffenburg S, Gillberg C 1988 Epilepsy in autism and autistic-like conditions. Arch Neurol 45: 666–668.

Osada H, Masuda K, Seki K, et al 2004 Multi-minicore disease with susceptibility to malignant hyperthermia in pregnancy. Gynecol Obstet Invest 58: 32–35.

Ouvrier RA, MacLeod JG, Conchin TE 1982 Friedreich's ataxia: early detection and progression of peripheral nerve abnormalities. J Neurol Sci 55: 137–145.

Ouvrier RA, MacLeod JG, Conchin TE 1987 The hypertrophic forms of hereditary motor and sensory neuropathy. A study of hypertrophic Charcot–Marie–Tooth disease (HMSN type I) and Dejerine–Sottas disease (HMSN type III) in childhood. Brain 110: 121–148.

Ouvrier RA, MacLeod JG, Pollard JD (eds) 1990 Peripheral Neuropathy in Childhood. Raven Press, New York.

Paetau R, Kajola M, Korkman M, et al 1991 Landau–Kleffner syndrome: epileptic activity in the auditory cortex. Neuroreport (Oxford) 2: 201–204.

Paladin F, Chiron C, Dulac O, et al 1989 Electroencephalographic aspects of hemimegalencephaly. Dev Med Child Neurol 37: 377–383.

Pampiglione G, Harden A 1984 Neurophysiological investigations in GM$_1$ and GM$_2$ gangliosidoses. Neuropediatrics (Suppl 15): 74–84.

Panayiotopoulos CP 2000. Benign childhood epileptic syndromes with occipital spikes: new classification proposed by the International League Against Epilepsy. J Child Neurol 15: 548–552.

Panayiotopoulos CP, Obeid T, Waheed G 1989 Differentiation of typical absence seizures in epileptic syndromes. A video EEG study of 224 seizures in 20 patients. Brain 112: 1039–1056.

Pareyson D 1999 Charcot–Marie–Tooth disease and related neuropathies: molecular basis for distinction and diagnosis. Muscle Nerve 22: 1498–1509.

Patwarl AK, Aneja S, Ravi RN, et al 1996 Convulsions in tuberculous meningitis. J Trap Pediatr 42: 91–97.

Paul DA, Goldsmith LS, Miles DK, et al 1993 Neonatal adrenoleukodystrophy presenting as infantile progressive spinal muscular atrophy. Pediatr Neurol 9: 496–497.

Pavone P, Bianchini R, Trifiletti RR, et al 2001 Neuropsychological assessment in children with absence epilepsy. Neurology 56: 1047–1051.

Payan J 1991 Clinical electromyography in infancy and childhood. In: Paediatric Neurology, 2nd edn (ed EM Brett). Churchill Livingstone, Edinburgh, pp 797–829.

Pearce J, Pitt M, Martinez A 2005 A neonatal diagnosis of congenital chronic inflammatory demyelinating polyneuropathy. Dev Med Child Neurol 47: 489–492.

Peltier AC, Russell JW 2006 Advances in understanding drug-induced neuropathies. Drug Safety 29: 23–30.

Percy AK, Kaback MM, Herndon RM 1977 Metachromatic leukodystrophy: comparison of early and late-onset forms. Neurology 27: 933–941.

Perlstein MA, Gibbs EL, Gibbs FA 1955 The electroencephalogram in infantile cerebral palsy. Am J Phys Med 34: 477–496.

Peterson BS 1995 Neuroimaging in child and adolescent neuropsychiatric disorders. J Am Acad Child Adolesc Psychiatry 34: 1560–1576.

Phelps M, Aicardi J, Vanier MT 1991 Late onset Krabbe's leukodystrophy: a report of four cases. J Neurol Neurosurg Psychiatry 54: 293–296.

Picard F, Baulac S, Kahane P, et al 2000 Dominant partial epilepsies. A clinical, electrophysiological and genetic study of 19 European families. Brain 123: 1247–1262.

Pietz J, Benninger Chr, Schmidt H, et al 1988 Long-term development of intelligence (IQ) and EEG in 34 children with phenylketonuria treated early. Eur J Pediatr 147: 361–367.

Pitt MC 1996 A system based study of the variation of the amplitude of the compound sensory nerve action potential recorded using surface electrodes. Electroencephalogr Clin Neurophysiol 101: 520–527.

Pitt M, Houlden H, Jacobs J, et al 2003 Severe infantile neuropathy with diaphragmatic weakness and its relationship to SMARD1. Brain 126: 2682–2692.

Plasmati R, Blanco M, Michelucci R, et al 1989 SEPs study in idiopathic infantile epilepsies. In: Reflex Seizures and Epilepsies (eds A Beaumanoir, H Gastaut, R Naquet). Editions Médécine et Hygiène, Geneva, pp 75–81.

Plassart-Scheiss E, Gervais A, Eymard B, et al 1998 Novel muscle chloride channel (CLCN1) mutations in myotonia congenita with various modes of inheritance including incomplete dominance and penetrance. Neurology 50: 1176–179.

Plaster NM, Tawil R, Tristani-Firouzi M, et al 2001 Mutations in Kir 2.1 cause the developmental and episodic electrical phenotypes of Anderson's syndrome. Cell 105: 511–519.

Polo A, Zanette G, Manganotti P, et al 1994 Spinal somatosensory evoked potentials in patients with tethered cord syndrome. Can J Neurol Sci 21: 325–330.

Pressler RM, Robinson RO, Wilson GA, et al 2005 Treatment of interictal epileptiform discharges can improve behavior in children with behavioral problems and epilepsy. J Pediatr 146:112–117.

Prior PF 2003 Head injury. In: Clinical Neurophysiology, Vol. 2, EEG, Paediatric Neurophysiology, Special Techniques and Applications (eds CD Binnie, R Cooper, F Mauguière, et al). Elsevier Science, Amsterdam, pp 355–370.

Prior PF, Maclaine GN, Scott DF, et al 1972 Tonic status epilepticus precipitated by intravenous diazepam in a child with petit mal status. Epilepsia 13: 467–472.

Pueschel SM, Findley TW, Furia J, et al 1987 Atlantoaxial instability in Down syndrome: roentgenographic, neurologic and somatosensory evoked potential studies. J Pediatr 110: 512–521.

Quane KA, Healy JMS, Keating KE, et al 1993 Mutations in the ryanodine receptor gene in central core disease and malignant hyperthermia. Nature Genet 5: 51–55.

Radermecker J 1956 Systématiques et électroencéphalographiques des encéphalites et encéphalopathies. Electroencephalogr Clin Neurophysiol (Suppl 5).

Ramaekers VTh, Lake BD, Harding B, et al 1987 Diagnostic difficulties in infantile neuroaxonal dystrophy. A clinicopathological study of eight cases. Neuropediatrics 18: 170–175.

Rapin I, Katzman R 1998 Neurobiology of autism. Ann Neurol 43: 7–14.

Rasmussen T 1978 Further observations on the syndrome of chronic encephalitis and epilepsy. Appl Neurophysiol 41: 1–12.

Rasmussen T, Olszewski J, Lloyd-Smith D 1958 Focal seizures due to chronic localized encephalitis. Neurology 8: 435–445.

Raymond GV 2001 Peroxisomal disorders. Curr Opin Neurol 14: 783–787.

Reye RDK, Morgan G, Babal J 1963 Encephalopathy and fatty degeneration of the viscera: a disease entity in childhood. Lancet ii: 749–752.

Robb SA, Harden A, Boyd SG 1989 Rett syndrome: an EEC study in 52 girls. Neuropediatrics 20: 192–195.

Robinson RJ 1984 When to start and stop anticonvulsants. In: *Recent Advances in Paediatrics* (ed R Meadow). Churchill Livingstone, Edinburgh, pp 155–174.

Robinson RO, Baird G, Robinson G, et al 2001 Landau–Kleffner syndrome: course and correlates with outcome. Dev Med Child Neurol 43: 243–247.

Rocha CT, Escolar DM 2004 Update on diagnosis and treatment of hereditary and acquired polyneuropathies in childhood. Clin Neurophysiol (Suppl 57): 255–271.

Roger J, Dravet C, Bureau M 1989 The Lennox–Gastaut syndrome. Cleve Clin J Med 56(Suppl Pt 2): S172–S180.

Roger J, Genton P, Bureau PM, et al 1992 Progressive myoclonus epilepsies in childhood and adolescence. *Epileptic Syndromes in Infancy, Childhood and Adolescence* (eds J Roger, M Bureau, C Dravet, et al). John Libbey, London, pp 381–400.

Rogers SW, Andrews PI, Gahring LC, et al 1994 Autoantibodies to glutamate receptor GluR3 in Rasmussen's encephalitis. Science 265(5172): 648–651.

Rothwell LJC, Obeso JA, Marsden CD 1984 On the significance of giant somatosensory evoked potentials in cortical myoclonus. J Neurol Neurosurg Psychiatry 47: 33–42.

Rowed DW, McClean JAG, Tator CH 1978 Somatosensory evoked potentials in acute spinal cord injury. Prognostic value. Surg Neurol 9: 203–210.

Roy N, Mahadevan MS, McLean M, et al 1995 The gene for neuronal apoptosis inhibitory protein is partially deleted in individuals with spinal muscular atrophy. Cell 80: 167–178.

Rubin DI, Patterson MC, Westmoreland BF, et al 1997 Angelman's syndrome: clinical and electroencephalographic findings. Electroencephalogr Clin Neurophysiol 102: 299–302.

Rudnik-Schoneborn S, Wirth B, Rohrig D, et al 1995 Exclusion of the gene locus for spinal muscular atrophy on chromosome 5q in a family with infantile olivopontocerebellar atrophy (OPCA) and anterior horn cell degeneration. Neuromuscul Disord 5: 19–23.

Rudnik-Schoneborn S, Goebel HH, Schlote W, et al 2003 Classical infantile spinal muscular atrophy with SMN deficiency causes sensory neuronopathy. Neurology 60: 983–987.

Rutherford MA, Heckmatt JZ, Dubowitz V 1989 Congenital myotonic dystrophy: respiratory function at birth determines survival. Arch Dis Child 64: 191–195.

Rutter M, Graham P, Yule W 1970 A neuropsychiatric study in childhood. Clin Dev Med 35/36: 1–265.

Ryan MM, Ouvrier R 2005 Hereditary peripheral neuropathies of childhood. Curr Opin Neurol 18: 105–110.

Ryan MM, Taylor P, Donald JA, et al 1999 A novel syndrome of episodic muscle weakness maps to xp22.3. Am J Hum Genet 65(4): 1104–1113.

Santoshkumar B, Radhakrishnan K 1996 Periodic electroencephalographic pattern in subacute sclerosing panencephalitis modified by preexisting damaged cerebral hemisphere. Electroencephalogr Clin Neurophysiol 99: 440–443.

Sasaki K, Tachi N, Shinoda M, et al 1992 Demyelinating peripheral neuropathy in Cockayne syndrome: a histopathologic and morphometric study. Brain Dev 14: 114–117.

Satterfield JH, Cantwell DP, Saul RE, et al 1974 Intelligence, academic achievement, and EEG abnormalities in hyperactive children. Am J Psychiatry 131: 391–395.

Satterfield JH, Schell AM, Nicholas T, et al 1988 Topographic study of auditory event-related potentials in normal boys and boys with attention deficit disorder with hyperactivity. Psychophysiology 25: 591–606.

Schalamon J, Singer G, Kurschel S, et al 2005 Somatosensory evoked potentials in children with severe head trauma. Eur J Pediatr 164: 417–20.

Scheffner D, Lipinski Chr, Holm Chr, et al 1978 Elektroenzephalographische Befunde bei friihbehandelten Kindern mit Phenyl-ketonurie 4 bis 10 Jahre nach Therapiebeginn. Monatsschr Kinderheilkd 126: 375–378.

Scheffer IE, Bhatia KP, Lopes-Cendes I, et al 1995 Autosomal dominant nocturnal frontal lobe epilepsy. A distinctive clinical disorder. Brain 118: 61–73.

Scher MS, Bergman I, Ahdab-Barmada M, et al 1986 Neurophysiological and anatomical correlations in neonatal nonketotic hyperglycinemia. Neuropediatrics 17: 137–143.

Schiffman R, Boespflüg-Tanguy O 2001 An update on the leukodystrophies. Curr Opin Neurol 14: 789–794.

Schmitt B, Seeger J, Jacobi G 1992 EEG and evoked potentials in HIV-infected children. Clin Electroencephalogr 23: 111–117.

Schneider S 1989 Usefulness of EEG in the evaluation of brain death in children: the cons. Electroencephalogr Clin Neurophysiol 73: 276–278.

Schwartz MS, Scott DF 1978 Pathological stimulus-related slow wave arousal responses in the EEG. Acta Neurol Scand 57: 300–304.

Selcen D, Anlar B, Renda Y 1996 Multiple sclerosis in childhood: report of 16 cases. Eur Neurol 36: 79–84.

Seyal M 1991 Cortical reflex myoclonus. A study of the relation between giant somatosensory evoked potentials and motor excitability. J Clin Neurophysiol 8: 95–101.

Shahwan A, Farrell M, Delanty N 2005 Progressive myoclonic epilepsies: a review of genetic and therapeutic aspects. Lancet Neurol 4(4): 239–248.

Shetty T 1973 Some neurological electrophysiological and biochemical correlates of the hyperactive syndrome. Pediatr Ann 29: 29–38.

Shibasaki H, Kuroiwa Y 1975 Electroencephalographic correlates of myoclonus. Electroencephalogr Clin Neurophysiol 39: 455–463.

Shibasaki H, Yamashita Y, Neshige R, et al 1985 Pathogenesis of giant somatosensory evoked potentials in progressive myoclonic epilepsy. Brain 108: 225–240.

Shorer Z, Philpot J, Muntoni F, et al 1995 Demyelinating peripheral neuropathy in merosin-deficient congenital muscular dystrophy. J Child Neurol 10: 472–475.

Shy GM, Magee KR 1956 A new congenital non-progressive myopathy. Brain 79: 610–621.

Shy GM, Engel WK, Somers JE, et al 1963 Nemaline myopathy; a new congenital myopathy. Brain 86: 793–810.

Simmons Z, Wald JJ, Albers JW 1997 Chronic inflammatory demyelinating polyradiculopathy in children: II. Long-term follow-up, with comparison to adults. Muscle Nerve 20: 1569–1575.

Singh R, Sutherland GR, Manson J 1999 Partial seizures with focal epileptogenic electroencephalographic patterns in three related female patients with fragile-X syndrome. J Child Neurol 14: 108–112.

Sladky JT, Brown HJ, Berman PH 1986 Chronic inflammatory demyelinating polyneuropathy of infancy: a corticosteroid-responsive disorder. Ann Neurol 90: 76–81.

Small JG 1975 EEG and neurophysiological studies of early infantile autism. Biol Psychiatry 10: 355–397.

Smith JMB, Kellaway P 1964 The natural history and clinical correlates of occipital foci in children. In: Neurological and Electroencephalographic Correlative Studies in Infancy (eds P Kellaway, I Petersen). Grune and Stratton, New York, pp 230–249.

Smith NJ, Prior PF 2003 Monitoring spinal cord function. In: Clinical Neurophysiology, Vol. 2, EEG, Paediatric Neurophysiology, Special Techniques and Applications (CD Binnie, R Cooper, F Mauguière, et al). Elsevier Science, Amsterdam, pp 801–813.

Smits MG, Gabreels FJ, Renier WO, et al 1982 Peripheral and central myelinopathy in Cockayne's syndrome. Report of 3 siblings. Neuropediatrics 13: 161–177.

Snoek JW, Minderhoud JM, Wilmink JT 1984 Delayed deterioration following mild head injury in children. Brain 107: 15–36.

So NK, Gloor P 1991 Electroencephalographic and electrocorticographic findings in chronic encephalitis of the Rasmussen type. In: Chronic Encephalitis and Epilepsy: Rasmussen's Syndrome (ed F Anderman). Butterworth-Heinemann, Boston, pp 37–45.

So N, Berkovic S, Andermann F, et al 1989 Electrophysiological studies in myoclonus epilepsy and ragged-red fibres (MERRF). 2: Electrophysiological studies and comparison with other progressive myoclonus epilepsies. Brain 112: 1261–1276.

Sonksen PM, Cottom DG, Harden A 1971 The evolution of the EEG in two patients with maple syrup urine disease (branched-chain ketonuria). Dev Med Child Neurol 13: 606–612.

Spiro AJ, Shy GM, Gonates NK 1966 Myotubular myopathy: persistence of fetal muscle in an adolescent boy. Arch Neurol 14: 1–14.

Sternberg D, Maisonobe T, Jurkat-Rott K, et al 2001 Hypokalemic periodic paralysis type 2 caused by mutations at codon 672 in the muscle sodium channel gene SCN4A. Brain 124: 1091–1099.

Stillerman ML, Gibbs EL, Perlstein MA 1952 Electroencephalographic changes in strabismus. Am J Ophthalmol (Suppl 35): 54–62.

Stöhr M, Dichgans J, Voigt K, et al 1983 The significance of somatosensory evoked potentials for localization of unilateral lesions within the cerebral hemispheres. J Neurol Sci 61: 49–63.

Stores G, Hart J, Piran N 1978 Inattentiveness in schoolchildren with epilepsy. Epilepsia 19: 169–175.

Strandburg RJ, Marsh JT, Brown WS, et al 1991 Reduced attention-related negative potentials in schizophrenic children. Electroencephalogr Clin Neurophysiol 79: 291–307.

Suakkonen A-L, Serlo W, von Wendt L 1990 Epilepsy in hydrocephalic children. Acta Paediatr Scand 79: 212–218.

Sue CM, Hirano M, DiMauro S, et al 1999 Neonatal presentations of mitochondrial metabolic disorders. Semin Perinatol 23: 113–124.

Surtees R 2000 Inherited ion channel disorders. Eur J Pediatr 159(Suppl 3): S199–S203.

Sussova J, Seidl Z, Faber J 1990 Hemiparetic forms of cerebral palsy in relation to epilepsy and mental retardation. Dev Med Child Neurol 32: 792–795.

Suzuki K, Rapin I, Suzuki Y, et al 1970 Juvenile GM2-gangliosidosis. Clinical variant of Tay Sachs disease or a new disease. Neurology 20: 190–204.

Talwar D, Baldwin MA, Horbatt CI 1995 Epilepsy in children with meningomyelocele. Pediatr Neurol 13: 29–32.

Task Force on Brain Death in Children 1987 Guidelines for the determination of brain death in children. Pediatrics 80: 298–300.

Tasker RC, Cole GF 1997 Acute encephalopathy of childhood. In: Paediatric Neurology, 3rd edn (ed EM Brett). Churchill Livingstone, Edinburgh, pp 691–729.

Tasker RC, Boyd SG, Harden A, et al 1988 Monitoring in non-traumatic coma. Part II: Electroencephalography. Arch Dis Child 63: 895–899.

Tasker RC, Boyd SG, Harden A, et al 1989 EEG monitoring of prolonged thiopentone administration for intractable seizures and status epilepticus in infants and young children. Neuropediatrics 20: 147–153.

Tasker RC, Boyd SG, Harden A, et al 1990 The cerebral function analysing monitor in paediatric medical intensive care: applications and limitations. Intens Care Med 16: 60–68.

Tassinari CA, Dravet C, Roger J, et al 1972 Tonic status epilepticus precipitated by benzodiazepines in five patients with Lennox–Gastaut syndrome. Epilepsia 13: 421–435.

Tassinari CA, Dravet C, Roger J, et al 1992 Epilepsy with myoclonic absences. In: Epileptic Syndromes in Infancy, Childhood and Adolescence (eds J Roger, M Bureau, C Dravet, et al). John Libbey, London.

Tassinari CA, Rubboli G, Volpi L, et al 2002 Electrical status epilepticus during slow wave sleep (ESES or CSWS) including acquired epileptic aphasia (Landau–Kleffner syndrome). In: Epileptic Syndromes in Infancy, Childhood and Adolescence, 3rd edn (eds J Roger, M Bureau, C Dravet, et al). John Libbey, Eastleigh.

Taylor D (ed) 1997 Paediatric Ophthalmology, 2nd edn. Blackwell Science, London.

Taylor DC, Falconer MA, Bruton CJ, et al 1971 Focal dysplasia of the cerebral cortex in epilepsy. J Neurol Neurosurg Psychiatry 34: 369–387.

Taylor MJ, Farrell EJ 1989 Comparison of the prognostic utility of VEPs and SEPs in comatose children. Pediatr Neurol 5: 145–150.

Taylor MJ, Robinson BH 1992 Evoked potentials in children with oxidative metabolic defects leading to Leigh syndrome. Pediatr Neurol 8: 25–29.

Taylor MJ, Boor R, Keenan NK, et al 1996 Brainstem auditory and visual evoked potentials in infants with myelomeningocele. Brain Dev 18: 99–104.

Thaker GK, Wagman AM, Tamminga CA 1990 Sleep polygraphy in schizophrenia: methodological issues. Biol Psychiatry 28: 240–246.

Tharp BR 1992 Unique EEG pattern (comb-like rhythm) in neonatal maple syrup urine disease. Pediatr Neurol 8: 65–68.

Thivierge J, Bedard C, Cote R, et al 1990 Brainstem auditory evoked responses and subcortical abnormalities in autism. Am J Psychiatry 147: 1609–1613.

Thomas PK 1987 Classification and electrodiagnosis of hereditary neuropathies. In: *Clinical Electromyography* (eds WF Brown, CF Bolton). Butterworth, Boston, MA, pp 179–207.

Thomas PK, Calne DB 1974 Motor nerve conduction velocity in peroneal muscular atrophy: evidence for genetic heterogeneity. J Neurol Neurosurg Psychiatry 37: 68–75.

Tinazzi M, Zanette G, Bonato C, et al 1996 Neural generators of tibial nerve P30 somatosensory evoked potential studied in patients with a focal lesion of the cervico-medullary junction. Muscle Nerve 19: 1538–1548.

Trauner DA, Stockard JJ, Sweetman L 1977 EEG correlations with biochemical abnormalities in Reye syndrome. Arch Neurol 34: 116–118.

Tristani-Firouzi M, Chen J, Mitcheson JJ, et al 2001 Molecular biology of K⁺ channels and their role in cardiac arrhythmias. Am J Med 110: 50–59.

Tsai LY, Tsai MC, August GJ 1985 Brief report of EEG diagnoses in the subclassification of infantile autism. J Autism Dev Disord 15: 339–344.

Tuchman R, Rapin I 2002 Epilepsy in autism. Lancet Neurol 1: 352–358.

Tuchman RF, Rapin I, Shinnar S 1991 Autistic and dysphasic children. II: epilepsy. Pediatrics 88: 1219–1225.

Uldall P, Alving J, Hansen LK, et al 2006 The misdiagnosis of epilepsy in children admitted to a tertiary epilepsy centre with paroxysmal events. Arch Dis Child 91: 219–221.

Upton A, Gumpert J 1970 Electroencephalography in the diagnosis of herpes-simplex encephalitis. Lancet i: 650–652.

Valeriani M, Restuccia D, Di Lazzaro V, et al 1997a The pathophysiology of giant SEPs in cortical myoclonus: a scalp topography and dipolar source modelling study. Electroencephalogr Clin Neurophysiol 104: 122–131.

Valeriani M, Restuccia D, Di Lazzaro V, et al 1997b Giant central N20–P22 with normal area 3b N20–P20: an argument in favour of an area 3a generator of early median nerve cortical SEPs? Electroencephalogr Clin Neurophysiol 104: 60–67.

Van der Meij W, Van Huffelen AC, Wieneke GH, et al 1992 Sequential EEG mapping may differentiate 'epileptic' from 'non-epileptic' Rolandic spikes. Electroencephalogr Clin Neurophysiol 82: 408–414.

Van der Meij W, Huiskamp GJ, Rutlen GJ, et al 2001 The existence of two sources in Rolandic epilepsy: confirmation with high resolution EEG, MEG and fMRI. Brain Topogr 13: 275–282.

van Huffelen AC 2003 Intracranial tumours. In: *Clinical Neurophysiology*, Vol. 2, *EEG, Paediatric Neurophysiology, Special Techniques and Applications* (eds CD Binnie, R Cooper, F Mauguière, et al). Elsevier Science, Amsterdam, pp 242–248.

Verduyn WH, Hilt J, Roberts MA, et al 1992 Multiple partial seizure-like symptoms following 'minor' closed head injury. Brain Inj 6: 245–260.

Verity CM, Greenwood R, Golding J 1998 Long-term intellectual and behavioral outcomes of children with febrile convulsions. N Engl J Med 338: 1723–1728.

Verma NP, Hart ZH, Nigro M 1985 Electrophysiologic studies in neonatal adrenoleukodystrophy. Electroencephalogr Clin Neurophysiol 60: 7–15.

Vigevano F, Fusco L, Granata T, et al 1996. Hemimegalencephaly: clinical and EEG characteristics. In: *Dysplasias of Cerebral Cortex and Epilepsy* (eds R Guerrini, AE Anderson, R Candau, et al). Lippincott-Raven, Philadelphia, PA, pp 285–294.

Vigevano F, Fusco L, Pachatz C 2001 Neurophysiology of spasms. Brain Dev 23: 467–472.

Vigliano P, Boffi P, Bonassi E, et al 2000 Neurophysiologic exploration: a reliable tool in HIV-1 encephalopathy diagnosis in children. Panminerva Med 42: 267–272.

Ville D, Kaminska A, Bahi-Buisson N, et al 2006 Early pattern of epilepsy in the ring chromosome 20 syndrome. Epilepsia 47: 543–549.

Vincent A, McConville J, Farrugia ME, et al 2004 Seronegative myasthenia gravis. Semin Neurol 24: 125–133.

Vohanka S, Zouhar A 1990 Benign posttraumatic encephalopathy. Acta Nervosa Superior (Prague) 32: 179–183.

Waldo MC, Cohen DJ, Caparulo BK, et al 1978 EEG profiles of neuropsychiatrically disturbed children. J Am Acad Child Psychiatry 17: 656–670.

Wallgren-Pettersson C, Sainio K, Salmi T 1989 Electromyography in Nemaline myopathy. Muscle Nerve 12: 587–593.

Watanabe K, Iwase K, Hara K 1973 The evolution of EEG features in infantile spasms: a prospective study. Dev Med Child Neurol 15: 584–596.

Weglage J, Demsky A, Pietsch M, et al 1997 Neuropsychological, intellectual, and behavioral findings in patients with centrotemporal spikes with and without seizures. Dev Med Child Neurol 39: 646–651.

White SR, Reese K, Sato S, et al 1993 Spectrum of EEG findings in Menkes' disease. Electroencephalogr Clin Neurophysiol 87: 57–61.

Wirrell EC, Camfield PR, Gordon KE, et al 1995 Benign Rolandic epilepsy: atypical features are very common. J Child Neurol 10: 455–458.

Wirrell EC, Camfield PR, Gordon KE, et al 1996 Will a critical level of hyperventilation-induced hypocapnia always induce an absence seizure? Epilepsia 37: 459–462.

Wohlrab G, Boltshauser E, Schmitt B 2001 Neurological outcome in comatose children with bilateral loss of cortical somatosensory evoked potentials. Neuropediatrics 32: 271–274.

Wulff CH, Trojaborg W 1985 Adult metachromatic leukodystrophy: neurophysiologic findings. Neurology 35: 1776–1778.

Wulff CH, Hoyer H, Asboe-Hansen G, et al 1985 Development of polyneuropathy during thalidomide therapy. Br J Dermatol 112: 475–480.

Yamada T, Young S, Kimura J 1977 Significance of positive spike burst in Reye syndrome. Arch Neurol 34: 376–380.

Yamada T, Stevland N, Kimura J 1979 Alpha-pattern coma in a 2-year-old child. Arch Neurol 36: 225–227.

Yamada S, Zinke DE, Sanders D 1981 Pathophysiology of 'tethered cord syndrome'. J Neurosurg 54: 494–503.

Yamada T, Graff-Radford NR, Kimura J, et al 1985 Topographic analysis of somatosensory evoked potentials in patients with well-localized thalamic infarctions. J Neurol Sci 68: 31–46.

Yamada T, Ishida T, Kudo Y, et al 1986 Clinical correlates of abnormal P14 in median SEPs. Neurology 36: 765–771.

Yaqub BA 1996 Subacute sclerosing panencephalitis (SSPE): early diagnosis, prognostic factors and natural history. J Neurol Sci 139: 227–234.

Yatziv S, Russell A 1981 An unusual form of metachromatic leukodystrophy in three siblings. Clin Genet 19: 222–227.

Zafeiriou DI, Kontopoulos EE, Tsikoulas I 1999 Characteristics and prognosis of epilepsy in children with cerebral palsy. J Child Neurol 14: 289–294.

Zerres K, Davies KE 1999 59th ENMC International Workshop. Spinal muscular atrophies: recent progress and revised diagnostic criteria. Neuromusc Disord 9: 272–278.

Index